Materials Crystal Chemistry

Relva C. Buchanan
Taeun Park
University of Cincinnati
Cincinnati, Ohio

MARCEL DEKKER, INC. NEW YORK • BASEL • HONG KONG

Library of Congress Cataloging-in-Publication Data

Buchanan, Relva C.
Materials crystal chemistry / Relva C. Buchanan, Taeun Park.
p. cm.
ISBN 0-8247-9798-1 (alk. paper)
1. Solid state chemistry. 2. Crystals. 3. Crystallography.
I. Park, Taeun. II. Title.
QD478.B83 1997
548—dc21

97-18615
CIP

The publisher offers discounts on this book when ordered in bulk quantities. For more information, write to Special Sales/Professional Marketing at the address below.

This book is printed on acid-free paper.

MARCEL DEKKER, INC.
270 Madison Avenue, New York, New York 10016
http://www.dekker.com

Current printing (last digit):
10 9 8 7 6 5 4 3 2 1

PRINTED IN THE UNITED STATES OF AMERICA

PREFACE

This book provides an overall description of crystal structure development in those materials (metals, ceramics, and polymers) that are used commonly in materials science and engineering. Emphasis is given to the relationships between developed structures and physical properties as well as to the design of new materials with improved properties.

The book is intended primarily for engineers, scientists, advanced undergraduate and graduate students in materials engineering, materials science and allied scientific disciplines, where understanding materials at the atomistic and structural levels is fundamental to their design, development and utilization. Practicing engineers and scientists who work on materials development as well as students and faculty in inorganic chemistry, geology or physics will find the book helpful. Accordingly, the text is designed to provide the reader with the general concepts of crystal chemistry and crystallography relating to the development of atomic structure, bonding, spatial arrangement, phase transition, symmetry and the unique structures developed in crystals. Quantum mechanical concepts are introduced early to explain the fundamentals of atomic structure, with symmetry and packing concepts used later to explain the evolution of crystal structures from simple unit cells. The structures are shown in great detail to aid in the understanding of their development. Concepts of crystallography and symmetry are presented to provide a framework for the description and analysis of the crystal structures by such techniques as X-ray diffraction and TEM (transmission electron microscopy) analysis.

The approach represents a unique synthesis of concepts relating to bonding, symmetry and crystallography. This allows for a more simplified analysis and explanation of crystal structure development evolution in relationship to properties observed. The approach contrasts with current texts, where only the first two of these important topics are likely to be treated in depth, thereby obscuring their combined influence on crystal structure evolution. Graphical representations of crystal structures in this text are presented in a new expanded form, developed through the aid of computer graphics. This treatment simplifies the visualization of these structures. In this way, important features in these very complex structures can be more readily understood. This is well illustrated for the more complex silicate structures, in which the modifying ions exercise a deterministic effect on their structural evolution.

The expert assistance of Dr. Aniruddha Thakur, Postdoctoral Associate (1995–1997) in the Department of Materials Science and Engineering at the University of Cincinnati, in the final editing of the text manuscript is gratefully acknowledged.

Relva C. Buchanan
Taeun Park

CONTENTS

INTRODUCTION : MATERIALS CRYSTAL CHEMISTRY

Any solid with an internal structure consisting of an ordered array of atoms or ions such that a unit of repetition—the unit cell— can be identified, is termed a crystal. This crystalline state can be contrasted to the amorphous state, typified by such materials as glasses, elastomers and most polymers for which no unit of repetition or long range order exists, even though the atoms or ions may be similarly bonded.

The state described as crystalline is generally the lowest energy state for the material or compound, compared to its liquid or gaseous existence. A given solid may, however, exist in different crystalline modifications or phases depending on the ambient temperature, pressure and the energy difference between states. Additionally, kinetic factors such as species mobility, bond breaking and changes in coordination of nearest neighbor atoms or ions can significantly influence phase stability. A mineral such as silica (SiO_2), for example, shows several crystalline modifications (α, β quartz, trydimite, cristobalite, fused silica) all of which, except for β quartz, can exist metastably at ordinary temperatures (25°C). This is due to the small energy difference between the phases and to a marked kinetic hindrance to phase transformation at such temperatures.

Analytical treatment of the crystalline state requires generally an understanding of the related disciplines of crystallography and crystal chemistry. Crystallography deals with the 3-dimensional arrangement of points in space as a mathematical concept. This arrangement is described by a space lattice which is generated by translation symmetry from lattice points. This aspect of crystallography can be recognized as structural and deals with the symmetry and internal arrangement of the atoms or ions in the crystal, as determined by x-ray or electron diffraction analysis techniques. These arrangements relate directly to the tensor (and other) properties of the crystal, including its thermal expansion, cleavage, elastic modulus, electrical conductivity, dielectric polarization and magnetization behavior.

Other important aspects of crystallography are concerned with the study of the relationships between the different faces or "forms" of the crystal. Before the advent of x-ray diffraction and with it the detailed analysis of internal structure, certain definite spatial relationships were found to exist between the crystal faces. The study of these relationships describes crystal morphology, which is of importance both in the identification and use of crystals. Optical crystallography deals with the interaction of visible electromagnetic radiation—particularly polarized light—with crystalline materials. This is an important analytical tool for mineral identification, by utilizing the unique differences in refractive index behavior. Electrooptic effects in some crystals are also made use of for light modulation and are important in various signal detection schemes.

Crystal chemistry deals with the nature of the atoms or ions at the lattice points, in relation to the chemical composition of the solid and the bonding which holds the structure together. Crystal chemistry seeks to explain why, from a given chemical composition, certain structural forms are favored. Required is a detailed understanding of atomic structure, chemical bonding, coordination theory and packing in terms of atom or ion sizes, ionic radius ratios and ion charge. These factors relate directly to the energy state of the crystal, to the bond strength, spatial arrangement of the bonds and hence to the type of structure formed. Properties such as density, melting point and hardness are a direct reflection of these factors. Also, intrinsic properties such as elastic compliance, electric susceptibility and magnetic susceptibility are directional and hence these are strongly related to crystal structures.

Table 1 summarizes the various interrelationships between the structure, symmetry and properties of crystalline materials.

One of the more important aspects of crystal chemistry relates to the presence of imperfections in the crystal lattice. As a practical matter, few crystalline lattices are perfect. Within the orderly arrangement of atoms or ions may be found vacant lattice sites, interstitial atoms or foreign atoms or ions on regular lattice sites. This latter may be accompanied by changes in valence states (ionic) or the formation of vacancies or interstitials. These so-called point defects provide the means for mass and charge transport within the solid and are, therefore, associated with some of the more interesting engineering properties of crystalline solids. Other line type defects such as dislocations and stacking faults also significantly influence mechanical deformation and crystal growth.

The broad objectives of this book, therefore, may be stated as follows:

a. To develop an understanding of concepts related to atomic structure, bonding, atomic packing, coordination theory, polymorphism, defect chemistry and symmetry, primarily in crystalline metallic and ceramic materials. Polymer structures, mainly amorphous, are also discussed.
b. To develop an understanding of simple crystallographic terms relating to planes, directions, symmetry, Bravais lattices, point groups and space groups.

c. To develop facility in the use of the Periodic Table in order to make intelligent interpretations of expected material behavior and as a guide in modifying this behavior to achieve desirable engineering properties. These concepts and the various factor which related crystal chemistry to the physical properties of crystal are illustrated in Fig.I-1.

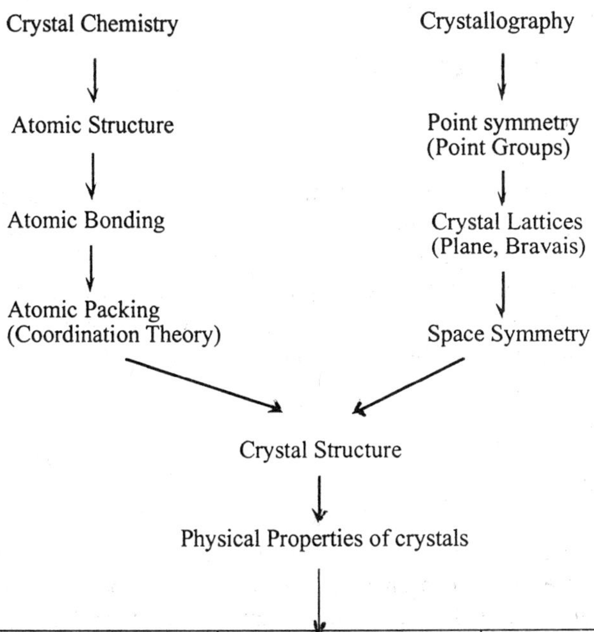

External conditions	Intrinsic Properties	Response
Temperature : T	Cp	S : Entropy
Stress : s	s	e : Strain
Electric Field : E	c	P : Dipole moment
Magnetic Field : H	m	M : Magnetic moment

Cp : heat capacity, s : elastic compliance,
m : magnetic susceptibility, c : electric susceptibility

Figure I-1 Crystal Chemistry Components and their effects on physical properties of crystals

Chapter 1

Fundamentals of Atoms, Molecules and Solids

SECTION 1 ATOMIC STRUCTURE

A. Classical Model

A useful starting point in the study of atomic structure is the concept of the atom developed by Rutherford and Bohr between 1911 and 1913. Rutherford proposed that the atom consisted of a positively charged nucleus containing essentially all of the mass surrounded by a negatively charged cloud containing most of the volume. This concept was extended to a system of planetary electrons orbiting the nucleus. Classically, the energy of the orbiting electrons is given by:

$$E = \frac{-e^2}{8\pi\varepsilon_o r} \tag{1}$$

where e = electron charge and r = electron orbit radius; ε_o = dielectric constant of vacuum. Since the electron is a charged particle it would be attracted to and gradually drawn into the nucleus emitting continuous radiation in the process, a process which was not observed to occur. Bohr's contribution was to provide an analytical background for the stability of such a planetary system for a one electron atom and, in particular, to provide a successful explanation for the observed spectral emission lines produced by hydrogen from a low pressure discharge. The Rutherford-Bohr atom is today mainly of historical interest being largely superseded by the quantum-mechanical model. Some concepts and terminology, however, are still valid. The background concepts on which the Bohr postulate are based include the following.

B. Background Concept

(i). Planck's Equation gives the fundamental relationship between frequency and the energy of light, viz:

$$E = h\nu \qquad (2)$$

h = Planck's constant

$$\nu = \frac{c}{\lambda}$$

ν = frequency of light
c = speed of light
λ = wavelength of light

$$\therefore \; E = h\nu = \frac{hc}{\lambda} \qquad (3)$$

(ii). The Einstein relationship between energy and matter is given by:

$$E = mc^2 \qquad (4)$$

From Equations (3) and (4)

$$mc^2 = \frac{hc}{\lambda}$$

$$\therefore \; mc = \frac{h}{\lambda} \qquad (5)$$

(iii). For two charges q_1 and q_2 separated by a distance r the force of attraction between them according to Coulomb's Law is given by:

q_1 •————————• q_2

$$F_a = \frac{q_1 q_2}{r^2} \bullet \frac{1}{4\pi\varepsilon_o} = \frac{q_1 q_2}{4\pi\varepsilon_o r^2} \qquad (6)$$

where $\frac{1}{4\pi\varepsilon_o}$ is a proportionality constant which, in the MKS(SI) system, is proportional to the dielectric constant of the medium separating the two charges. In the CGS unit system, the proportionality constant is 1. Note that q = ze where z = charge on the ion and e = electronic charge. If q_1 and q_2 are both of the same sign the force is repulsive. For q_1 and q_2 of opposite sign, the force is attractive.

(iv). For a body of mass m moving in a circular orbit of radius r with velocity v, the Newtonian centripetal force is given by:

$$F = \frac{mv^2}{r} \tag{7}$$

with the positive direction (by convention) being that of increasing r.

(v). The angular momentum for an electron traveling in a circular orbit about the nucleus is given by:

$$L = mvr = I\,\omega$$

where I = moment of inertia = mr^2 and ω = angular velocity = $\frac{v}{r}$

$$\therefore \quad I\omega = mr^2 \frac{v}{r} = mvr \tag{8}$$

Table 1-1 gives the value of the constants appropriate to these relationship.

C. Bohr's Postulate

Bohr pictured the Rutherford model of the atom as in Fig. 1-1. By this concept, the electron exists in a stationary orbit around the nucleus, and is kept in orbit by a balance between the centrifugal force arising from the circulation of electrons around the nucleus with velocity, v, and centripetal force from Coulombic forces between electrons and the nucleus.

Table 1-1 List of constants

h	=	6.625×10^{-34} J. sec. (Planck)
e	=	1.609×10^{-19} Coulombs (electron charge)
eV	=	1.602×10^{-19} J = 23 kcal/mole
ε_o	=	8.854×10^{-12} C/v-m (farad) - [permittivity of free space]
M_e	=	9.109×10^{-31} kg (electron mass)
c	=	2.998×10^{8} m/sec (speed of light)
μ_o	=	$4\pi \times 10^{-7}$ H/m (Magnetic permeability of free space)

Figure 1-1. Rutherford-Bohr model of a atom

r = electron orbital radius
m = electron mass
v = electron velocity
e^- = electronic charge
M = mass of proton (nucleus)
Ze = +ve charge on nucleus
Z = atomic # (# of protons)

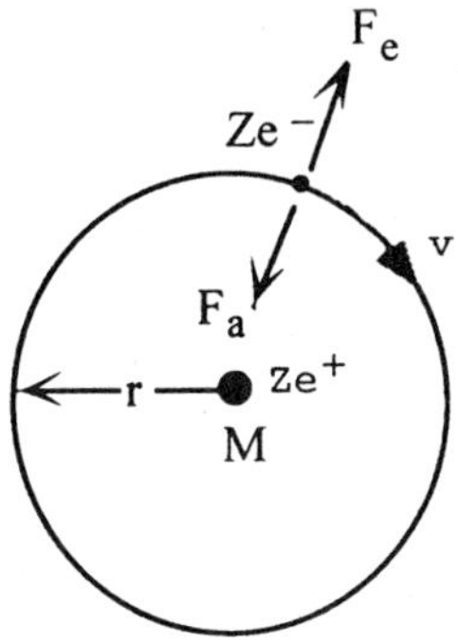

Although the experimental facts pointed to this model, it must fail since eletromagnetic theory predicts that an alternating charge, such as an electron orbiting around the nucleus, would lose its energy and, therefore, finally fall into the positive nucleus. However, this collapsing of electrons toward the nucleus does not occur. Bohr overcame this theoretical problem in the nucleus-electron planetary model of Rutherford by proposing the following postulate; viz:

> The only stationary orbits available to the electron are those for which the angular momentum L for the electron is given by the quantum condition:

$$L = mvr = \frac{nh}{2\pi} \tag{9}$$

where n = 1, 2, 3,...(principal quantum number)

This postulate can be visualized by considering the wave nature of the electron. The waves associated with electrons must be standing waves as shown in Fig. 1-2(a) since nonstanding waves result in destructive interferences and eventually lose their energies. For the wave to be a standing wave, the condition $n\lambda = 2\pi r$ must be satisfied, where n is a integer. Since $\lambda = \frac{h}{p}$ according to de Broglie relation, the necessary condition for the standing wave of electrons is $\frac{nh}{p} = 2\pi r$, which is the same as the Bohr's postulate.

D. Bohr's Radii

From Eqs. (6) and (7)

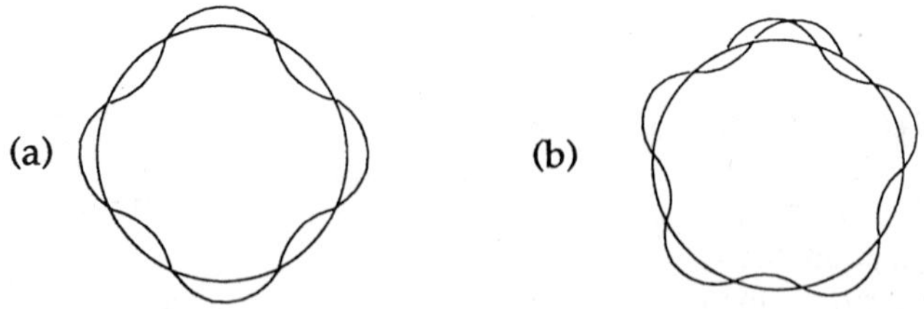

Figure 1-2 (a) Standing wave and (b) nonstanding wave about a circular orbit

$$F_a = F = \frac{mv^2}{r} = \frac{q_1 q_2}{r^2}$$

therefore, $v^2 = \dfrac{e^2}{mr}$ (10)

From the Bohr's postulate $mvr = \dfrac{nh}{2\pi}$

$$\therefore \quad v = \frac{nh}{2\pi mr} \text{ and } v^2 = \frac{n^2h^2}{4\pi^2 m^2 r^2} \tag{11}$$

Equating (10) and (11)

$$v^2 = \frac{-e^2}{mr} = \frac{n^2h^2}{4\pi^2 m^2 r^2}$$

Therefore: $\boxed{r = \dfrac{n^2h^2}{4\pi^2 me^2} = \dfrac{h^2}{4\pi^2 me^2} n^2}$ (12)

For values of n = 1, 2, 3, 4 … we have

n	shell	r_i	Bohr radius
1	K	$r_1 = \dfrac{\varepsilon_o h^2}{\pi e^2 m}$	$r_1 = 5.292 \times 10^{-11}$m
2	L	$r_2 = r_1 \times 2^2$	$r_2 = 2.117 \times 10^{-10}$m
3	M	$r_3 = r_1 \times 3^2$	$r_3 = 4.763 \times 10^{-10}$m
4	N	$r_4 = r_1 \times 4^2$	$r_4 = 8.467 \times 10^{-10}$m

E. Bohr Energy

For any orbit, the total energy of the electron, E_n, is made up of potential and kinetic energies, i.e.:

$$E_n = V + T$$

where $T = \frac{mv^2}{2}$ = kinetic energy (Newton)

Since $v^2 = \frac{e^2}{mr}$ from Eq.(10), then:

$$T = \frac{e^2}{2r} \qquad (13)$$

The potential energy V is defined as the work done by the electron in moving from infinity (zero energy) to r.

i.e. $|V| = \left| \int_0^r F\, dr \right|$

$$V = \int_0^r F\, dr = \{\text{work done by electron}\}$$

where $F = -\frac{e^2}{r^2}$

Integrating: $V = -\frac{e^2}{r}$ (14)

Therefore $E_1 = \frac{e^2}{2r} - \frac{e^2}{r} = -\frac{e^2}{2r}$

Substituting for $r = \frac{h^2}{4\pi^2 me^2} n^2$ according to Eq.(12)

$$E_n = -\frac{2\pi^2 me^4}{h^2} \cdot \frac{1}{n^2} \qquad (15)$$

For $n = 1$; $E_n = -13.6$ eV (Hydrogen)

$$n = 2;\ E_n = -13.6 \cdot \frac{1}{2^2} = -3.41\ \text{eV}$$

$$n = 3;\ E_n = -13.6 \cdot \frac{1}{3^2} = -1.51\ \text{eV}$$

Equations (12) and (15) reveal, therefore, that electrons are distributed in discrete orbits whose radii and energy levels are quantized.

F. Electron Energy Transitions

As shown in the Bohr atom, energy levels in the atom are quantized. Hence the energy loss or gain from the transition from one orbit to another has a discrete value. The total energy change during this transition in the Bohr's atom can be expressed as:

$$\Delta E = E_f - E_i$$

$$\Delta E = E_f - E_i = -\frac{2\pi^2 me^4}{h^2} \cdot \frac{1}{n_f^2} - -\frac{2\pi^2 me^4}{h^2} \cdot \frac{1}{n_i^2}$$

$$= -\frac{2\pi^2 me^4}{h^2}\left(\frac{1}{n_f^2} - \frac{1}{n_i^2}\right) \tag{16}$$

Thus if $n_i = 3$ and $n_f = 2$, the electron would emit an energy (quanta). The electron would absorb energy if $n_i = 2$ and $n_f = 3$. If a transition occurs by emitting light, the frequency or wave length of the light is determined by the energy difference between two orbits. Therefore the frequency of the light can be expressed according to Planck Equation as follows:

$$E_{light} = h\nu = \frac{hc}{\lambda}$$

$$\Delta E = E_f - E_i$$

$$\Delta E = \frac{hc}{\lambda} \tag{17}$$

Combining Eqs. (16) and (17)

$$\frac{1}{\lambda} = \frac{\Delta E}{hc} = -\frac{2\pi^2 me^4}{h^3 c}\left(\frac{1}{n_f^2} - \frac{1}{n_i^2}\right)$$

$$\text{i.e.}\ \frac{1}{\lambda} = R\left(\frac{1}{n_i^2} - \frac{1}{n_f^2}\right) \tag{18}$$

where $R = \dfrac{2\pi^2 me^4}{h^3 c}$ which is equivalent to Rydberg constant - 1.0974×10^7/m

Eq.(18) is the basis for the discrete line spectrum emitted during the discharge of atomic hydrogen. This also explains the formation of a series of lines in the emission spectrum of the hydrogen atom. These series of lines in the regions of ultraviolet, visible, and infrared light are called Lyman, Balmer and Paschen series, respectively, according to the discoverer's names. In relation to Eq.(18), these line series arise with the same n_i and can be classified as follows:

$n_i = 1$, $n_f = 2,3,4$—Lyman Series
$n_i = 2$, $n_f = 3,4,5$—Balmer Series
$n_i = 3$, $n_f = 4,5,6$—Paschen Series

A schematic energy diagram for the electron transition for each of these line series is illustrated in Fig. 1-3. As shown in Fig. 1-3(b), the series of lines are called K, L, and M series. This designates values for n_i of 1, 2 and 3, respectively. Also each line in a series is called by α, β and γ which relates to $\Delta n(n_i - n_f)$ of 1, 2 and 3, respectively.

G. Summary of the Bohr's Model

The overall picture of the Bohr atom, therefore, is that of a positive nucleus (containing most of the mass) surrounded by electrons in stationary orbits, as illustrated in Fig. 1.3(a).

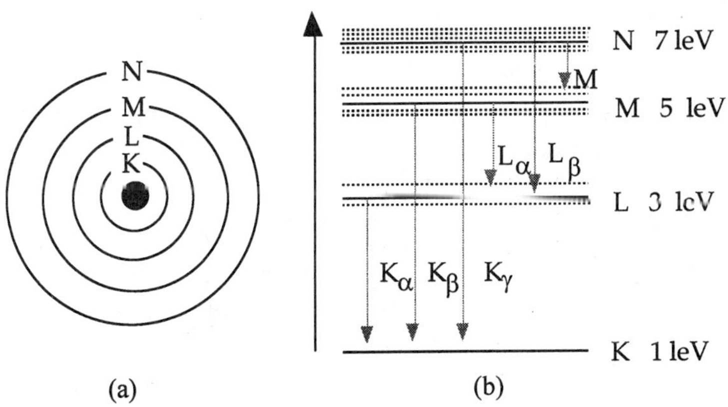

Figure 1-3 (a) Bohr Atom and (b) its characteristic energy levels

Transitions occur when an inner orbit electron is knocked out, say by x-rays. Outer orbit electrons falling into the inner orbit, emit photons of characteristic wavelength and energy. More generally we can write for a multi-electron atom:

$$\frac{1}{\lambda} = RZ^2\left(\frac{1}{n_i^2} - \frac{1}{n_f^2}\right) \tag{19}$$

where λ represents the wavelength of the emitted photon. $Z = 1, 2, 3$, and 4 for H, He^+, Li^{2+}, and Be^{3+}, respectively.

H. Quantum Mechanical Concept

The Bohr model introduced quantum concepts into atomic structure and successfully explained the stability and spectra of the hydrogen atom. The theory, nevertheless, suffered from certain limitations, namely:

a. The Bohr theory could not predict the spectra of other elements such as helium or more complex atoms, and certainly not the spectra of molecules.

b. It could also not explain the splitting of some spectral lines in the presence of a magnetic field — the Zeeman effect. The Bohr theory would predict that all spectral lines would split, since the energy states all have angular momentum.

c. Bonding of atoms could also not be explained and, finally

d. The assumption of stationary states was arbitrary. It provided no real understanding of the quantization process.

These objections were overcome by the evolution of Quantum Mechanics in the early part of the 1900s. The difference between quantum mechanics and classical mechanics is that, in quantum mechanics, the position of an object, including small particles such as electrons, cannot be unambiguously specified. Rather the position is represented by a probability function. The reason for this difference was formulated by Heisenberg in his "uncertainty principle" which states that there is a limit in accuracy to the simultaneous determination of the position and momentum of an object. This principle is expressed by a mathematical form as

$$\Delta p_x \Delta x \geq \frac{h}{2\pi} \tag{20}$$

In the macroscopic world, any dynamic parameters can be described or predicted from classical Newtonian mechanics, represented by Eq.(21), where the position and momentum of the object are measurable within a certain accuracy.

$$F = m\frac{dv}{dt} \tag{21}$$

In contrast, the Heisenberg uncertainty principle, Eq.(20), illustrates that, in the microscopic world, the magnitude of error in the measurement of position and momentum is not negligible. Classical mechanics, therefore, cannot accurately describe the motion of microscopic particles, since the position and momentum are not accurately known. Hence, in quantum mechanics, a definite location for the particles cannot be assigned. Rather, the location is described by a function, say Ψ, which expresses not the certainty that the particle is at a given point, but only the probability that it is at that point. This uncertainty, along with wave character of small particles, underlies the Schrödinger Equation which represents the foundation of Quantum Mechanics.

The following derivation of Schrödinger's Equation for an electron illustrates the underlying principles of quantum mechanics. First, to describe the motion of an electron, the wave function for the motion of a photon can be used. This wave function, according to the Maxwell Equation, is given as:

$$\frac{\partial^2 W}{\partial^2 x} + \frac{\partial^2 W}{\partial^2 y} + \frac{\partial^2 W}{\partial^2 z} = \frac{1}{c^2}\frac{\partial^2 W}{\partial t^2} \tag{22}$$

where c is the velocity of light and W is the amplitude of the photon. In a similar way, the motion of the electron can be described by a wave Equation as

$$\frac{\partial^2 \mathcal{W}}{\partial^2 x} + \frac{\partial^2 \mathcal{W}}{\partial^2 y} + \frac{\partial^2 \mathcal{W}}{\partial^2 z} = \frac{1}{\upsilon^2}\frac{\partial^2 \mathcal{W}}{\partial t^2} \tag{23}$$

where v is the velocity of the electron wave and W is the amplitude of the electron wave. As in Bohr's model, it is assumed that the wave is a standing wave. Then $\mathcal{W}$ is expressed as, $\Psi\exp(-i\nu\pi t)$, where ν is the frequency of the electron wave, and Ψ is the wave function of coordinates x, y and z. By replacing $\mathcal{W}$ by $\Psi\exp(-i\nu\pi t)$, Eq.(23) becomes

$$\frac{\partial^2 \Psi}{\partial^2 x} + \frac{\partial^2 \Psi}{\partial^2 y} + \frac{\partial^2 \Psi}{\partial^2 z} = \frac{4\pi^2 \nu^2}{\upsilon^2}\frac{\partial^2 \Psi}{\partial t^2} \tag{24}$$

Since $v = \nu\lambda$ and $p = \frac{h}{\lambda}$ according to de Broglie Equation, Eq.(24) becomes

$$\frac{\partial^2\Psi}{\partial^2 x} + \frac{\partial^2\Psi}{\partial^2 y} + \frac{\partial^2\Psi}{\partial^2 z} = -\frac{4\pi^2 p^2}{h^2}\Psi \qquad (25)$$

With $E_{total} = V + T$ and $T = \frac{p^2}{2m}$, therefore:

$$\frac{\partial^2\Psi}{\partial^2 x} + \frac{\partial^2\Psi}{\partial^2 y} + \frac{\partial^2\Psi}{\partial^2 z} + \frac{8\pi^2 m}{h^2}(E_{total} - V)\Psi = 0 \qquad (26)$$

Eq.(26) is the well known Schrödinger Wave Equation (SWE). The derivation of Schrödinger Wave Equation above illustrates characteristics of quantum mechanics such as the probability concept and the wavelike behavior of microscopic particles such as electrons. From this similarity between Maxwell Equation and SWE, the physical meaning of Ψ can be pursued in the same way. In the Maxwell Equation, W is the amplitude(phase) of light at a certain point in space. A measurable quantity is obtained by the square of W which represents the intensity of the light. Likewise, Ψ in SWE is the phase of the electron wave at a point in space while its square (Ψ^2) represents the probability of electron to be found in that position. Since the wave function Ψ describes the condition for one electron, the probability of finding the electron in whole space must be 1. This condition of wave function is expressed as follows:

$$\int \Psi^* \Psi dV = \int \Psi^2 dV = 1 \qquad (27)$$

Here the integration is over whole space, and the procedure to make the integration be 1 is called normalization of the wave function. In comparison with Maxwell Equation - Eq.(23), the normalization allows the amplitude of the wave functions to have a actual value. A normalized wave function is different from the unnormalized one by a constant, such as :

$$\Psi_{normal} = c\,\Psi_{unnormal} \qquad (28)$$

Hence unnormalized wave functions show the same phase as the normalized ones.

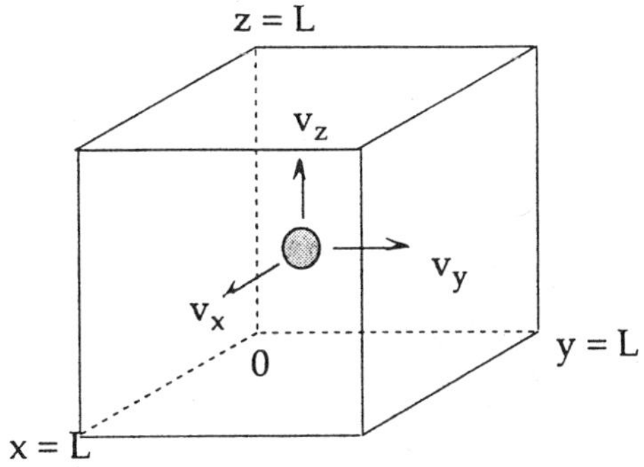

Figure 1-4 A free particle in a Box

Particle in a Box model

The following example illustrates the quantum concept of energy levels being *quantized* in the microscopic world, where the motion of a particle is described by the Schrödinger Equation.

Fig. 1-4 depicts a particle confined in a box. First, consider the motion of the particle in one direction, say along the x axis. For the particle to be confined in the box, the potentials, V_x, at x = 0 and L are infinite. Conversely V = 0 at 0 < x < L, since the particle moves freely. The motion can be described by the SWE as

$$\frac{\partial^2\Psi}{\partial x^2} + \frac{8\pi^2 m}{h^2}(E-0)\Psi = 0 \tag{29}$$

One possible solution for this Equation is $\Psi = A\sin\frac{2\pi}{h}(2mE)^{\frac{1}{2}}x$. Since the particle is confined in the box, $\Psi = 0$ at x = 0 and L. To satisfy this boundary condition, the expression $\frac{2\pi L}{h}(2mE)^{\frac{1}{2}} = n\pi$ must hold where n is an integer. Therefore, $E = \frac{n^2h^2}{8mL^2}$ and the energy of the particle is quantized.

A schematic energy diagram for the particle is shown in Fig. 1-5 which illustrates the quantum energy levels and shapes for the wave functions with the quantum numbers n = 1, 2 and 3. The motion of the particle in the y and z directions are independent from that in the direction of x axis. Therefore, the energy levels in the y and z axes can be described in the same way. Hence the total energy of the particle in a box is expressed as $E = h^2(n_x^2 + n_y^2 + n_z^2)/8mL^2$

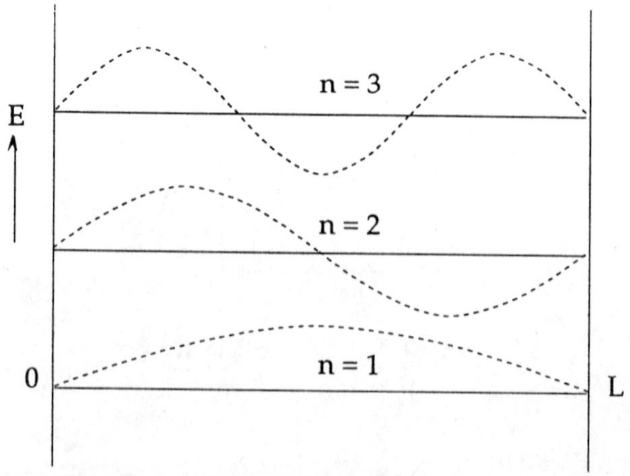

Figure 1-5 Allowed wave functions and their energy levels in one direction for a particle in a box

where n_x, n_y, and n_z are quantum numbers in x, y , and z directions. This example of a particle in a box illustrates why, in the quantum mechanics approach, physical properties such as energy are quantized.

Hydrogen atom

The simplest of atomic structures is that of the hydrogen atom, which consists of a nucleus of charge +e and mass M, and an electron of charge -e and mass m. To simplify the solution to the Schrödinger Equation, it is assumed that the position of the nucleus is fixed at the origin of the coordinates. Then the Schrödinger Equation can be expressed as:

$$\nabla^2\Psi + \frac{8\pi^2 m}{h^2}\left(E + \frac{e^2}{\sqrt{(x^2 + y^2 + z^2)}}\right)\Psi = 0 \qquad (30)$$

where x, y and z are the coordinates of the electron in the rectangular coordinate system. Actually m must be replaced by μ, the reduced mass $\frac{mM}{m+M}$ in which M and m represent the masses of nucleus and electron, respectively. Since $M >> m$, μ can be approximated by m. Hence, in the following equations for the electron of the hydrogen atom, m is used for μ without difficulty. If the coordinate system is converted to the spherical coordinate system, Eq.(30) can be expressed as:

$$\nabla^2\Psi + \frac{8\pi^2 m}{h^2}\left(E+\frac{e^2}{r}\right)\Psi = 0 \tag{31}$$

If $\nabla^2\Psi\left(=\frac{\partial^2\Psi}{\partial x^2}+\frac{\partial^2\Psi}{\partial x^2}+\frac{\partial^2\Psi}{\partial x^2}\right)$ is expanded in this spherical coordinate system which is shown in Fig.1-6, Eq.(31) will become

$$\frac{1}{r^2}\frac{\partial}{\partial r}\left(r^2\frac{\partial\Psi}{\partial r}\right)+\frac{1}{r^2\sin\theta}\frac{\partial}{\partial\theta}\left(\sin\theta\frac{\partial\Psi}{\partial\theta}\right)+\frac{1}{r^2\sin\theta}\frac{\partial^2\Psi}{\partial\varphi^2} + \frac{8\pi^2 m}{h^2}\left(E+\frac{e^2}{r}\right)\Psi = 0 \tag{32}$$

Eq.(32) reveals that variables can be separated and Ψ can be expressed in the form of $R(r)Y(\theta,\varphi)$ in which R is a function of r only and Y are functions of θ and φ only.

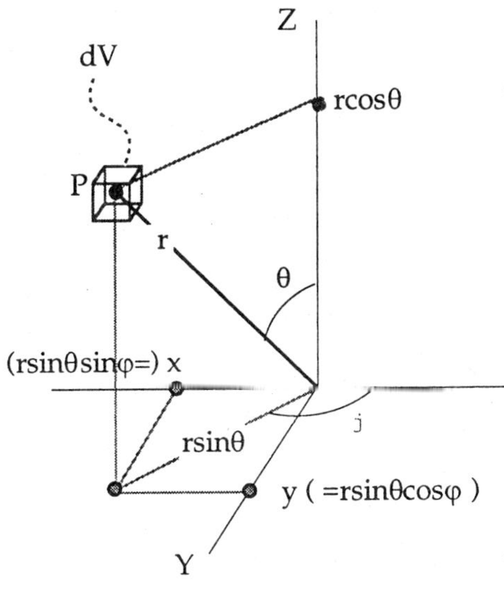

$$dV = dx\,dy\,dz = r^2\sin\theta\,dr\,d\theta\,d\varphi$$

$$x = r\sin\theta\cos\varphi$$

$$y = r\sin\theta\sin\varphi$$

$$z = r\cos\theta$$

Figure 1-6 Illustration of spherical coordinates with rectangular coordinates system

By putting R(r)Y(θ,φ) for Ψ, Eq.(32) becomes

$$\frac{1}{R}\frac{\partial}{\partial r}\left(r^2\frac{\partial R}{\partial r}\right)+\frac{8\pi^2 m}{h^2}\left(E+\frac{e^2}{r}\right)r^2$$
$$=-\frac{1}{Y\sin\theta}\frac{\partial}{\partial\theta}\left(\sin\theta\frac{\partial Y}{\partial\theta}\right)-\frac{1}{Y\sin^2\theta}\frac{\partial^2 Y}{\partial\varphi^2} \quad (33)$$

Since R(r) and Y(θ,φ) are independent from each other, both sides of Eq.(33) must be a constant. Let's assume that this constant is $\ell(\ell+1)$. By putting this constant into both sides of Eq.(33) and solving it, the solution obtained is Ψ = $R_{n,\ell}(r)\, Y_{\ell,\pm m}(\theta,\varphi)$ where

$$R_{n,\ell}=-\sqrt{\left(\frac{2}{na_o}\right)^3\frac{(n-\ell-1)!}{2n(n+\ell)!}}\left(\frac{2r}{na_o}\right)^\ell \exp\left(-\frac{2r}{na_o}\right)L^{2\ell+1}_{n-\ell-1}\left(\frac{2r}{na_o}\right) \quad (34)$$

and

$$Y_{\ell,\pm m}=-\sqrt{\frac{2\ell+1}{4\pi}\frac{(\ell-|m|)!}{(\ell+|m|)!}}\,P^m(\cos\theta)^\ell \exp(\pm m\varphi) \quad (35)$$

Here $a_o=\dfrac{h^2}{4\pi^2\mu e^2}$ is the Bohr radius which is ≈ 0.53A°, L^β_α represents the associated Laguerre polynomials, and P^m_ℓ is the associated Legendre polynomials. For the derivation of the solution, refer to the appendix. There is, however, a restriction on the solution of the SWE for the hydrogen atom, which is the inequality relationship between n, ℓ, and m as $n > \ell > m$. Table 1-2 gives the calculated solution of the radial function R(n,ℓ) for the hydrogen atom.

The two special functions in the solution are expressed in mathematical form as

$$L^\beta_\alpha=\sum_{m=0}^{\alpha}(-1)^m\frac{(\alpha+\beta)!}{(\alpha-m)!(m+\alpha)!m!}x^m \text{ , } (k > -1) \quad (36)$$

$$P^m_\ell(x)=\frac{d^m}{dx^m}\left[\frac{1}{2^\ell \ell!}\frac{d^\ell}{dx^\ell}\left(x^2-1\right)^\ell\right] \quad (37)$$

As in the case of a particle in a box with quantum numbers, n_x, n_y, and n_z, the solution, Ψ, to the SWE of a electron in the hydrogen atom requires 3 quantum numbers n, ℓ, and m. The required conditions for these quantum numbers are $n \geq \ell + 1$, and $-\ell \leq m \leq \ell$. The quantum numbers n, ℓ, and m are related to the energy, total angular momentum, and the z-directional angular momentum, respectively. In fact, the energy is represented by $E_n = -\frac{2\pi^2 me^4}{n^2h^2}$. Total angular momentum is $\ell(\ell+1)$. The z-directional angular momentum is m. The energy of the Bohr atom in Eq.(15) is the same as that derived from the Schrödinger wave equation for the hydrogen atom.

A wave function with discrete quantum numbers, n, ℓ, and m, describes a unique electron motion. However, it is known that there are two electrons having the same quantum numbers, n, ℓ, and m. To separate the two electrons in the orbital with the same quantum numbers, a fourth quantum number (QN), 's', arises in the solution of the SWE. In terms of the spatial coordinates, there is no difference between these two electrons. The difference arises when relativistic effects are taken into account. 's' represents the spin of the electron about its own axis and has the value, $\pm\frac{1}{2}$. If one electron has a spin value of $+\frac{1}{2}$, then the other has $-\frac{1}{2}$. Conventionally this difference is expressed by up and down arrows, as illustrated in following sections where the electron filling in the atomic wave functions is presented.

The solutions to the SWE for the hydrogen atom are listed in Table 1-2 for the radial function, R(r), and in Table 1-3 for the angular function, $Y(\theta,\varphi)$. The radial function consists of two parts - an exponential term and a polynomial term with degree of n (primary quantum number), the variable is given as $\rho = \frac{2r}{a_o}$. Since the exponential term decreases monotonously with r, the variation of the radial function depends primarily on the polynomial term. Note that the number of terms in the polynomial part is equal to n – 1 and that the sign of each term alternates with the order of ρ. For a function of polynomials of x with such alternating signs, in general, there must be a point other than x = 0 at which the value of function is '0'. This point is called node and represents the point where the value of the radial function is zero. The number of nodes for the radial function of an electron in the hydrogen atom is equal to $n - \ell - 1$. For radial functions with the same primary quantum number, therefore, the one with lower angular momentum has the larger number of nodes. This wavelike aspect of the radial function is illustrated in Fig.1-7 which shows a plot of the electron density distribution for the hydrogen atom along the radial direction which is represented by $r^2R(r)^{2*}$.

---

* $R(r)^2$ is the probability of finding the electron at r. In real 3-dimensional space, this probabilty, $R(r)^2$, must be multiplied by $4\pi r^2$ since any point on

The angular part of the wave function of the hydrogen atom consists of L(θ) and M(φ) which are solely functions of the polar angle(θ) and the azimuth angle(φ), respectively. L(θ) is expressed as a polynomial of sinθ and cosθ. The order of the polynomial is equal to the angular quantum number, ℓ. Since the angular range is $0 < \theta < 180°$, sinθ is always positive. Hence the sign of L(θ) follows the 'cosθ' part. The collection of points which make L(θ) equal to zero forms a conical cylinder around the z axis since r and φ do not affect the value of L(θ). The characteristic of L(θ) is that it is a function of odd or even order of cosθ. This implies that L(θ) = L(180°–θ) which, in consequence, means the existence of twins of conical cylinders across the xy plane. In relation to the azimuthal quantum number, *m*, the number of these conical cylinders is equal to ℓ - *m* if the twins of conical cylinders are counted as one.

In contrast to L(θ), M(φ) has a simpler form which is expressed as cos*m*φ or sin*m*φ. This is listed in Table 1-3. Since M is a function of sine or cosine of φ only which ranges from 0° to 360°, this results in a nodal plane or planes being formed on which M(φ) = 0. The number of nodal planes is equal to the azimuthal quantum number, *m*. The concept of a node as derived from the radial function or conical cylinder of nodes will be illustrated when the shape of orbitals is discussed.

s, p, d, f, g notation for atomic wave functions

It is not easy to represent graphically the wave functions for an atom for the full expressions listed in Tables 1-2 and 1-3. However, as shown in Eq.'s (34) and (35), wave functions can be simply represented by the 3 quantum numbers (n, ℓ, m) instead of by the completer Equations. The orbital, in turn, is usually expressed by $\Psi_{n,\ell,m}$. This notation for the wave function is further simplified by using an alphabetical notation, which is simply called the *s p d f* notation, and is related to the wave function as expressed by angular quantum numbers in Table 1-3. The terms *s* , *p* , *d* and *f* are used to represent total angular momentum of 0, 1, 2 and 3, respectively. Since the angular momentum, ℓ, is less than the primary quantum number, *n* , only s and p orbitals are allowed in the case of primary quantum number 2. In the case of n = 1, only s orbital is allowed. *m* also is an alphabetical term which designates azimuthal momentum.

As the number of ℓ is restricted by $\ell \leq n$, the azimuthal angular momentum, *m* , is limited in relation to total angular momentum. The relationship is that the absolute value of *m* is equal to or less than ℓ. Hence the

---

the sphere with radius r has the same probability and the surface area of the sphere is $4\pi r^2$.

Table 1-2 Radial Functions($R_{n,\ell}$) of Hydrogen-like Atom

R_{10}: 1s	$2\left(\frac{Z}{a_o}\right)^{3/2} e^{-\rho}$
R_{20}: 2s	$\frac{1}{2\sqrt{2}}\left(\frac{Z}{a_o}\right)^{3/2}(2-\rho)e^{-\frac{\rho}{2}}$
R_{21}: 2p	$\frac{1}{2\sqrt{6}}\left(\frac{Z}{a_o}\right)^{3/2}\rho\, e^{-\frac{\rho}{2}}$
R_{30}: 3s	$\frac{2}{81\sqrt{3}}\left(\frac{Z}{a_o}\right)^{3/2}\left(27-18\rho-2\rho^2\right)e^{-\frac{\rho}{2}}$
R_{31}: 3p	$\frac{4}{81\sqrt{6}}\left(\frac{Z}{a_o}\right)^{3/2}\left(6\rho-\rho^2\right)e^{-\frac{\rho}{3}}$
R_{32}: 3d	$\frac{4}{81\sqrt{30}}\left(\frac{Z}{a_o}\right)^{3/2}\rho^2 e^{-\frac{\rho}{3}}$
R_{40}: 4s	$\frac{1}{768}\left(\frac{Z}{a_o}\right)^{3/2}\left(192-144\rho+24\rho^2-\rho^3\right)e^{-\frac{\rho}{4}}$
R_{41}: 4p	$\frac{1}{256\sqrt{15}}\left(\frac{Z}{a_o}\right)^{3/2}\left(80\rho-20\rho^2+\rho^3\right)e^{-\frac{\rho}{4}}$
R_{42}: 4d	$\frac{1}{768\sqrt{5}}\left(\frac{Z}{a_o}\right)^{3/2}\left(12\rho^2-\rho^3\right)e^{-\frac{\rho}{4}}$
R_{43}: 4f	$\frac{1}{768\sqrt{35}}\left(\frac{Z}{a_o}\right)^{3/2}\rho^3 e^{-\frac{\rho}{4}}$
R_{50}: 5s	$\frac{2}{3\times 5^6\times\sqrt{5}}\left(\frac{Z}{a_o}\right)^{3/2}\left(9375-7500\rho+1500\rho^2-100\rho^3+2\rho^4\right)e^{-\frac{\rho}{5}}$
R_{51}: 5p	$\frac{4}{3\times 5^6\times\sqrt{30}}\left(\frac{Z}{a_o}\right)^{3/2}\left(3570\rho-1125\rho^2+90\rho^3-2\rho^4\right)e^{-\frac{\rho}{5}}$
R_{52}: 5d	$\frac{2\sqrt{2}}{5^6\times\sqrt{35}}\left(\frac{Z}{a_o}\right)^{3/2}\left(525\rho^2-70\rho^3+2\rho^4\right)e^{-\frac{\rho}{5}}$
R_{53}: 5f	$\frac{2}{3\times 5^6\times\sqrt{70}}\left(\frac{Z}{a_o}\right)^{3/2}\left(40\rho^3-2\rho^4\right)e^{-\frac{\rho}{5}}$
R_{54}: 5g	$\frac{8}{9\times 5^6\times\sqrt{70}}\left(\frac{Z}{a_o}\right)^{3/2}\rho^4 e^{-\frac{\rho}{5}}$
R_{60}: 6s	$\frac{1}{2^4\times 3^8\times 5\times\sqrt{6}}\left(\frac{Z}{a_o}\right)^{3/2}\left(174960-145800\rho+32400\rho^2-2700\rho^3+90\rho^4-\rho^5\right)e^{-\frac{\rho}{6}}$

Table 1-3 Angular wave functions $L_{l,m}(\theta)$ and $M_m(\varphi)$ for different l and m

l	m	s, p, d, f	$L_{l,m}(\theta)$	m	$M_m(\varphi)$
0	0	s	$\frac{\sqrt{2}}{2}$	0	$\frac{1}{\sqrt{2\pi}}$
1	0	p_0	$\frac{\sqrt{6}}{2}\cos\theta$		
	±1	$p_{\pm 1}$	$\frac{\sqrt{3}}{2}\sin\theta$	1	$\frac{1}{\sqrt{\pi}}\cos\varphi$
2	0	d_0	$\frac{\sqrt{10}}{4}\left(3\cos^2\theta-1\right)$	-1	$\frac{1}{\sqrt{\pi}}\sin\varphi$
	±1	$d_{\pm 1}$	$\frac{\sqrt{15}}{2}\cos\theta\sin\theta$		
	±2	$d_{\pm 2}$	$\frac{\sqrt{15}}{4}\sin^2\theta$	2	$\frac{1}{\sqrt{\pi}}\cos 2\varphi$
3	0	f_0	$\frac{3\sqrt{14}}{4}\left(\frac{5}{3}\cos^3\theta-\cos\theta\right)$	-2	$\frac{1}{\sqrt{\pi}}\sin 2\varphi$
	±1	$f_{\pm 1}$	$\frac{\sqrt{42}}{8}\sin\theta\left(5\cos^2\theta-1\right)$		
	±2	$f_{\pm 2}$	$\frac{\sqrt{105}}{5}\sin^2\theta\cos\theta$	3	$\frac{1}{\sqrt{\pi}}\cos 3\varphi$
	±3	$f_{\pm 3}$	$\frac{\sqrt{70}}{8}\sin^2\theta$	-3	$\frac{1}{\sqrt{\pi}}\sin 3\varphi$

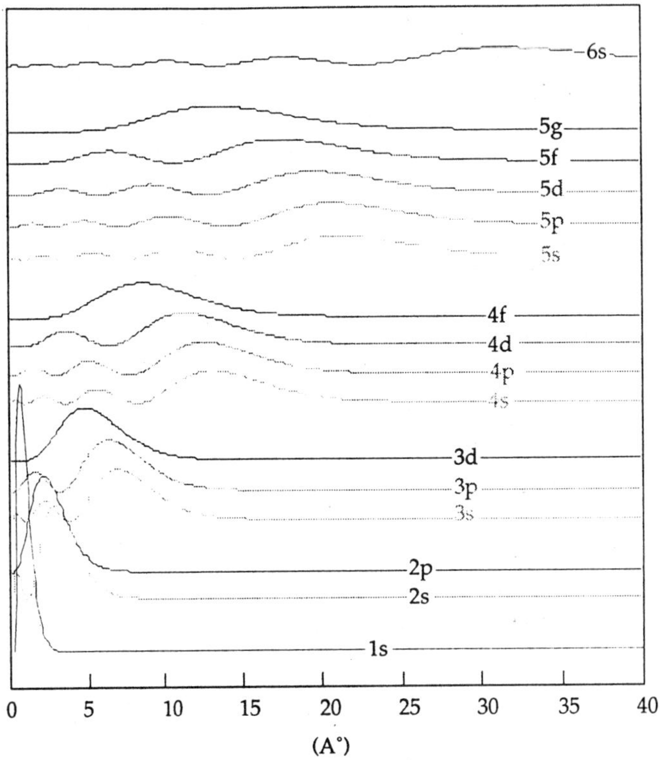

Figure 1-7 Plot electron density distribution with r by $r^2R(r)^2$

number of parameters representing m is equal to $2\ell + 1$. For example, the p orbital ($\ell = 1$) can have azimuthal angular momentum of +1, 0 and -1. p orbitals with m of +1, 0 and -1 are designated by p_+, p_o, and p_-. For the purpose of convenience, these are converted to p_x, p_y and p_z, where

$$p_x = \frac{p_+ + p_-}{\sqrt{2}},\ p_y = -\ i\ \frac{p_+ - p_-}{\sqrt{2}},\ \text{and}\ p_z = p_o \qquad (38)$$

The reason for this conversion is that it is not convenient to graphically depict p^+, p^0, and p^- and distinguish between them. Hence, in the s,p,d,f notation, parameters representing the azimuthal angular momentum do not represent the actual value. Rather they represent the absolute value of the azimuthal angular momentum. In the case of the p orbital, z denotes $|m| = 0$, and x and y both denote $|m| = 1$. Similarly, in the case of d orbital, x^2-y^2 and xy denotes $|m| = 2$

while xz and yz for $|m| = 1$ and z^2 for $|m| = 0$. These relationships are summarized in Table 1-4. The conversion of d_{+2}, d_{+1}, d_0, d_{-1}, and d_{-2} into alphabetical notation are as follows:

$$|m| = 2 \rightarrow \quad d_{xy} = -i\,\frac{d_{+2} - d_{-2}}{\sqrt{2}},\; d_{x^2-y^2} = \frac{d_{+2} + d_{-2}}{\sqrt{2}}$$

$$|m| = 1 \rightarrow \quad d_{yz} = -i\,\frac{d_{+} - d_{-}}{\sqrt{2}},\; d_{yz} = -i\,\frac{d_{+} - d_{-}}{\sqrt{2}}$$

$$|m| = 0 \rightarrow \quad d_{z^2} = d_o$$

I. Energy Levels in Atom

The solution to the SWE for the hydrogen atom was developed without considering the interaction between electrons since there is only one electron. If the atomic number is larger than 2, interactions between electrons must be taken into consideration as well the interactions between spin moment and angular moments. To obtain analytical solutions for the SWE similar to Eq.(26) for atoms other than hydrogen, therefore, is very difficult, but with simplifying

Table 1-4 s, p, d, f notation for orbitals with quantum number n of 3

n	ℓ	m	number notation	$\|m\|$	alphabetical notation
3	0	0	3s	0	3s
3	1	1	$3p_1$	1	$3p_x$*
3	1	-1	$3p_{-1}$		$3p_z$,$3p_y$
3	1	0	$3p_o$	0	
3	2	2	$3d_2$	2	$3d_{xy}$, $3d_{x^2-y^2}$
3	2	-2	$3d_{-2}$		
3	2	1	$3d_1$	1	$3d_{yz}$, $3d_{xz}$
3	2	-1	$3d_{-1}$		
3	2	0	$3d_0$	0	$3d_{z^2}$

* For p_x, p_y, d_{xy}, d_{x2-y2}, d_{xz} and d_{yz} , m is not the actual azimuthal angular momentum. Rather they connote the absolute value of m (azimuthal angular moment).

assumption the wave functions calculated for the hydrogen atom can be utilized to describe the actual atomic wave functions for other atoms. However, this simplified description of other atoms by the hydrogen atomic wave function does not consider those interactions other than electron-nucleus interaction and consequently the energy levels for hydrogen atomic orbitals in other atoms may have different order. The aufbau principle is a schematic description of energy levels of hydrogen atomic orbitals in other atoms without referring to detailed analytical description of energy levels of wave functions in a particular atom.

The aufbau principle describes the order in which the atomic orbitals are sequentially filled with electrons. This principle embodies three basic concepts, namely (a) the lowest energy state(Energy Principle); (b) electron occupation of the orbitals (Pauli Principle) and (c) electron spin orientation in degenerate orbitals (Hund's rule).

1. The Energy Principle → Order of Filling

This principle deals with the order of filling and states that electrons will go into the lowest available energy level first. In the hydrogen-like atom, the primary quantum number n determines the energy level of the orbital. Therefore electrons fill from the shell of lowest n. The order of the electron filling in the hydrogen-like atom should be as follows:

$$1s < 2s\ ,\ 2p < 3s\ ,\ 3p\ ,\ 3d < 4s\ ,\ 4p\ ,\ 4d\ ,\ 4f < 5s\ ,\ 5p\ ,\ 5d\ ,\ 5f$$

This ordering of energy levels is calculated from the one-electron system, where the energy of an orbital is due solely to the interaction between the electron and the nucleus. In more complex atoms, however, the energy levels of orbitals are determined not only by the interaction of electrons with the nucleus but also by the interaction between electrons themselves. When these interactions are considered, the ordering of the energy levels and of the electron fitting of the orbitals is changed to the following sequence:

$$1s < 2s < 2p < 3s < 3p < 4s \sim 3d < 4p < 5s \sim 4d < 5p < 6s < 5d \sim 4f < 6p < 7s < 6d < 5f$$

2. Pauli Exclusion Principle → Orbital Filling

This principle deals with electron filling in the same orbital and states that no two electrons can have all four QN's(n, ℓ, m, s) identical . This means that an orbital, defined by permitted values for (n, ℓ, m) can contain two electrons but only if s = +1/2 or −1/2, i.e. an orbital can contain a maximum of two electrons but with opposite spins.

3. Hund's Rule

While the Energy principle is concerned with the filling order for electrons going into orbitals with different primary quantum numbers, Hund's rule deals with the filling order for electrons going into orbitals with the same primary and angular quantum numbers. For these degenerate orbital, Hund's rule states that the lowest energy configuration is obtained by the electrons being in their highest spin states. Thus, two or more electrons in equivalent orbitals (i.e. degenerate orbitals having the same n with different ℓ) will exhibit parallel spins, because this is the state of lowest energy. Nitrogen illustrates how electrons fill orbitals according to Hund's rule. The nitrogen atom has seven electrons. The first four electrons fill 1s and 2s orbitals according to the Energy principle above. The next 3 electrons go into three 2p orbitals. Since there are three 2p orbitals, two filling modes of 3 electrons into the 2p orbitals are possible. They are $2p^2\,2p^1$ or $2p^1\,2p^1\,2p^1$ (= $2p^3$). The Energy principle does not distinguish between these configurations with respect to energy level. However, according to Hund's rule, the latter is a more stable status than the first. Therefore, the electron configuration for the nitrogen atom is as follows

N (Z=7); electron configuration: $1s^2 2s^2 2p^3$

In some elements the effect of Hund's rule on electron filling is more dominant than the Energy principle. Since half shells are particularly stable (Hund's Rule), some distortion in the filling of the atomic orbitals does occur, although these violate the Energy principle. According to Energy principle, 4s orbitals should be filled before 3d orbitals since the 4s energy level is lower than the 3d. However, this sequence

Table 1-5 Electronic configuration of transition metals

Element	Atomic Number	Electronic configuration
Sc	21	$[Ar]3d^1 4s^2$
Ti	22	$[Ar]3d^2 4s^2$
V	23	$[Ar]3d^3 4s^2$
Cr	24	$[Ar]3d^5 4s^1$
Mn	25	$[Ar]3d^5 4s^2$
Fe	26	$[Ar]3d^6 4s^2$
Co	27	$[Ar]3d^7 4s^2$
Ni	28	$[Ar]3d^8 4s^2$
Cu	29	$[Ar]3d^{10} 4s^1$

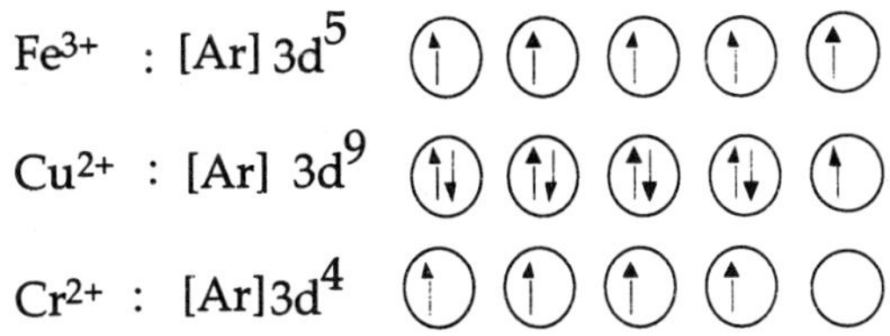

Figure 1-8 Electron configurations for ions of transition metal elements

is broken in Cr where the electron configuration is $[Ar]3d^54s^1$ not $[Ar]3d^44s^2$. This deviation in electron filling from the Energy principle is attributed to a particularly stable d^5 configuration, which energy gain easily compensates for the added energy needed to fill the 3d orbital. Like half filled shells in the case of Cr, fully filled shells also are very stable. In the case of Cu, because of the greater stability of the $3d^{10}$ (filled subshells), the 4s level gives up an electron to maintain this stability, resulting in a configuration of $[Ar]3d^{10}4s^1$ instead of $[Ar]3d^94s^2$.

Similar phenomena showing a tendency toward half filled or fully filled shells occur in the ionization process of some transition metal elements. For example, the 4s electrons rather than 3d electron are removed in the ionization for Fe^{3+}, Cu^+, and Cr^{2+}. Thus we have what are shown in Fig.1-8.

J. Shell, Subshells, Orbitals

This is another scheme illustrating the energy level of atomic wave functions. In the section G, the energy of atomic the wave function was seen to depend not only on the primary quantum number but also on the angular moment quantum number. For this reason, wave functions are first categorized into groups called shells. Each shell contains wave functions with the same primary quantum number. The shell is further divided into a smaller group called a subshell in which the wave functions have the same angular moment quantum number, ℓ. The lowest level in this scheme is the orbital which is equivalent to the wave functions characterized by n, ℓ, and m. In this sense, the orbital is the same as the wave function for each electron.

Since the quantum number n largely determines the energy level, orbitals in a shell of higher primary QN have a higher energy level than orbitals in a shell of lower primary QN. A subshell, therefore, is a group of orbitals with the same angular quantum number, ℓ, in the same shell. As with the shell, the angular QN determines the energy level of the subshell within the shell. The higher the angular QN, the higher the energy level. The subshell is further degenerated into orbitals which have different azimuthal angular momentum. Since, in general, orbitals within the same subshell have the same energy level,

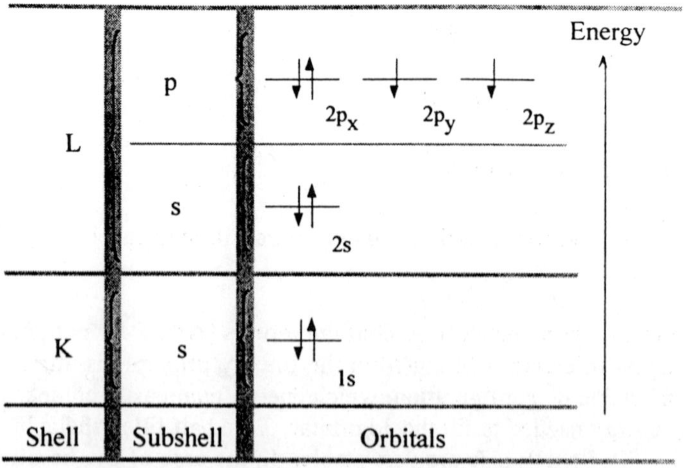

Figure 1-9 Illustration of *s,p,d,f* notation by the electron configuration of Nitrogen

the orbitals in the subshell are called degenerate orbitals. Each orbital can accommodate two electrons with opposite spin states. Fig.1-9 shows the configuration of electrons of nitrogen atom illustrating the energy scheme by shell structure along with *s,p,d,f* notation of N: $1s^2 2s^2 2p^3$.

K. Shapes of Orbitals in Hydrogen-like atoms

The accurate description of orbital is to display electron density at each point in space. However, it is not possible to indicate electron densities for each point in the whole space. Therefore, the concept of shape is utilized to characterize the orbitals. The shape of an orbital is not the same as the concept of a solid material which has an outer boundary. Rather, the shape of the orbital is a theoretical definition to characterize the overall change of electron density in space. There are two distinctive ways for describing the shape of orbitals. One is contour method and the other is functional method. In the contour method, contours for a constant electron density are mapped. This is shown in Fig. 1-10 for $2p_z$ orbital. Electrons can be depicted by either Ψ^2, or Ψ. The contour itself expressed by Ψ is not different from that of Ψ^2. The difference is that the phase is expressed by (+) and (–) signs on each contour surface displayed by Ψ^2.

The functional method shows the change of electron density with coordinates. In 3-dimensional space, a function with only two variables can be

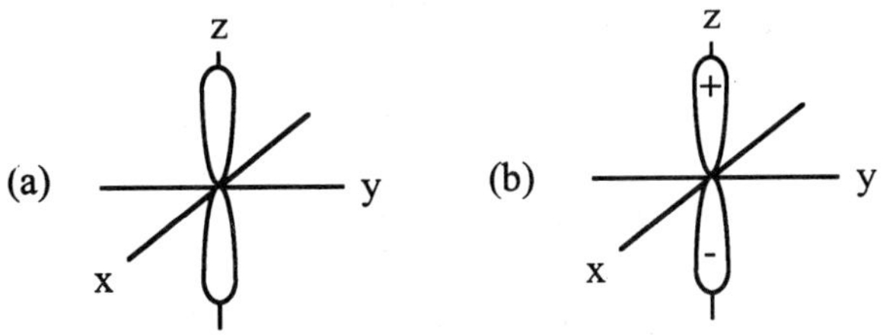

Figure 1-10 Contour map of $2p_z$ by (a) Ψ^2 and (b) Ψ

displayed. The radial function of the hydrogen atom is a illustration of orbitals with one variable (r) which is distance from the center of the nucleus.

Usually, in the contour method, the contour of an orbital forms one or several closed surfaces. If an orbital has more than one closed contour surfaces, they are bounded by a nodal plane. The nodal plane is defined as a plane where electron density is zero. From the contour method, the nodal planes can be easily seen. For example, in $2p_z$ orbital shown above, one closed surface is formed above xy plane and the other below the xy plane which is a nodal plane. In contrast to $2p_z$, the contour of 2s orbital shows only one closed surface which is a sphere. If nodal plane is not assumed to be a flat plane, there is a nodal plane in 2s orbital, too. This nodal plane cannot be depicted in contour method since the other contour is hid by the outer one. This nodal plane can be inferred from the radial function of 2s orbital. In Fig.1-7, 2s orbital shows one node in the radial direction. This node connotes the sphere in space. Hence the nodal plane of 2s orbital is a sphere with radius equal to the radial node. In a wide sense, the nodal plane can be divided into spherical and nonspherical. The spherical nodal planes are formed due to the nodes in radial wave function and the nonspherical nodal planes are formed due to the angular wave functions. d_{z2} exhibits the nonspherical nodal planes which are conical planes. They are two cones above and below xy plane as shown in Fig.1-10. If the nodal plane is defined in this wide meaning, the number of nodal plane is equal to the primary QN – 1.

$$n - 1 = N_f + N_s \quad (39)$$
$$N_f = \ell$$
$$N_s = n - \ell - 1$$

Here N_{ns} and N_s represents the number of nonspherical (flat) nodal plane and spherical nodal plane, respectively. Electron density at $r = \infty$ is 0. Another spherical nodal plane can be drawn at infinity. If this nodal plane is

Table 1-6 The relationship between n, ℓ, N_f and N_s

Orbitals	n	ℓ	N_f	N_s
1s	1	0	0	0
2s	2	0	0	1
2p		1	1	0
3s	3	0	0	2
3p		1	1	1
3d		2	2	0
4s	4	0	0	3
4p		1	1	2
4d		2	2	1
4f		3	0	0
5s	5	0	0	4

included in the number of nodal planes, the total number of nodal plane is equal to the primary QN. Hence Eq(39) can be expressed as

$$n = N_f + N^*_s \tag{40}$$

Here N^*_s is the total number of spherical nodal planes including the one at $r = \infty$. s orbitals are spherically symmetric and, therefore, the probability of finding the electron at a given distance r from the nucleus is identical in all directions. Regarding nodal plane, N_f is 0 for s orbitals. Other orbitals has nonzero N_f which means they are not spherical in shape. Unlike s orbital, p orbitals are dumbbell shaped and are determined by the quantum numbers l and m. The d and f orbitals are more complex but are also nonspherical with respect to the nucleus, and exhibit directionality similar to p orbitals.

The contours of s, p, and d orbitals are shown in Fig.1-11. As shown in Fig.1-7, within the same primary QN, the maximum peak moves toward the nucleus as l increases. Hence the positions of maximum peaks are in the ascending order of g < f < d < p < s in the same shell. The maximum peak moves outward with increase of primary QN for the same kind subshells. For example, the positions of maximum peaks for s orbitals are in the order of 1s < 2s < 3s < 4s < 5s < 6s.

L. Hydrogen-like Atoms

1. Limitation of Hydrogen-like Atoms

The SWE for atoms with atomic number larger than 2, must be expressed as:

$$\nabla^2\Psi + \frac{8\pi^2 m}{h^2}\sum_{i,j(i\neq j)}^{Z}\left(E - \frac{Ze^2}{\sqrt{(x^2+y^2+z^2)}} + \frac{e^2}{\sqrt{(x_{ij}^2+y_{ij}^2+z_{ij}^2)}}\right)\Psi = 0 \quad (41)$$

The energy term for the interaction between electron and the nucleus is corrected by inserting Z since the charge of the nucleus is Ze. Also the energy term for interaction between electrons are supplemented since there are more than one electron. However, the analytical solution of Eq.(41) is not obtainable like the SWE for hydrogen atom. The approach to the real solution is based on the assumption that the wave function obtained from SWE without considering the interaction between electrons can describe the real solution of Eq.(41). It is called hydrogenlike atom whose wave functions have the solutions of Eq.(42). Hydrogenlike means that there is no other electrons to be considered in solving the SWE for the electron.

$$\nabla^2\Psi + \frac{8\pi^2 m}{h^2}\left(E - \frac{Ze^2}{\sqrt{(x^2+y^2+z^2)}}\right)\Psi = 0 \qquad (42)$$

The hydrogenlike atoms have the same angular wave functions as the hydrogen atom. The only difference is the radial functions are modified by replacing the nucleus charge with Ze. The radial electron densities are plotted in Fig.1-12 for atomic numbers 11, 19, 24, and 74 which corresponds to Na, K, Cr, and W, respectively. Since the electron densities in Fig.1-12 are not real densities, they are indicated by atomic numbers rather than by names. To compare the hydrogenlike atoms with the real atoms, the electron density is plotted along the radial direction. If atomic radius is approximately defined as the position of maximum electron peak, the atomic radii in the hydrogenlike atoms are always smaller than real atoms. In Fig.1-12, the real atomic radii are indicated by solid arrows and the ionic radii by gray arrows, neglecting interactions between electrons. In actual atoms, electrons interact repulsively and the outer electrons are pushed away from the nucleus by the repulsive interactions with the inner electrons. The atomic radii in consequence increases due to this repulsive force between electrons. In the case of Na, as shown in Fig.1-12, the atomic radius defined as the point where the electron density is maximum, is about the half the actual value indicated by dark arrow. This discrepancy is due to the omission of the interactions between electrons. Fig.1-12 shows that, in general, the atomic radii predicted by the hydrogenlike atomic wave functions assuming the radius is the point where electron density is the maximum, are even smaller than actual ionic radii which is measured even after stripping off valence electrons. In order to make closely approximate the real

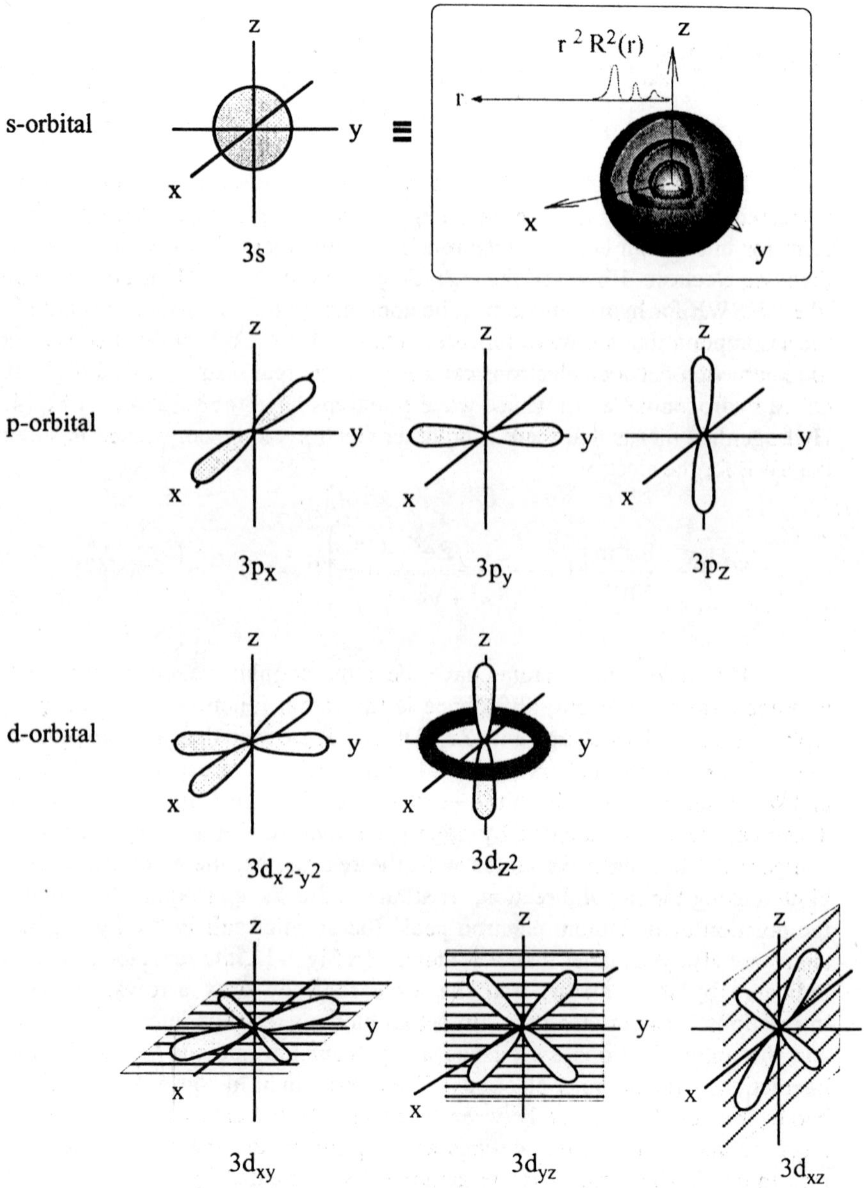

Figure 1-11 Shape of s, p and d orbitals

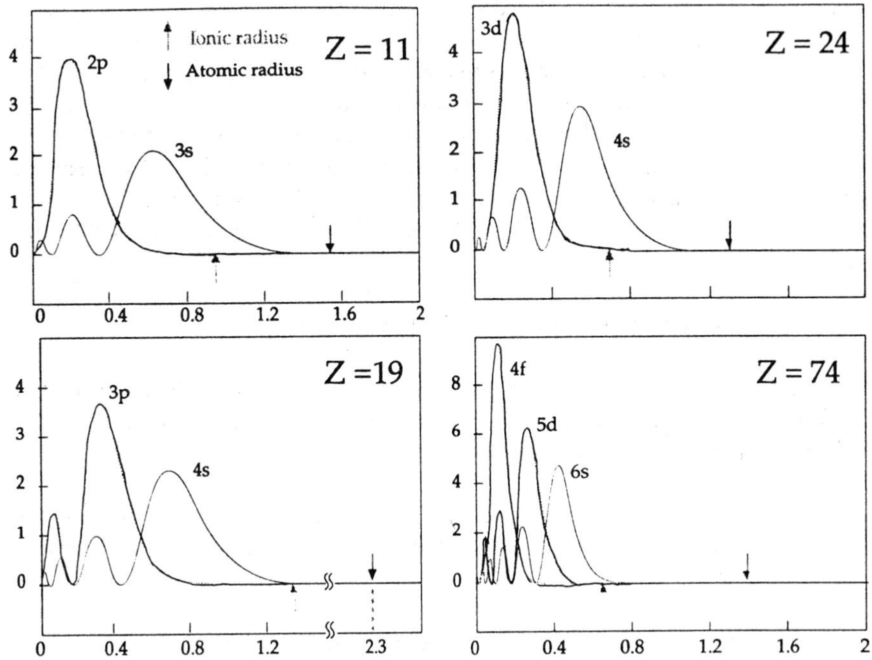

Figure 1-12 The electron density distribution calculated by the $r^2R(r)^2$ for hydrogenlike atom

value for the wave functions, Slater proposed the effective nucleus charge which takes into account the repulsive interactions between inner electrons and outer electrons.

2. Slater's Model

To compensate the effect of electron-electron repulsive interaction on the wave function of the hydrogenlike atoms, Zener proposed the empirical equation given in Eq.(43). This expansion is similar to the radial wave functions of hydrogenlike atoms in Table 1-2 except that there are no nodes.

$$R'(r) = r^{n^*-1} e^{-(Z-Z_0)r/a_0 n^*} \qquad (43)$$

Here n* is the modified primary QN for the orbital and 'Z_0' is a constants representing the shielding effect of inner electrons on the electrons in the orbital.

Originally Zener applied this function for lightweight atoms upto F. Later Slater extended this analysis for the heavy atoms. By introducing a shielding constant and modifying the primary QN, this equation approaches the real wave functions with respect to electron-electron interactions. Slater provided also some schematic rules by which determine the shielding constant and effective primary QN. Using this schematic, the relative of electron-electron interactions in a atom can be estimated without requiring a solution to the SWE for the atom. This function with an effective primary QN and shielding constant, satisfactory values for the atomic size and potential energy for each electron could be readily estimated.

2-1. Slater's Rules for Estimation of Shielding Constant(Z_0)

Slater's Rules are empirical guidelines based on the average behavior of the various electrons in an atom, but they are simple estimates to be made of the shielding phenomenon, as follows:

(i). Write out the electronic configuration for the each element, grouping s and p electrons in the same shell. d and f are separated from the s, p group, viz.

(1s) (2s,2p) (3s,3p) (3d) (4s,4p) (4d) (4f) (5s,5p) etc.

Let's assume the electron being considered is in the n-th shell, then:

(ii). Electrons in the group right to that of this electron contribute nothing to Z_0.

(iii). Electrons in the same group as the electron being considered, shield the electron to an extent of 0.35 each. For 1s group, 0.3 instead of 0.35.

(iv). Electrons in the group with (n-1)th shell and to the left of the electron in the nth shell shield the electron to an extent of

0.85 if the electron considered is a s,p electron
1.00 if the electron considered is not a s,p electron

(v). Electrons in the (n-2)th shell shield completely. i.e. to an extent of 1.0, regardless of whether is an s, p, d, or f electron

Example 1 Nitrogen ($_7N$)

a. Rule 1 : $(1s)^2$ $(2s, 2p)^5$
b. Rule 3 : $4\times 0.35 = 1.40$ - Shielding constant from 2s, 2p electrons

c. Rule 4 : $2 \times 0.85 = 1.70$ - Shielding constant from 1s electrons

The total shielding constant for 2s,p electron from other electrons is $Z_o = 3.1$

Example 2 Nitrogen ($_{30}Zn$)

a. Rule 1 : $(1s)^2$ $(2s,2p)^8$ $(3s,3p)^8$ $(3d)^{10}$ $(4s,4p)^2$
b. Rule 3 : $(2\text{-}1) \times 0.35 = 0.35$- Shielding constant from 4s, 4p electrons
c. Rule 4 : $(8 + 10) \times 0.85 = 15.30$ - Shielding constant from 3s, 3p, 3d electrons
d. Rule 5 : $(2 + 8) \times 1.0 = 10$ - Shielding constant from 1s, 2s, 2p electrons

The total shielding constant for 4(s,p) electron from other electrons is

$$Z_o = 25.65$$

2-2. Atomic Size Estimation with Shielding Constant(Z_o) and Effective Primary QN(n^*)

The net nuclear charge which is effective on an electron can be expressed as

$$Z_{eff} = Z - Z_o \tag{44}$$

As seen in the above examples, the effective nuclear charge is much smaller than the real nuclear charge as electrons fill the outer orbitals. This illustrates why the atomic radii calculated from the hydrogenlike wave functions are smaller than the real values. As indicated atomic radius is approximately defined as the position of maximum electron density along the radial direction. With reference to Eq.(43), this maximum peak occurs at the position where the derivative of $r^2R'(r)^2$ is equal to 0, ie:

$$r^2R'(r)^2 = r^2 \times (r^{n^*-1}e^{-Z^*r/a_o n^*})^2 = r^{2n^*}e^{-2Z^*r/a_o n^*}$$

$$\frac{\partial}{\partial r} r^{2n^*}e^{-2Z^*r/a_o n^*} = \frac{2nr^{2n^*-1}}{e^{2Z^*r/a_o n^*}} - \frac{2r^{2n^*}Z^*}{a_o e^{2Z^*r/a_o n^*} n} = 0 \tag{45}$$

$$r^* = \frac{(n^*)^2}{Z^*} a_o \quad [\, a_o \text{ is Bohr's radius} \approx 0.53\text{A}^\circ \,] \tag{46}$$

Since the electron in the outermost orbital determines atomic size, Eq.(46) applies to the outermost electron. Along with the shielding constants, Slater proposed corrections to the primary QN, as follows:

n	1	2	3	4	5	6
n*	1	2	3	3.7	4.0	4.2

Using the n* values above and shielding constants, the radii of atoms such as Li, Na, K through Zn are listed in Table 1-7. In general the calculated atomic radii are larger than the actual values. The largest deviation is in Cr by 90% and Li shows the smallest with less than 10%. Although the calculated radii deviate from the actual values, they do illustrate the variation of atomic size with atomic number. From K through Zn, n* does not change while the effective nuclear charge continuously increases. According to Eq.(46), this trend in n* and Zeff with atomic number brings about an decrease in atomic size as atomic number increases from K through Zn. This can be explained by differences in the rates of increase of n* and Z_{eff} with atomic number. Since n* increases more rapidly than Z_{eff}, the atomic radii consequently increases with atomic number.

Table 1-7 Electronic configuration and shield constants of atoms Li, Na and K through Zn

	Z	1s	2s	2p	3s	3p	3d	4s	Z_o	Z_{eff}	r_{cal}	r
Li	3	2	1						1.7	1.3	1.63	1.55
Na	11	2	2	6	1				8.8	2.2	2.17	1.90
K	19	2	2	6	2	6		1	16.8	2.2	3.3	2.35
Ca	20	2	2	6	2	6		2	17.2	2.85	2.55	1.97
Sc	21	2	2	6	2	6	1	2	18	3	2.42	1.62
Ti	22	2	2	6	2	6	2	2	18.9	3.15	2.3	1.47
V	23	2	2	6	2	6	3	2	19.7	3.3	2.2	1.34
Cr	24	2	2	6	2	6	5	1	21.1	2.95	2.46	1.30
Mn	25	2	2	6	2	6	5	2	21.4	3.6	2.02	1.35
Fe	26	2	2	6	2	6	6	2	22.3	3.75	1.93	1.26
Co	27	2	2	6	2	6	7	2	23.1	3.9	1.86	1.25
Ni	28	2	2	6	2	6	8	2	24	4.05	1.79	1.24
Cu	29	2	2	6	2	6	10	1	25.3	3.7	1.96	1.28
Zn	30	2	2	6	2	6	10	2	25.7	4.35	1.67	1.38

M. The Periodic Chart

The periodic chart is a grouping of the elements in terms of the systematic variation of the valence electrons in atoms of different atomic numbers. The valence electrons are those electrons in the outermost occupied shell which give the chemical and bonding characteristics of the atom. Elements in the same period have the same shell (same primary quantum number). Table 1-8 and Figure 1-13 illustrate the grouping of elements in the periodic chart.

1. Short-Hand Notation

Since each period ends in a rare gas with a closed shell configuration, this fact may be used as a starting point for succeeding elements and only the valence electrons need be written in this short hand notation, thus:

Na: [Ne]$3s^1$ Mg: [Ne]$3s^2$ Cl: [Ne]$3s^2 3p^5$
K : [Ar]$4s^1$ Ca : [Ar]$4s^2$ Cr: [Ar]$3d^5 4s^1$

2. Classification of the Elements

The elements are typically classified based on the orbital filling in accordance with the Energy Principle for atomic orbitals. The order of energy level in atomic orbitals is as follows:

$$\underbrace{1s}_{1st} < \underbrace{2s < 2p}_{2nd} < \underbrace{3s < 3p}_{3rd} < \underbrace{4s \sim 3d < 4p}_{4th} < \underbrace{5s \sim 4d < 5p}_{5th} < \underbrace{6s < 5d \sim 4f < 6p}_{6th} < \underbrace{7s < 6d < 5f}_{7th\ group}$$

The orbitals can be combined in groups with the same primary quantum number. Within each group the lowest energy orbital is the s orbital which starts each group. The 3d orbital belongs to the 4th group instead of the 3rd

Table 1-8 Periodical Chart Grouping

Period #	n	Shell	# of elements in shell	Period Ending	Rare Gas configuration
1	1	K	2	$1s^2$	He
2	2	L	8	$2p^6$	Ne
3	3	M	8	$3p^6$	Ar
4	4	N	18	$4p^6$	Kr
5	5	O	18	$5p^6$	Xe
6	6	P	32	$6p^6$	Rn
7	7	Q		—	

group although the primary quantum number is 3. Likewise 4d orbital belongs to the 5th group and the 4f and 5f orbitals belong to the 6th and 7th groups, respectively. A characteristic of this grouping is that the s and p orbitals are located at the beginning and the end of each group, respectively, with the d and f orbitals located in between s and p orbitals. If elements are arranged by a atomic number in a table in which the first group occupy the first row, the second group takes the second low and so on, it will be like the table in Fig. 1-13. Another characteristic of this table is that elments with s orbital valence electrons only are arranged in the left columns and elements with p orbital valence electrons are grouped and arranged in the right columns, repectively. Those elements with d or f orbital electrons as valence eletrons are grouped and arranged in between s and p orbital groups. Beside this arrangement, there is a special group called the rare gas elements. In a strict sense this group belongs to the p orbital group but it is treated seperately since elements in this group have completely filled p valence orbitals. In Fig. 1-13, elements in the same column have the same configuration of valence electrons. For example, the electronic configuration of Na and K are $[Ne]3s^1$ and $[Ar]4s^1$ in the short notation. Both atoms have only one s-orbital valence electron. Due to the same configuration of valence electrons, elements in the same column of the periodic table have very similar chemical chracteristics and can be grouped as a family. Similar chemical properties repeat as the atomic number finishes one row in the periodic table and reaches the same column in the next row. It is this repetition in the mode or row by row characteristics which gives to the Table the name 'periodic table of elements'. Each row is also called a period. The first row is called the first period and the second row the second period and so on.

i. Type I Elements - Rare gas block

These are the elements with rare gas configuration. Because of the closed shell, they exhibit high ionization energies and rarely enter into any chemical reaction. They are extremely stable, and most exist as gases.

ii. Type II Elements - s and p block

These are s-block and p-block elements consisting of atoms with only p and/or s valence electrons outside a closed shell. By losing or gaining electrons they easily acquire a rare gas configuration. These elements tend to form stable ions of known valence.

s-orbitals: Na, K, Mg, Ca, Ba, etc.

p-orbitals: Cl, O, N, Al, Si, etc.

iii. Type III Elements - d block

The energy level of d orbital is very close to s orbital. This closeness results in the easy trasition of electrons from d to s or vice versa. Hence this block is usually called transition series. Transition series elements in which the outer s orbitals are occupied and the inner d orbitals are being progressively filled. Also, variable valency states develop and in the oxide state many are brightly colored because the energy levels in the d states are close together.

e.g. Fe, Co, Ni, Cr, Mn, etc.

iv. Type IV Elements - f block

These are Rare Earth Series Elements with filled outer *n*s and *n*p orbitals and there is a progressive filling of the inner *(n-2)*f orbital takes place after *n*s and *n*p orbitals are filled. Again, because of similarities in the energies of the f orbitals with *n*s and *n*p orbitals and also because of the good shielding by the outer orbitals, valencies and chemical properties tend to be similar to transition series elements in which the inner *(n-1)*d orbital is progressively filled after outer *n*s and *n*p orbitals are filled.

N. Magnetic Properties of Atoms

The spin state of electrons in an atom determines the magnetic property of the atom. Depending on this spin state of the electrons, atoms may be categorized into diamagnetic and paramagnetic.

(i). Diamagnetic atoms (or ions) are those in which there are no uncoupled electron spins, e.g. Mg: [Ne]$3s^2$

(ii). Paramagnetic atoms or ions are those with uncoupled electron spins in the orbitals giving rise to a net magnetic spin or moment.

The magnetic moment of an uncoupled electron is expressed in Bohr magnetons, μ_B. One Bohr magneton designates the magnetic moment of one uncoupled electron. Therefore, the net magnetic moment for the Fe^{3+} atom is five Bohr magnetons (μ_B) since it has five uncoupled electrons as shown in Fig. 1-8.

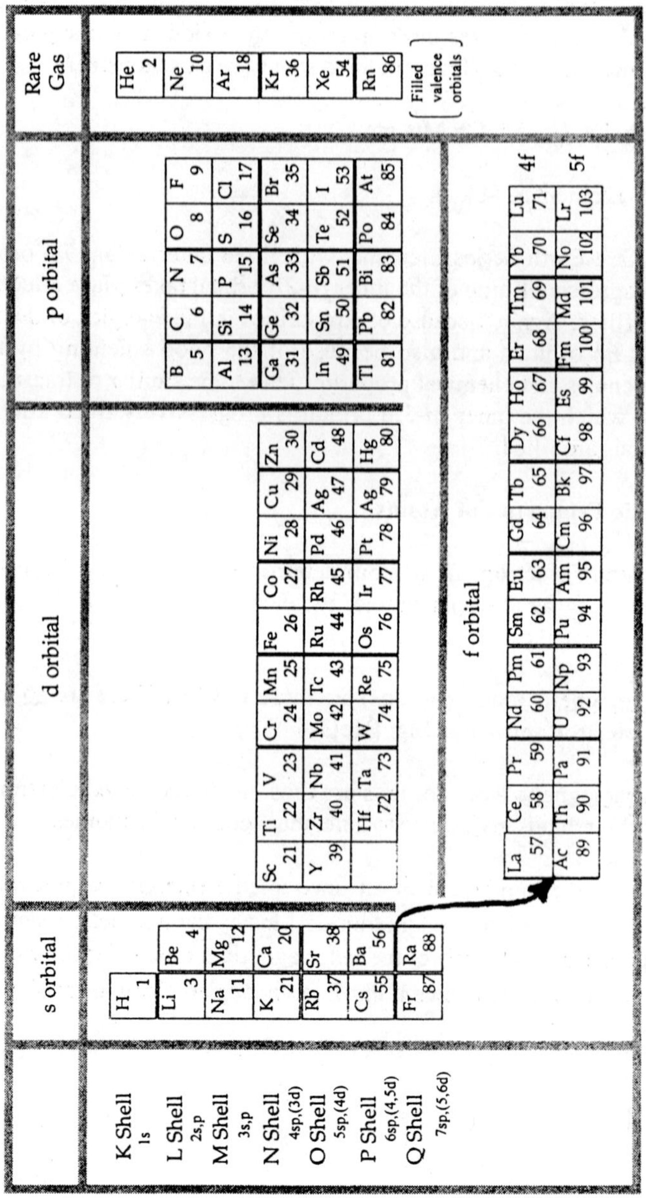

Figure 1-13 Periodic table of the elements

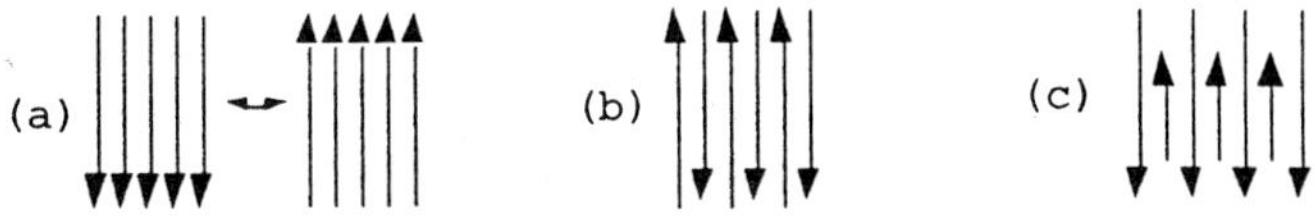

Figure 1-14 Schematic illustration of magnetic moment structures in (a)ferro, (b) antiferro and (c) antiferrimagnetic solids

Although paramagnetic atoms have net magnetic moments, the overall magnetic moment of a crystal may be zero due to interaction of the atoms in the structure. Depending on the nature of this interaction, atoms may be classified as ferromagnetic, antiferromagnetic or ferrimagnetic. A crystal is called a ferromagnetic if consisting atoms are paramagnetic and their directions are aligned in one direction and switchable in the opposite direction. Antiferromagnetic crystal is one whose total magnetic moment is zero because magnetic moment of atoms aligned in one direction is compensated by other atoms whose magnetic moments are aligned in the opposite direction. In ferrimagnetic crystal, magnetic moments of atoms are arranged in a similar way to antiferromagnetic crystal. However, if the magnetic moment in one direction is larger than the other direction, the result is a non-zero overall magnetic moment. This classification of crystals of paramagnetic atoms are illustrated in Fig.1-14. Even in ferromagnetic crystal, both up and down magnetic moment coexist. The difference between ferromagnetic and ferrimagnetic is that up and down moments coexist intrinsically within the crystal for ferrimagnetic while they are formed by two regions (domains) of crystal in ferromagnetic. Therefore, only one direction of magnetic moment exists in a domain of ferromagnetic crystals.

SECTION 2 MOLECULAR STRUCTURE

A. Valence Electrons

Elements do not exist as atoms except for the rare gas elements. Rather they form molecules or multi-atomic compounds which may be molecules, crystals, or glassy compounds. By forming the multi-atomic compound, atoms get lower energy than as separated from other atoms. As atoms approach one another, the interaction between these are concerned with those electrons in the outermost orbitals. An electron in the outermost orbital is called a valence electron. Since the valence electron possesses the highest energy, this electron is more easily removed or modified from the atomic state than inner orbital electrons. The inner electrons do not change their characteristics by bondings. These electrons, therefore, primarily determine the nature of the chemical bonding of the atom. There have been used several characteristic features of atoms which are helpful to describe the chemical bonding in a macroscopic way without referring to analytical calculation of quantum mechanics. Those are ionization potential, electron affinity, and electronegativity. Since quantum mechanical description of molecules seem even more difficult than atoms, these are utilized to help the understanding of the chemical bondings for molecules and solids in a simpler way.

Chemical bonds can be said that they are redistribtution of electrons from two different atoms and lower the total energy by binding them. Since the inner electrons are too strongly bound to the each nucleus, the last sentence can be rephrased as the redistribution of *valence* electrons from two atoms. Chemical properties of molecules or solids are determined by this redistribution of electrons around the nuclei. Due to the redistribution of valence electrons, the chemical properties of molecules are different from that of atoms. In other words, the redistribution can be represented by giving up by one atom and taking up of electrons by the other atom. In this respect, ionization energy measures how easily electron is removed from the nucleus. Electron affinity measures how strongly the nucleus attracts an electron. Electronegativity is a relative scale to measure the electron affinity or ionization energy.

B. Ionization of Atoms

Ionization refers usually to the removal of the valence electrons from the attractive field of the nucleus since the valence electrons have the higher enrgy than the inner electrons. The removal of electron leaves the atom (ion) with a net positive charge. The ionization can occur not only in an neutral atom but also in ions. Hence it can produce positives charges on the nucleus more than +1. For the semi-quantitative description of ionization energy, Slater's rule is used for the calculation of effective nucleus charges.

1. Hydrogen Atom

Removal of electron means that distance of electron from the nuclues changes from $r_n \rightarrow r_\infty$. For hydrogen atom, this corresponds to a transition of an electron from n = 1 to n = ∞. Since the energy of an electron in the nth shell is $-\frac{2\pi^2 me^4}{h^2}\cdot\frac{1}{n^2}$, the ionization energy is

$$IE = -\frac{2\pi^2 me^4}{h^2}\cdot\frac{1}{n^2} \qquad : \quad Z=1 \text{(H atom)} \tag{47}$$

2. Multiple Electron Atom

Let's assume the valence electron of a multiple electron atom is in the n-th shell. The energy of the valence electron is $-\frac{2Z^2\pi^2me^4}{h^2}\cdot\frac{1}{n^2}$ on the basis of the hydrogenlike atom. Considering the shielding effects of inner electrons upon the attractive force of nucleus on the valence electron, Z is replaced by Zeff which is equal to Z - Z_o. The energy of multiple electron atom is expressed as :

$$E_n = -\frac{2(Z-Z_o)^2\pi^2me^4}{h^2}\cdot\frac{1}{n^2} \tag{48}$$

where: E_n = energy of electron in n^{th} shell
n = 1, 2, 3, 4...; primary QN.
Z = charge on nucleus
Z_o = shielding factor (Z_{eff} = Z - Z_o = effective charge)
m = electron mass (9.1×10^{-31} kg)
e = electron charge (1.6×10^{-19}) Coulomb (C)
h = Planck's Const. (6.625×10^{-34} J. sec)

Ionization scheme for the Na atom is illustrated in Fig.1-15.

3. Second and Third Ionization Energy (IE)

Ionization energy is defined as the energy needed to remove an electron from the valence orbital of isolated atom in its gaseous state. Since this energy must be

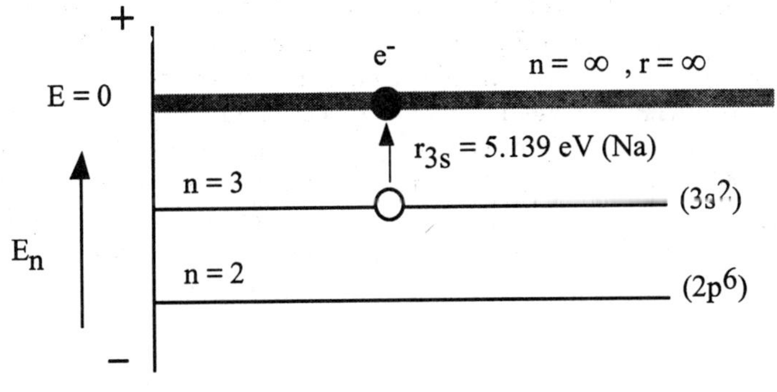

Figure 1-15 Model of electron removal for Na

supplied it is considered a positive energy. It may be possible to remove several valence electrons in turn, and the successive energies required would be designated as the first, second, third ionization energys. Thus for cation M the first, second and third ionization energiess are expressed as:

1) $M(g) \rightarrow M^+ (g) + e^-$; I_1 : 1st IE
2) $M^+ (g) \rightarrow M^{2+} (g) + e^-$; I_2 : 2nd IE
3) $M^{2+}(g) \rightarrow M^{3+} (g) + e^-$; I_3 : 3rd IE

Generally $I_3 > I_2 > I_1$. For example, the first, second and third IE of Al are 5.98, 18.82 and 28.40eV, respectively. With the short notation for the ionization of Al, Al^{+3} shows the electon configuration of Ne.

$$Al \rightarrow Al^{3+} : [Ne]3s^2 3p^1 \rightarrow [Ne]3s^2 \rightarrow [Ne]3s^1 \rightarrow [Ne]$$

4. Ionization Energy Predicted by Slater's Model

Ionization energies for elements in the second period of the periodic table are calculated according to Eq.(48). For this, effective nucleus charges are calculated according to Slater's rule. IE's calculated in this way are listed in Table 1-9 and the first ionization energies are plotted in Fig.1-16. A distinctive aspect shown in this figure is that ionization energy increases with atomic number. This increasing trend suddenly changes when elements jumps into the next period such as from Ne to Na.

(i) Effect of closed shell configuration on IE : Na and Ne

In a short hand notation Na is expressed by [Ne] $2s^1$. The valence electron of Na is fully shielded by the electrons in the closed shell shown as [Ne]. Hence the effective charge Z_{eff} in Eq.(48) for the valence electron of Na is 2.2. In contrast the electrons in the closed shell are not shielded by themselves. Therefore the effective charge for the inner electron in the closed shell is much greater than that for the valence electron and the ionization energy for the inner electron of Na is much higher than that of inner electrons. Since Ne has the closed shell configuration, its outermost electrons are like the inner electrons of other elements. The calculated Z_{eff} of outermost electron of Ne is 5.85. This value is much higher than that for the valence electron of Na.

Table 1-9 Ionization energies for Li through Na with effective nucleus charge within brackets.
(The first ionization energy is indicated by bold squares.)

	$1s_1$	$1s_2$	$2s_1$	$2s_2$	$2p_1$	$2p_2$	$2p_3$	$2p_4$	$2p_5$	$2p_6$	$3s_1$
Li	9.00	7.29	**0.42**	0.95							
	(3)	(2.7)	**(1.30)**								
Be	16.00	13.69	1.32	**0.95**	1.60						
	(4)	(3.7)	(2.30)	**(1.95)**							
B	25.00	22.09	2.72	2.18	**1.69**	2.25					
	(5)	(4.7)	(3.30)	(2.95)	**(2.60)**						
C	36.00	32.49	4.62	3.90	3.24	**2.64**	2.90				
	(6)	(5.7)	(4.30)	(3.95)	(3.60)	**(3.25)**					
N	49.00	44.89	7.02	6.13	5.29	4.52	**3.80**	3.55			
	(7)	(6.7)	(5.30)	(4.95)	(4.60)	(4.25)	**(3.90)**				
O	64.00	59.29	9.92	8.85	7.84	6.89	6.00	**5.18**			
	(8)	(7.7)	(6.30)	(5.95)	(5.60)	(5.25)	(4.90)	**(4.55)**			
F	81.00	75.69	13.32	12.08	10.89	9.77	8.70	7.70	**6.76**		
	(9)	(8.7)	(7.30)	(6.95)	(6.60)	(6.25)	(5.90)	(5.55)	**(5.20)**		
Ne	100.0	94.09	17.22	15.80	14.44	13.14	11.90	10.73	9.61	**8.56**	
	(10)	(9.7)	(8.30)	(7.95)	(7.60)	(7.25)	(6.90)	(6.55)	(6.20)	**(5.85)**	
Na	121	114.5	21.62	20.03	18.49	17.02	15.60	14.25	12.96	11.73	**0.54**
	(11)	(10.7)	(9.30)	(8.95)	(8.60)	(8.25)	(7.90)	(7.55)	(7.20)	(6.85)	**(2.20)**

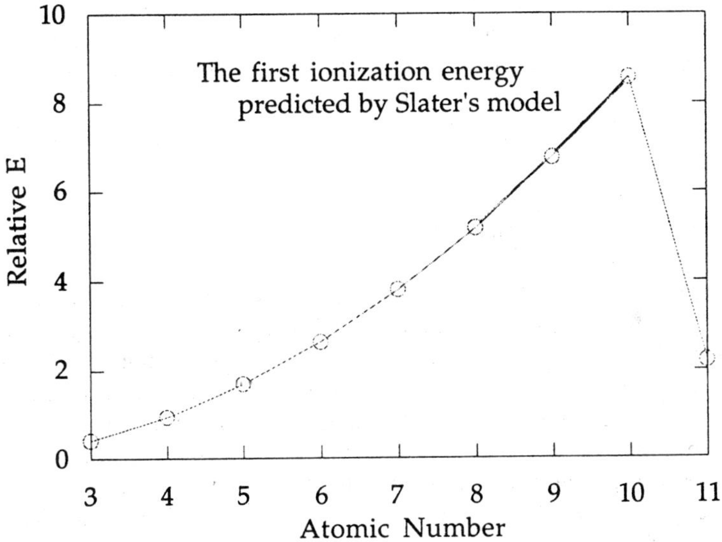

Figure 1-16 Plot of the first ionization energy calculated by Slater's model according to Eq.(48)

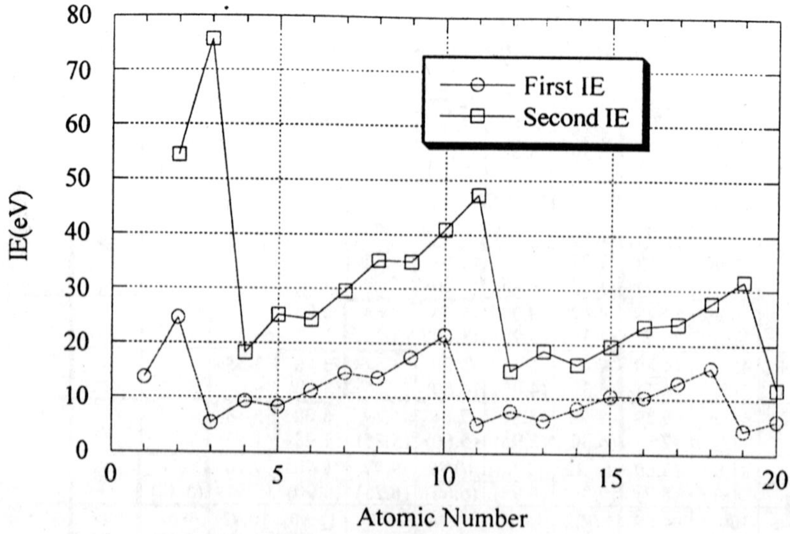

Figure 1-17 Variation of Ionization Energy with atomic number

This also accounts for the large difference between the first and second ionization energies of Na. The actual ionization of Ne and Na agree well with this calculations. Those data of ionization energies of Ne and Na are as follow:

$$I_{1,Na} = 5.14, \; I_{2,Na} = 47.3$$
$$I_{1,Ne} = 21.56$$

(ii) Variation of IE within the same group

There is a periodic trend in ionization energy. Ionization energy (IE) decreases within a family (in the same column of the periodic chart) with increasing Z and atomic weight. This is shown in Table 1-10. The reason for this trend in a family is the decrease in the ratio of Z_{eff}/n in Eq.(48) as atomic number increases in a family. For example, the Z_{eff} for Li, Na and K in the same family are 1.33, 2.22 and 2.22 according to Slater's rule while their primary QN's are 2, 3, and 4. Hence IE decrease as atomic number increase in the same family. Actual data of ionization energies of elements are listed in Table 1-11. Similarly, the sharp decrease in IE's from Ne to Na, Ar to K, Kr to Rb and Xe to Cs can be attributed to the same reason. In a transition from rare earth element in the n-th period to alkali element in the next period, the primary QN changes from n to

Table 1-10 Variation of ionization energy in the same group of the period table

Alkali Family		Alkali Earth Family		Chalcogenide Family		Halogen Family	
Li	5.4 eV	Be	9.3	O	13.6	F	17.4 eV
Na	5.1	Mg	7.6	S	10.4	Cl	12.9
K	4.5	Ca	6.1	Se	9.75	Br	11.8
Rb	4.2	Sr	5.7	Te	9.0	I	10.45
Cs	3.9	Ba	5.2				

n+1. However, the effective nucleus charge, Z_{eff}, does not change as much as the primary quantum number. Therefore the sharp decrease in IE arises due to rapid decrease in the ratio of Z_{eff}/n when the position of an element changes from one row to the next row below in the periodic table, which means that the primary qunatum number changes from n to n+1.

5. Effect of Exchange Energy and Spin Pairing on IE

Slater's model predicts the monotonic increase of IE with atomic number in the same period. The actual data of the first and second ionization energies are plotted in Fig.1-17. The actual data also exhibit the same trend as Slater's model in that IE increases with atomic number in the same period and sudden decrease of IE when atomic number jumps into the next period. However, some elements exhibit lower ionization energies than the nearest element with lower atomic number. These elements are such as B, O, Al and S in the second and third periods of the periodic table. Due to these elements, the variation of IE in a given period is stepwise rather than monotonic increase. In Slater's model, there is no difference between s and p orbitals. Also exchange energy described by Hund's rule or stabilization energy by spin pairing of electrons in the same orbital are not considered in Slater's model. Hence, taking these two factors into consideration, the stepwise increase of IE within a given period can be explained.

Exchange Energy

Pauli exclusion prinicple says that electrons in this $\frac{1}{2}$ filled subshell gains extra energy by switching their positions in the subshell with degenerate orbitals such as p_x, p_y and p_z. This gain of extra energy from switching

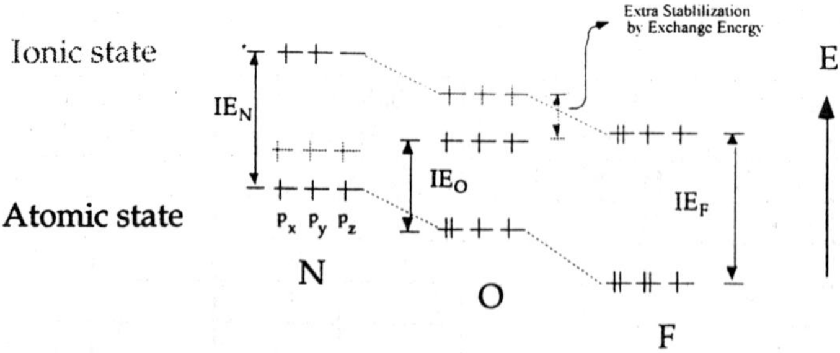

Figure 1-18 Schematic diagram showing extra stabilization energy for O^+ by exchange energy

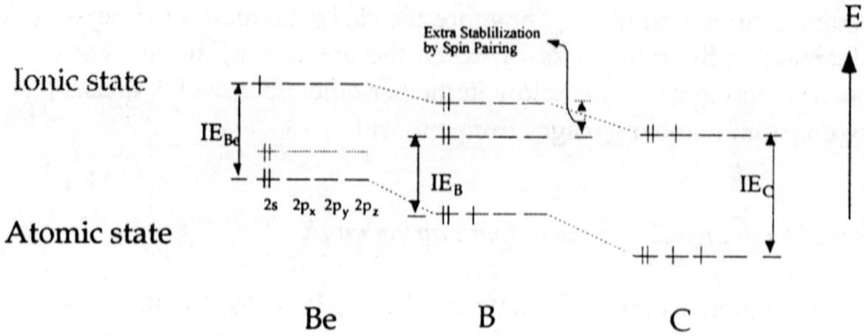

Figure 1-19 Schematic diagram showing extra stabilization energy for B^+ by spin pairing

electrons in the subshell is called exchange energy. Since N has $\frac{1}{2}$ filled subshell of p orbitals, it acquires more stabilization energy than C (partially empty-$2p^2$) or O(partially complete-$2p^4$). In the same token, O^+ enjoys the more stabilization energy by the exchange energy than N^+ (same configuration as C) or F^+ (same as O). This effect of exchange energy on the energy level of nitrogen atom and oxygen ion results in the lower IE of oxygen than F and even N. The ionization energies for N, O and F are 14.5, 13.6 and 17.4 eV, respectively.

This effect of exchange energy is illustrated in Fig.1-18 in a schematic way. For simplicity, only p orbitals are shown. The dotted line indicates the change of averaged atomic energy with atomic number.

Spin pairing

Pauli Principle predicts that electrons with the same spins cannot occupy the same orbital. By coupling spins of electrons in the same orbital, the atom obtains extra stabilization energy. Like exchange energy, spin pairing has a similar effect on the change of IE. This lowers atomic energy level than expected by Slater's model. A schematic drawing is shown in Fig.1-19 for the case of B whose IE is lower than that of Be. Since Be and B^+ have the same electronic configurations as $[He]2s^2$, spin pairing contributes to lowering energy level of atomic state of Be and ionic state of B^+ resulting lower IE of B than even Be.

6. Measurement of IE and Characterization of Atomic Structrues

Electrons can be removed by incident electrons from an external source or photon (light). Photon is a conceptual particle of light due to its particle-like properties. Each photon has a energy of hv. In relation to de Broglie equation, the moementum of photon is $\frac{h\nu}{c} = \frac{h}{\lambda}$. quantized concept for the particle In either case, energy conservation law is obeyed. If the incident one has the energy E_i, which must be larger than the ionization energy of the atom, electrons will be removed. The kinetic energy, E_k, of the electron removed must be E_i - IE accronding to energy conservation law. Hence, the ionization energy is calculated by measuring E_i and E_k. The following example illustrates the measurement of IE for Li atom with photo excitation. Since the energy of incident photon (light) is equal to hv, the ionization energy is expressed as :

$$IE = h\nu - \frac{1}{2}mv^2$$

for the photo excitation of

$$Li(g) + h\nu \rightarrow Li^+(g) + e^-$$

Here it must be distinguished between ionization energy and the energy

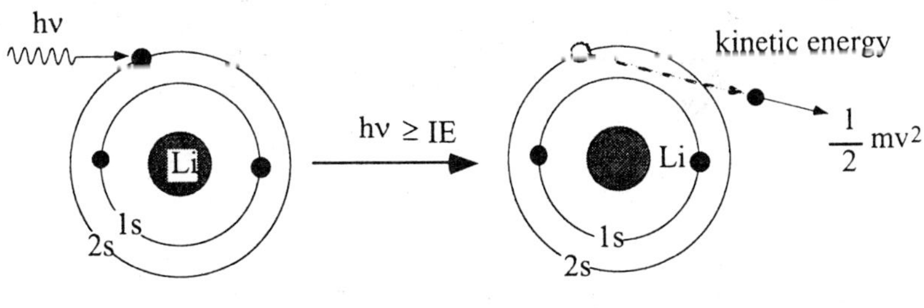

Figure 1-20 Interaction of photon(light) with electrons

required to remove an electron if that electron is not the outermost electron with the highest energy. For example, the second ionization energy of Li represents the energy required to remove an 1s electron from Li^+ not from the neutral Li atom. Hence the electron energy in a neutral atom is not equal to ionization energy needed to remove that electron.

X-ray or photon generation generation

This is the reverse phenomena of the ionization of an atom by photo excitation. If an inner electron is removed by energetic particles such as other electrons from an external source, outer electron would move to the empty position of the inner electron removed since the energy of the latter is lower than that of the former. During this transition, photons are emitted. The photon emission is schematically depicted in Fig.1-21 for LI. Here the wavelength of the photon emitted must be equal to the energy difference between outer electron and inner electron. The energy difference(ΔE) between 2s and 1s of Li atom is 4.34 eV. Hence the photon emitted during the transiton from 2s to 1s orbital must have a wavelength of $\frac{hc}{\Delta E}$ since the energy of a photon is $h\nu$ or $\frac{hc}{\lambda}$. The wavelength corresponding to 4.34eV is calculated as follows :

$$\lambda = \frac{6.625 \times 10^{-34}\,\mathrm{J\,sec} \times 3.0 \times 10^{8}\,\mathrm{m/sec}}{4.34\mathrm{eV} \times 1.6 \times 10^{-19}\,\mathrm{J/eV}}$$

$$= 2.862 \times 10^{-7}\mathrm{m}$$

Since there are several electron orbitals in an atom whose energy levels are discretely separated, spectra with discrete wavelengths are formed like the spectra of hydrogen atom as shown in Fig.1-3. From these spectra, the electronic energy levels of atoms can be characterized.

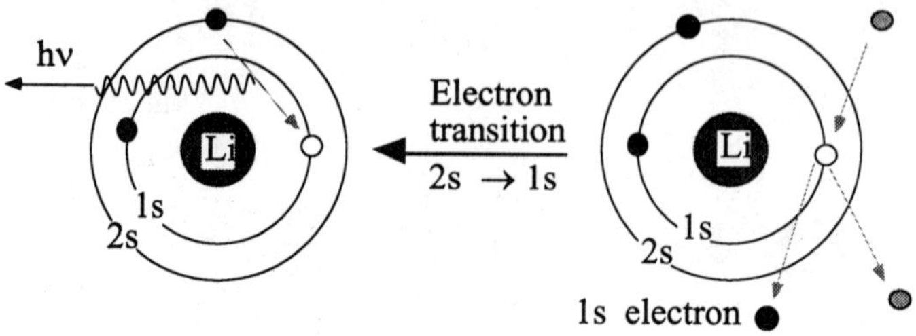

Figure 1-21 Schematic drawing for photon emission during electron transition from outer(2s) to inner(1s) orbital of Li atom

Table 1-11 Ionization energies

	I	II	III	IV	V	VI	VII	VIII
1 H	13.598							
2 He	24.587	54.416						
3 Li	5.392	75.638	122.451					
4 Be	9.322	18.211	153.893	217.713				
5 B	8.298	25.154	37.930	259.368	340.217			
6 C	11.260	24.383	47.887	64.492	392.077	489.981		
7 N	14.534	29.601	47.448	77.472	97.888	552.057	667.029	
8 O	13.618	35.116	54.934	77.412	113.896	138.116	739.315	871.387
9 F	17.422	34.970	62.707	87.137	114.240	157.161	185.182	953.886
10 Ne	21.564	40.962	63.45	97.110	126.21	157.93	207.27	239.09
11 Na	5.139	47.286	71.64	98.914	138.39	172.15	208.47	264.18
12 Mg	7.646	15.035	80.143	109.24	141.26	186.50	224.94	265.90
13 Al	5.986	18.828	28.447	119 99	153.71	190.47	241.43	284 59
14 Si	8.151	16.345	33.492	45 141	166.77	205.05	246.52	303.17
15 P	10.486	19.725	30.18	51.37	65.023	230 43	263.22	309.41
16 S	10.360	23.33	34.83	47.30	72.68	88.049	280.93	328.23
17 Cl	12.967	23.81	39.61	53.46	67.8	98.03	114.193	348.28
18 Ar	15.759	27.629	40.74	59.81	75.02	91.007	124.319	143.456
19 K	4.341	31.625	45.72	60.91	82.66	100.0	117.56	154.86
20 Ca	6.113	11.871	50.908	67.10	84.41	108.78	127.7	147.24
21 Sc	6.54	12.80	24.76	73.47	91.66	111.1	138.0	158.7
22 Ti	6.82	13.58	27.491	43.266	99.22	119.36	140.8	168.5
23 V	6.74	14.65	29.310	46.707	65.23	128.12	150.17	173.7
24 Cr	6.766	16.50	30.96	49.1	69.3	90.56	161.1	184.7
25 Mn	7.435	15.640	33.667	51.2	72.4	95	119.27	196.46
26 Fe	7.870	16.18	30.651	54.8	75.0	99	125	151.06
27 Co	7.86	17.06	33.50	51.3	79.5	102	129	157
28 Ni	7.635	18.168	35.17	54.9	75.5	108	133	162
29 Cu	7.726	20.292	36.83	55.2	79 9	103	139	166
30 Zn	9.394	17.964	39.722	59.4	82.6	108	134	174
31 Ga	5.999	20.51	30.71	64				
32 Ge	7.899	15.934	34.22	45.71	93.5			
33 As	9.81	18.633	28.351	50.13	62.63	127.6		
34 Se	9.752	21.19	30.820	42.944	68.3	81.70	155.4	
35 Br	11.814	21.8	36	47.3	59.7	88.6	103.0	192.8
36 Kr	13.999	24.359	36.95	52.5	64.7	78.5	111.0	126
37 Rb	4.177	27.28	40	52.6	71.0	84.4	99.2	136
38 Sr	5.695	11.030	43.6	57	71.6	90.8	106	122.3
39 Y	6.38	12.24	20.52	61.8	77.0	93.0	116	129
40 Zr	6.84	13.13	22.99	34.34	81.5			
41 Nb	6.88	14.32	25.04	38.3	50.55	102.6	125	
42 Mo	7.099	16.15	27.16	46.4	61.2	68	126.8	153
43 Tc	7.28	15.26	29.54					
44 Ru	7.37	16.76	28.47					
45 Rh	7.46	18.08	31.06					
46 Pd	8.34	19.43	32.93					
47 Ag	7.576	21.49	34.83					
48 Cd	8.993	16.908	37.48					
49 In	5.786	18.869	28.03	54				
50 Sn	7.344	14.632	30.502	40.734	72.28			

Table 1-11 Continued

	I	II	III	IV	V	VI	VII	VIII
51 Sb	8.641	16.53	25.3	44.2	56	108		
52 Te	9.009	18.6	27.96	37.41	58.75	70.7	137	
53 I	10.451	19.131	33					
54 Xe	12.130	21.21	32.1					
55 Cs	3.894	25.1						
56 Ba	5.212	10.004						
57 La	5.577	11.06	19.175					
58 Ce	5 47	10.85	20.20	36.72				
59 Pr	5.42	10.55	21.62	38.95	57.45			
60 Nd	5.49	10.72						
61 Pm	5.55	10.90						
62 Sm	5.63	11.07						
63 Eu	5.67	11.25						
64 Gd	6.14	12.1						
65 Tb	5.85	11.52						
66 Dy	5.93	11.67						
67 Ho	6.02	11.80						
68 Er	6.10	11.93						
69 Tm	6.18	12.05	23.71					
70 Yb	6.254	12.17	25.2					
71 Lu	5.426	13.9						
72 Hf	7.0	14 9	23.3	33.3				
73 Ta	7.89							
74 W	7.98							
75 Re	7.88							
76 Os	8.7							
77 Ir	9.1							
78 Pt	9.0	18.563						
79 Au	9.225	20.5						
80 Hg	10.437	18.756	34.2					
81 Tl	6.108	20.428	29.83					
82 Pb	7.416	15.032	31.937	42.32	68.8			
83 Bi	7.289	16.69	25.56	45.3	56.0	88.3		
84 Po	8.42							
85 At								
86 Rn	10.748							
87 Fr								
88 Ra	5.279	10.147						
89 Ac	6.9	12.1						
90 Th		11.5	20.0	28.8				
91 Pa								
92 U								
93 Np								
94 Pu	5.8							
95 Am	6.0							

Numerical values in this Table are expressed in electron volts. The conversion factor used in converting the spectral data to eV was 1 eV = 8065.73 cm^{-1}.
From Moore, C. E., Analyses of Optical Spectra,
NSRDS-NBS 34, Office of Standard Reference Data,
National Bureau of Standards, Washington, D.C.

In a reverse way, if we know the energy levels of electrons in an atom, the wavelength of emitted photon can be calculated. By choosing an appropriate element, a photon with the desired wavelength can be generated. Also the ionization energy can be calculated from these atomic spectrum data along with other experimental data such as thermochemical data. The ionization energies are listed in Table 1-11 obtained in this way.

Table 1-12 Electron affinity of non-metallic elements

Atom	Orbital Electronic configuration	EA(eV)	Orbital Electronic configuration of Anion
H	1s1	0.754	[He]
F	[He]$2s^22p^2$	3.34	[Ne]
Cl	[Ne]$3s^23p^5$	3.61	[Ar]
Br	[Ar]$4s^23d^{10}4p^5$	3.36	[Kr]
I	[Kr]$5s^24d^{10}5p^5$	3.06	[Xe]
O	[He]$2s^22p^4$	1.47	[He]$2s^22p^5$
S	[Ne]$3s^23p^4$	2.08	[Ne]$3s^23p^5$
Se	[Ar]$4s^23d^{10}4p4$	2.02	[Ar]$4s^23d^{10}4p^5$
Te	[Kr]$5s^24d^{10}5p^4$	1.97	[Kr]$5s^24d^{10}5p^5$
N	[He]$2s^22p^3$	0±0.2	[He]$2s^22p^4$
P	[Ne]$3s^23p^3$	0.77	[Ne]$3s^23p^4$
As	[Ar]$4s^23d^{10}4p^3$	0.80	[Ar]$4s^23d^{10}4p^4$
C	[He]$2s^22p^2$	1.27	[He]$2s^22p^3$
Si	[Ne]$3s^23p^2$	1.24	[Ne]$3s^23p^3$
Ge	[Ar]$4s^23d^{10}4p^2$	1.20	[Ar]$4s^23d^{10}4p^3$
B	[He]$2s^22p^1$	0.24	[He]$2s^22p^2$
Al	[Ne]$3s^23p^1$	(0.52)	[Ne]$3s^23p^2$
Ga	[Ar]$4s2^3d^{10}4p^1$	(0.37)	[Ar]$4s23d^{10}4p^2$
In	[Kr]$5s^24d^{10}5p^1$	(0.35)	[Kr]$5s^24d^{10}5p^2$
Be	[He]$2s^2$	≤ 0	[He]$2s^22p^1$
Mg	[Ne]$3s^2$	≤ 0	[Ne]$3s^13p^2$
Zn	[Ar]$4s^23d^{10}$	≤ 0	[Ar]$4s^23d^{10}4p1$
Cd	[Kr]$5s^24d^{10}$	≤ 0	[Kr]$5s^24d^{10}5p1$
Ti	[He]$2s^1$	0.62	[He]$2s^2$
Na	[Ne]$3s^1$	0.55	[Ne]$3s^2$
Cu	[Ar]$4s^23d^9$	1.28	[Ar]$4s^23d^{10}$
Ag	[Kr]$5s^14d^{10}$	1.30	[Kr]$5s^24d^{10}$
Au	[Xe]$4f^{14}5d^{10}6s^1$	2.31	[Xe]$4f^{14}5d^{10}6s^2$
He	[1s2]	≤ 0	$1s^22s^12p^1$
Ne	[He]$2s^22p^6$	≤ 0	[He]$2s^22p^63s^1$

Source: E. N. Chen and W. E. Wentworth, " The Experimental Values of Atomic Electron Affinities," *J. Chem. Educ.* 52: 486(1975). These values have been rounded to two significant figures after the decimal

C. Electron Affinity (EA)

Like the ionization energy, electron affinity can characterize the valence electron strucutre. Electron affinity is the energy released when electron is added to an atom. Thermodynamically this energy should be expressed as a negative one since the reaction is exothermic. However, EA is conventially expressed as a positive one although it is contrary to the thermodynamical principle. To overcome this confusion, it may be defined by an energy required to remove electron from negative ion in gasseous state.

$$A^-(g) \rightarrow A(g) + e^- : \; EA = \Delta H$$

Usually EA defined as above is positive since $A^-(g)$ is more stable than A(g). If ionization energy is important to metallic elements, EA is important to non-metallic elements such as F, Cl and Br. Table 1-12 lists the data of EA for some elements. Data in a bracket is not an experimental but theoretical data. For some elements such as He-, Ne- , EA values vary due to metastable phase between atomic and anionic states. To measure EA of an element, first there must be an ion source of that element. Since this source is limited for the whole elements, EA data is limited. Due to this limitation, it is not easy to measure electron affinity. Some of them are not possible to get experimental data and theoretial values are listed in a bracket in Table 1-12.

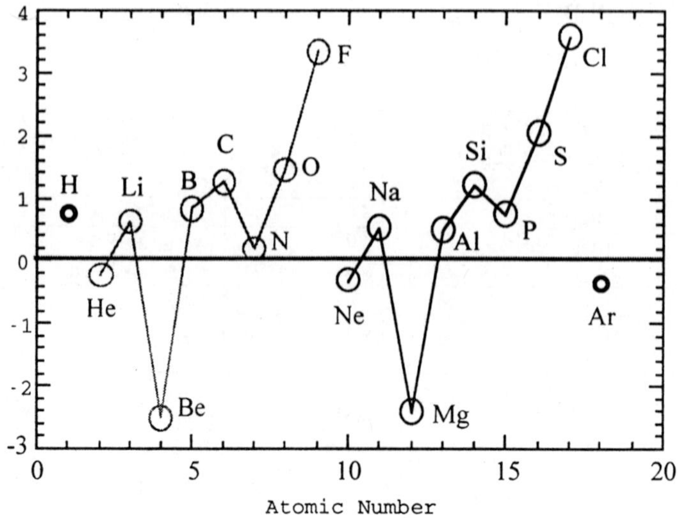

Figure 1-22 Plot of EA for elements of 1st, 2nd and 3rd period with atomic number

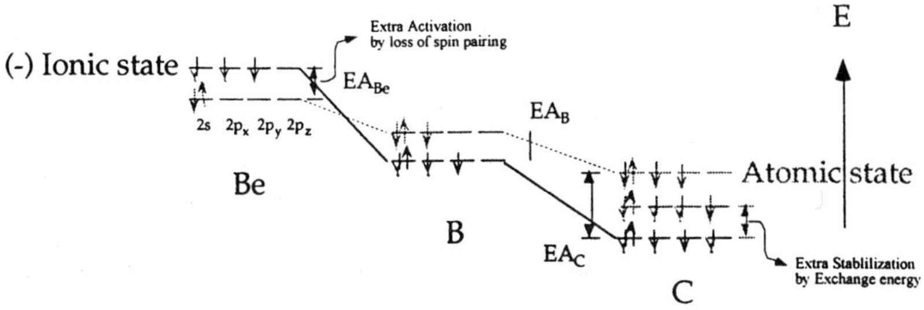

Figure 1-23 Electronic configurations and their energy levels of Be, B and C

As shown in Fig.1-17 for IE of elements, EA increases with atomic number. It exhibits the same trend as ionization energies except that the trend shifts to the right by one atomic number since EA represent ionization energy of the anion which has the same electronic configuration as the next one on the right in periodic table. Also there are some elements such as Be, Mg, N and P which show lower EA than the nearest element to the left in the periodic table. This phenomena also can be attributed to the effect of spin pairing and exchange energy.

As in ionization energy, the electron affinity is generally positive and shows the general trend in the variation of EA with atomic number. It is noticeable that those elements in the alkali earth metal have extremely low EA compared to other elements even rare gas group elements. Also these elements have been reported with several values for EA. The severals values for EA of elements such as He and Be are due to their metastable transition states. As shown in Table 1-12, $1s^2 2s^1$ is not the true ionic state. Rather $1s^1 2s^1 2p^1$ is the true ionic state which was identified experimentally. Although the definintion of ionic(-) state is varied, let's assume that the true ionic state is stabler than others. With this respect, the added electron in 2s orbital in the ionic state of He as $1s^2 2s^1$, is loosely bound compared to the electron in 2p orbital in another ionic state of He as $1s^1 2s^1 2p^1$. The following illustrates the effective nucleus charge for each electron of these ionic states calculated according to Slater's rule.

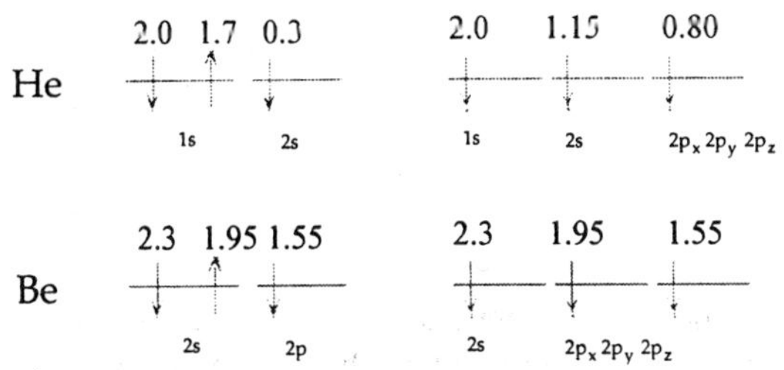

There is no difference in Z_{eff} for both ionic states of Be. Then it seems natural to ask why $2s^12p^2$ is the true ionic state rather than $2s^22P^1$. As shown in ionization energy, Slater's rule does not take the difference of s and p orbitals into account. The repulsive interaction between electrons in s and p orbitals in the same shell are averaged as 0.35. This repulsive interaction between electrons in 2s and 2p orbitals is lateral interaction compared to the vertical interaction between inner(1s) and outer(2s or 2p) electrons along radial direction resulting in shielding effects. This can partially explain why two p orbital electrons in carbon are distributed in p_x and p_y rather than in p_x orbital with spin pairing. Geometrically p_z and p_y are separated by 90° as shown in Fig.1-11. This geometrical separation reduces the electronic repulsive interaction. Likewise, innitrogen, three p orbital electrons are distributed in p_x, p_y and p_z. In Be^-, therefore, this lateral interaction between electrons of 2s and 2p orbitals can be reduced by promoting one of 2s orbital electrons into p orbital although the promoted state has higher energy. Then another question may arise concerning this lateral interaction. Why does the promotion such as in Be^- not happen in B which is next to Be in the periodic table? If the binding energy of an electron is assume to be equal to the ionization energy of that electron, two s orbital electrons in B has higher binding energy than those of Be. Simply referring to Table 1-12, these values are 1.32 and 0.95 for Be and 2.72 and 2.18 for B. Therefore the extra electron added to make B^- does not disturb the 2s orbitals due to high binding energies of these electrons and occupies p_y orbital. The qualtative energy diagram of electronic states of He and Be are schematically illustrated in Fig.1-24.

Summary

Ionization energy and electron affinity are important physical properties for characterization of valence electrons and, consequently, for other chemical factors such chemical bond, reactivity, etc. Also they can reveal the atomic structure and provide information concerning quantum mechanical aspects such as Hund's rule, spin pairing and stability of closes shell structure.

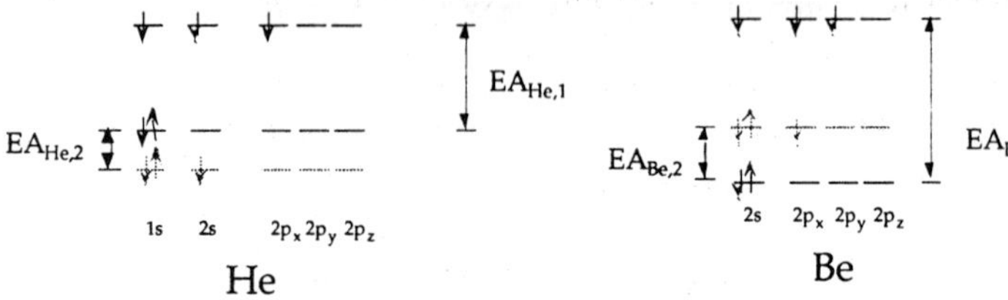

Figure 1-24 Schematic energy diagram for atomic and ionic states with possible metastable states which is shown as gray color

D. Electronegativity (EN)

Although ionization energy and electron affinity may represent the power to release electrons from atoms and attract electrons to atoms, respectively, they can not be easily used as factors to determine the tendencies of atoms to attract or release electrons in a molecule. In this respect, electron negativity provides the simple relative scale for the releasing or attracting power of atoms relative to one another. Therefore, electronegativity descirbes how these atomic properties are related to bonding scheme in a molecule while ionization potential and electron affinity are characteristics of individual atoms. Thus, electronegativity is not a property of the isolated atom but rather the property of the atom in a molecule, namely in the environment and under the influence of surrounding atoms.

The empirical scale of electronegativity was first formulated by Pauling from thermochemical data. As Pauling described, electronegativity is the relative ability or power of an atom in a molecule to attract electrons to itself. In Pauling's scale, each atom is assigned a value regardless of surrounding atoms in a molecule. Later, Mulliken interpreted the physical meaning of electronegativity calculated by Pauling. According to Mulliken, electronegativity combines elements of the ionization energy and electron affinity of a atom within the molecule which are influenced by other atoms in the molecule. As will be seen in the next section (I-3 Molecular structure), the electronegativity is a simple tool describing the bonding nature between atoms within a molecule. In valence bond theory, electrons are shared by two atoms and the shared electrons are locating between two atoms. Quantum mechanics predicts that the shared electrons are not always in the same distance from these two atoms. Rather electrons are shifted toward more negative atoms. The advantage of electronegativity is that it characterizes this uneven distribution of shared electrons between two atoms in a simple way compared to quantum mechanics. In this simple aspect of electronegativity, Pauling's method is more widely used compared to Mulliken's method due to its simplicity. Therefore, electronegativity is a powerful concept in determining the bonding tendencies of atoms without referring to quantum mechanics.

1. Pauling

The first scale for electronegativity was developed by Pauling. It had been observed that bonds between dissimilar atoms were almost always greater (stronger) than the covalent bonds formed between the like atoms, thus:

Ex:	1)	Cl:Cl	Cl:F	F:F
		(Cl_2)	(ClF)	(F_2)
		242	255	153(kJ/mol)

2)	H:H	H:F	F:F
	(H_2)	(HF)	(F_2)
	103	135	37(kcal/mol)

Pauling suggested that bonds formed from dissimilar atoms would be stabilized by an <u>ionic resonance energy</u> which he considered as arising from resonance between the ionic and covalent charge configurations:

$$Cl^-F^+ \leftrightarrow Cl{:}F \leftrightarrow Cl^+F^-$$
(ionic) (covalent) (ionic)

This ionic resonance energy produces excess bonding energy to that of purely covalent bond in which electrons are supposed to be evenly distributed in between. This excess bonding energy is expressed as

$$\Delta_{AB} = E_{AB} - (E_{AA} + E_{BB})/2 \qquad (48\text{-}1)$$

His calculation showed that there was a correlation between values of Δ_{AB} for various diatomic molecules and certain parameters for constituent atoms of the molecule which were later defined as electronegativities. This correlation found between excessive resonance energy and the parameters was as follows;

$$\Delta_{AB} = (\chi_A - \chi_B)^2 \qquad (48\text{-}2)$$

Here Δ_{AB} is expressed in electron volts as unit and χ_A and χ_B are values of electronegativity for atoms A and B, respectively, from Pauling's scale. As shown above, the calculation of electronegativity is empirical rather than theoretical. Also, strictly speaking, there is no fixed value χ_A or χ_B for any isolated atom as defined above since the electronegativity is a characteristic within a molecule not an atom. Pauling, however, calculated a value χ as a measure of its electronegativity from thermochemical data (bonding energies) and assigned it to each atom. These χ values calculated by Pauling are listed in Table 1-13.

With respect to wave function for the molecule AB, the ionic character of the bond is represented by λ as:

$$\Psi_{AB} = (1-\lambda)\Psi_{cov.} + \lambda\Psi_{ionic} \qquad (49)$$

Here λ is related to bonding excessive energy and to the electronegativity

Table 1-13 Electronegativity of elements in the periodic Table

I	II	III	II	II	II	II	II	II	II	I	II	III	IV	III	II	I
H 2.20																
Li 0.98	Be 1.57											B 2.04	C 2.55	N 3.04	O 3.44	F 3.98
Na 0.93	Mg 1.31											Al 1.61	Si 1.90	P 2.19	S 2.58	Cl 3.16
K 0.82	Ca 1.00	Sc 1.36	Ti 1.54	V 1.63	Cr 1.66	Mn 1.55	Fe 1.83	Co 1.88	Ni 1.91	Cu 1.90	Zn 1.65	Ga 1.81	Ge 2.01	As 2.18	Se 2.55	Br 2.96
Rb 0.82	Sr 0.95	Y 1.22	Zr 1.33		Mo 2.16			Rh 2.38	Pd 2.20	Ag 1.93	Cd 1.69	In 1.78	Sn 1.96	Sb 2.05		I 2.66
Cs 0.79	Ba 0.89	La 1.10			W 2.36			Ir 2.20	Pt 2.28	Au 2.54	Hg 2.00	Tl 2.04	Pb 2.33	Bi 2.02		
		Ce 1.12	Pr 1.13 (III)	Nd 1.14 (III)		Sm 1.17 (III)		Gd 1.20 (III)		Dy 1.22 (III)	Ho 1.23 (III)	Er 1.24	Tm 1.25 (III)		Lu 1.27 (III)	
					U 1.38 (III)	Np 1.36 (III)	Pu 1.28 (III)									

Note: Roman numerals refer to the oxidation numbers of the atoms in the molecules used in the calculation of atomic electronegativity values

difference between A and B, $|\chi_A - \chi_B|$. Also λ illustrates the degree of ionicity of the molecule. Thus for single bond between atoms A and B, an electronegativity difference can be calculated and related to the ionic component of the bonding. Thus the exponential relationship between λ and electronegativity difference ($\Delta\chi$) was proposed as Eq.(50) and the plot of λ vs $\Delta\chi$ is shown in Fig. 1-25.

$$\lambda = 1 - \exp\left(-\frac{(\chi_A - \chi_B)^2}{4} \right) \qquad (50)$$

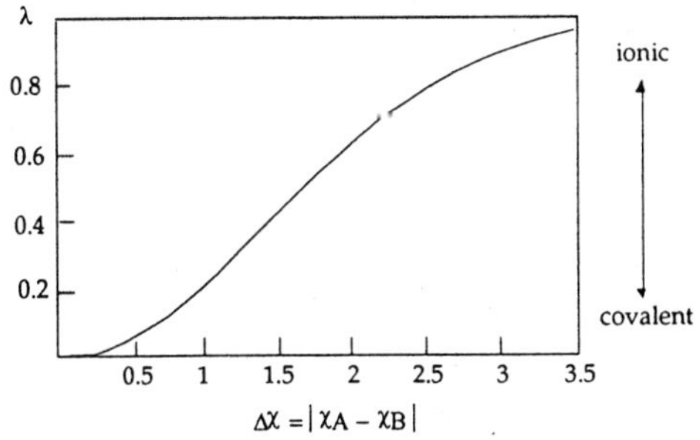

Figure 1-25 Variation of ionic character of bond with EN($\Delta\chi$)

2. Mulliken

Shortly after Pauling proposed his method, Mulliken suggested an absolute scale of electronegativity. His derivation of this scale can be illustrated in a diatomic molecule as the following.

$$A^-B^+ \leftrightarrow \quad AB \quad \leftrightarrow \quad A^+B^- \qquad (51)$$
(ionic) (covalent) (ionic)

$$\Psi_{AB} = \gamma\psi_{AB} + \alpha\psi_{A^-B^+} + \beta\psi_{A^+B^-} \qquad (52)$$

Even a covalent bond does not mean there is no ionic character. Hence, in a diatomic molecule, the wave function of the molecule can be depicted as a combination of purely covalent and two ionic wave functions as in Eq.(52). In order for AB to be a covalent bond, the necessary condition is that the two ionic states in Eq.(52) are probable in the same magnitude. In other words, the coefficients of the wave function for two ionic states are the same, namely $\alpha = \beta$. This condition is met when the energies of the two ionic states are same. The total energy of the ionic state can be divided into three energy terms which are electrostatic energy between positive and negative atoms, electron affinity for the negative atom and ionization potential for the positive atoms. Hence the total energy is expressed as

$$E_{total}(A^-B^+) = E_{electrostatic}(A^-B^+) + EA(A^-) - IE(B^+)$$
$$E_{total}(A^+B^-) = E_{electrostatic}(A^+B^-) + EA(B^-) - IE(A^+)$$

In a trial simplification, Mulliken assumed that interatomic distance does not change regardless of the sign of electric charge on either atom. Hence the electrostatic energy are assumed to be same for A^-B^+ and A^+B^-. On this assumption the necessary condition for the bond to be covalent is reduced to the sum of electron affinity and ionization potential as:

$$E_{total}(A^-B^+) = E_{total}(A^+B^-)$$

$$E_{electrostatic}(A^-B^+) + EA(A^-) - IE(B^+)$$
$$= E_{electrostatic}(A^+B^-) + EA(B^-) - IE(A^+)$$

$$EA(A^-) - IE(B^+) = EA(B^-) - IE(A^+) \qquad (53)$$

Rearranging the above Equation

$$EA(A^-) + IE(A^+) = EA(B^-) + IE(B^+) \tag{54}$$

Therefore, a bond is covalent if the sum of electron affinity and ionization potential for both atoms in a diatomic molecule are same. In the light of Eq.(52), it is apparent that the sum of electron affinity and ionization potential is related to the ionicity of AB. If electron affinity for A^- is high, the coefficient α in Eq.(52) is expected to be high since the energy of the ionic state A^-B^+ is low. Also, if ionization potential for A^+ is high, the energy of the ionic state A^+B^- becomes high. Hence, the coefficient β in Eq.(52) is expected to be low. Consequently, the sum of electron affinity and ionization potential for an atom indicates the magnitude favoring the ionic state with negative charge on that atom in the molecule. This is the basis of Mulliken's electronegativity scale. Actually his scale of electronegativity is expressed as an average of the ionization energy and electron affinity, given by:

$$EN = \frac{1}{2}(IE + EA) \tag{55}$$

EA = Electron Affinity

EN = Electronegativity = ξ_A for atom A

Here the electron affinity and ionization potential are those calculated for atoms within molecular environment. These are not values shown in Table 1-11. This difference can be illustrated in HCl as an example. Like Eq.(52), the molecular wave function of HCl can be expressed as:

$$\Psi_{HCl} = \gamma\psi_{HCl} + \alpha\psi_{H^-Cl^+} + \beta\psi_{H^+Cl^-} \tag{56}$$

The actual spin moment of HCl is 0 which means that HCl is in a singlet state. Hence the three wave functions on the right side of Eq.(38) must be in singlet states. The spin conditions for ψ_{HCl} and $\psi_{H^+Cl^-}$ are singlet since H^+ has no electron and Cl^- has filled electron orbitals whose spin state is a singlet. Also the purely covalent wave function, ψ_{HCl}, is an average of H^+ and Cl^-, both of which are singlet states. In H^-Cl^+, H^- is in singlet state. However, the spin state of Cl^+ with its ordinary $3s^23p^4$ electronic structures does not match the singlet condition since $3p^4$ is not singlet states. Rather it must be in $3\pi^23\pi^2$ whose spin state is singlet. (Here we do not mention in detail about $3\pi^23\pi^2$ because it is beyond the scope of this book.) This is different from the ordinary atomic orbital notation of $3p^4$. Therefore the sum of electron affinity and ionization potential in Eq.(55) must be expressed as

Table 1-14 Comparison of Mulliken's and Pauling's electronegativity

	I	II	III	IV
	ξ (=EA+IE)	$\xi-\xi_H$ Relative to H	$(\xi-\xi_H)/2.72$ Mulliken	$\chi - \chi_H$ Pauling
H	7.12	0.00	0.00	0.00
F	12.7	5.54	2.05	2.00
Cl	9.81	2.69	1.00	0.94
Br	9.06	1.94	0.72	0.75
I	8.23	1.11	0.41	0.40

$$\begin{aligned} &EA + IE \\ &= EA(3s^23p^6 \rightarrow 3s^23p^5) + IE(3s^23\pi^23\pi^2 \rightarrow 3s^23p^5) \end{aligned} \qquad (57)$$

As shown above, the electron affinity and ionization potential must be those calculated considering molecular structures not calculated from atomic orbitals as shown in Tables 1-11, and 1-12. In Mulliken's scale, IE and EA are for valence electron within a molecule not for a valence electron in an atom. In Table 1-14, values of electronegativity of Pauling's(χ) and Mulliken's (ξ) are compared and exhibits $\frac{\xi - \xi_H}{2.72} = \chi - \chi_H$. Mulliken provides the physical meaning to electronegativity values calculated by Pauling since he calculated the electronegativity from the theoretical basis without thermochemical data and their data shows a good agreement.

3. Dipole Moment

Dipole moments are measured from the degree of movement or orientation of the molecule (usually in the solution) in an electric field. The dipole moment for a molecule is a measure of the charge difference or ionicity between its constituent atoms or ions. Therefore it seems natural to infer that the electronegativity is directly related to dipole moment since electronegativity measures the magnitude of shifting of electrons toward one atom in a bond. It has been found that the dipole moment is linearly proportional to electronegativity difference between two atoms. This is illustrated in Table 1-15 for hydrogen halogenide compounds. However it is difficult to generalize this linear relationship between dipole moment(μ) and electronegativity difference($\Delta\chi$) since the dipole moment calculated from this relationship shows discrepancies from observed values for many molecules.

Table 1-15 Dipole moments of hydrogen halogenide compounds and their ionic characters

Compound	$\Delta\chi = \lvert\chi_A - \chi_B\rvert$	μ	Bonding
H–H	0.00	0.00	0% ionic
H–Br	0.76	0.78	13% ionic
H–Cl	0.94	1.03	20% ionic
H–F	1.78	2.00	55% ionic

$$\Delta\chi = \lvert\chi_A - \chi_B\rvert \approx \mu \text{ (dipole moment)} \tag{58}$$

4. Electronegativity Trends of the Elements

Electronegativity is related to IE and EA as shown by Mulliken although Eq(55) requires IE and EA within molecules. With respect to atomic IE and EA, elements with high IE or EA have high electronegativities and those with low IE or EA have low electronegativity except elements in rare gas group which have high IE and low EA. Hence elements in the left side of the periodic table has low electronegativities and those on the right side have high electronegativities. The actual data for electronegativity of elements are listed in Table 1-17. Also those of ions are listed.

As shown in IE, electronegativity decrease as atomic number increases in the same family(column) of the periodic table. As will be shown in the next section about chemical bonding, this trend in electrognegativity results in a division of the periodic table into two gropus, metallic and nonmetallic groups if rare gas group is handled separately. Elements in metallic group form ionic compound with elements in nonmetallic group. Elements in non-metallic group form covalent compounds between themselves.

Table 1-16 Electronegativity range of elements

Metallic	Akali, Alkali earth Transition metal	< 2.0	Na, K, Cs, etc. Mg, Ca, Sr
Nonmetallic	Moderate EN	~1.8–3.0	C, N, Si, P, etc.
	Chalcogenides	2 ~ 3.5	O, S, Se
	Halogens	> 3.0	Cl, F, etc.
Rare Gas	Rare gases	~ 0	He, Ne, Ar, Kr...

Table 1-17 Atomic Electonegativity

Element	Pauling	Sander-son	Allred-Rochow
1 . H	2.20	2.31	2.20
2. He			5.50
3. Li	0.9 8	0.86	0.97
4. Be	1.57	1.61	1.47
5. B	2.04	1.88	2.01
6. C	2.55	2.47	2.50
7. N	3 .04	2.9 3	3 .07
8. 0	3.44	3.46	3.50
9. F	3.98	3.92	4.10
10. Ne		4.50	4.84
11. Na	0.93	0.85	1.01
12. Mg	1.31	1.42	1.23
13. Al	1.61	1.54	1.47
14. Si	1.90	1.74	1 74
15. P	2.19	2.16	2.06
16. S	2.58	2.66	2.44
17. Cl	3.16	3.28	2.83
18. Ar		3.31	3.20
19. K	0.82	0.74	0.91
20. Ca	1.00	1.06	1.04
21. Sc	1.36	1.09	1.20
22. Ti(II)	1.54	1.13	1.32
23. V(II)	1.63	1.24	1.45
24. Cr(II)	1.66	1.35	1.56
25. Mn(II)	1.55	1.44	1.60
26. Fe(II)	1.83	1 47	1.64
Fe(III)	1.96		
27. Co(II)	1.88	1.47	1.70
28. Ni(II)	1.91	1.47	1.75
29. Cu(I)	1.90	1 74	1 75
Cu(II)	2.00		
30. Zn(II)	1.65	1.86	1.66
31. Ga(III)	1.81	2.10	1.82
32. Ge(IV)	2.01	2.31	2.02
33. As(III)	2.18	2.53	2.20
34. Se	2.55	2.76	2.48
35. Br	2.96	2.96	2.74
36. Kr	3.00	2.91	2.94
37. Rb	0.82	0.70	0.89
38. Sr	0.95	0.96	0.99
39. Y	1.22	0.98	1.11
40. Zr(II)	1.33	1.00	1.22
41. Nb	1.60	1.12	1.23
42.Mo(II)	2.16		
Mo(III)	2.19		
Mo(IV)	2.24	1.24	1.30
Mo(V)	2.27		
Mo(VI)	2.35		
43. Tc	1.90	1.33	1.36
44. Ru	2.20	1.40	1.42
45. Rh	2.28	1.47	1.45
46. Pd	2.20	1.57	1.35
47. Ag	1.93	1.72	1.42
48. Cd	1.69	1.73	1.46
49. In	1.78	1.88	1.49
50. Sn(II)	1.80	1.58	
Sn(IV)	1.96	2.02	1.72

Element	Pauling	Sander-son	Allred-Rochow
50. Sn(II)	1.80	1.58	
Sn(IV)	1.96	2.02	1.72
51. Sb	2.05	2.19	1.82
52. Te	2.10	2.34	2.0
53. I	2.66	2.50	2.21
54. Xe	2.60	2.34f	2.40
55. Cs	0.79	0.69	0.86
56. Ba	0.89	0.93	0.97
57. La	1.10	0.92	1.08
58. Ce	1.12	0.92	1.08
59. Pr	1.13	0.92	1.07
60. Nd	1.14	0.93	1.07
61. Pm		0.94	1.07
62. Sm	1.17	0.94	1.07
63. Eu		0.94	1.
64. Gd	1.20	0.94	1.11
65. Tb		0.94	1.10
66. Dy	1.22	0.94	1.10
67. Ho	1.23	0.96	1.10
68. Er	1.24	0.96	1.11
69. Tm	1.25	0.96	1.11
70. Yb		0.96	1.06
71.Lu	1.27	0.96	1.14
72. Hf	1.30	0.98	1.23
73. Ta	1.50	1.04	1.33
74. W	2.36	1.13	1.40
75. Re	1.90	1.19	1.46
76. Os	2.20	1.26	1.52
77. Ir	2.20	1.33	1.55
78. Pt	2.28	1.36	1.44
79. Au	2.54	1.72	1.42
80. Hg	2.00	1.92	1.44
81. Tl(I)	1.62	1.36	
Tl(III)	2.04	1.96	
82. Pb(II)	1.87	1.61	
Pb(IV)	2.33	2.01	1.55
83. Bi	2.02	2.06	1.67
84. Po	2.00		1.76
85. At	2.20		1.90
86. Rn			2.06
87. Fr	0.70		0.86
88. Ra	0.90		0.97
89. Ac	I .10		1.00
90. Th	1.30		1.11
91. Pa	1.50		1 14
92. U	1.70		1.22
93. Np	1.30		1.22
94. Pu	1.30		1.22
95. Am	1.30		(1.2)
96. Cm	1.30		(1.2)
97. Bk	1.30		(1.2)
98. Cf	1.30		(1.2)
99. Es	1.30		(1.2)
100. Fm	1.30		(1.2)
101. Md	1.30		(1.2)
102. No	1.30		(1.2)

E. Chemical Bonding

1. Introduction

Two or more atoms which coalesce to form a molecule may be said to form a chemical bond. The driving force for the atomic coalescence or bond formation is the overall lower energy state of the system.

In a simpler way this can be said as that atoms tend to get electron configuration of noble gases(He, Ne, Ar etc.). Usually the valence orbitals of noble gas consist of s and p orbitals whose electrons amount to 8. The trend to fill the valence orbital is called as 8N rule (octet rule) for this reason. In this regard, metallic elements lose their electrons to more electronegative elements such as halogen(F, Cl, Br, I) resulting in electronic state like noble gas. And they form ionic bonding with these electronegative elements. In non-metallic elements which have similar electronegativity, atoms shares valence electrons one another to follow the octet rule.

Bonds are formed primarily by the interaction of valence electrons and valence orbitals on the bonding atoms. Each atom in the reaction will seek to acquire a stable (filled or half-filled shell) configuration by electron transfer or sharing. This leads to formation of these primary type bonds: ionic, covalent, and metallic. Van der Waals and hydrogen bonds are considered to be secondary bonds. Half shell configurations are often stabilized by the interaction of different orbitals e.g. (s, p or d), leading to the formation of the mixed or hybrid bonds. Carbon, for example, forms four sp^3 hybrid bonds according to the scheme:

$$1s^2 2s^2 2p^2 \longrightarrow 1s^2 2s^1 2p^3$$

These bonds are often directional. The type bonds which are likely to form can be predicted by consideration of the electronegativity differences between the reacting atoms, and the proximity of the valence electrons to a filled or half-filled shell. However, bonding between atoms are rarely completely of one type, there being some covalent character to even the most ionic bonds.

2. Bonding Types

Due to this spherical shape, s orbitals tend to form ionic bonds which has little or no directionality. These dumbbell shaped orbitals (including p) result in poorer shielding and typically form directional covalent type bonds.

Atoms consist of nucleus, neutron and electrons. Nucleus and neutron are located at center of the atom surrounded by electron shells or clouds. In terms of quantum mechanics, electrons orbit in discrete orbitals which have different potential energies. For stable configuration, electrons are filled from the lowest energy orbital. Usually the highest energy orbitals are not filled with complete electrons since the number of electrons must be the same as the number of protons. On formation of bonds, the outer shells are filled with complete electrons which is more stable configuration than in unbonded atoms. Bonding , therefore, take place among the electrons in the highest energy level (called valence electrons). Depending on the method to fill the outer shells, bonds can be divided into 3 primary groups described as covalent, ionic, and metallic bonds. These are illustrated in Figure 1-26 and discussed in further detail in the pages following. Other than the primary bonds above, secondary bonds(hydrogen bonding and van der Waals) develop from dipolar fluctuations or from the dipolar nature of such molecule as H_2O.

2-1 Primary Bonding

Typically, also, these orbitals ae associated with mixed bonding, involving outer s orbitals and partially filled d or f orbitals.

(i). Ionic bonding

For ionic bonding, at least one valence electron form one atom is essentially transferred to the other atom with greater electronegativity resulting in positive charge on the donating atom and negative charge on the accepting atom. These opposite charges induce the electrostatic force between the atoms. Since the electrostatic force developed has no directionality, the positive charge can be surrounded by as many of atoms of negative charge as space considerations allow.
Ionic bonds are typically formed between atoms with large electronegativity differences. Such atoms can readily loose or gain an electron to form cations(Na+, Ca+2, Al+3,...) or anions (F-, Cl-, O2-) with a closed shell or rare gas configuration. Conditions for forming an ionic bond are summarized below:

Between atoms (ions) of low and high EN

$\Delta X \geq 1.7$

Essential Electron Transfer

Ex: $Na + Cl \rightarrow Na^{+}Cl^{-}$

$\Delta X = 2.1$ (67% ionic) ions formed

(ii). Covalent bonding

As illustrated in Fig. 1-26, covalent bonds occur when the valence electrons are shared between bonding atoms. In terms of electronegativity principles, covalent bonds occur between atoms with moderate to high electronegativities, but with essentially Equal and strong attraction for electrons. The molecule formed is symmetrical with no significant dipole moment. Such bonds may be formed from atoms with either closed or half shell configuration. Conditions for covalent bond formation are summarized below:
With atoms of moderate to strong EN

$$\Delta X = |X_a - X_b| < 0.7 \text{ (12\% ionic)}$$

Ex: SiC
$\Delta X = 0.7$ Electrons Shared

(iii). Polar covalent bonding

Polar covalent bonds are typically formed between atoms of moderate to high electronegativities but with disparate attraction for the electron. On forming a bond, the electronegativity difference results in displacement of the electrons towards the more electronegative atom. The molecule formed is non-symmetric and polar with a distinct dipole moment. Polar covalent bonds are commonly formed between hydrogen and highly electronegative atoms such as Cl, F, O, and N. Secondary bonding within such compounds (HCl, H_2O) often leads to molecular association and a relatively high boiling point. The conditions for the formation of a polar covalent bond are summarized below:

Ex. Polar Covalent - Dissimilar but moderately strong EN

$$0.7 \leq \Delta X < 1.7$$

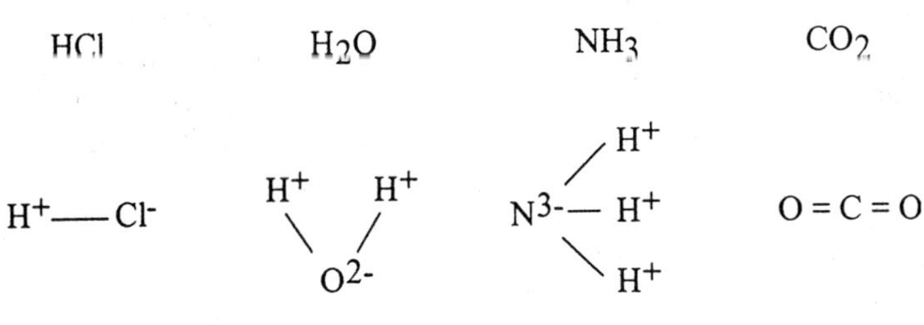

Figure 1-27 Partial transfer of electron to more EN atom.

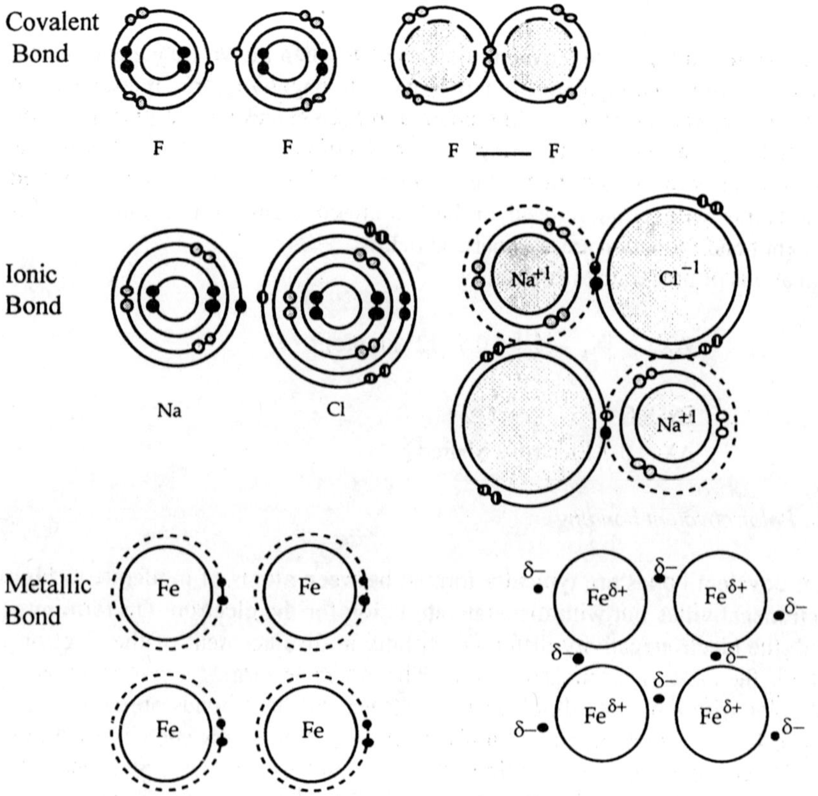

Figure 1-26. Schematic drawing of molecular structures of primary bonds

(iv). Metallic bonding

Metallic bonds are typically formed between cluster of atoms with weak low electronegativities. since the atoms attract only weakly, the valence electrons are essentially "free" and shared by all atoms. Actual bonding occurs by localization of (negative) electron charge clouds between atomic nuclei. Bond strengths depend on the orbitals involved, with s-orbitals alone, bonding is typically weak(Na- Na, Ca-Ca, ...) whereas very strong bonds may be formed (Fe-Fe, Cr-Cr,...) with orbital mixing (s-d, or s-p orbitals). Conditions for metallic bond formation are summarized below:

Ex. Atoms of low EN

$\Delta X < 0.5$

Na-Na Ca:Ca etc.
Essentially "free" electron since bonding is weak.

2-2 Secondary Bond

(i). Hydrogen bonding

Hydrogen bonding occurs between strongly electronegative atoms such as F, Cl, N, O and H. The strong attraction causes displacement of the electrons leaving a net positive charge on the electronegative atom.

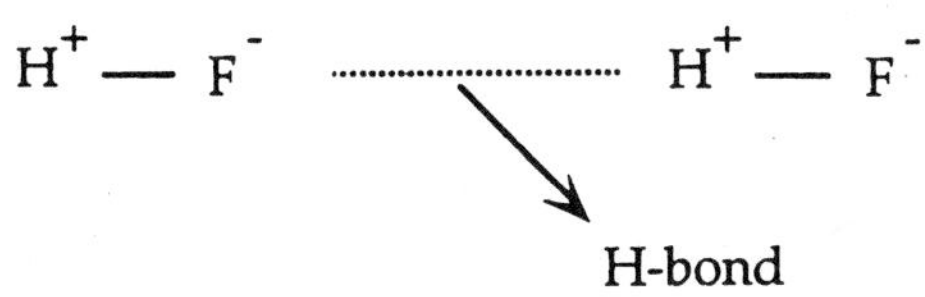

Because of this charge displacement association of the molecules takes place resulting in a weakly ionic bonded (0.25 eV) material. Because of this association, hydrogen bonding often results in stable, higher boiling point liquids. Hydrogen bonding also occurs in solids such as clays—between O^{2-} and OH^{-} layers.

(ii). Van der Waals bonding

Random fluctuation within an atomic or ionic structure can lead to distortion of electron clouds surrounding the nuclei. Such fluctuation may be induced by thermal or electromagnetic vibration. The asymmetric displacement of electron shell induces the formation of momentary dipoles. Interaction between these dipoles produce weak attractive forces known as London forces. These generate weak Van der Waals bonds which are electrostatic in nature. Such dipole fluctuation is common to all solids, but Van der Waals bonding is dominant in only a few compounds. Conditions for van der Waals bond formation are summarized below:

> There is no electron transfer between atoms or sharing electrons. Electronegativity is not involved to characterize the bonding. Consequently bonding is in nature dipolar. More or less this bonding occurs in all solids.

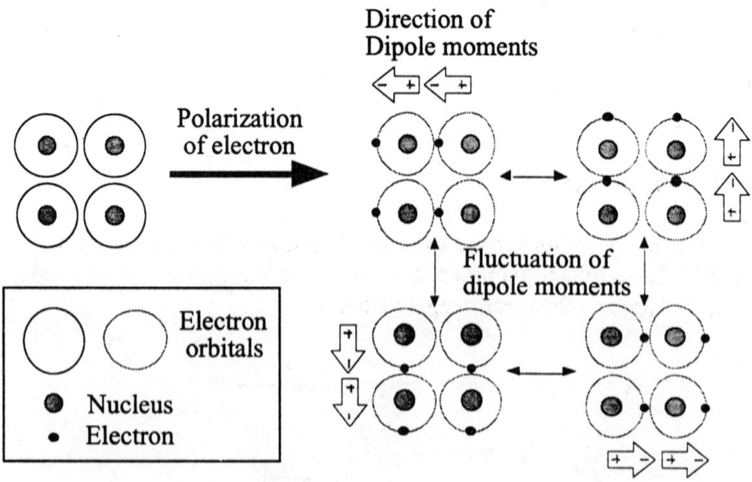

Figure 1-28 Dipole-Dipole interaction in van der Waals bonding

3. Bonding Forces

In order to understand the physical properties of solids, one must develop an understanding of the bonding forces which hold atoms(together). These binding forces are of 2 kinds, namely

a) an attractive force which holds the tom or ions together
b) a repulsive force which prevents the atoms or ions from coalescing.

Coions with the same kind charges would of course have an electrostatic repulsion between them. Therefore such coions do not form bonds. The following energy diagram can illustrate the variation of attractive and repulsive forces with interatomic distance.

3-1. Components of Potential Energy - V

Both V_a and V_r expressed as f(r). Also V must be negative for bonding to take place i.e. the process must result in a net gain of energy.

$$V = V_a - V_r = \left(-\frac{a}{r^n} + \frac{b}{r^m} \right)$$

where: V = Total potential energy

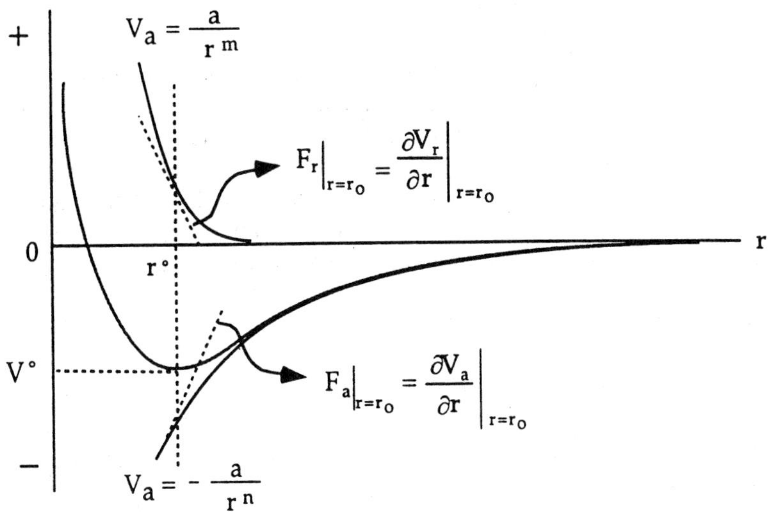

Figure 1-29 Schematic diagram for potential energy V with attractive and repulsive energies

$V_a = -\frac{a}{r^n}$: Attractive energy, keeps atoms together.

$V_r = \frac{b}{r^m}$: Repulsion term, prevents coalescence of atoms.

a, b, m and n = constants, r = interionic separation distance

Typically m > n , so that V for bonding is a negative value. The potential energy at interatomic distance r is the sum of attractive energy and repulsive energy as $V = V_a + V_r$. At equilibrium position, the potential energy, V_o, must be the minimum. r_o is the equilibrium interatomic separation where V_o is the minimum.

Hence, $\frac{\partial V}{\partial r}\Big|_{r=r_o} = 0 \quad \frac{\partial V_a}{\partial r} = -\frac{\partial V_r}{\partial r}\Big|$ and $F_r = F_a$ at Equilibrium distance r_o.

For diatomic molecule, $\frac{r_o}{2}$ is equal to the atomic radius. In Fig.1-29, the attractive force, F_a and repulsive force F_r are the slopes of tangent lines of curves for V_a and V_r at r_o, respectively.

3-2. Nature of the Repulsive Force F_r

The repulsive force is the same for all solids. It arises from overlap of inner atomic orbitals and too close approach of the nuclei as the atoms(ions) coalesce. To degree the repulsive forces can be considered coulombic due to mutual repulsion of the +ve charged nuclei. However the overlap of the inner atomic orbitals introduces quantum mechanical effects as the orbitals are compressed, and cause a momentum increase in the electrons comprising the orbitals - Heisenberg uncertainty principles. The repulsive force is defined by Born as $F_r \propto -\frac{1}{r^{n+1}}$.

3-3. Nature of Attractive Force F_a

The attractive forces vary for different solids, depending on the type of bonding (ionic, covalent, metallic)

i) For ionic solids

Ionic solids consist of cations and anions. The bonding force, therefore, is the electostatic forces between these cations and anions. According to Coulomb's law, the bonding force is quadratic regarding interactomic distance.

$$F_a = \frac{q_1 q_2}{4\pi\varepsilon_o r^2}$$

Coulomb's Law

where q_1 and q_2 are electric charges of cation and anions, respectively, and r is distance between cation and anion. i.e. the attractive forces or coulombic or electrostatic in nature

ii) For covalent bonding solids

The attractive force are quantum mechanical in nature i.e. they are due to the overlap of atomic orbitals which causes localization of the electron cloud between nuclei- thus lowering the potential energy. The localized -ve charges tend to repel similarly *localized cells* so that the bonding is directional. The magnitude of the localization of electrons between two nuclei is measured by the energy difference of the bonding orbital, which is formed by combining two atomic orbitals, from that of atomic orbitals. This difference is expressed as ΔE in Fig.1-30 in the case of the hydrogen molecule. The greater the energy difference, DE, is, the more overlap of electron orbtials is accompanied.

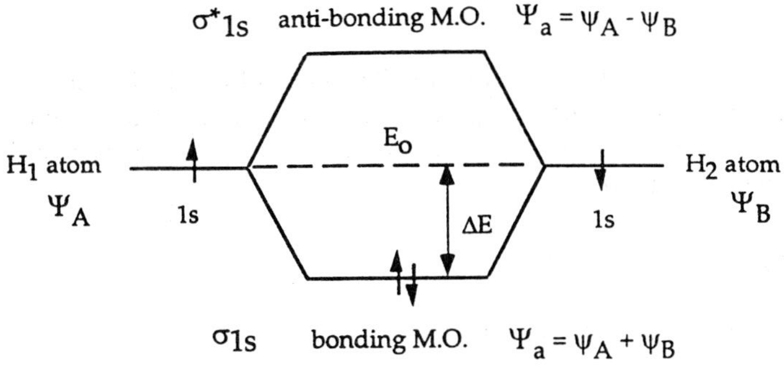

Figure 1-30 Energy diagram for H_2 molecule
[Here, E_o and E are energy levels for the isolated atoms and bonding molecular orbital, respectively. $\Delta E = E - E_o$ = Exchange or Bonding Energy (-ve)]

iii) Metallic bonding

Metals cannot be characterized simply as exchange or pairing of electrons nor as dipole interaction. Rather they are considered as collection of nuclei bonded by free electrons. In the 'free electron model' for metals, free electrons are distributed throughout the metals and form a electron cloud where nuclei of metal atoms are implemented as shown in Fig.1-31. Here the "free" electrons perform the function of neutralizing the mutual repulsion of the +ve nuclei. In return the +ve ions (nuclei) provide a low potential wall for the electrons. It is this reduction in the potential energy of the electrons which provides the attractive force for bonding. More than 2 atoms (> 10) needed to define the metallic state. Potential Energy of system lowered by juxtaposition of atoms relative to the isolated state

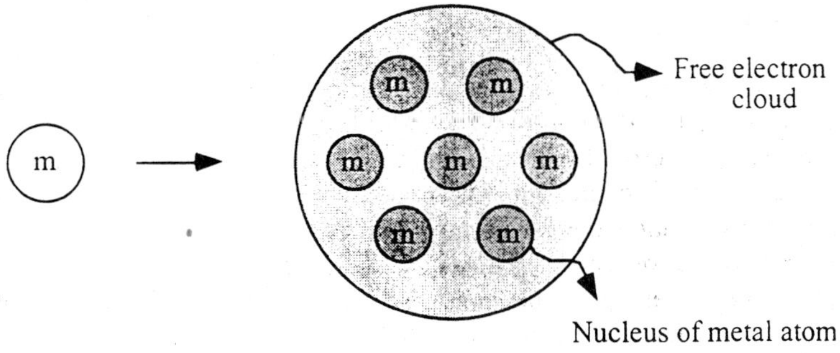

Figure 1-31 Free electron model for bonding in metals

iv) Van der Waals

Van der Waals attractive forces for bonding derives from the London dispersion forces i.e. the dipolar fluctuations associated with atoms in an assembly. The dipoles formed (-ve and +ve ends of atoms) provides the attractive force for bonding which varies roughly

$$F_a \propto r^{-7}$$

3-4. Strength of Primary Bond Types

In ionic crystal, the attractive force is the Coulombic electrostatic force which is proportional to $\frac{1}{r^2}$ and the repulsive force is the interactive force between electron cloud surrounding nuclei which is proportional to $\frac{1}{r^{m+1}}$, m = 5 – 12. Coions with the same kind charges would of course have an electrostatic repulsion between them. Therefore such coions do not form bonds. The following energy diagram can illustrate the variation of attractive and repulsive forces with interatomic distance.

In other types of crystals, it is not easy to express the potential energy like in ionic crystals. Only for molecular crystals which are bonded by van der Waals force, the mathematical expression is developed. The van der Waals force is explained by a synchronized movement of electrons as shown in Fig.1-28. This synchronization of movement of electrons is equivalent to a rotating electric dipoles. The potential energy between two dipoles is proportional to $\frac{1}{r^6}$. Hence the attractive force is proportional to $\frac{1}{r^7}$. This attractive force is less than that of electrostatic force in ionic crystals by orders of 4 for 'r'. Hence the bonding energies for molecular crystals are much less than those for ionic crystals.

Since analytical expressions for the other two bonding types, covalent and metallic bondings, are not available, it is difficult to compare the relative strengths of materials having different bonding types. However, the experimental data provide a rough estimation for the relative magnitude of potential energies of bonding types. The data in Table 1-18 illustrates the trend in the bond dissociation energies of typical materials of different bonding types. Since the dissociation energy is directly related to the potential energy of a bond, the potential energy of bonds can be estimated from the data in Table 1-18. From these data, in general, covalent bond show the highest bond energies. Ionic bonds show bond strengths close to the covalent bonds. van der Waals bond is much lower even than metallic bonds. The bond energy of W(tungsten)

Table 1-18 Enthalpies of sublimation for selected solids at 25°C

Bond Type	Substance	$\Delta H_{sm,25^\circ C}$	Bond Type	Substance	$\Delta H_{sm,25^\circ C}$
van der Waals	Ar	7.5	Ionic	LiF	582
	O_2	7.5		NaCl	588
	CO_2	25		MgO	619
	CH_4	18		CaF_2	438
Hydrogen	H_2O	61	Metallic	Na	107
	NH_3	35		Mg	148
Covalent	C(diamond)	698		Al	326
	Si	456		Cu	338
	SiC	798		Ag	285
	InSb	375		W	849

[Enthalpy of sublimation (ΔH_{sm},25°C) from the Journal of Physical and Chemical Ref. Data, Vol. 11, 1982, Supplement No 2.]

is exceptionally high among metals. This large bond energy has been attributed to a covalent character of W(tungsten).

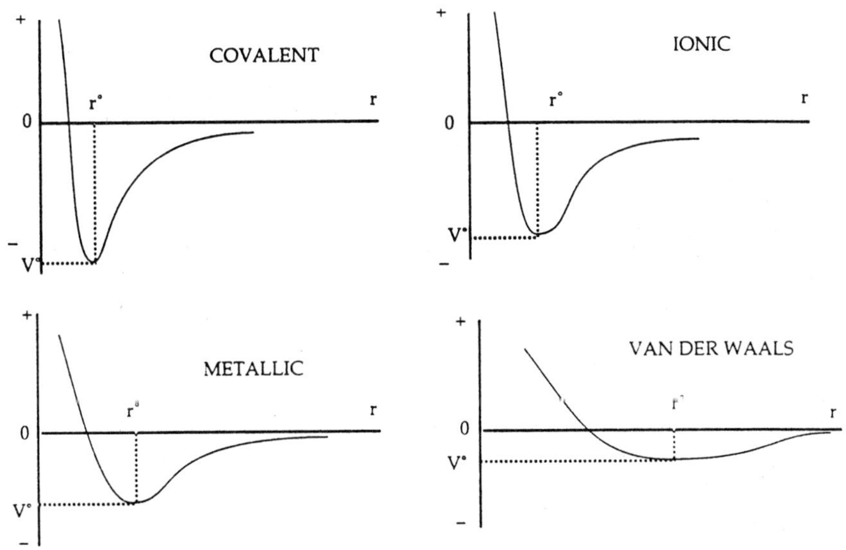

Figure 1-32 Potential curves for covalent, ionic, metallic and van der Waals bonds

The equilibrium distance between two atoms involved in a bond is proportional to the bond strength since greater bonding force means greater attractiveness of the atoms resulting in a shorter distance. Regardless of bonding types, the magnitude of attractive force is estimated how close the nuclei are approached until it is compensated by repulsive force. This closeness of nuclei is sketched by the change of atomic radii during bond formation. If atoms or ions are assumed as hard spheres and the interactomic distance at equilibrium is assumed as the sum of radii of two atoms or ions, the atomic radii in Table 1-32 of later section in this chapter will be helpful. The trend of atomic size agrees the trend in the potential energies in Table 1-18 since different types of radii follows the order of covalent radii < metallic radii < van der Waals radii if there are more than two types of atomic radii for an element. In this respect, the potential energy curves for bonding types can be schematically drawn as in Fig.1-32. Covalent bond has the deepest potential well and shortest interatomic distance while van der Waals bond has the most shallow potential well and largest bond distance.

4. Bonding in Polymer

(i). Introduction

The term polymer denotes a molecule made up by the repetition of simple units, the mer or monomer. Polymers are typically organic, although polymeric type structure can be formed in metalloorganic solution precursors of ceramics, glasses and metals. The repetition of these units in polymers is one dimensional, compared to crystals in which the repetition unit is 3 dimensional. Bonds between atoms in the monom er as well as the bonds linking monomer mus be strongly covalent in order to make linear polymeric structure. This configuration requires free bonds available on opposite sides of bonds. The condition is satisfied by very few atoms mainly carbon and silicon. Compared to carbon, silicon forms less strong bond between them. Therefore silicon needs other atoms such as oxygen and carbon to make strong covalent bonds. Hence polymeric structure are predominantly organic. The basis for the wide varieties of organic polymers is that carbon has 4 unpaired valence electrons. Two of these typically form the backbone of the polymer and the other two can be variously modified to give different monomeric structures.

The fundamental structure of organic polymers is that of polyethylene with ethylene(CH_2CH_2) as repeating unit or monomer). The backbone of polyethylene is a linear chain of carbons as shown in the below.

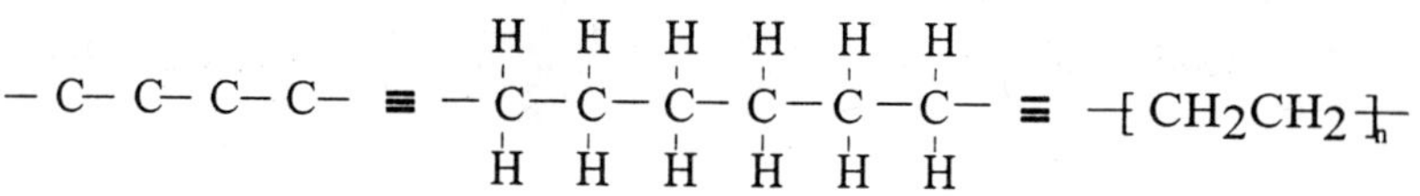

For polymers, the monomer or repeating unit is typically determined from consideration of precursor origin rather than the chemical formula of the finished polymer. For example the chemical reaction for synthesis of polyethelen can be depicted as below where the ethylene monomer forms the basics of the reaction.

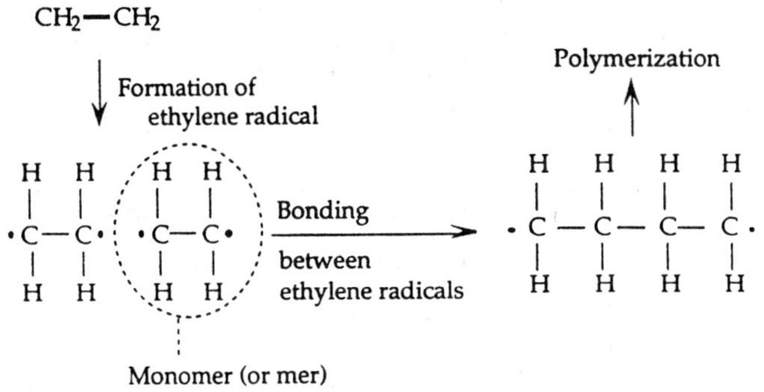

As shown in the above, the monomer for polyethylene is ethylene not methylene which is very difficult to make and polymerize. Hence polyethylene is not called as polymethylene although its basic repeating unit can be considered as methylene.

Various polymers can be considered as derivatives of polyethylene by replacing H in ethylene(CH_2CH_2) by other atoms such as Cl, F, and O or large carbon groups such as methyl(CH_3) and benzene(C_6H_5). Also carbon backbones of polyethylene can be modified by substitution of ethylene group of polyethylene with other longer carbonic group such as isoprene and butadiene as shown in the third column of Table 1-19. Some important derivatives of polyethylene are listed in Table 1-19.

(ii) Structural features of polymers

Since the polymer is a repeated sequence of monomers, it can be represented as:

$$-\left[R \right]_n-$$

where R is the repeating unit and n is degree of polymerization, DP. Polymers with more than one repeating unit are called copolymers. These can be further classified as random copolymer, regular copolymer, block copolymer, and graft copolymer, depending on how each kind of monomer is arranged. These are illustrated in Figure 1-33.

Table 1-19 Important organic polymers

Polymers with vinyl backbone		Polymers with non-vinyl backbone
Carbon branch	Non-carbon branches	
Polyethylene $-(CH_2CH_2)_n-$	polyacrylonitrile $-(CH(CN)CH_2)_n-$	polyisoprene $-(C(CH_3){=}CHCH_2CH_2)_n-$
polypropylene $-(CH(CH_3)CH_2)_n-$	poly(vinyl chloride) $-(CH(Cl)CH_2)_n-$	polybutadiene $-(CH{=}CHCH_2CH_2)-$
polyisobutylene $-(C(CH_3)_2-CH_2)_n-$	poly(vinyl alcohol) $-(CH(OH)CH_2)_n-$	polyformaldehyde $-(OCH_2)_n-$
polystyrene 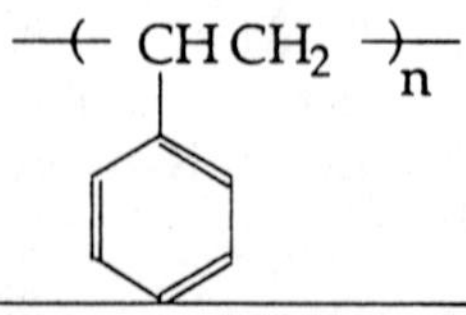	poly(vinyl acetate) $-(CH(OOCCH_3)CH_2)_n-$	poly(ethylene oxide) $-(OCH_2CH_2)_n-$
	poly(vinylidene fluoride) $-(CF_2-CH_2)_n-$	Poly(phenylene oxide) 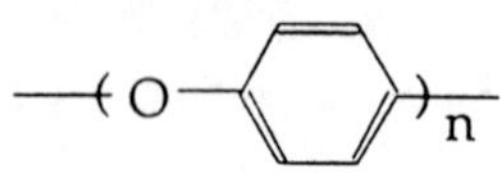
	poly(tetrafluoroethylene) $-(CF_2CF_2)_n-$	poly(vinyl butyral) 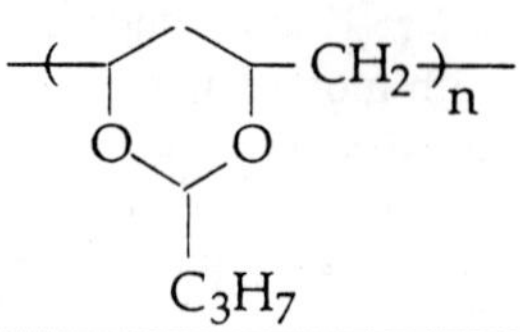

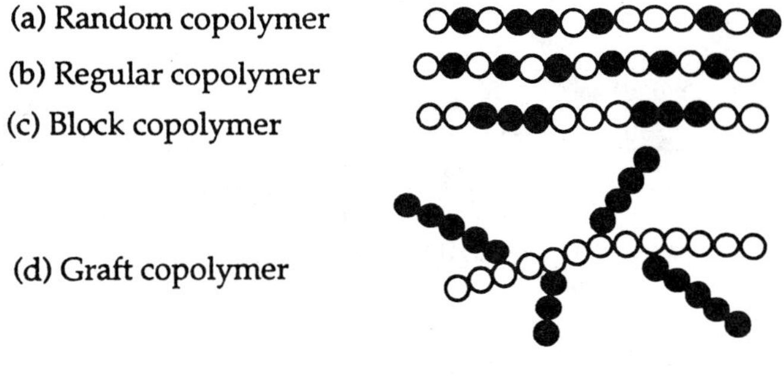

Figure 1-33 Arrangement of copolymers

Polymers with one type repeating monomer having different side groups can be classified into 3 groups depending on how the side groups are arranged. These groups are described as atactic, isotactic, and syndiotactic. An isotactic polymer is a type of polymer in which the side group is arranged on only one side of the polymer backbone. In atactic type, side groups are positioned irregularly in either side of polymer backbone and in syndiotactic polymers there is a zig-zag type arrangement of side groups along the polymer backbone. These are illustrated schematically in Figure 1-34 using simple vinyl polymer.

(ii) Energetics in polymers

The difference of polymer from other molecules is that its average molecular weight is 10,000 ~ 1,000,000. Although the polymer backbone is formed from

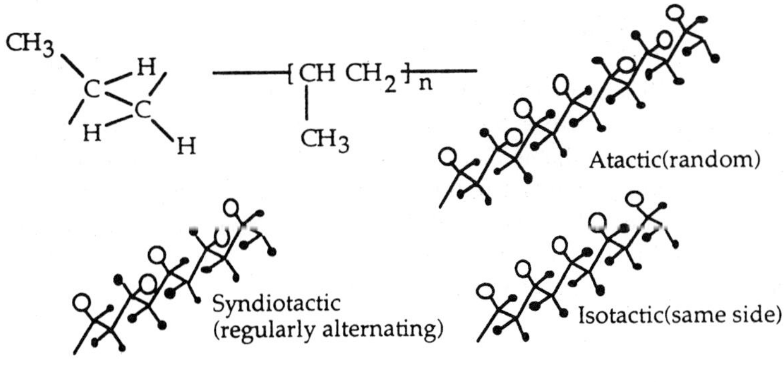

Figure 1-34 Stereo isomerism of polypropylene

Table 1-20 List of weighting factor for some organic groups[Dunkel]

R	Y_i(cal)	R	Y_i	R	Y_i
— CH_3	1780	— O —	1630	— COOH	8970
— CH_2 —	990	— NH_2	3530	— Cl	3430
		— CO—	4270	— F	2060
		— OH	7250	— Br	4300

covalent bonds, the intermolecular forces between polymer chains are the secondary bonding forces such as Van der Waals, hydrogen, and dipolar bonding. These secondary bonds are much weaker than covalent bonds. Due to the huge molecular weight, however, its intermolecular force per each molecule (= chain) is not small. This large intermolecular force makes polymers easily solidify at room temperature and have large cohesive energy per mole (= chain). This cohesive energy of polymer is additive with respect to molecular weight and repeating unit. The cohesive energy can be expressed as a linear sum of contributions of each characteristic segment of the polymer.

$$E_{cohesive} = \sum_i Y_i \tag{59}$$

Here Yi is a coefficient for i-th segment of the polymer. Table 1-20 list the values of Y_i for some important organic group(R).

Among organic groups, methylene(-CH_2-) has the lowest coefficient. Therefore polyethylene whose backbone consists of methylene groups is expected to have the lowest cohesive energy if chain length is the same. As hydrogen atom in polyethylene is replaced by larger alkyl group or non-carbonic elements such as N, O and Cl, its cohesive energy increases. The coefficient for hydroxyl group(-OH) and carboxylic group(-COOH) is much higher than for other organic groups. These two groups illustrates the effect of hydrogen bonding. The reason for this is that Van der Waals force is proportional to r^{-6} and hydrogen bonding force is proportional to r^{-1} to r^{-3} where r designates the distance between atoms. If there is hydrogen bonding from high polarized group such as -OH or -COOH, cohesive energy increases more rapidly than without it. Therefore hydrogen bonding results in rigid structure of polymer.

(ii). Polymer crystals

Compared to conventional crystalline materials, polymer has two characteristic features regarding crystal structure. Those are bonding forces and locality of crystalline structure. The binding forces in the polymer crystals are secondary

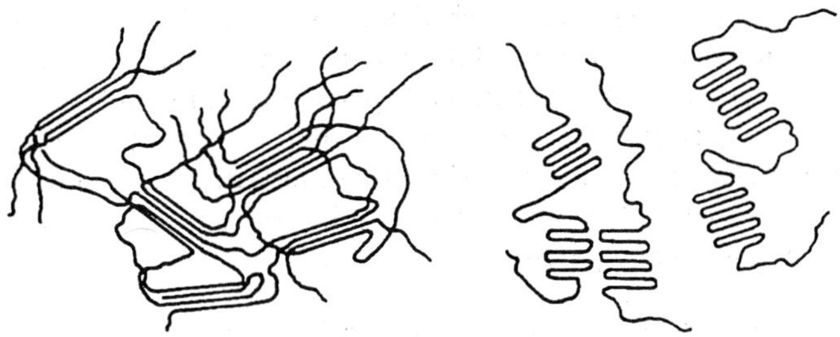

Figure 1-35 (a) Fringed miscelle structure and (b) Folded-chain structure of crystallite in polymers

molecular forces in which is little or no sharing or exchange of electrons between atoms. Usually these secondary forces are not sufficiently strong to make crystalline structures of simple molecules. As shown above, however, these secondary forces can hold polymer chains together in crystalline forms if the polymer chains are arranged in parallel ways as shown in Fig. 1-35 to maximize the accumulated bonding forces. These arrangements of polymer chains are described as fringed miscellar structures and folded chain structures, respectively.

As can be seen in Fig. 1-35, repeating unit in crystallites must be considered as long chain not the monomer. Polymers in liquid states can freely rotate around C-C bond in polymer chain. The arrangement of polymer chains in the melt state is not packed in a parallel way as in Fig. 1-35. Since polymer is long chain, it is very difficult to arrange these whole length of polymer chains in a parallel packing to form a crystalline one during solidification. Therefore, polymer is in general combination of amorphous and crystalline structures in solid states.

The crystallinity of polymers is related to the kinetics of parallel packing of the polymer chains. This is dependent on the linearity of polymer chain, and the arrangement of the side groups. In general, linear structures more

Table 1-21. The relationship between crystallinity and irregularity of polymers

Irregularity	Polymer	Typical percent of crystallinity
Linearity	Linear polyethylene	90
	Branched polyethylene	40
Tacticity(Side groups)	Isotactic polypropylene	90
	Atatic	none

readily form crystalline areas than branched or network structures, since the side groups hinder dense parallel packing of the polymer chains. Therefore branched polymers exhibit less crystallinity than linear polymers without branches. Also syndiotactic and isotactic type of polymer exhibit higher crystallinity than atactic since irregular arrangements of side groups hinders regular packing in a crystalline form. Table 1-21 shows the effect of linearity and side groups on the formation of crystallite in polymers.

(iv) Structural effects of polymers on thermal mechanical properties

(a) Thermosetting and thermoplastic polymers

Thermal motion of polymers denotes vibration of each atoms in polymer and the movement polymer chains against each other. Hence the thermal properties of polymers are strongly dependent on the molecular structure of the monomer(repeating unit) and configuration of the polymer chain. With respect to monomer, polymers are classified into two groups. They are thermoplastic and thermoset polymers. Thermoplastic polymers are those which show reversible shape change with heat treatment within a certain temperature range. In contrast, thermosetting polymers are those which have no reversible shape change with heat treatment. For thermosets, the decomposition temperature is lower than the softening temperature. Therefore thermosets starts decomposing before they soften. In terms of microstructure, these thermal properties are due to the microstructure of monomers. These are illustrated in Table 1-22. The main characteristic of thermosetting polymers is that they have C-X bonding in the *backbone* where X is a non-carbon atom which can result in cross-link between polymer chains. Therefore thermosetting polymers form cross-links when they are heated to certain temperatures. If heated further, they decompose rather than melt since the bond strength of cross-link is stronger than C-C bond in the polymer backbone. In thermoplastic polymers, they do not have C-X bonding in the backbone which is needed for cross linking of polymer chains. Therefore thermoplastic polymers melt if heated.

Table 1-22 List of typical thermosetting and thermoplastic polymers

Thermosetting polymers	Thermoplastic polymers
Epoxides	Polyethylene
Polyesters	Polypropylene
Ureas	Polystyrene
Melamines	Polyamides
Phenolics	Polyvinyl chlorides

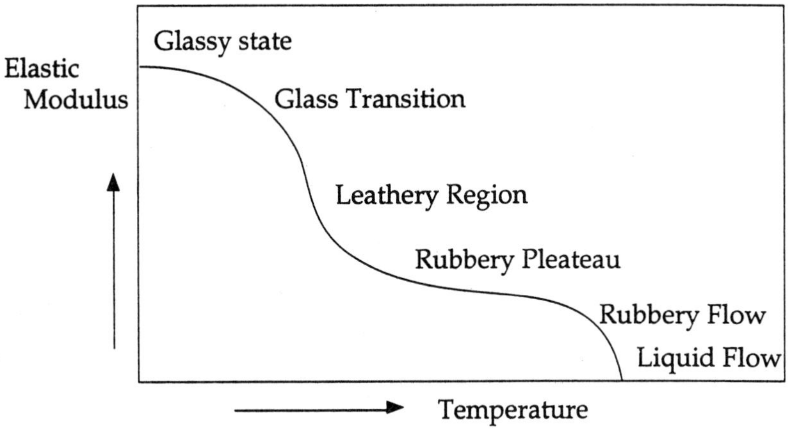

Figure 1-36 Elastic modulus change with temperature

Thermoplastic polymers typically exhibit the modulus change with temperature shown in Fig. 1-36. The curve can be divided into 3 major regions identified as glassy, rubbery, and liquid region. These regions can be explained in terms of microstructural features. In the glassy region, only bending and stretching of polymer chain contribute to the modulus since the energy barrier for the rotation of polymer chain is higher than thermal energy, kT (k is Boltzmann constant and T is absolute temperature). It exhibits Hookean elastic behavior. During glass transition, side groups and main chain can partly rotate contributing to lowering the modulus. In the rubbery region, rotation of the main chain is greatly extended resulting in some mobility of the chain. In the liquid region, polymer chains slips against each other and shows liquid like properties.

(b) Glass transition temperature(T_g)

This is a very important parameter which is related to the mechanical properties of polymer. The modulus changes over this temperature is approximately by a factor of 1000. If Tg is lower than room temperature, polymer can be described as elastomer or rubbery polymer. If Tg is higher than room temperature, it is brittle rigid polymer.

Although theoretical explanations for glass transition phenomena are not perfect, glass transition temperatures can be measured experimentally. One method used to determine Tg is plotting variation of relative volume vs. temperature. At glass temperature, the slope for the variation of relative volume suddenly increases as shown in Fig. 1-37.

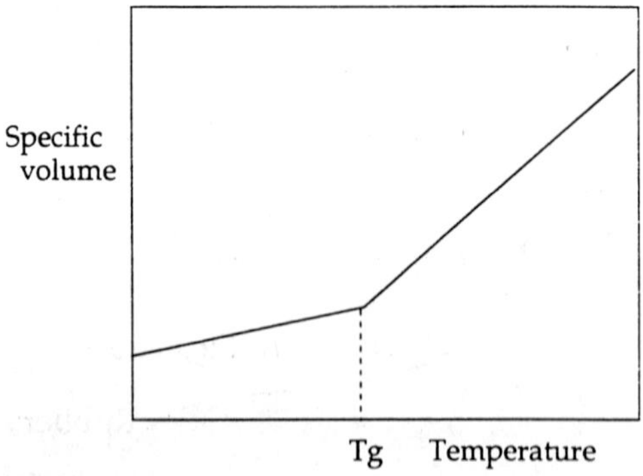

Figure 1-37 Determination of Tg by change in the specific volume

As explained above, reason for increase of relative volume with temperature below Tg is bond length increase due to increase in amplitude of stretching and bending mode of polymer bondings. Above Tg, rotation of polymer chain is not restricted and polymer molecule produces free space for this rotation. Not only vibration but also rotation of polymer molecules contribute to increase of relative volume above Tg. Therefore there is a sudden increase in relative volume over Tg.

The rotation of polymer chain is related to its flexibility which in turn is dependent on structural features such as linearity of the polymer backbone as well as to the length and bulkiness of the side groups.

Table 1-23 illustrates the structural effects on Tg. With the same side group, polyacrylates have lower Tg than polymethacrylates. The additional methyl group(CH_3) in the backbone of the polymethacrylates reduces the flexibility since it makes more difficult the rotation around C-C bond in the backbone than without it. In this respect, the effect of side group attached to the backbone of polymer chain increases as its size increases in the direction of raising Tg. This is, however, not always the case. Increased side group may show the reverse as in Table 1-23 where increase in size of the side group lowers Tg by 10 - 20°C per each CH_2 in alkyl group attached to —COO—. This reversed effect of side groups on Tg of polyacrylates and polymethacrylates can not be explained by the hindering effect of side groups on rotation of polymer chain. On the other hand, the increased side group induces the void space between polymer chains and hinders dense packing of polymer chains since the side group can rotate more freely than the backbone of polymer chain. This void space in turn gives rise to easier rotation of the backbone of polymer chain. Hence the steric effect of side groups on the rotation of polymer chain is compensated or surmounted by the negative effect

Table 1-23 Glass transition temperatures of acrylate and metacrylates with diffenrent side groups

	Acrylates	Methacrylates	
Side group(R)	Tg	Tg	V_{Tg}*
Methyl(CH_3)	3	105	2.4
Ethyl (C_2H_5)	-22	65	2.2
Propyl(C_3H_7)	-44	35	2.1
Butyl(C_4H_9)	-56	21	2.0
Repeating unit	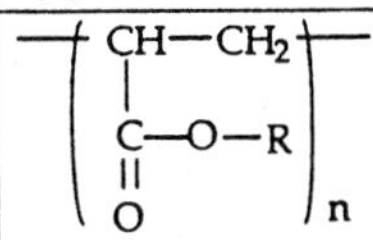	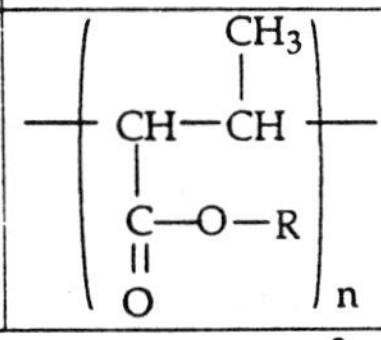	

* Specific volume of polymer at Tg. Its unit is cm^3/g

of void space resulting in lowered Tg. This is shown in the final column where the increase of specific volume increases with increase in size of side group.

In a similar way to cohesive energy, Tg can be estimated in a additive way using the coefficients of each organic group in polymers. As shown above, however, Tg can not be estimated in a linear addition of coefficients of each group in polymers like Eq.(60). Rather it is estimated by considering other effects such as void space and interactions between adjacent groups. General expression of Tg is

$$Tg = \frac{\sum_i s_i T_{g,i}}{\sum_i s_i} \qquad (60)$$

where Tg,i is a coefficient of i-th group which is closely related to Yi in Eq.(60) and si is a weighting factor for Tg,i. For copolymers, Tg can be estimated as

$$T_g = v_1 T_{g,1} + v_2 T_{g,2} \quad \text{or} \qquad (61\text{-}1)$$

$$\frac{1}{Tg} = \frac{W_1}{T_{g,1}} + \frac{W_2}{T_{g,2}} \qquad (61\text{-}2)$$

where $T_{g,\ 1}$ and $T_{g,\ 2}$ are glass transition temperatures of homopolymers with monomer 1 and monomer 2, respectively. W_1 and W_2 weight fractions of monomer 1 and monomer 2, respectively. (v_1 and v_2 volume fractions of monomer 1 and monomer 2, respectively.)

For the difference of syndiotatic and atactic types in homopolymers,

$$Tg(syndiotactic) - Tg(atactic) \approx 120^{\circ}C$$

5 Molecular Shape

Among the primary bondings only the covalent bonding has directionality since the electrons for the bonding are localized in between two nuclei of atoms associated with the bonding. Hence the skeleton of a molecule is determined by the character of the bonding electrons. In contrast the bonding electrons are delocalized in metallic bonding and, therefore, do not constrain the geometry of the bonding. In ionic bonding, the bonding electrons are localized and distributed in a more electronegative atom. This delocalization of bonding electrons in ionic bonding also does not constrain the positions of the two atoms associated with the bonding. In a strict sense, the geometry of metallic and ionic bondings can be described by packing of each atoms in a regular structure. The pure ionic or metallic compounds lacks the concept of molecule which has an identity differentiating it from atoms and other molecules of same kind. Hence the electronic structure of molecule can be started with the covalent bonding type where electrons are localized between atoms. There are two approaches to describe bonding electrons. The first one is Valence bond theory. The other one is molecular orbital theory. Although the molecular orbitals was developed to describe the molecular structure with covalent bonding, it can be further developed to describe the metallic and ionic bonding.

3-1. Valence Bond Theory

Bonds are formed by overlap of valence orbitals or localization of negative charges between the positive nuclei. Electrons are shared between atoms and not localized on either of atoms. There is not net charge developed on either atom. In Lewis scheme, which describes valence electrons with dots, it is represented as A: A or A — A where : and — illustrates the bond. There are certain rules for the bond to form between atoms. These are as follow:

a. Valence orbitals must not be filled
b. Valence orbitals must overlap
c. Electron spins in the two valence orbitals must be opposite direction
d. Valence orbitals must be filled to fulfill the octet rule by bonding

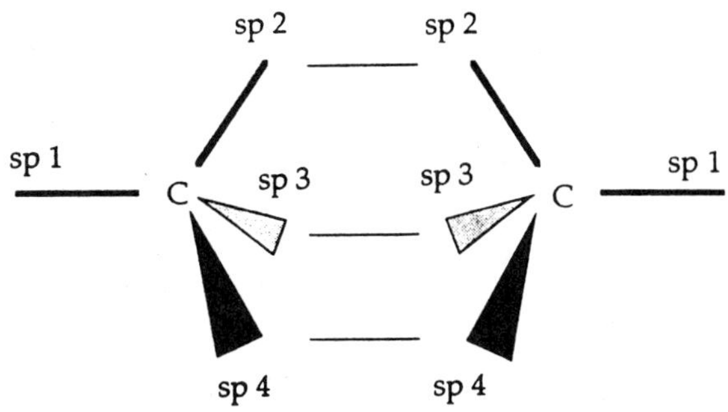

Figure 1-38 Overlap of valence orbitals of carbon atoms in the configuration of acetylene (C_2H_2)

From these rules, stable molecular structure can be predicted. The relative significance of these rules is in the order of a to d. Therefore, it can be stated that 'a' is the most important for the stability of diatomic molecules and 'd' is the least. Table 1-12 depict the formation of diatomic molecule in the first 10 elements in the periodic Table. He and Ne violate the rule 1 because their valence orbitals are filled which is against the rule 'a'. Hence they do not form diatomic molecule regardless of fulfilling of other rules. Among these elements, nitrogen, oxygen and flourine fulfill all the rules from 'a' through 'd'. Hence they exist as stable diatomic molecules. Diatomic molecules of other elements do not follow the octet rule in 'd'.

The relative significance of each rule can be illustrated in the case of carbon. For carbon, its diatomic molecule has bonding order of 2 instead of 4 although it has 4 valence electrons. The reason is due to the greater effect of rule 'b' than rule 'd'. To illustrate the reason, the geometry of acetylen can be referred to, which is shown in Fig.1-38. In acetylene, carbon atom satisfy the octet rule by forming bondings with 4 hydrogens. To obtain this structure, valence electrons are spreaded into the 4 sp^3 hybrid obitals in a tetrahedral configuration and take overlaps with the 1s hydrogen obital. Acetylene satisfies all the rules, 'a' to 'd'. With this geometry of electron configuration for carbon. overlapping of all the electron orbitals is not possible leaving one unpaired valence electron in the diatomic molecule of carbon as shown in Fig.1-38. Hence electron configuration of carbon atom in the diatomic molecule is [$1s^2 2s^2 2p^2$] instead of [$1s^2 2sp^4$] as in acetylene and bondings are formed by overlapping two 2p electrons.

(i) Bond geometry

Bond geometry refers to the spatial arrangement of atoms around a central atom.

Table 1-24 Bonding Schemes for valence electron of the first 10 elements

Element	Configuration of Electron for bonding		Diatomic Molecule	Bonding structure of diatomic molecule		Bond type
1. H	$1s^1$	↑	H – H	$1s^1 - 1s^1$	$1\ \sigma_s$	1 σ bond
2. He	$1s^2$	↑↓		$1s^1 - 1s^1\ (1\sigma)$ $1s^1 - 1s^1\ (1\sigma^*)$	1σ $1\sigma^*$	no net bonding
3. Li	$1s^2\ 2s^1$	↑↓ ↑	Li – Li	$2s^1 — 2s^1$	$1\ \sigma_s$	1 σ bond
4. Be	$1s^2\ 2s^2$	↑↓ ↑↓		$2s^1 — 2s^1\ (2\sigma)$ $2s1 — 2s1\ (2\sigma^*)$	$2\ \sigma s$ $2\ \sigma_s^*$	no net bonding
5. B	$1s^2\ 2s^2\ 2p_x^{\ 1}$	↑↓ ↑↓ ↑		$2p_x 1 — 2p_x^{\ 1}$	π_p	
6. C	$1s^2\ 2s^2\ 2p_x^{\ 1}\ 2p_y^{\ 1}$	↑↓ ↑↓ ↑ ↑		$2p_x^{\ 1} — 2p_x^{\ 1}$ $2p_y^{\ 1} — 2p_y^{\ 1}$	$2\ \pi_p$	
7. N	$1s^2\ 2s^2\ 2p_x^{\ 2}\ 2p_y^{\ 1}\ 2p_z^{\ 1}$	↑↓ ↑↓ ↑ ↑ ↑	N≡ N	$2p_x^{\ 1} — 2p_x^{\ 1}$ $2py^1 — 2p_y^{\ 1}$ $2p_z^{\ 1} — 2p_z^{\ 1}$	σ_p $2\ \pi_p$	1 σ and π double bonds
8. O	$1s^2\ 2s^2\ 2p_x^{\ 2}\ 2p_y^{\ 1}\ 2p_z^{\ 1}$	↑↓ ↑↓ ↑↓ ↑ ↑	O = O	$2py^1 — 2p_y^{\ 1}$ $2p_z^{\ 1} — 2p_z^{\ 1}$	σ_p $1\ \pi_p$	1 σ and π bond
9. F	$1s^2\ 2s^2\ 2p_x^{\ 2}\ 2p_y^{\ 2}\ 2p_z^{\ 1}$	↑↓ ↑↓ ↑↓ ↑↓ ↑	F – F	$2p_z^{\ 1} — 2p_z^{\ 1}$	$1\ \sigma_p$	σ bond
10. Ne	$1s^2\ 2s^2\ 2p_x^{\ 2}\ 2p_y^{\ 2}\ 2p_z^{\ 2}$	↑↓ ↑↓ ↑↓ ↑↓ ↑↓				

Ex) a) BeCl Cl-Be-Cl Linear

b) BF 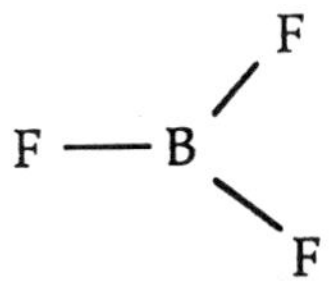 Triangular

Figure 1-39 Molecular shape of $BeCl_2$ and BF_3

These configurations can be predicted by reference to Lewis scheme and Valence-shell electron repulsion rule (VSEPR). In the valence bond theory for molecules, orbitals have fixed spatial configuration. As shown in Fig. 1-8 the s orbitals are circular. p-orbitals are mutually directed at 90° to each other. d-orbitals have fixed directions. Even hybrid orbitals of s, p, and d have fixed spatial configuration, too.

Knowing the orbitals involved, spatial geometry can be predicted. VSEPR predicts that electron pairs around a central atom will adopt configuration to minimize pair repulsion, i.e. each atom will seek to be as far away as possible spatially. Hence the configuration of electron pairs is characteristic depending on the number of electron pairs. This relationship between # of electron pairs and their geometry is shown in Fig. 1-40.

Also it says that lone pair repulsion is stronger than bond-pair. The order of repulsion between electron pair is

$$\begin{pmatrix}\text{lone}\\\text{pair}\end{pmatrix}-\begin{pmatrix}\text{lone}\\\text{pair}\end{pmatrix} > \begin{pmatrix}\text{lone}\\\text{pair}\end{pmatrix}-\begin{pmatrix}\text{bond}\\\text{pair}\end{pmatrix} > \begin{pmatrix}\text{bond}\\\text{pair}\end{pmatrix}-\begin{pmatrix}\text{bond}\\\text{pair}\end{pmatrix}$$

This ordering in the repulsion between lone pairs and bond pairs can be seen in the cases of CH_4, NH_3 and H_2O where the bond angle is the measure of the repulsion. This is shown in Fig. 1-41.

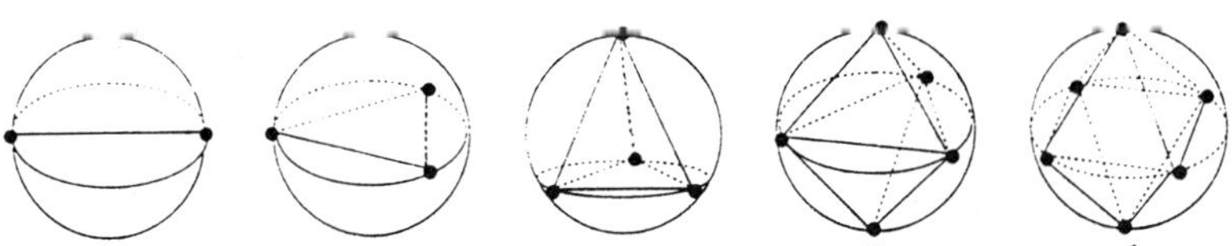

Figure 1-40 The arrangement of electron pair surface of a sphere

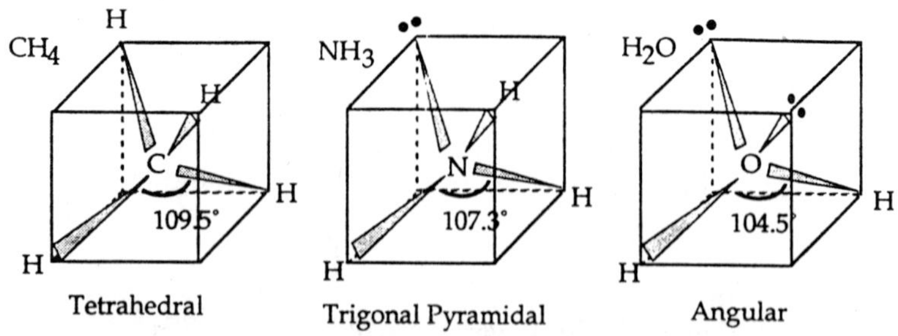

Figure 1-41 Molecular geometry of CH_4, NH_3 and H_2O

The difference of repulsion of lone pair and bond pair makes the configuration of electron pairs a little different from those shown in Fig. 1-40. Considering the VSEPR, molecular structures of some some compounds can be constructed by Lewis scheme as in the following Fig. 1-42.

3-2 Molecular Orbital Theory

(i) Molecular orbitals

In the valence bond theory bonding is formed by overlap of valence electron clouds. In molecular orbital theory electrons are not confined between the atoms rather it is distributed throughout the molecule. The difference between valence bond and molecular orbital theory can be described in terms of orbital. For separated atoms A and B, the wave function for electrons is described as

$$\Psi = \Psi_{A(1)} \cdot \Psi_{B(2)} \tag{62}$$

where A and B represent atoms A and B and 1 and 2 represent valence electrons. After bonding, the wave function is expressed from the point of view of valence bond theory as

$$\Psi = \Psi_{A(1)}\Psi_{B(2)} + \Psi_{A(2)}\Psi_{B(1)} \tag{63}$$

The wave function in Eq.(63) illustrates the fact that valence electrons can be switched between two atoms. This switching results in the exchange energy. Simply the exchange energy can be considered as the bonding energy for the molecule AB.

In molecular orbital theory, wave function does not exclude the possibility of net charge development on either atom A or B. Hence the wave function can be described as:

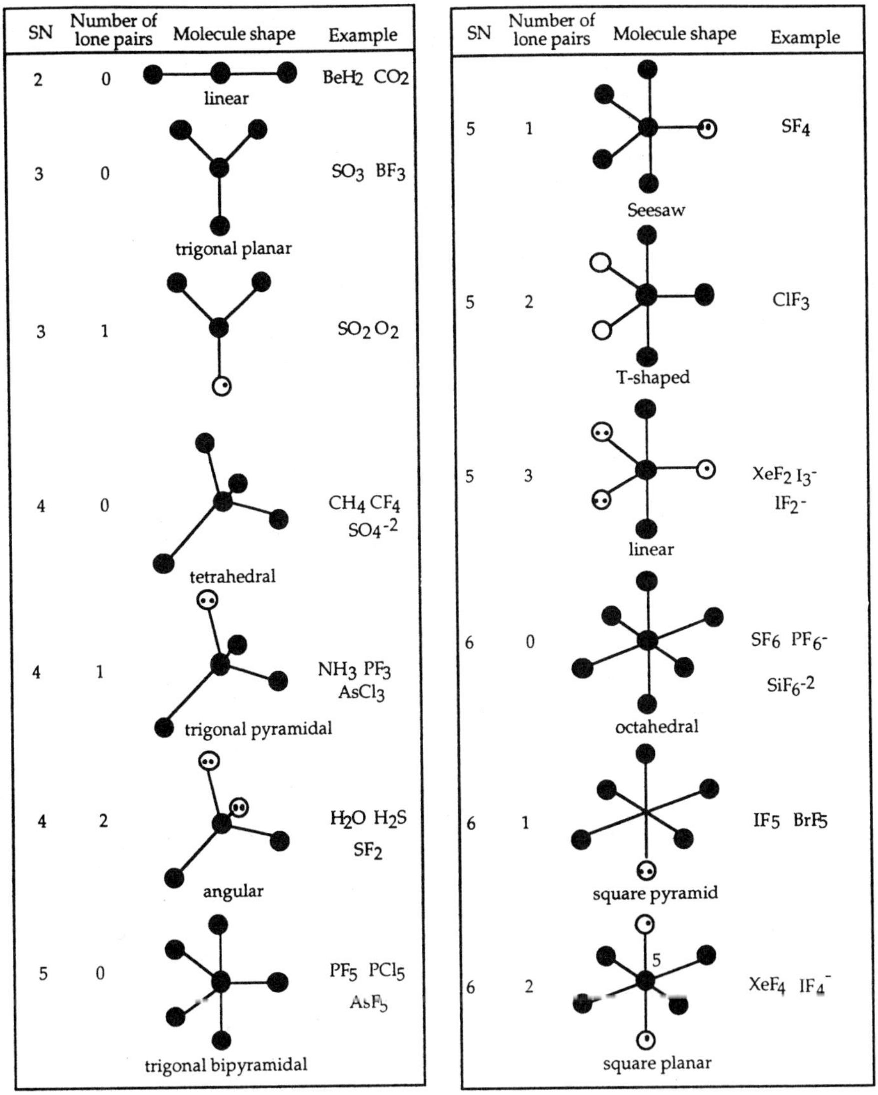

Figure 1-42 Shape of molecules with number of lone pair electrons

$$\Psi = \Psi_{A(1)}\Psi_{B(2)} + \Psi_{A(2)}\Psi_{B(1)} + \lambda_A\Psi_{A(1)}\Psi_{A(2)} + \lambda_B\Psi_{B(2)}\Psi_{B(1)} \qquad (64)$$

Here λ_1 and λ_2 exhibit the extent of localization of bonding electrons on atoms A and B, respectively. The Eq. (64) is the actual molecular orbital for bonding electron pair. The actual wave function must be obtained by solving the Schroedinger Equation with the potential energy term expressed as

$$V = -\frac{e^2}{4\pi\varepsilon_o}\left(\frac{1}{r_{A,1}} + \frac{1}{r_{A,2}} + \frac{1}{r_{B,1}} + \frac{1}{r_{B,2}} - \frac{1}{r_{1,2}} - \frac{1}{r_{A,B}}\right) \qquad (65)$$

Here $r_{x,y}$ represent the distance between nuclei and electrons as shown in the following figure.

For simple hydrogen atom the potential energy can be described by the coordinate of the electron. The potential in Eq.(65) is not able to calculate and is unknown. Therefore, it seems impossible to solve the Schroeding Equation for the wave function with the potential energy expressed by Eq.(65). To overcome this problem, linear combination of atomic orbital (LCAO) method is used. In this method it is assumed that molecular orbital can be approximately described by a linear combination of the atomic orbitals for bonding electrons. The justification for this approximation is that electrons are controlled by one or the other of the two nuclei and molecular orbital looks like the atomic orbital.

Since the square of atomic orbital is the probability to find a electron in a certain position, the integration of the square of the atomic orbital over the entire space is 1. Likewise the integration between different orbital is zero. This condition for the atomic orbital is called othogonality of the wave function for the atomic orbitals. In mathematical form this is expressed as

$$\int_v \Psi_n^* \Psi_m dv = 0 \text{ if } n \neq m,\ 1 \text{ if } m=n \qquad (66)$$

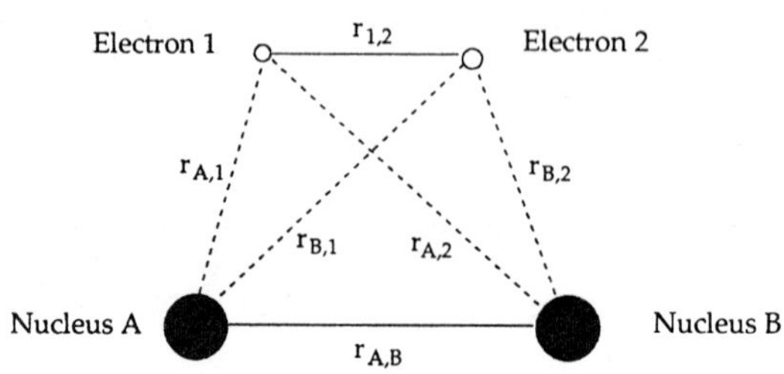

Figure 1-43 Schematic diagram of bonding electron configuration
(Solid line shows the repulsion. Dotted line shows attraction.)

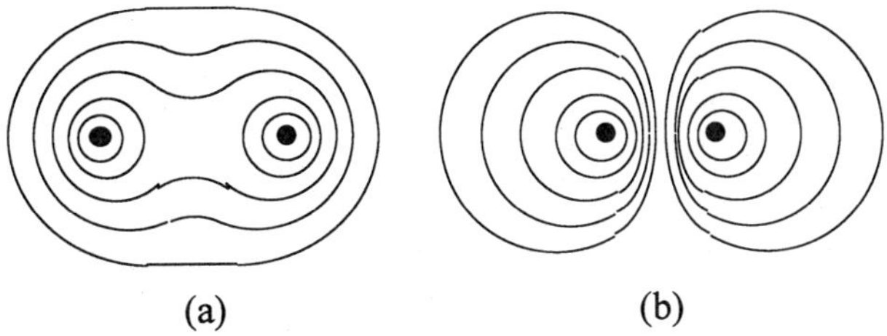

Figure 1-44 Electron contour of (a) bonding and (b) antibonding molecular orbital of H_2^+ ion

In LCAO (Linear Combination of Atomic Orbitals) method, the resulting molecular orbitals from the combination of atomic orbitals must be orthogonal to each other. Hence the molecular orbitals are expressed according to LCAO method as

$$\Psi_{MO,\,b} = \Psi_A + \Psi_B \tag{67}$$

$$\Psi_{MO,\,a} = \Psi_A - \Psi_B \tag{68}$$

Here MO imply molecular orbital. b and a represent bonding and antibonding. Ψ_A and Ψ_B represent atomic orbital of atoms A and B, respectively. Bonding and antibonding characteristic of molecular orbitals derived from LCAO method in the above can be compared with the actual electron contour for H_2^+ ion in Fig. 1-44.

With respect to energy, $\Psi_{MO,\,b}$ has lower energy than Ψ_A and Ψ_B by say ΔE if Ψ_A and Ψ_B are the same kind of atomic orbital. Also $\Psi_{MO,\,a}$ has higher energy than Ψ_A and Ψ_B by the same amount of energy. This is schematically drawn in Fig. 1-45.

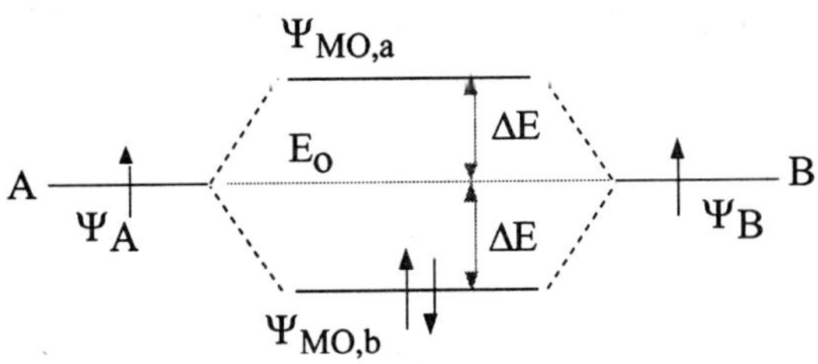

Figure 1-45 Energy diagram of molecular orbitals derived from LCAO

(ii) *Geometry of Molecular Orbitals*

In valence bond theory the bond shape is determine by VSEPR rule. In molecular orbital theory, bond geometry is determined by the shape of molecular orbitals. Like spdf notation in the atomic orbitals, the molecular orbitals are designated by the $\sigma\pi\delta\phi$ notation. σ bond does change phase in circumferential direction around the bonding axis. This means that there is no nodal plane parallel with the bonding axis although it has nodal plane perpendicular to the bonding axis. π molecular orbital has one nodal plane parallel with the bonding axis. δ molecular orbital has two nodal plane. Fig. 1-46 illustrates the general drawing of molecular orbitals.

(iii) *Rules for Molecular Orbital Filling*

Two atomic orbitals combine to form two molecular orbitals that have a lower energy bonding orbital and a higher energy antibonding orbital. Bonding occurs only where an excess of bonding over anti bonding orbitals exists. Filling of MOs follows the same Aufbau Principles for atomic orbitals. Each molecular orbital can accommodate a maximum of two electrons (Pauli exclusion principle). Lowest energy orbitals fill first. Hund's Rules are obeyed for degenerate orbitals π, δ, ψ. Electrons first assume parallel orientation in MOs. σ bonds generally stronger than π bonds.

Example:

Filling of molecular orbitals of N_2 molecule

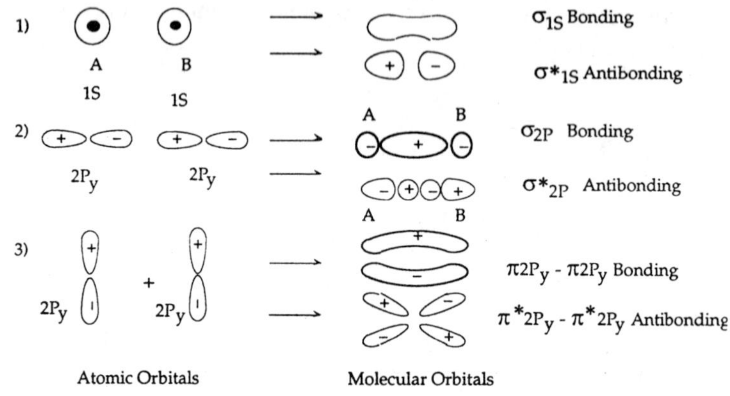

Figure 1-46 Formation of molecular orbitals from atomic orbitals s and p

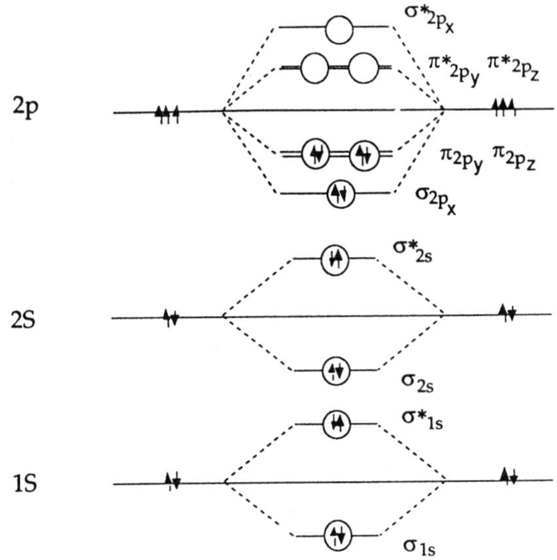

Figure 1-47 configuration of molecular orbitals of N_2

N_2: $2(1s^22s^22p^3) \rightarrow$ 14 Valence electrons
N_2: $(\sigma_{1s})^2(\sigma^*_{1s})^2\,(\sigma_{2s})^2(\sigma^*_{2s})^2(\sigma_{2p_x})^2\,(\pi_{2p_y}=\pi_{2p_z})^4$

(iv) *Bond Order and Bond energy*

Bond order is the number of bonds formed for a molecule. It is defined by $\frac{1}{2}$ $(N_b - N_a)$ where N_b is number of bonding electrons and N_a is number of antibonding electrons. In Fig. 1-47 N_2 has bond order of 3 since its N_b and N_a are 10 and 4, respectively. The bond order is the number of bond in relation to valence bond theory. Table 1-25 list the bond order and bonding energy for the diatomic molecules. One noticeable point of molecular theory is for oxygen molecule. Experimentally it exhibits paramagnetic property. In valence bond theory bond electrons must be pair. Therefore, it predicts the diamagnetic property of oxygen molecule which is not the fact. The electron configuration of oxygen in molecular orbital theory shows that oxygen molecule has unpaired two electrons in π antibonding and explains that oxygen molecule is paramagnetic. This reveals that molecular orbital theory is more close to the fact.

Table 1-25 configuration of molecular orbitals of diatomic molecules

Element	Molecular Orbitals	B.O.	Bond Energy
N_2	$\sigma_{1s}^2\,\sigma^*{}_{1s}^2\,\sigma_{2s}^2\,\sigma^*{}_{2s}^2\,\sigma_{2p}^2$ $\pi_{2p_y}{}^2\,\pi_{2p_z}{}^2$	$\frac{10-4}{2}=3$	9.76 eV (942 kJ/mol)
O_2	$\sigma_{1s}^2\,\sigma^*{}_{1s}^2\,\sigma_{2s}^2\,\sigma^*{}_{2s}^2\,\sigma_{2p}^2$ $\pi_{2p_y}{}^2\,\pi_{2p_z}{}^2\pi^*{}_{2p_y}{}^1\,\pi^*{}_{2p_z}{}^1$	$\frac{10-6}{2}=2$	5.11 eV
F_2	$\sigma_{1s}^2\,\sigma^*{}_{1s}^2\,\sigma_{2s}^2\,\sigma^*{}_{2s}^2\,\sigma_{2p}^2$ $\pi_{2p_y}{}^2\,\pi_{2p_z}{}^2\pi^*{}_{2p_y}{}^2\,\pi^*{}_{2p_z}{}^2$	$\frac{10-8}{2}=1$	1.6 eV
Ne_2	$\sigma_{1s}^2\,\sigma^*{}_{1s}^2\,\sigma_{2s}^2\,\sigma^*{}_{2s}^2\,\sigma_{2p}^2$ $\pi_{2p_y}{}^2\,\pi_{2p_z}{}^2\pi^*{}_{2p_y}{}^2\,\pi^*{}_{2p_z}{}^2\,\sigma^*{}_{2p}{}^2$	$\frac{10-10}{2}=0$	0 eV

(v) *Ionic character of molecular orbital*

Using the bonding molecular orbital as Eq(63), the bonding electrons can be described as

$$
\begin{aligned}
&\Psi(1)\cdot\Psi(2)\\
&= [\Psi_A(1)+\Psi_B(1)][\Psi_A(2)+\Psi_B(2)]\\
&= \Psi_A(1)\Psi_B(2)+\Psi_A(2)\Psi_B(1)+\Psi_A(1)\Psi_A(2)+\Psi_B(1)\Psi_B(2) \quad (69)
\end{aligned}
$$

In comparison of the above Equation with Eq.(63) from the valence bond theory, the molecualr orbital expressed in Eq.(69) stresses too much about the ionic character. Also for heteronuclear molecule whose atoms have different electronegativity, the molecular orbital in Eq.(69) stresses the covalent character of the bonding. To correct this, molecular orbitals are modified as

$$\Psi_{MO,\,b} = a\Psi_A + b\Psi_B \quad (70)$$

$$\Psi_{MO,\,a} = b\Psi_A - a\Psi_B \quad (71)$$

These two molecular orbitals satisfies the orthogonal rule for orbitals. Let's suppose a > b, which means atom A is more electronegative than atom B. In bonding molecular orbital in Eq.(70), electrons are drawn closer to atom A than atom B. Conversely in antibonding molecular orbital, electrons are drawn to closer to atom B than atom A. With respect to the energy diagram, the following examples of HF and NaF illustrates characteristics of ionic character of molecular orbitals. For HF(a > b), F is more electronegative than H.

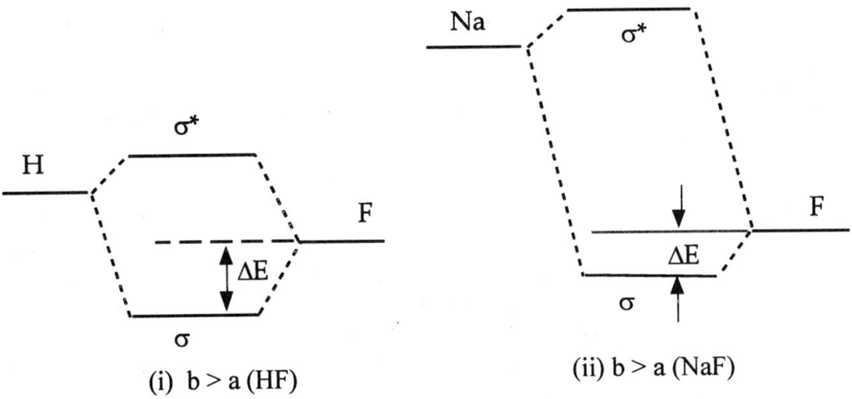

Figure 1-48 Schematic drawing of energy levels of atomic orbitals for (i) HF and (ii) NaF

Therefore, the atomic orbital from F for the bonding is at a lower energy level than that from H and F will attract the electron more. For NaF, this energy gap is much higher than HF. This energy gap means that the overlep of two atomic orbitals are small. Also this large energy gap results in small stabilization energy ΔE. This does not imply that the bonding strength is weak when ΔE is small. Rather ΔE represents the stabilization energy with respect to covalent bonding character.

(vi) *Summary*

Molecular structure is easily described by valence bond theory and molecualr orbital theory without referring exact solution of Schroeding Equation for the molecules. Advantage of valence bond theory is its simple approach to the geometry of molecule. However, more details of molecular structrue can be obtained by molecular orbital theory. In molecular orbital theory, LCAO method is adopted. This method illustrates that hybridization of orbitals, in a strict sense, atomic orbitals is efficient way to describe the molecular system otherwise it is very complex. In this section, the hybridization concept is utilized to describe the molecular structure. Also the hybridization concept can be extende to describe the electronic structure of solids.

3-4. Atomic Radii

Radii for individual atoms or ions are representation of the relative size assuming their shapes are spherical. However, this volumetric definition of atomic radii cannot be rigorously defined since the outer orbital electron orbitals have no defined limits and electron density is zero only at infinity. Hence a term, the atomic radii, is inevitably vague concept. Also, in molecule with more than two atoms, it is more difficult to assume spherical shape. This is well illustrated in Fig.1-44 in which the bonding molecular orbital of hydrogen molecule shows accumulation of electrons between two hydrogen atoms, resulting in nonspherical electron density distribution. With this nonspherical shape of electron density the concept of atomic radii lose its original meaning defined from the spherical shape of the atoms. Viewed in this light, it is more definitively related to the atomic bonding characteristic rather than it is a characteristic of volumetric definition for atoms. In this simple case of hydrogen molecule, the atomic radius is defined as one half the bond distance between two atoms. Since the bond distance is 0.74A°, the resulting atomic radius of hydrogen atom, therefore, is 0.37A°. Because of this bond-related characteristics, prefixes such as covalent-, ionic-, metallic- or van der Waals- must be attached in order to correctly describe the atomic radii for given bonding environment. Conversely, for a given element, the magnitude of atomic radii directly is related to the nature of the bonding. For example, the element oxygen has 1.66, 1.40 and 0.68A° as van der Waal, ionic and covalent radius, respectively.

(i). van der Waals radii

Electrons in an atoms are described as cloud surrounding the nucleus. More or less this electron cloud is compressible. When two atoms are brought together, on the assumption that there is no electron exchange or sharing, the resulting distortion of electron cloud will lead to fluctuating dipoles resulting in van der Waals bonding. The half of the distance between two atoms at this condition is defined as the van der Waals radius of the atom. Examples of van der Waals radii are illustrated by half the distance between two adjacent atoms in crystals of rare gas element and by half the distance between two adjacent atoms of different molecules in molecular crystals. These van der Waals radii are calculated from the crystallographic data of corresponding crystals for rare gas elements. For other elements, the van der Waals radii are induced from the crystallographic data of molecular organic crystals which contains the element concerned. This is illustrated in Fig.1-49.

Bondi calculated the van der Waals of elements using crystallographic data of organic compounds along with rare gas elements. A few of them are listed in Table 1-26. This radii variation with atomic number agrees with the prediction of Eq.(46) although the radius of oxygen is larger than that of nitrogen. These values also are dependent on other factors such as extent of chemical bond other than van der Waals bonding and effect of adjacent atoms on the radii. Some caution must be exercised in using these values since the values

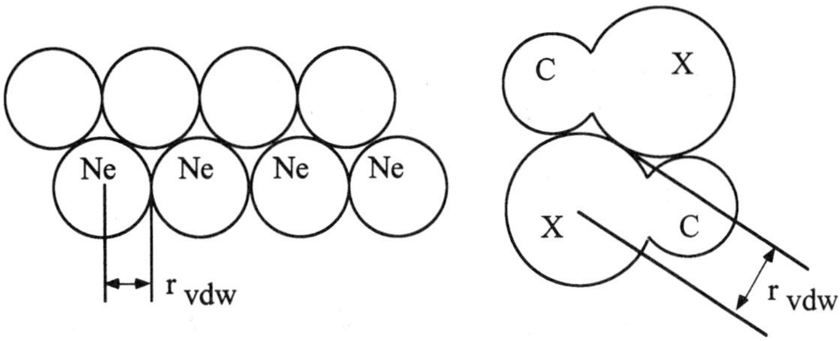

Figure 1-49 van der Waals radius (a) rare gas element and
(b) of elements X in organic compound containing X

Table 1-26 van der Waals radii of lightweight nonmetallic element

Element	B	C	N	O	F	Ne
r_{vdw}	2.13	1.82	1.61	1.66	1.47	1.32

calculated from other methods or by using other compounds may vary from those values in Table 1-26.

(ii). Covalent radii

Simple example of covalent radii is that of diatomic molecules of identical atoms like H_2, N_2, O_2, and F_2. The covalent radius is, therefore, defined as half the bond length between two atoms. However, the covalent radius cannot be interpreted to mean that the atom is a sphere of that size. Rather it gives only bond length between atoms joined by covalent bonds and tells nothing about r_0 distances with the same atom not so joined. The following example of Cl_2 illustrates this aspect of covalent radii.

Bond length of Cl_2 in the gaseous state (Cl — Cl) = 1.988 Å

$r_{cov,Cl} \sim 1/2$ bond length = 0.99 Å

Atomic distance between adjacent Cl atoms in the molecular crystal of Cl_2 is 3.6A°

$r_{Cl,vdw} = 1.8$ Å (Van der Waal radius of Cl)

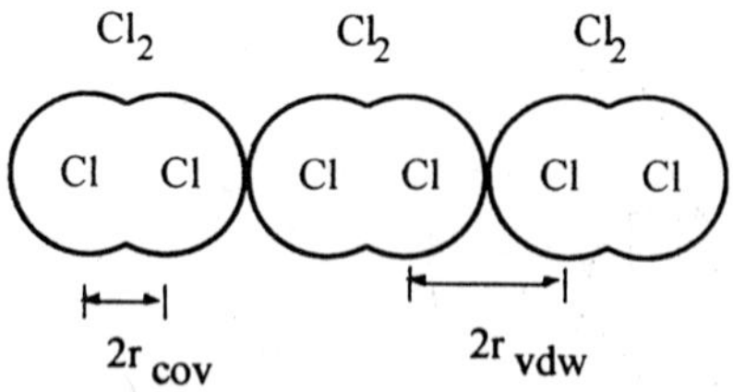

Figure 1-50 Illustration of covalent radius and van der Waals radius in the Cl_2 crystal

Compared to van der Waals radii, covalent radii are shorter since two atoms share valence electrons. Covalent bond lengths are a function of the bond order that is of single, double or triple bonds. The ionization energy is proportional to $r^* = \frac{(n^*)^2}{Z^*} a_o$.. which is in turn proportional to the radius of the orbital which electron occupies. Likewise, bond length can be considered as the orbital radius of bonding electrons. Hence there is a good relationship between bond length and binding energy.

Table 1-27 Covalent Bond Lengths and Associated Binding Energies

Bond	Bond Length (nm)	Binding Energy (kJ/mol of bonds)
H—F	0.092	561
H—O	0.096	464
H—N	0.101	389
H—C	0.109	414
O═O	0.121	494
N≡N	0.110	942
C—C	0.154	347
C═C	0.134	611
C≡C	0.120	837
F—F	0.144	154
Cl—Cl	0.202	239
Br—Br	0.228	190
I—I	0.267	149
H—H	0.074	436
H—Cl	0.128	428
H—Br	0.142	362
H—I	0.161	295

Table 1-27 lists bond length of various covalent molecules and their binding energies illustrating this direct relationship between bond length and binding energies.

(iii). Ionic radii

Ions can be considered as hard spheres stripped of valence electrons (cations) or as hard spheres fully filled with valence electrons (anion). The ionic bond is formed by a electrostatic attractive force between these two hard shells with opposite charges. Hence bond length for an ionic bond represents the distance between centers of two ions. That is, the radius and bond length are related by :

$$r_o = \underset{\text{(cation radius } r_c\text{)}}{r_c} + \underset{\text{(anion radius } r_a\text{)}}{r_a}$$

Here r_o represents the ionic bond length. This simple additivity of ionic radii for the bond length is very important for crystallographers, inorganic chemists and geologists. Once these ionic radii are known, crystal structure for ionic crystals can be predicted. Also crystallographic data can be easily interpreted from these ionic radii data. Based on this need, tables of ionic radii have been developed from experimental data. Among these, Goldschmidt, Pauling, and Shannon-Prewitt have reported systematic list of ionic radii. Since experimental data gives only r_o, the ionic radii must be calculated with reference to a standard ions such as oxygen (O^{2-}) or fluorine (F^-). From the ionic radii of these anions, ionic radii of cations are calculated from observed ionic bond lengths, r_o. Ionic radii of other anions, in turn, can be calculated from ionic radii of these cations. Reported values for ionic radii are empirical ones regressed from data obtained on large number of molecules and crystal structures. In the case of Shannon-Prewitt, more than one thousand structural data were analyzed and their data show good agreements with additive rule, $r_o = r_c + r_a$. However, it is inevitable that different values of ionic radii have been reported, since the ionic radii of O^{2-} and F^- referenced in the above procedure are not absolute values. Rather the values assumed for O^{2-} and F^- ranged from 1.32 - 1.40A°. Table 1-28 illustrates the discrepancies between these values.

Table 1-28 Ionic radii with coordination number of 6

	Goldschmidt	Pauling	Shannon-Prewitt	Zhdanov
Li	0.78	0.60	0.74	0.68
Na	0.98	0.95	1.02	0.98
K	1.33	1.33	1.38	1.33

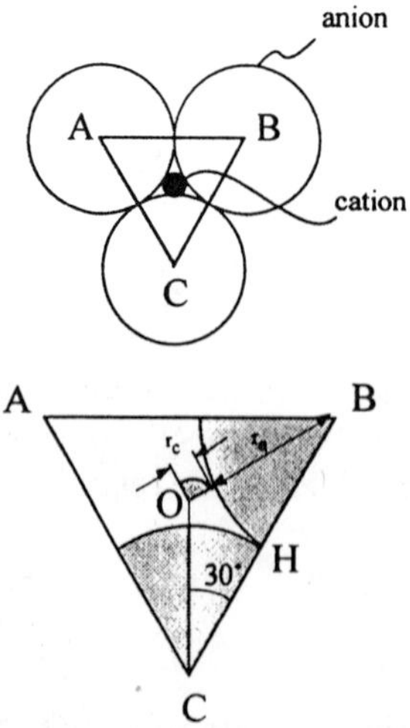

$$\overline{OC} = r_a + r_c$$

$$\overline{OC}\ \cos 30^\circ = \overline{CH} = r_a$$

$$(r_a + r_c)\cos 30^\circ = r_a$$

$$(1 - \cos 30^\circ)\ r_a = \cos 30^\circ\ r_c$$

$$\frac{r_c}{r_a} = \frac{\cos 30^\circ}{1 - \cos 30^\circ} = \frac{\frac{\sqrt{3}}{2}}{1 - \frac{\sqrt{3}}{2}} = 0.155$$

Figure 1-51 Derivation of the minimum ratio of r_c / r_a for trigonal structure of a ionic crystal

The eventual purpose of calculating the ionic radii is to establish their relationship with crystal structures and other characteristics of crystals such as ionic diffusion in crystals. Pauling's coordination theory well illustrates the usefulness of ionic radii regarding crystal structures. His coordination theory for ionic crystals utilizes the concept that ionic crystals consist of hard spheres of anions and cations with defined radii. Unlike covalent crystals, binding force is electrostatic between cations and anions with no directionality. From these facts, Pauling's rules simply states that ionic crystal structures adopt a structure in which the anions form a packed framework with the cations fitting into the interstitial positions. Hence the crystal structure adopted by a ionic crystal is determined by the ratio of cationic radius over anionic radius and by their respective charges. Derivation of the ratio for trigonal structure is shown in Fig.1-51.

As shown in Fig.1-51, if the ratio of r_c/r_a is smaller than 0.155, the cation (sphere) is separated at least from one of the anions and the repulsion between anions overcomes the attractive interaction between the cation and anions and make the trigonal structure unstable. Hence the ratio of rc/ra must be larger than 0.155 to make trigonal configuration in which the cation can make contacts with all of anions while anions are separated from one another. If this ratio, however, becomes 0.225, one more anion can be added to make the coordination of cation with anions four. With this ratio, the cation takes tetrahedral configuration with 4 anions around it. Therefore, the range of the ratio rc/ra for trigonal configuration is $0.155 < r_c/r_a < 0.225$. Table 1-29 illustrates the relationship

Table 1-29 Ionic radii ratio for the criteria of polyhedron in ionic crystals

Polyhedron	coordination #	R (= r_c/r_a ratio)	Example
Trigonal	3	**0.155** < R < 0.225	BF_3*
Tetrahedron	4	**0.225** < R < 0.414	SiO_2
Octahedron	6	**0.414** < R < 0.732	NaCl
Cube (Square Antiprism)	8	**0.732** < R < 1.000	CsCl
Cubo-octahedron	12	**1.000** < R	$BaTiO_3$**

* In a strict sense, BF_3 does not form a ionic crystal. With trigonal configuration, it seems not possible to build up 3 dimensional crystal lattice.

** Configuration of Ba^{+2} with O^{-2} is concerned. This will be illustrated in Chapter 3.

between crystal configurations and the range of ratio, r_c / r_a, in ionic crystals. The configuration defined in Table 1-29 describes how a cation is coordinated to anions according to the ratio of rc/ra. This does not describe how these polyhedra are connected in three dimensional space. Different modes of connectivity between ionic polyhedra can result in many different crystal structures. This is best illustrated in diverse crystal structures of SiO_2 which are called polymorphs. Polymorphs are crystals with same composition but different crystal structures. The actual crystal structures of SiO_2 are shown in Chapter 4.

The prediction of crystal structure by the ratio of ionic radii is straightforward. It is so simple that it has been widely used as an introduction to ionic crystals in relation to atomic properties. As any rule, there are exceptions from the criteria in Table 1-29. LiF, LiCl, LiBr and LiI all have NaCl structure in which Li has 6 coordinations. The ionic radii proposed by Pauling does not predict the crystal structures of these lithium halide as NaCl structure. Rather Pauling's own ionic radii predict the crystal structures of lithium halide are like SiO_2 with 4 coordination of Li. Compared to Pauling's values, those ionic radii proposed by Shannon-Prewitt predict NaCl structure for lithium halide according to the criteria in Table 1-29. This discrepancy between reported radii by different person indicates validity of ionic radii. Although there is no absolute criteria for the values of ionic radii, the values provided by Shannon and Prewitt are more realistic and widely accepted since they are derived from extensive experimental data. Even their data give different values for coordination numbers and for different spin states of transition metal ions.

However, in this text, atomic radii by Zhdanov are listed in Table 1-32(a). This gives not only ionic radii but also covalent and metallic radii. Also his correction table for ionic radii and metallic radii regarding coordination number and covalent radii regarding multiplicity of bonds gives some insight for the nature of crystals in a simple way. Hence the purpose of giving this table of Zhdanov is to get the overall change with atomic number and change within the same period and the same group.

Valence electrons are loosely bound to the nucleus and usually occupy more than half of the atomic volume. Atoms become cations by losing more than half of their volumes whereas anions are not expected less than 1% volume change from atomic states. Hence cations normally are smaller than anions in a given period (row in the periodic table). Due to this big change in sizes of cations, the ionic(cationic) radii exhibit descending order with atomic number in a given row of the periodic table. However, within a given vertical group of the periodic table, ionic radii increase with increasing atomic number which is the same trend as in sizes of neutral atoms. In ionic crystals, the size effect on the crystal structure is more apparent than covalent crystals since the directionality of bonding orbital is more important than the size of atoms in covalent crystals. Due to this restriction by bond geometry, covalent bond results in more empty space and low density than ionic crystals.

(v). Metallic radii

The metallic radii are defined in the same way as ionic crystals except for no difference between atoms as cation and anion .

$$r_{metal} = \frac{1}{2} d_{M\text{-}M}$$

Here, r_{metal} is metallic radius of an atom and $d_{M\text{-}M}$ is the bond length between the nearest atoms in a metal. Metals consist of elements with low electronegativity and their valence orbitals are usually filled less than half. Due to these facts, metal atoms have more than 6 coordinations to fill the valence orbitals from the covalent bonding point of view. For example, Na has 1 valence electron expressed as $3s^1$. In order to fill the valence orbital, it needs at least seven coordination on assumption that each coordination provides one electron. In the light of covalent bonding, this coordination scheme between Na atoms indicates the bond strength is less than one. Hence the bond length is inevitable larger than that resulting from pure covalent bond. In this respect, the metallic radii exhibit how loosely metal atoms are bonded together compared to covalent and ionic crystals. If the van der Waals radii of atoms are assumed to represent the size of atoms with no bonding, the metallic radii are about 0 - 20% smaller than the van der Waals radii. They are twice the ionic radii. This is illustrated

Table 1-30 Comparison of metallic, ionic, van der Waals radii of some metallic elements

	Li	Na	K	Cu	Zn
r_{ion}(A°)	0.68	0.98	1.33	0.80	0.83
r_{vdw}(A°)	1.82	2.27	2.75	1.4	1.39
r_{metal}(A°)	1.55	1.89	2.36	1.28	1.39
$\frac{r_{metal} - r_{vdw}}{r_{vdw}}$	-0.15	-0.17	-0.14	-0.09	0.00

in Table 1-30 for alkali metal elements along with Cu and Zn.

In a single element metal, there is no such difference as cations and anions. Hence atomic size has nothing to do with crystal structures as in Pauling's rule for ionic crystals. Rather the crystal structures of single element metals can be considered in terms of efficient space filling with hard homogenous atomic spheres. Majority of metals, therefore, take closest packing such as FCC and HCP. Also some of them takes BCC(Body Centered Cubic) structure which has a similar packing density to FCC or HCP.

In alloys, the size effect of metals atoms can be considered along with others such as electrochemical factors and valence electron concentration as pointed out by Hume-Rothery. Unlike ionic crystals in which charge balance determines the component ratio, more than one phase is possible for alloys at given concentration of multi-elements. This variation of component ratio in alloys may result in a different crystal structure. Compared to ionic crystals, the role of atomic size seems to be less apparent in alloy system. Although it is true, there have been efforts to reveal the relationship between crystal structures and simple parameters such as atomic radii. This will be given in Chapter 3 in detail.

In one way, atomic radii are thought to affect the crystal structure. In the other way, crystal structures also affect the atomic radii or atomic sizes. In ionic crystals, a cation is surrounded by anions. If the number of anions increases, electrostatic repulsion between anions increases. Hence anions are expected to decrease in size. Regarding cations, if the number of anions increases, the bond strength decreases and the size of the cation is expected to increase. The same thing can be said about anions and metallic atoms in metals. In Table 1-32, correction factors for coordination numbers in metallic and ionic radii illustrate the extent of the effect of crystal structures on the radii. In the same token, multiplicity in covalent bond can be regarded as a factor which represent the effect of crystal structures on atomic radii. If multiplicity is 3, the number of nearest atoms decrease by 3. Since the coordination number decreases with multiplicity, it is expected that there is a inverse relationship between multiplicity and covalent radii. The correction factor for covalent radii with multiplicity reveals this relationship.

(vii). Discrepancies in radii

a. Hybridization

The atomic radii are defined with respect to chemical environment in which a chemical bond is formed. In a strict sense, there is no absolute ionic, covalent or metallic bond. Rather the chemical bond is more or less a mixture of these bond characters. Considering this, the covalent bond is refined with ionic characters represented by electronegativities as:

$$r_o(AB) = r_A + r_B - 0.09\,|\,x_A - x_B\,|$$

where r_A, r_B = covalent radii for A & B and $|\,x_A - x_B\,|$ is electronegativity difference between two atoms. Hybridization of ionic character with covalent character decreases the normal covalent bond distances.

b. Crystal Field Theory

The hybridization of ionic and covalent characters in a bond shown above illustrates that chemical environment surrounding a bond can be a mixture of covalent, ionic and metallic ones. In actual bonds, this hybridization may even further occur with a different magnitude with respect to spatial direction of a bond . The best example is the change of bond length in ZrO_2 crystals regarding Zr – O. At high temperature above 2100°C, Zr is surrounded by 8 oxygens and their bond lengths are all the same. As temperature is lowered below 2100°C, the bond length are not all the same. Rather they degenerate into two sets of Zr – O with different bond lengths. Zr has bond length of 2.065Å with four oxygens and has 2.455Å of bond length with the other four oxygens. This variation of bond length of Zr atom with different oxygens is further developed when the temperature is lowered below about 1150°C. The coordination number of Zr with oxygen changes from 8 to seven and bond lengths are all different ranging from 2.05 – 2.28Å. This difference in bond length signifies that the interaction between Zr and O is different in magnitude depending on the relative position of oxygen with respect to Zr. In the case of ZrO_2, atomic radii cannot be applied to calculate the different bond lengths of Zr with oxygens. There are many crystals which show different bond lengths although the bonds are formed between the same two atoms. TiO_2(Brookite) and $BaTiO_3$ (Tetragonal structure) are the examples. The structures of these crystals are illustrated in chapter 4.

One of the simple theoretical ways to explain why the bond length of one atom with adjacent atoms changes with spatial direction of the bonds can be the crystal field theory. It describes the interaction of metal ion with anions surrounding it. Main concern of the crystal field theory is d-orbitals of transition metals. Usually non-transition metal ions such as Na^+ and Ca^{+2} have the electronic structure of the rare gas in which orbitals are completely filled. In these cases, electrons of metal ions are distributed in a spherically symmetric way and the metal ions can be treated as a point charge. However, in the case of transition metal ions, d-orbitals are not completely filled and cannot be treated as being spherically symmetric. Consequently, electrons in d-orbitals interacts differently with anions depending on the geometry of their corresponding orbitals and on the configuration of anions surrounding metal ions. Regarding energy level of d-orbitals, five d-orbitals of transition metal ions are not the same if d-orbitals are not filled with anions around metal ion. Due to the splitting in energy level of d-orbitals caused by anions, d-orbitals with lower energy are filled before the others are filled if the effect of anions is bigger than the stabilization energy which is gained with higher spin states according to Hund's rule. In the atomic structure, electron filling follows the Hund's rule which states that high spin state has lower energy than low spin state. In the light of crystal field theory, this does not hold under the influence of anions on metal ions within crystals which splits energy levels of d-orbitals. Hence, in a crystal structure, the energy state of transition metal ion varies with surrounding environment such as coordination and subsequent non-spherical distribution of electrons results in uneven bond length and distortion of $M^{+a}X^{-b}{}_n$ polyhedra. The geometry of MX_n polyhedra is directly related to the coordination number, *n*. With coordination number 4, the geometry of anions takes the tetrahedra. Coordination number 6 and 8 connote octahedra and cubic geometries, respectively. Due to this relationship between coordination number and geometry of anions, coordination number may reverse the spin state from higher to lower spin and, in consequence, the atomic radii.

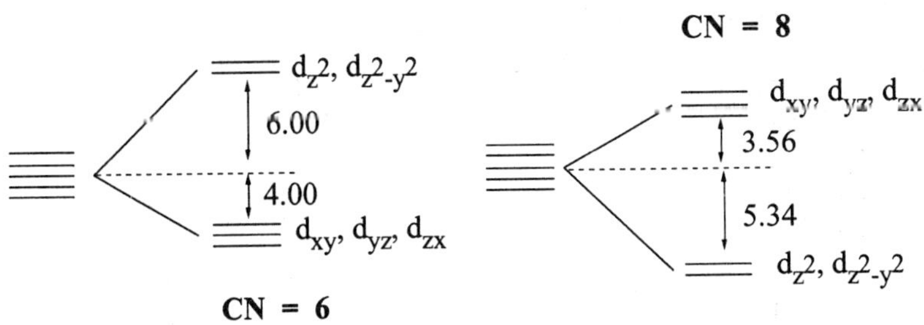

Figure 1-52 Splitting of d-orbitals in a octahedral (CN=6) and cubic (CN=8) structure[Zuckerman]

Table 1-31 Ionic radii of Fe^{2+} and Fe^{3+} with CN of 6 and 8

CN	6	8
Fe^{2+}	r = 0.61 LS (Low Spin)	r = 0.77 HS (High Spin)
Fe^{3+}	r = 0.55 LS	r = 0.64 HS

If the result of the crystal field theory is simply borrowed, the d-orbitals of metal ion are split into two groups under coordination number 6 and 8 as in Fig.1-52. For example, Fe^{+2} and Fe^{+3} have d^6 and d^5 electronic configurations. Without the influence of anions, the spin states follow the Hund's rule resulting high spin state which first fills all five d-orbitals for Fe^{+3} resulting in 5 unpaired electrons and then fill again one of them for Fe^{+2} resulting in 4 unpaired electron. However, for coordination number 6, Fe^{+2} and Fe^{+3} both first fill d_{xy}, d_{yz}, and d_{zx} resulting in one and none unpaired electron for Fe^{+2} and Fe^{+3}, respectively, under the influence of anions according to crystal field theory as shown in Fig.1-52. With coordination number 8, electrons fill d_{z2} and d_{x2-y2} and then d_{xy}, d_{yz} and d_{zx} resulting in unpaired electrons of 2 and 1 for Fe^{+2} and Fe^{+3}, respectively. Hence, coordination number 8 produces higher spin states than CN = 6. Since high spin state distributes electrons in a wider space than low spin state, it make ionic radii bigger than low spin state. The ionic radii of Fe^{+2} and Fe^{+3} are listed in Table 1-31 with CN of 6 and 8.

The variation of spin states illustrates the correlation between atomic radii and crystal structure. In other words, it shows the effect of crystal structure on the atomic radii. Hence the atomic radii also should be understood as a simple guide to understanding of bond structure and characters like electronegativity. As in the monoclinic form of ZrO_2, the bond lengths are 2.04, 2.10, 2.15, 2.16, 2.18, 2.26, and 2.26A°. When these are averaged, the bond length of Zr – O in monoclinic form is 2.16A°. The calculated ionic radius of Zr^{+4} in Table 1-32(a) is 0.82A°. The ionic radius of O-2 is 1.36A°. The predicted value of Zr – O bond length is 2.18. This calculated value is very close to the average distance between Zr – O. Although there is a variation of bond length with the same cation and anion combination, the close relationship between average value and calculated value illustrates the significance of the valued in Table 1-32(a) for atomic radii. Hence these values can be used as approximate values. The atomic radii given in Table 1-32(a) are drawn by representation of circles in Table 1-32(b) and (c) for atomic number from 1(hydrogen) to 36(krypton) with oxidation states. General trend of atomic radii can be easily seen with graphical representation of radii by circles. From H to Cl, each element has a unique cationic or anionic state. Their radii change in a monotonic way for covalent, ionic and metallic radii. For any metallic element of atomic number less than 17, the order of radii is metallic > covalent > ionic. For non-metallic elements, the order changes as ionic (anion) > covalent > ionic

Table 1-32(a) Atomic radii of 103 elements in the periodic table

Correction for Multiplicity of the Covalent Bond	
Multiplicity of Bond	Bond length Percent
1	100
2	86
3	78

Correction for Coordination No.	Radius Percent	
C.N.	Ionic	Metallic
12	112	100
8	103	98
6	100	96
4	94	88

Na 11		
+1 0.98	→ Ionic radii :	Plain face with charge state
	(van der Waals radii :	Underlined)
1.54	→ Covalent radii :	**Bold face**
1.89	→ Metallic radii :	*Italic*

H 1 +1 0.00 -1 2.08																	He 2 1.12
Li 3 +1 0.68 **1.23** *1.52*	Be 4 +2 0.31 -1 1.95 **0.89** *1.11*											B 5 +3 0.20 +1 0.35 **0.88** *0.83*	C 6 +4 0.15 -4 2.60 **0.77**	N 7 +5 0.11 +3 0.13 +1 0.25 -3 1.71 **0.70**	O 8 -2 1.40 **0.66**	F 9 -1 1.36 **0.64** *0.72*	Ne 10 1.60
Na 11 +1 0.95 **1.57** *1.54*	Mg 12 +2 0.65 **1.36** *1.60*											Al 13 +3 0.50 **1.25** *1.43*	Si 14 +4 0.41 +1 0.65 -1 3.84 -4 2.71 **1.17**	P 15 +5 0.34 +3 0.42 -3 2.12 **1.10**	S 16 +6 0.29 +4 0.37 -2 1.84 **1.04**	Cl 17 +7 0.26 -1 1.81 **0.99**	Ar 18 1.91
K 21 +1 1.33 **2.02** *2.27*	Ca 20 +2 0.99 **1.74** *1.97*	Sc 21 +3 0.81 **1.44** *1.60*	Ti 22 +4 0.68 +3 0.77 +2 0.90 +1 0.96 **1.32** *1.45*	V 23 +5 0.59 +4 0.60 +3 0.74 +2 0.88 **1.22** *1.32*	Cr 24 +6 0.52 +4 0.56 +3 0.64 +2 0.84 **1.17** *1.27*	Mn 25 +7 0.46 +4 0.54 +3 0.62 +2 0.80 **1.17** *1.30*	Fe 26 +3 0.64 +2 0.76 **1.17** *1.29*	Co 27 +3 0.63 +2 0.74 **1.16** *1.25*	Ni 28 +3 0.62 +2 0.72 **1.15** *1.25*	Cu 29 +2 0.72 +1 0.96 **1.17** *1.28*	Zn 30 +2 0.74 +1 0.88 **1.25** *1.33*	Ga 31 +3 0.62 **1.25** *1.22*	Ge 32 +4 0.53 +2 0.70 -4 2.72 **1.22** *1.22*	As 33 +5 0.47 +3 0.58 -3 2.22 **1.21** *1.25*	Se 34 +6 0.42 +4 0.69 -1 2.32 -2 1.98 **1.17**	Br 35 +7 0.39 +5 0.47 -1 1.96 **1.14**	Kr [illegible] 1.98
Rb 37 +1 1.48 **2.16** *2.48*	Sr 38 +2 1.13 **1.92** *2.15*	Y 39 +3 0.93 **1.62** *1.81*	Zr 40 +4 1.09 **1.45** *1.60*	Nb 41 +5 0.70 +4 0.74 **1.34** *1.43*	Mo 42 +6 0.62 +4 0.66 **1.29** *1.36*	Tc 43 +7 0.58 +4 0.72 +2 0.95 *1.36*	Ru 44 +8 0.54 +4 0.63 +3 0.77 **1.24** *1.33*	Rh 45 +4 0.67 +3 0.75 +2 0.86 **1.25** *1.35*	Pd 46 +4 0.64 +2 0.86 **1.28** *1.38*	Ag 47 +2 0.97 +1 1.26 **1.34** *1.44*	Cd 48 +2 0.97 +1 1.14 **1.41** *1.49*	In 49 +3 0.81 +1 1.32 **1.50** *1.63*	Sn 50 +4 0.71 +2 1.02 -1 3.70 -4 2.94 **1.40** *1.41*	Sb 51 +5 0.62 +3 0.90 +1 0.89 -3 2.45 **1.41**	Te 52 +6 0.56 +4 0.81 -1 2.50 -2 2.21 **1.37** *1.43*	I 53 +7 0.50 +5 0.62 -1 2.16 **1.33**	Xe 54 2.18
Cs 55 +1 1.69 **2.35** *2.65*	Ba 56 +2 1.35 **1.98** *2.17*		Hf 72 +4 0.78 **1.44** *1.56*	Ta 73 +5 (0.7) **1.34** *1.43*	W 74 +6 0.65 +4 0.68 **1.30** *1.37*	Re 75 +7 0.60 +6 0.61 +4 0.72 **1.28** *1.37*	Os 76 +8 0.53 +6 0.60 +4 0.65 +2 0.89 **1.26** *1.34*	Ir 77 +6 0.56 +4 0.64 +3 0.75 +2 0.89 **1.26** *1.36*	Pt 78 +4 0.70 +2 0.85 **1.29** *1.38*	Au 79 +3 0.91 +2 1.05 +1 1.37 **1.34** *1.44*	Hg 80 +2 1.10 +1 1.27 **1.44** *1.60*	Tl 81 +3 0.95 +1 1.44 **1.55** *1.70*	Pb 82 +4 0.84 +2 1.20 -4 2.15 **1.54** *1.75*	Bi 83 +5 0.74 +3 0.96 -3 2.13 **1.52** *1.55*	Po 84 +6 0.56 +4 0.65 -2 2.30 **1.53** *1.67*	At 85 +7 0.51 +5 0.57 -1 2.27	Rn 86 2.20
Fr 87	Ra 88 +2 1.40 *2.20*																

La 57 +3 1.06 **1.69** *1.87*	Ce 58 +4 0.92 +3 1.03 **1.65** *1.83*	Pr 59 +4 0.90 +3 1.01 **1.65** *1.83*	Nd 60 +3 1.00 **1.64** *1.82*	Pm 61 +3 0.98 *1.81*	Sm 62 +3 0.96 +2 1.11 **1.66** *1.80*	Eu 63 +3 0.95 +2 1.12 **1.85** *2.04*	Gd 64 +3 0.94 **1.61** *1.80*	Tb 65 +4 0.84 +3 0.92 **1.59** *1.78*	Dy 66 +3 0.91 **1.59** *1.77*	Ho 67 +3 0.89 **1.58** *1.77*	Er 68 +3 0.88 **1.57** *1.76*	Tm 69 +4 0.94 +3 0.87 **1.56** *1.75*	Yb 70 +2 1.13 +3 0.86 **1.70** *1.94*	Lu 71 +3 0.85 **1.56** *1.73*
Ac 89 +3 1.11 *1.88*	Th 90 +4 0.99 +3 1.08 *1.80*	Pa 91 +5 0.90 +4 0.96 +3 1.05 *1.61*	U 92 +6 0.83 +5 0.87 +4 0.93 +3 1.03 *1.39*	Np 93 +6 0.82 +5 0.88 +4 0.92 +3 1.01 *1.31*	Pu 94 +6 0.81 +5 0.87 +4 0.90 +3 1.00 *1.51*	Am 95 +6 0.80 +5 0.86 +4 0.89 +3 0.99 *1.84*	Cm 96 +4 0.88 +3 0.99 +2 1.19	Bk 97 +4 0.87 +3 0.98 +2 1.18	Cf 98 +4 0.86 +3 0.98 +2 1.17	Es 99 +4 0.85 +3 0.98 +2 1.16	Fm 100 +4 0.84 +3 0.97 +2 1.15	Md 101 +4 0.84 +3 0.96 +2 1.14	No 102 +4 0.83 +3 0.95 +2 1.13	Lw 103 +4 0.83 +3 0.94 +2 1.12

Table 1-32(b) Illustration of atomic and ionic sizes of elements of the first, second and third row in the periodic table

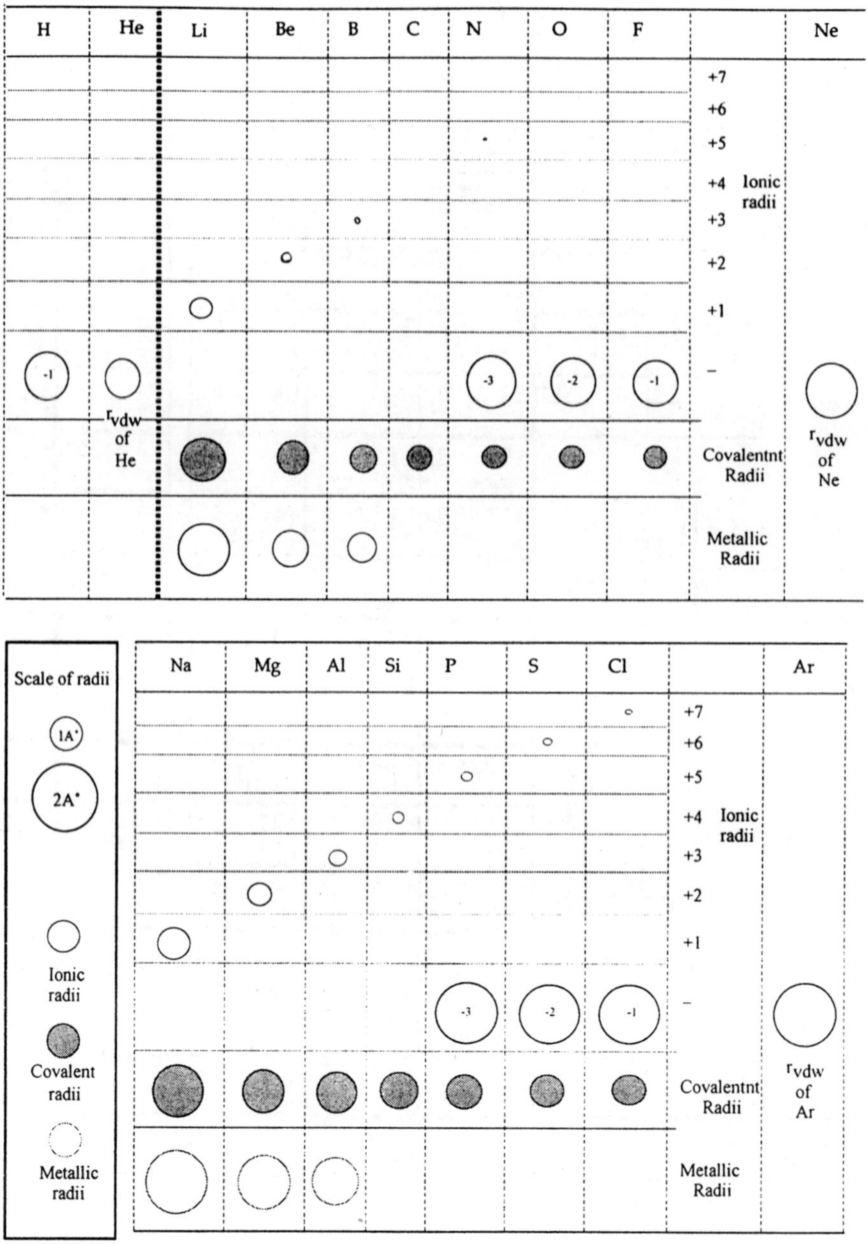

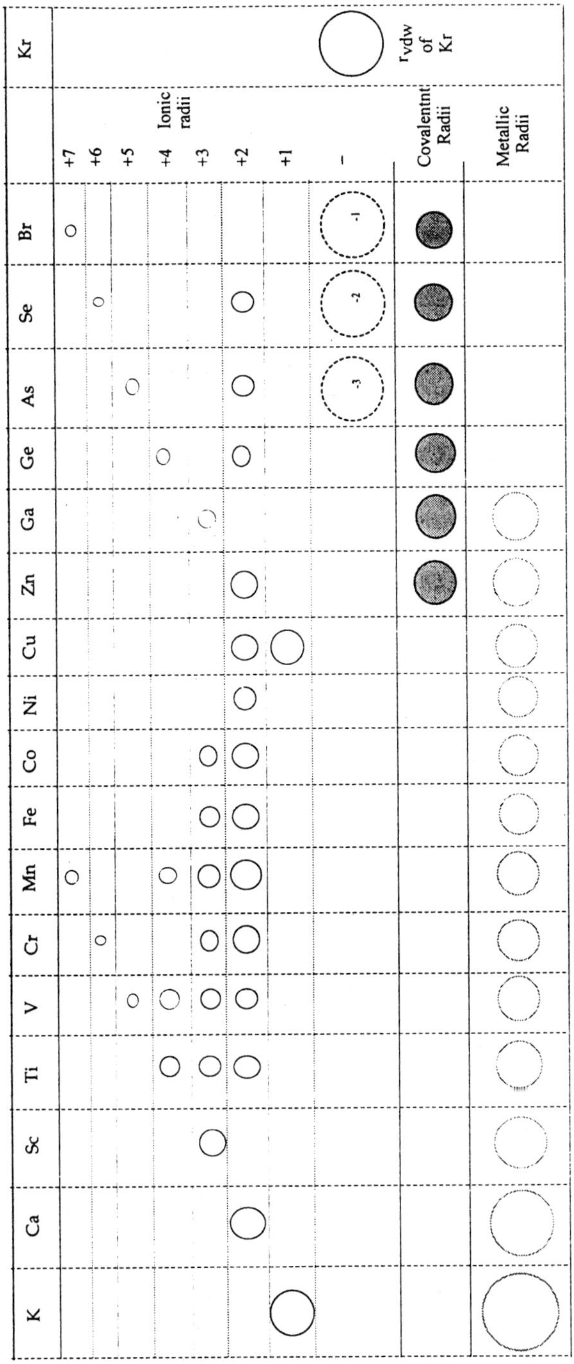

Table 1-32(c) Atom c and ionic radii of elements in the fourth period (cation). The same trend of atomic radii can be stated about the transition metal elements from Sc through Zn. These transition metal elements have more than one oxidation states although Sc, Ni, and Zn show only one oxidation state. For each transition metal element, ionic radii decrease with oxidation number. Another noticeable point for transition metal elements is that they show irregular change of ionic radii for the same oxidation states, especially +2 oxidation state. This irregularity is due to the unfilled d-orbitals and subsequent change in the spin state as shown above for Fe^{+2} and Fe^{+3}.

Here the crystal field theory is introduced regarding atomic radii. More detailed discussion of the crystal field theory such as distortion of MX_n polyhedra and splitting of d-orbitals with respect to other geometries than octahedral and cubic structures of anions surrounding metal cation will be presented in Chap. 4 when crystal structures of metal oxides are discussed.

SECTION 3 ENERGETICS AND ELECTRONIC STRUCTURES OF CRYSTALS

A. Introduction

Atoms/ions in a combined state are held together by a net attractive force. This condition represents a lower energy state than the uncombined atoms or ions. Bond strength is a measure of the bond dissociation energy, i.e. the energy needed to separate the bonded pair into its gaseous constituents at infinite separation.

Examples:

$$HCl_{(g)} \longrightarrow H_{(g)} + Cl_{(g)}; \quad H_d = 433 \text{ kJ/mol}$$
$$NaCl_{(g)} \longrightarrow Na_{(g)} + Cl_{(g)}; \quad H_d = 229.3 \text{ kJ/mol}$$

But,

$$NaCl_{(s)} \longrightarrow Na_{(g)} + Cl_{(g)}: \quad H_d = 640.2 \text{ kJ/mol}$$

Here H_d is the enthalpy of dissociation. The dissociation of crystalline NaCl is three times that of gaseous NaCl. This large difference in dissociation energies for the gaseous and solid states of NaCl reflects the much greater number of bonding interactions per atom in solid. Components which must be considered in analyzing bond strengths in solids are:

a. Bond length (atom/ion separation)
b. Bond type (eg. covalent, ionic)
c. Number of electrons involved in the bonding and
d. The coordination state of the atoms/ions

Due to difficulty in quantitative description of potential energy in general for covalent and metallic bonding, bonding strength with respect to microstructure will be described for ionic bonding solids in a quantitative way.

Energy of a molecule is the energy gained by forming the molecule compared to the energy of atoms. During the formation of the molecule, only the valence electrons are concerned since the inner electrons of atoms are not changed much from the atomic states. Only the interaction of valence electrons with two atoms bonded is important. In a crystal, this concept of molecule is extended through the entire crystal. Electrons are not any more confined within a single atom or between two atoms and wave equations for electrons extend from one end to the other of the crystal. Therefore the energy of an electron in a crystal cannot be calculated by considering the interaction between two atoms or just nearest atoms as in molecule. Therefore the binding energy for crystals is a collective concept which goes beyond the two atoms bonded together as in a molecule.

Lattice energy is the potential energy of a *solid*. Hence lattice energy is not the potential energy between two adjacent atoms. Rather it is a potential energy between all atoms in a solid modified considering the long range interaction across the solid. This long range interaction is best described in ionic solid. The parameter which shows this long range interaction in ionic solids is Madelung constant which will be described later.

In ideal ionic solid it is assumed that no valence electron overlap is developed as in covalent solids since the bonding is developed by electrostatic forces between cations and anions. Therefore it is not necessary to include the interaction between electrons of different atoms in calculating the bonding energy. Hence the lattice energy of a ionic solid is a measure of the cohesive strength which needs to be provided to dissociate a mole of solid into its constituent ions not neutral atoms.

The lattice energy can be determined in one or two ways. The first approach, which is somewhat more theoretical, employs the concept of a net potential energy which results from the interaction of an attractive force and a repulsive force. These forces arise when two separated ions or atoms approach each other to form a bond. Bonding occurs at an Equilibrium separation distance (r_0) which represents the sum of the radii of the two entities. Corresponding to r_0 is the net potential energy V_0. The lattice energy can be defined in terms of the molar summation of the potential energy V_0, a structure factor called the Madelung constant (which increases the overall energy) and r_0. While these quantities are perhaps more difficult to evaluate, the potential energy approach provides more useful insights into the structure and physical properties of the solid.

The second approach, the Born-Haber cycle, uses an experimental approach involving a net summation of the several molar thermal energies which constitute subsets of the reactions required to form the solids. These energy terms include the heat of formation, dissociation energy, sublimation energy, the ionization energy and the electron affinity, all readily measurable, except perhaps for the electron affinity.

B. Calculation of Lattice Energy in Ionic Solids

1. Estimation of Bond Energies

The bond energy (E_b) for an ionic crystal is obtainable by calculating the following equation in which the energy term U is concerned since EA and IE are all known as electron affinity and ionization energy of corresponding ions. Refer to the appendix for the derivation of Eq.(72).

$$E_b = \sum_{I,J}\left(EA_I - IE_J\right) + U \tag{72}$$

Here I and J indicate the anion and cations, respectively, and U is called lattice energy and defined as the potential energy arising from interactions between cations and anions which are assumed as hard sphere and electrons are confined within ions. Hence U depends on crystal structure especially arrangements of cations and anions. U is a potential energy arising from Coulombic force between two electric charges, i.e.

$$U = \int_{\infty}^{r_o} \frac{q_1 q_2}{4\pi\varepsilon_o r^2} dr = \frac{q_1 q_2}{4\pi\varepsilon_o r_o},$$

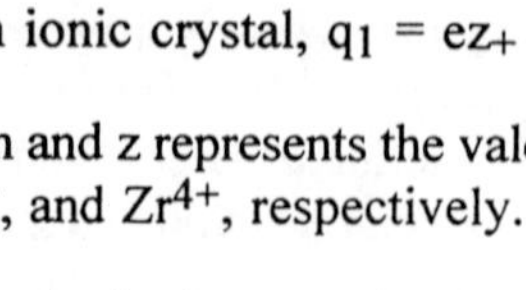

According to Coulomb's Law, the force between two charges separated by a distance of 'r' is $F = \frac{q_1 q_2}{4\pi\varepsilon_o r^2}$. In ionic crystal, $q_1 = ez_+$ and $q_2 = ez_-$ in which 'e' is electric charge of an electron and z represents the valence of ions like $z_+ = 1, 2, 3$, and 4 for Na^+, Mg^{2+}, Al^{3+}, and Zr^{4+}, respectively. Therefore $U = \frac{z_+ z_- e^2}{4\pi\varepsilon_o r_o}$. In deriving Eq.(72), ions in the ionic crystal are assumed as point charges. Actual ions more or less, however, have a certain volume. Most of the volume is occupied by electrons while the nucleus takes a very little volume at the center of the ion. Therefore, when ions of different charges make contacts, repulsive force arises due to the penetration of electron clouds in addition to the Coulombic attractive force predicted by Coulombic law for point charges. This additional repulsive force between ions is given as $\frac{b}{r^{m+1}}$ according to the

Born's model. Here 'b' and 'm' are empirical constants which compensate the repulsive force between ions. The potential energy including this repulsive force is then

$$U = \frac{z_+ z_- e^2}{4\pi\varepsilon_o r} + \frac{1}{m}\frac{b}{r^m} \tag{73}$$

Since the potential energy at equilibrium position of ions should be the minimum, the derivative of the energy is equal to zero.

$$\left.\frac{\partial U}{\partial r}\right|_{r=r_o} = 0$$

$$\frac{\partial U}{\partial r} = \frac{z_+ z_- e^2}{4\pi\varepsilon_o r^2} + \frac{b}{r^{m+1}}$$

$$\frac{z_+ z_- e^2}{4\pi\varepsilon_o r^2} = -\frac{b}{r^{m+1}} \tag{74}$$

Substituting Eq.(74) into Eq.(73) for V_r, the potential energy at Equilibrium is

$$V_o = \frac{z_+ z_- e^2}{4\pi\varepsilon_o r_o}\left(1 - \frac{1}{m}\right) \tag{75}$$

This Equation has been derived from a pair of a cation and an anion. In a solid there are many pairs of cations and anions. In the light of Eq.(72), the summation goes beyond the adjacent ions which make a contact with a ion concerned. These interactions depend on the arrangement of cations and anions, and is defined by a structural constant, Madelung constant. The following shows the derivation of Madelung constant for a one dimensional crystal

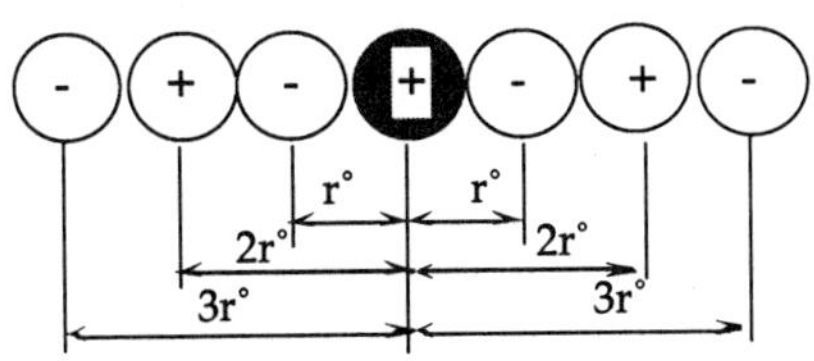

Figure 1-53 One dimensional ionic crystal (MX) with interatomic distance r_o

There are anions (X^-) at distance of r_o, $3r_o$, $5r_o$, and so on. Other cations (M^+) are located at distances of $2r_o$, $4r_o$, $6r_o$ and so on. The Coulombic potential for the cation marked as black in Fig.1-53 with other ions has the general form:

$$U = \frac{z_+z_-e^2}{4\pi\varepsilon_o r_o}\left(\frac{2}{1}-\frac{2}{2}+\frac{2}{3}-\frac{2}{4}+\frac{2}{5}-\frac{2}{6}+....\right)\left(1-\frac{1}{m}\right)$$

The Madelung constants, M, is the summation of fractional numbers in the first bracket of the above equation and defined as

$$A = \left(\frac{2}{1}-\frac{2}{2}+\frac{2}{3}-\frac{2}{4}+\frac{2}{5}-\frac{2}{6}+....\right) \tag{76}$$

The Coulombic potential for a ion in a crystal structure with the Madelung constant is

$$U = \frac{Az_+z_-e^2}{4\pi\varepsilon_o r_o}\left(1-\frac{1}{m}\right) \tag{77}$$

Since the bond energy is defined as the energy per 1 mole of chemical formula (MX_n) for ionic crystal, the energy given in Eq.(89) is multiplied by Avogadro number, $N_o(= 6.03 \times 10^{23})$, which is the number of molecules in one mole. Cations and anions have the electronic configuration of rare gases whose m value is about 9. The only variable in Eq.(77) which depends on the crystal structure is the Madelung constant since r_o is calculated from the ionic radii of consisting ions as $r_o = r_c + r_a$. Here r_c and r_a are radii of the cation and anion, respectively. Hence Eq.(77) is the basis for the theoretical calculation of bonding energy of ionic crystals although it gives a approximate value due to some assumptions made for electronic structures of ionic crystals.

The Madelung constant for the crystal of NaCl which is shown in Fig.1-54 is calculated as follows. Since, with reference to central Na^+ ion in NaCl, there are 6 Cl ions at r_o, 12 Na ions at $\sqrt{2}\ r_o$, 8 Cl ions at $\sqrt{3}\ r_o$, 6 Na ions at $\sqrt{4}\ r_o$, 24 Cl ions at $\sqrt{5}\ r_o$, and so on. The electrostatic energy for the crystal of NaCl is given as

$$U = \frac{N_o z_+z_-e^2}{4\pi\varepsilon_o r_o}\left(\frac{6}{1}-\frac{12}{\sqrt{2}}+\frac{8}{\sqrt{3}}-\frac{6}{\sqrt{4}}+\frac{24}{\sqrt{5}}+....\right)\left(1-\frac{1}{m}\right) \tag{78}$$

When the summation in Eq.(90) extends to infinity, the Madelung constant for the NaCl structure becomes. A = 1.7476.

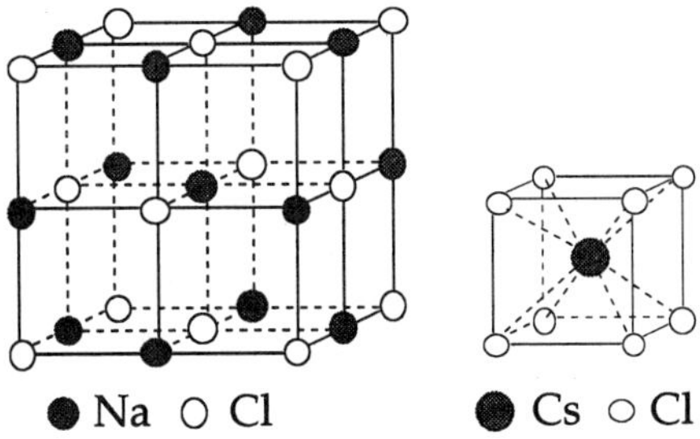

Figure 1-54 Crystal structure of NaCl and CsCl

The bonding energy per mole of crystal NaCl is then

$$U = \frac{N_0 A z_+ z_- e^2}{4\pi\varepsilon_0 r_0}\left(1 - \frac{1}{m}\right)$$

$$= \frac{6.03 \times 10^{23} \times 1.7476 \times (+1)(-1) \times \left(1.602 \times 10^{-19}\,\text{Coulomb}\right)^2}{4 \times \pi \times \left(8.854 \times 10^{-12}\,\text{Coulomb}/_{\text{Volt}\bullet\text{m}}\right) \times 2.79 \times 10^{-10}\,\text{m}} \times \left(1 - \frac{1}{9}\right)$$

$$= 774 \text{ kJ/mole}$$

Table 1-33 Madelung constants for ionic crystals* and lattice energies calculated

Crystals	z^+z^-	r_0**	A	U_0	Crystal structure
NaCl	1	2.79	1.7456	774	FCC
CsCl	1	3.46	1.7627	630	BCC
ZnS(Blende)	4	2.65	1.6381	3059	FCC
ZnS(Wurtzite)	4	2.65	1.641	3064	Hexagonal
CaF_2	2	2.37	2.519	2629	Cubic
TiO_2	8	2.00	2.400	11875	Tetragonal
CdI_2	2	3.19	2.355	1826	Hexagonal
MgO	4	2.10	1.7456	4113	FCC
CaO	4	2.40	1.7456	3599	FCC
SrO	4	2.56	1.7456	3374	FCC
Al_2O_3	6	1.93	4.170	16036	Rhombohedral

* For the crystal structures of these crystals, refer to Chap.3.
** Calculated from the ionic radii in Table 1-32

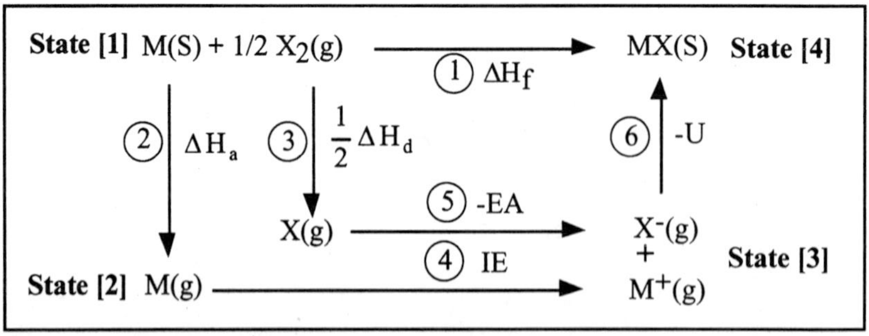

Figure 1-55 Born-Haber cycle for ionic solid

The Madelung constants and energies calculated using these Madelung constants for other ionic crystals are listed in Table 1-33. MgO, CaO and SrO have the same value for the Madelung constant as NaCl since they have the NaCl structure. In FCC structure, the Madelung constant does not depend on the length of unit cell length. Rather the Madelung constant depends on the relative position of ions. However, in other crystal systems, the Madelung constant also depends on the size of the unit cell. This aspect of the Madelung constant with a crystal structure will be explained in more detail in Chap. 4.

2. Experimental Measurement of Bond energy : Born-Haber Cycle

The basis for the measurement of bonding energy in ionic crystal is Born-Haber cycle as shown in Fig.1-55. The bonding energy measured from this scheme can

Table 1-34 Energy terms in Born-Haber cycle

Chemical Reaction		Description
$M(s) + \frac{1}{2}X_2(g) \xrightarrow{\Delta H_f^o} MX(s)$	ΔH_f^o	Heat of Formation of crystal from elements in their standard states
$M(s) \xrightarrow{\Delta H_a} M(g)$;	ΔH_a	Heat of sublimation
$\frac{1}{2}X_2(g) \xrightarrow{\frac{1}{2}\Delta H_d} X(g)$	ΔH_d	Dissociation energy for 2 atoms (X_2)
$M(g) \xrightarrow{I} M^+(g)$	IE	Ionization energy
$X(g) \xrightarrow{-EA} X^-(g)$	EA	Electron affinity
$M^+(g) + X^-(g) \xrightarrow{-U} MX(s)$	U	Lattice energy

be compared with that calculated according to Eq.(77). With this comparison the importance of the Madelung constant as a factor determining the bonding energy can be tested and the assumption made for the deriving Eq.(77) can be evaluated.

The energy terms in Fig.1-55 can be measured in experiments except U. The derivation of U is based on Hess' Law which states that heat of reaction to produce a given end product is the same regardless of reaction path or number of steps involved. Simply it is expressed in relation to reactions shown in Fig.1-55 as

$$\Delta H_{f(1\rightarrow4)} = \Delta H_{f(1\rightarrow2\rightarrow3\rightarrow4)}$$

$$\text{and } \Delta H_f = I - EA + \Delta H_a + \frac{1}{2}\Delta H_d - U$$

$$U = \Delta H_f - IE + EA - \Delta H_a - \frac{1}{2}\Delta H_d \qquad (79)$$

For example, NaCl has the following reaction heats.

ΔH_f^o	ΔHa	ΔHd	IE	EA	U
–411	108	121	496	–349	787kJ/mole

Therefore the lattice energy of NaCl is

$$U = -411 - 108 - \frac{1}{2} \times 121 - 496 + 349 = -787\text{kJ/mole}$$

Hence U is derived from these energy terms which can be measured experimentally according to Eq.(91). These energy terms are listed and described along with the chemical reaction in Table 1-31. In comparison with the schematic drawing in Fig.1-55, reactions (4), and (5) correspond to the atomic state(state[2]) to ionic state(state[3]). Reaction (6) corresponds to the transition from ionic state to ionic crystal state(state[4]).

Here it should be mentioned that the energy terms in Eq.(79) are not all enthalpy terms. Hess' law is stated in terms of heat generated or absorbed in the reactions. In thermodynamics, heat is represented by enthalpy which is defined as E + PV. Here E is total non-translational energy of a system excluding translational kinetic energy. P and V are pressure and volume of a system, respectively. The lattice energy is a good example which can explain the difference of enthalpy with internal energy.

The lattice energy represented as U in Eq.(72) is the energy which is taken up when MX is separated into M^+ and X^- at *rest* in gaseous state. In the light of Eq.(72), this energy arises solely due to the Coulombic electrostatic attraction. However, the lattice energy, U, shown in the Born-Haber cycle in Fig.1-55 represents the energy when MX is separated into M^+ and X^- at the reaction temperature in gaseous state which are *not at rest*. Therefore to compare the calculated lattice energy, E', with the lattice enthalpy, U, there must be a correction for the kinetic energy of M^+ and X^- in the gaseous state. The kinetic energy molecules in gaseous state is equal to PV. Hence the relation between lattice energy and lattice enthalpy is as U = E' + PV. The correction of the PV in lattice enthalpy is about 2.9kJ/mole at 25°C. This is less than 1% of the lattice energy. Therefore U in Eq.(72) can be treated as the lattice energy, U in the Born Harber cycle.

If there is no change in PV during the reaction, the energy and the enthalpy of the reaction are the same. This is the case of ionization energy and electron affinity in which only electron is transferred without changing the kinetic energy of atoms. Hence energy terms, IE and EA, are used in Born-Haber cycle in Fig.1-55.

For FCC, alkali halides, the Madelung constant and electric charge are the same. Therefore lattice energy is proportional to r_o^{-1} only. Hence, between K and Na halides, r_o is the only significant variable although, for larger sized ions, covalency effects distort this picture. For the FCC alkali earth oxides, values for z_+z_- are twice of alkali halides. CaF_2 and BaF_2 have lattice energies which are intermediate between alkali halides and alkali earth oxides since r_o and

Table 1-35 Comparison of observed and calculated lattice energies for ionic crystals along with melting points

Compound	z_+z_-	r_o	A	U (Experimental)	E' (calculated)	Melting Point(°C)
NaF	1	2.31	1.7456	923(kJ/mol)	936	996
NaCl	1	2.79	"	786	775	801
NaBr	1	2.94	"	747	735	747
NaI	1	3.18	"	704	680	660
KF	1	2.66	"	821	813	858
KCl	1	3.14	"	715	688	771
KBr	1	3.29	"	682	657	734
KI	1	3.53	"	649	612	681
MgO	4	2.10	"	3791	4117	2800
CaO	4	2.40	"	3401	3603	2580
SrO	4	2.56	"	3223	3378	2430
CaF_2	2	2.37	2.37	2630	2629	1418
BaF_2	2	2.71	"	2352	2300	1368

z_+z_- are intermediate although the Madelung constant is largest. The melting point data of these halide or oxide compounds well illustrate their relationship with the lattice energy.

Variable parameters are those in Eq.(77)

a. (z_+z_-)—Product of the charges of the cation and anion

b. $\frac{1}{r_o}$—inverse interatomic separation distance

$$r_o = \underset{\text{cation}}{r_c} + \underset{\text{anion radius}}{r_a}$$

c. m = compressibility factor
- constant for inert gas configuration
(cation and anion have the iner gas configuration)

Therefore : $U_o \sim \frac{Az_+z_-}{r_o}$

d. Madelung Constant : Depends on Crystal structure which must be known

Li and K halides have NaCl crystal structures. Hence the only variable determining the potential energy is the interionic distance which is the sum of radii of cation and anion. The melting point and compressibility of these alkali halides are plotted in Fig.1-56. Melting point is expected to be proportional to

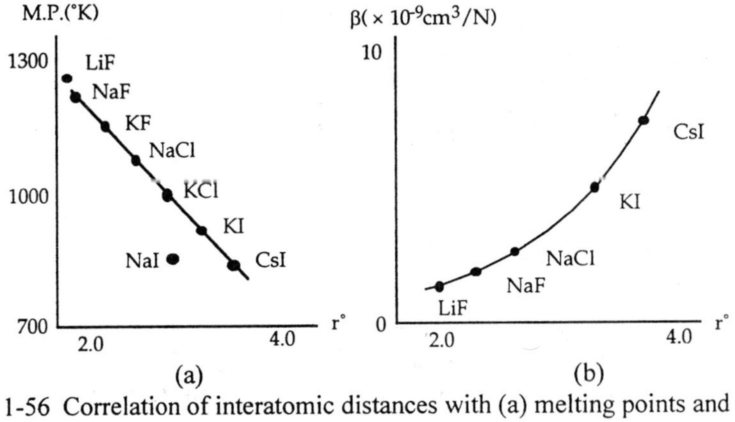

Figure 1-56 Correlation of interatomic distances with (a) melting points and (b) compressibility of alkali halides

of the potential energy. Since the potential energy is a function of r_0^{-1}, the melting point will be inverse to r_0 and the compressibility is proportional to r_0. These are well illustrated in Fig. 1-56. Since $\beta \propto (\text{hardness})^{-1}$, it is expected that m.p. and hardness are correlated for most ionic compounds. the potential energy and the compressibility is to be proportional to the inverse

C. Thermal Expansion of Materials

Thermal expansion is the result of atomic motion in a crystal. Fig.1-57 describes the oscillating motion of nuclei around the equilibrium position. At 0°K, atoms are arranged in the crystal without any motion. As temperature increases, the positions of atoms are shifted from their equilibrium positions at 0°K due to kinetic energy of nuclei which makes nuclei vibrate from its equilibrium position. Since the kinetic energy needed to make the nuclei jump from the ground state of vibration to the next level, however, is very large, the position of the most nuclei, even at room temperature, does not shift to the next level of oscillation. Only small portion of them is activated to the next level of the oscillation. As temperature increases upto significantly high temperature, the population of nuclei in the higher level of oscillation is increased. As shown in Fig.1-57(a), the potential well with interatomic distance, 'r', is not symmetric around the equilibrium position, r_0(0°K). If the potential well is a regular parabolic shape, the oscillation does not induce a expansion of a crystal. This asymmetric shape of potential well results in increase of interatomic distance as amplitude of oscillation increases with temperature. This change of equilibrium

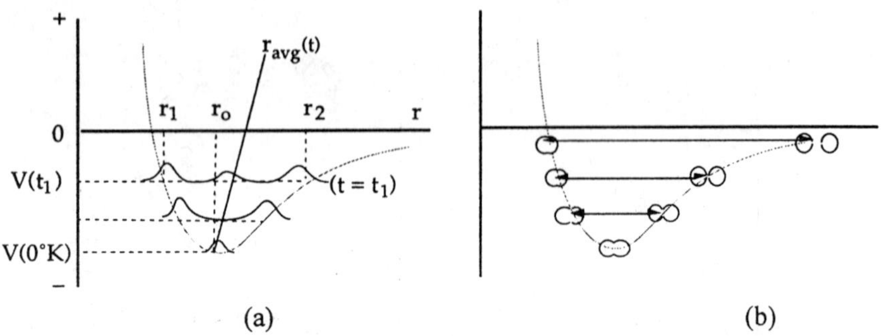

Figure 1-57 Vibrations of nuclei in the potential well at t = 0°K and t = t_1
(a) represents the change of probability to find the nuclei at different temperature and (b) illustrates the shape of atoms at different potentials with the vibrations

position of nuclei is shown in Fig.1-57(a) with a straight line for $r_{ave}(t)$. The slope of this line indicates the rate of change of 'r_{avg}'. Since the thermal expansion is related to this slope according to $\frac{\partial r_{avg}}{\partial T} = \frac{\partial r_{avg}}{\partial V}\frac{\partial V}{\partial T}$, it can be stated that the thermal expansion coefficient of a crystal depends on the gradient of potential with interatomic distance, $\frac{\partial r_{avg}}{\partial V}$, if the gradient, $\frac{\partial V}{\partial T}$, is constant for crystals.

Total energy (U) of a crystal is the sum of kinetic energy(T) and potential energy(V). T is difficult to evaluate for solids. In contrast, V is relatively easy to measure. Hence kinetic energy can be evaluated if the potential energy is known. At 0°K, total energy is equal to the potential energy since no kinetic energy exists. At t_1, the kinetic energy is not '0'. However, it becomes '0' at a certain point like r_1 and r_2 at t_1. Hence the total energy(U) at t_1 is equal to the potential energy, $V(t_1)$, at $r = r_1$ or r_2 due to zero kinetic energy at these points In relation to the oscillation of nuclei at t_1, there is no other energy supply from outside. At $r = r_1$ and r_2, the kinetic energy is '0' since the velocity of nuclei is '0'. Hence the lattice energy (U) at t_1 is $V(t_1)$ according to $U = T + V$. At the point ro,t1, the potential energy is V(ro). The kinetic energy is derived from $T = U - V(r_o) = V(r_1) - V(r_o)$. In a vibration, the kinetic energy is proportional to the square of displacement from the equilibrium position.

$$T = V(r_1) - V(r_o) = \frac{1}{2}k\left(r - r_o\right)^2 = \frac{1}{2}k\Delta r^2$$

Here Δr is the amplitude of the oscillation. The change in radial separation with increase in temperature from t_o to t_1 results in increased amplitude of vibration of the ions(atoms). This increased separation leads to expansion of the solid since $r_o(\ t = 0°K\ ) < r_o(\ t = t_1\ )$ with the asymmetric shape of the potential well. In relation to depth of potential well which is equal to potential energy, the asymmetry is exaggerated as the potential becomes shallow. This aspect is shown in Fig.1-58 in which potential wells are schematically drawn for SiC, MgO, Al and H_2O. From the potential curve for H_2O, it is expected that the same increase in temperature push the oscillation motion of nuclei into the asymmetric region. Thus the expansion coefficient, young's modulus, melting point, sublimation and dissociation temperature) are related to the depth and width or shape of the potential energy well. For typical covalent and ionic crystals, this well is deep and narrow. For a metal, it is shallow and wide although W(tungsten) would have a deep potential according to the data in Table 1-18. The difference in thermal expansion coefficients for SiC, MgO and Al is illustrated in Fig.1-59 in which the potential curves are those in Fig.1-58.

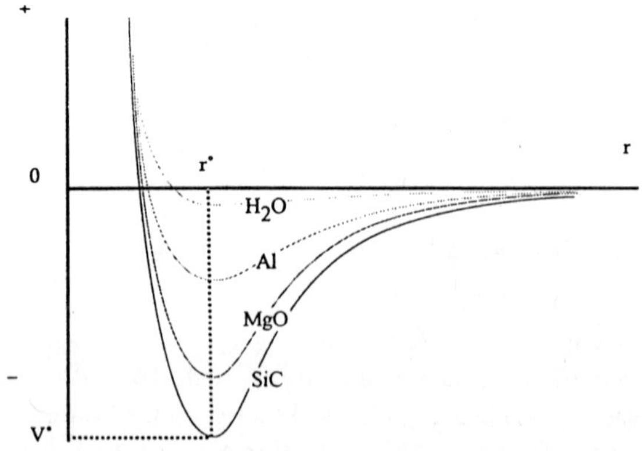

Figure 1-58 Schematic drawing of potential curves with values in Table 1-18 for SiC(covalent), MgO(ionic), Na(metallic) and H_2O(Hydrogen bonds)

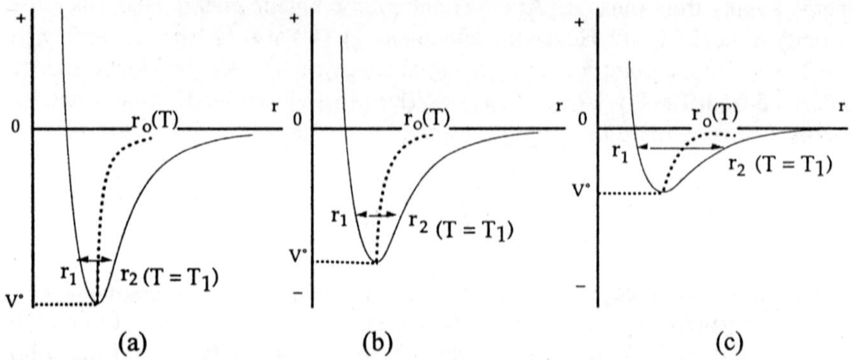

Figure 1-59 Illustration of the relationship of thermal expansion coefficient for (a) SiC (b) MgO and (c) Al representing covalent, ionic and metallic bonding , respectively. Potential curves are those shown in Fig.1-68

D. Molecular Orbital Theory for Crystals

1. Electronic Structure of Solids

The same kind method which was adopted to depict the molecular structure can be used to construct the electronic structure of solids. In molecular structure only two atoms were considered to form molecular orbitals for the bonding. In

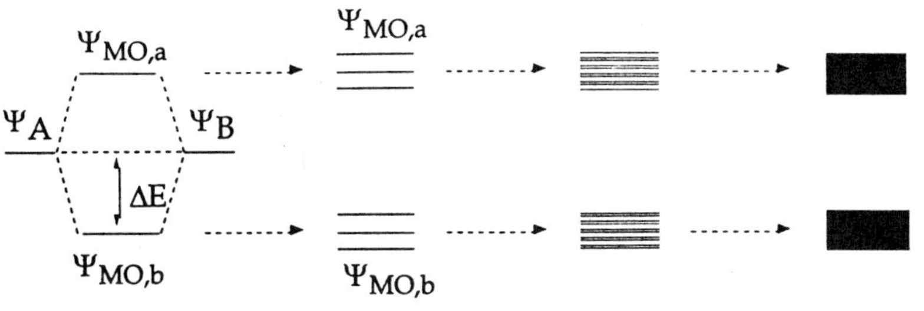

Figure 1-60 Schematic energy diagram for molecular orbitals from (a) two (b) six and (c) many atomic orbitals and (d) for solid

solids, consideration is exteneded to whole atoms in the solid. To simplify the electronic structure of solid, it will be treated as a very large molecule. The formation of the characteristic orbitals in solid from atomic orbitals is depicted in Fig. 1-60(d) as in the molecular orbital theory. Since the number of atomic orbitals are the same as the number of molecular orbitals, the number of orbitals in a solid must be the same as the number of atomic orbitals involved in the formation of molecular orbitals. In a small molecule, the energy levels of resulting molecular orbitals are well separated from one another. Like in solids which have a huge number of atoms, the molecular orbitals are rather very closely spaced as shown in Fig. 1-60(d). The spacing in energy level of molecular orbitals is about 10^{-18}eV if there are 1 mole of atomic orbitals in the solid. These energy states for electrons of solid can be considered as a continuous group of molecular orbitals. Due to the continuous characteristic, this group of molecular orbitals are usually called as a x-band. Here x represents the atomic orbital which constructs the band orbital. If the atomic orbital is 1s orbital, the group is called as 1s band.

As in a molecule, filling of the orbital bands is of interest because filling or emptiness of the orbital bands determines the physical properties of solids such as chemical bonding and electrical conduction.

2. Electrical Conduction in Solids

Electrical conduction in solids is determined by the existence of electrons which are to move in an electrical field. If band orbitals are empty, no electrical conduction occurs because there is no electon. Reversely if band orbitals are filled, electrons are confined within the band orbitals. Hence electrons are not affected by an electric field and there is no conduction. The necessary condition

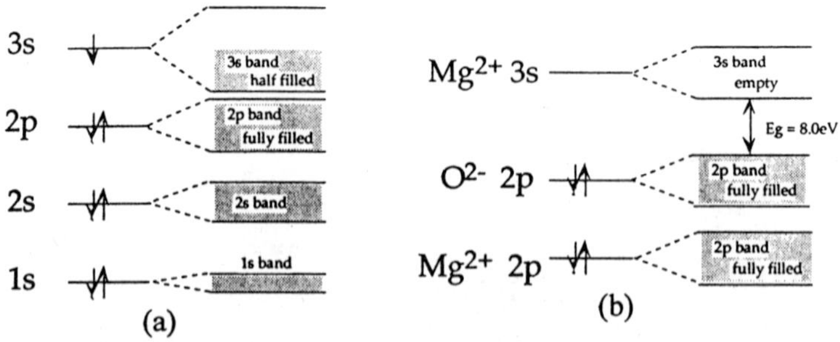

Figure 1-61 Band structure of (a) Na metal and (b) MgO

for the conduction to occur is that the band orbitals are half filled or partially filled.

The sequence of filling orbitals are same in the solids as in molecules. Electrons first fills the lowest energy state and then the next upper level state. The orbital band of interest in electrical conduction is the filled highest energy orbital band regardless of fully or partially filled since the orbital bands of lower energy than this must be fully filled. If this highest orbital band is fully filled like MgO in Fig. 1-61(b), it is an insulator. If it is half filled as Na metal, it is metallic.

Since the filled highest orbital band is important for electrical conduction, this orbital band is used to illustrate the characteristics of solids regarding electric conduction. According to this scheme, solids are classified into metal, insulator, semi metallic and semiconductor in Fig. 1-62. In insulator, the filled highest energy band orbital is called as valence band and the lowest energy empty band orbital is called as conduction band. The difference in energy between the valence band and conduction band determines the electrical resistivity of insulators. This is called as band gap. If band gap is larger than 1.5eV, the solid is insulator. If band gap is less than 1.0eV, it is semiconductor.

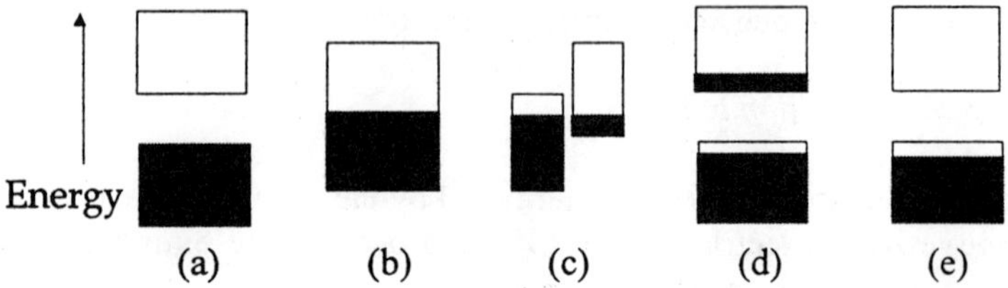

Figure 1-62 Filling of electrons in orbital bands in (a) insulator, (b) metal, (c) semimetal,(d) n-type semiconductor and (e) p-type semiconductor

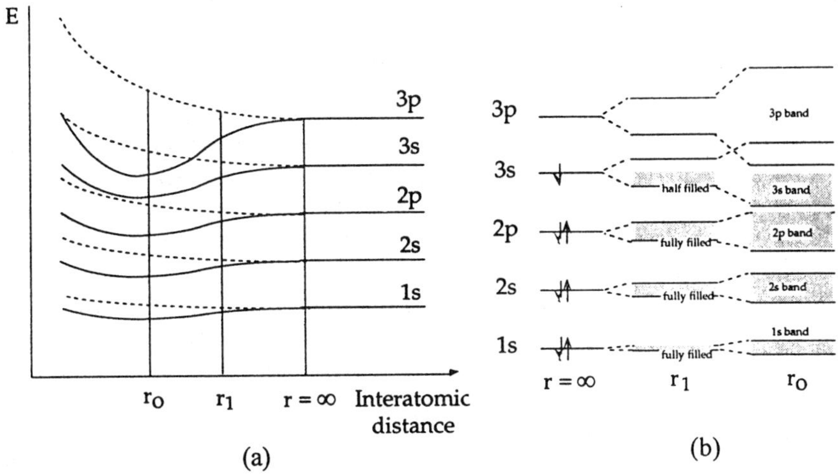

Figure 1-63 (a) Pertubation of atomic orbitals of Na metal with internuclear distance and (b) variation of energy levels of orbital bands at r_o, r_1 and $r = \infty$

The band gap of MgO is 8.0eV which is very large. This high band gap results in the high resistivity of MgO.

n-type semiconduction can be considered as thermally excited insulator and p-type is a insulator with electron holes which are produced by adding impurites. Semimetallic compound is different from others in the fact that it has overlapping orbital bands. Obital band of lower energy is partially empty and orbital band of higher energy is partially filled.

As shown in molecular structure, molecular orbitals usually follow the order of energy level of atomic orbitals. However in solid this order may be reversed. Fig. 1-63 reveals this phenomena. When atoms are separated in infinite distance there is no molecular orbitals formed. Therefore each atom retains the atomic structure. At interactomic distance r_1, Na metal forms the electronic structure as in Fig. 1-63(b) which follows the order of energy level of atomic orbitals. Once it reaches the Equilibrium distance, r_o, 3s orbital band overlaps with 3p orbital band. The lowest energy level of 3p orbital band is lower than the highest energy level of 3s orbital band. If Mg with one more electron than Na has the same band structure, its band structure must be semimetallic since both 3p and 3s orbital bands would be partially filled like in Fig. 1-62(c).

E. Band Theory and Cohesive Energy

1. Introduction

Crystals are different from atoms and molecules in the fact that their properties are not determined simply by interactions between the nearest atoms. The interaction goes far beyond the nearest atoms. This is shown in the calculation of binding energy of ionic crystal, NaCl in Eq.(77). The cohesive energy in solid is a collective concept which must include contribution of every electron. In the example of NaCl binding energy shown in Eq.(77), each electron confined to Cl^- and Na^+ stripped of an electron contributes to the binding energy. If the energy contribution of each Cl^- or Na^+ is E_i and the number of Cl- or Na- in the same distance from the origin is $N(E_i)$, then the binding energy of NaCl crystal can be expressed as

$$E = \frac{\sum_i E_i N(E_i)}{\sum_i N(E_i)} = E_o \frac{\sum_i k_i N(E_i)}{\sum_i N(E_i)} \tag{82}$$

where $E_o = \frac{q^+ q^-}{\pi \varepsilon_o}$ and $k = \frac{1}{r_i}$. In comparison of the above equation with Eq.(77), Madelung constant can be expressed as

$$A = \frac{\sum_i k_i N(E_i)}{\sum_i N(E_i)}$$

Hence the Madelung constant for ionic crystals represents a factor which averages energies from different electronic states in ionic crystals. Since electronic states of ionic crystals are represented by point charges as expressed by delta function in Eq.(77), energy band structures are constructed by geometrical configuration of atoms in a crystal lattice. In real crystals, electrons are not confined in a point as described by a point charge. Rather electrons are distributed in a way as a wave function describes. Unlike delta function, the real wave function of an electron has a shape and sometimes directionality. These geometric factors of wave functions along with atomic arrangement in crystal lattice determine the energy states distribution. This is shown in energy band structure of Fe metal in Fig.2-xx in which d orbitals form different energy band structrues. Since the band structrue is formed vs reciprocal lattice, the detailed description will be given in Chapter 2. Only the concept of band structure will be presented in the next with simple model.

2. Energy Level Distribution and Cohesive Energy in Crystals

In the section 1-2, molecular orbital theory introduced the concept of band theory for crystals. However, the molecular orbital theory does not provide any information about distribution of energy states. To illustrates how of energy states are distributed, a purely hypothetical 2-dimensional model is proposed. The model as shown in Fig.1-64 consists of N atoms arranged in a stright line with a constant interval, 'a'.

Here several assumptions are supposed.

(i). Each atom has one electron in p orbital. Molecular obitals are built by LCAO method (Linear Combination of Atomic Orbital) and is expressed as :

$$\Psi_k = \frac{1}{\sqrt{N}} \sum_i p_{ki}, \quad (81)$$

Here p_{ki} is p orbital of i-th atom for k-th molecular orbital, Ψ_k.

(ii). Another assumption for the molecular orbital is that each atomic orbital, $p_i(x)$, vanishes when x goes beyond 0.5a to the left and right of the corresponding atomic position. In other words, $p_i(x)$ is 0 when $(i-1)a < x$ or $x > ia$.

(iii). Since there are N atoms with single p orbital, N molecular orbitals are formed. Fig.1-66 shows the first and the k-th molecular orbitals. Other molecular orbitals are generated by rotating with a certain angle, θ. With the second assumption, rotating direction must alternate clockwise and anticlockwise in every other atoms in order to make the molecular orbital continuous across the cell boundary (x = na). Also, to make quantized N molecular orbitals, however, it is assumed that the rotating angle for k-th molecular orbital is equal to $\frac{360k}{N}$ degree.

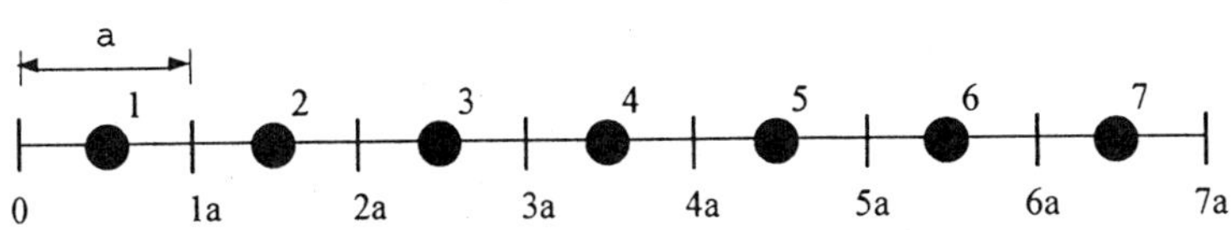

Figure 1-64 2-dimensional lattice

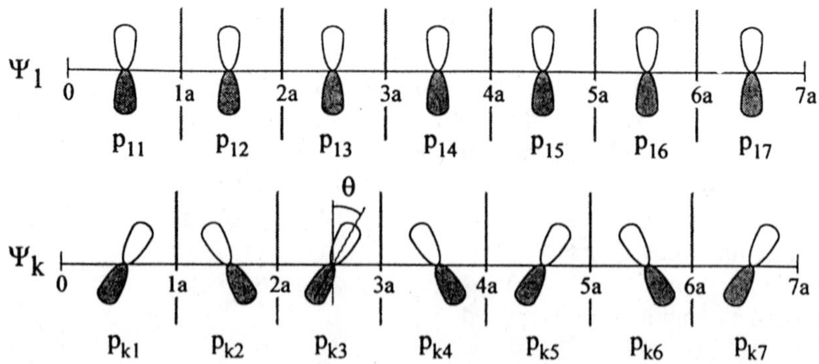

Figure 1-65 Hypothetical molecular orbitals from atomic p orbitals of N

(iv). Repulsive interactions between electrons and nuclei are assumed to be averaged as E_{rep}. Binding energy arises from the attractive interaction between electrons and the nearest nuclei. The calculation of binding energy is based on the assumption that electron in porbital is equivalent to a pole of length 0.5a with half of electron charge on both ends.

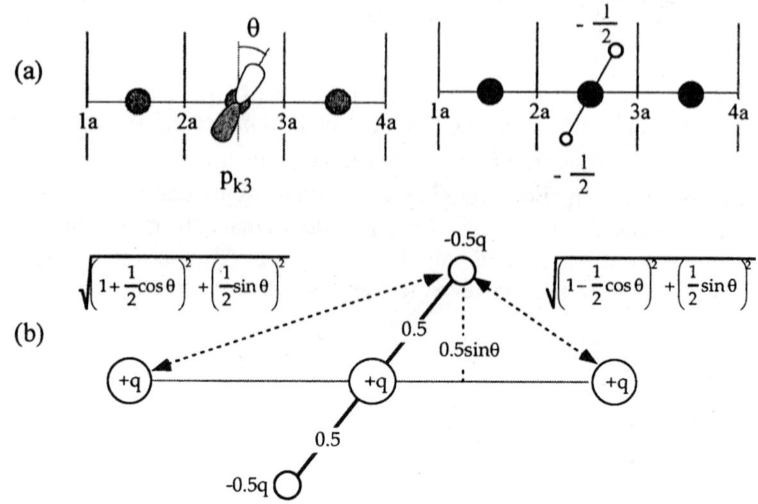

Figure 1-66 (a) Equivalence between p orbital and electron pole and (b) Corresponding dimensions

By considering the interaction of electron charge on the both ends of the pole with the nearest nuclei, the binding energy is as follows :

$$E = \frac{-0.5q^2}{\sqrt{\left(1+\frac{1}{2}\cos\theta\right)^2+\left(\frac{1}{2}\sin\theta\right)^2}} + \frac{-0.5q^2}{\sqrt{\left(1-\frac{1}{2}\cos\theta\right)^2+\left(\frac{1}{2}\sin\theta\right)^2}} + E_{rep} \qquad (82)$$

Lowest energy occurs at $\theta = 0°$. The highest energy occurs at $\theta = 90°$. Energy levels are most crowded around $\theta = 0°$, and 90° and least crowded around $\theta = 45°$. Let's define $N(E_o)$ is the number of energy states with the range of $E_o < E < E_o + \Delta E$. Since, in real crystal, N is very large, the discrete energy level in Fig.1-60 turns into the band structure as shown in Fig.1-67. Electrons fill from the bottom of the band structure. E_{ave} is calculated as in Eq.(80). $E_{cohesive}$ is the difference between E_{ave} and E_{atom}. E_F is known as the Fermi energy level which corresponds to the highest energy level filled with electrons. In this model, the band structure is half-filled since there is only one electron in each atom. If each atom has two electrons, the band structure should be filled. although it changes the direction of p orbitals relative to the lattice axis. In contrast, k in Eq(83) of Block theorem shows not only dependency of molecular

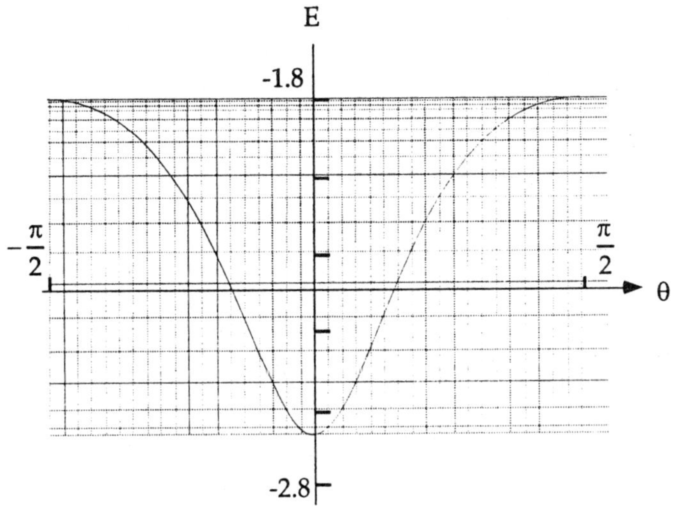

Figure 1-67 Variation of binding energy of single p orbital with rotating angle, θ

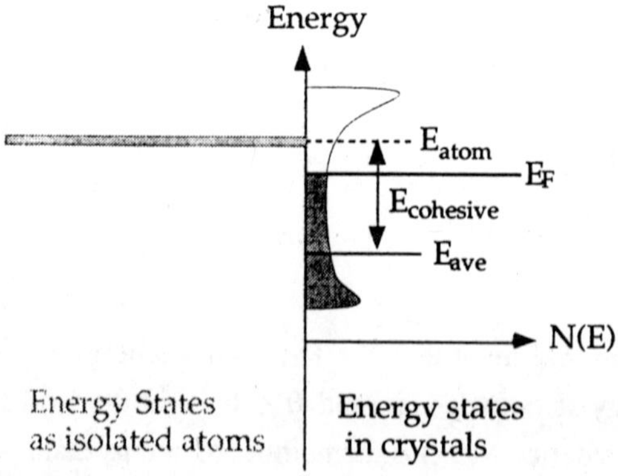

Figure 1-68 Energy states distribution for the binding energy expressed as Eq(82)

orbital, ψ_k, on the direction but also on the magnitude of k. Hence the band structure of energy level of a crystal must show this double dependency with the direction of vector k and its magnitude. Fig.1-69 is a plot of band structure of Ni which shows the variation of energy level of ψ_k from d orbital vs different direction of vector k.

3. Block Equation

The base of band theory in solid state physics is Block theorem which descibes the wave function of electron in a periodic potential. In crystal, the position of nuclei are assumed to be stationary in a certain position which forms crystal lattice while electrons moves around whole nuclei. Since this theorem includes the reciprocal lattice, detailed explanation will be given in Chapter 2. Here it is compared with the proposed model.

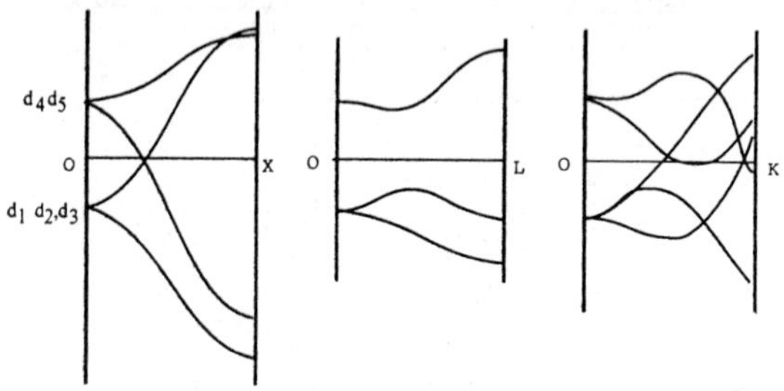

Figure 1-69 Energy band structure of d orbitals in Ni with 3 different direction of vector k (OX, OL, and OK) [Fletcher]

$$\psi_k(r) = e^{ik\cdot r} u_k(r) \quad (83)$$

Here ψ_k is a wave function for an electron which satisfies SWE for a crystal and k is a (reciprocal lattice) vector. ψ_k is equivalent to a molecular orbital in Eq.(81). In the proposed model, θ does not change the magnitude of p orbitals.

The energy band structure of Ni illustrates how molecular orbitals from atomic d orbital are energetically separated with vector k. When k = 0, which means that there is no difference of d orbitals in Ni crystal from atomic states, d_1, d_2, and d_3 have all same energy level. Likewise, d_4 and d_5 have the same energy level.

In the proposed model, energy level repeats with θ in a way that $n\pi + \theta$ result the same energy level as that of θ. Hence the energy level are illustrated within the range of $0 < \theta < \pi$ (or $-\pi/2 < \theta < \pi/2$). Likewise, energy band structures for crystals are illustrated within a certain range of vector k. In Fig.1-70, this range of vector in the direction of OL is $0 < k < \pi$. Beyond this range of vector k, band structure repeats as in Fig.1-67. This repeating property of (reciprocak lattice) vector k on the band structure defines a (reciprocal lattice) cell.

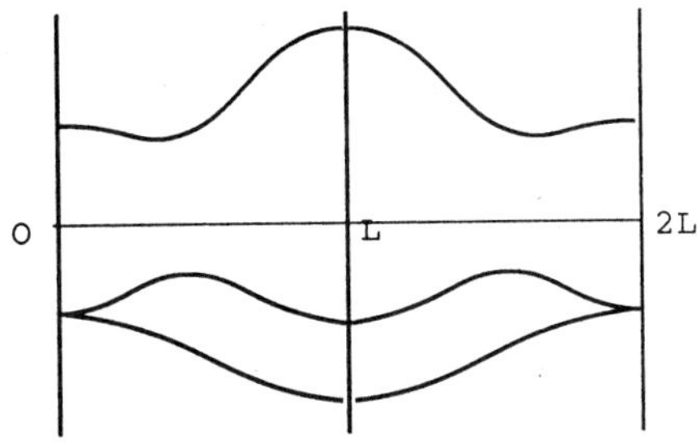

Figure 1-70 Repetition of band structure with the magnitude of vector k

CHAPTER 2

FUNDAMENTALS OF CRYSTALLOGRAPHY

SECTION 1 CRYSTAL STRUCTURE STATES

A. Introduction

The crystalline state can be defined as the most stable and lowest energy state for a solid under a given environmental condition of temperature and pressure. This energy state is represented by the Gibb's free energy function G=(H – TS) at constant pressure or by the Helmholtz function H=(U – TS) at constant volume. Here H, T, S and U represent enthalpy, temperature, entropy, and internal energy, respectively. Lowest energy state is achieved when an imaginary line that bisects any free surface of a crystal would have the energy state on each side of the line the same, as illustrated in Fig.2-1(b) and Fig.2-2. Any deviation from this condition would, therefore, represent a noncrystalline state, as illustrated In Fig.2-1(a).

This energetical approach to crystal definition implies that the crystalline structure must be ordered. The unique feature concerning the crystalline state in a crystal is its periodic structure or pattern. Due to this periodic structure, crystals can be defined by simple building blocks with defined stacking directions. In the two dimensional case, the stacking block is a closed plane outlined by two non-parallel vectors. Similarly, in three dimensions, the block is a polyhedron and three non-parallel directions are needed to define the stacking directions.

If the building block is regarded as a point, the resulting array of points that are arranged by the stacking directions, forms a lattice. Since within the space lattice, each lattice point bears a fixed, geometric relationship to every other point and the stacking direction is equivalent to a translational vector, the entire array can be treated and analyzed as a mathematical entity.

A crystal can be described as a collection of smallest blocks stacked in defined directions. This non-divisible smallest block is called the "unit cell" in crystallography. Thus far, atomic arrangement within the unit cell has not been a concern. However, if this atomic arrangement is taken into consideration, the stacking direction is no longer arbitrary. For example, if the block is filled with homogeneous media, there are an infinite number of stacking directions for the blocks as illustrated in Fig. 2-3. Since the stacking block is, however, filled with discrete atoms, stacking directions will be limited to those that make continuos connections between atoms in adjacent blocks. This relationship between the stacking direction and the atomic arrangement within the block is illustrated in Fig.2-4. Here, three sets of directions are possible due to the hexagonal

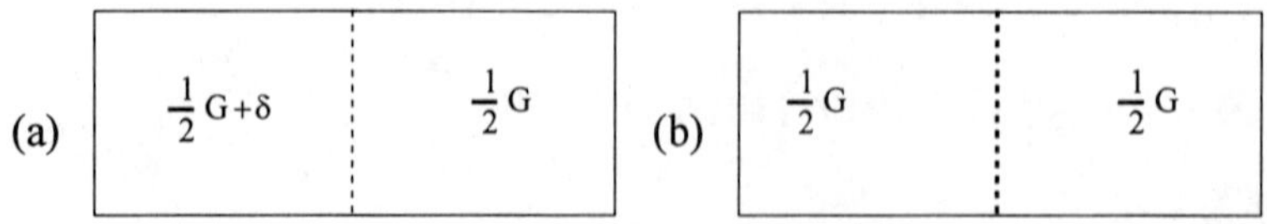

Figure 2-1 Energy distribution in (a) non-crystalline state and (b) crystalline state

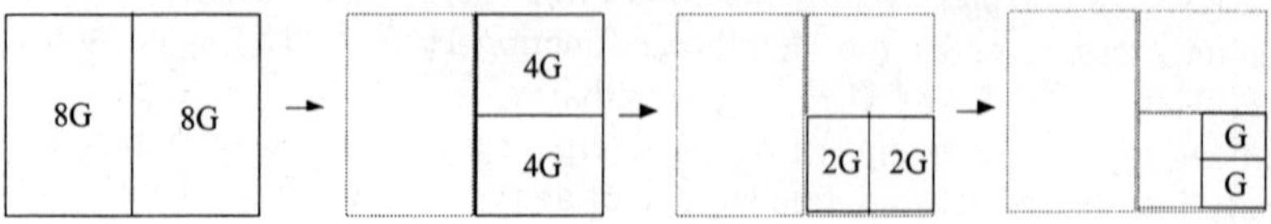

Figure 2-2 Partitioning of crystals into the smaller blocks with the same energy state

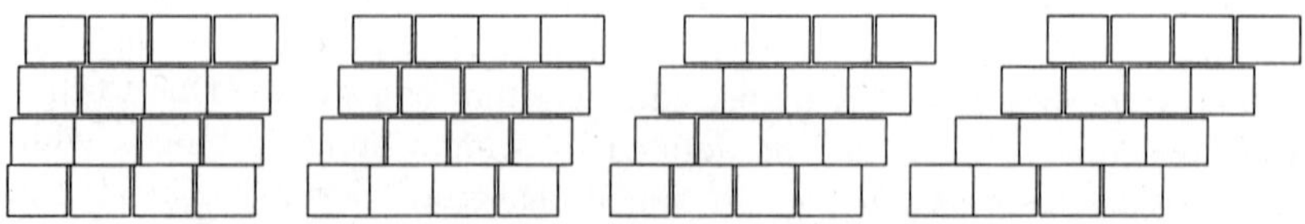

Figure 2-3 Continuous change of stacking direction for blocks filled with continuous medium

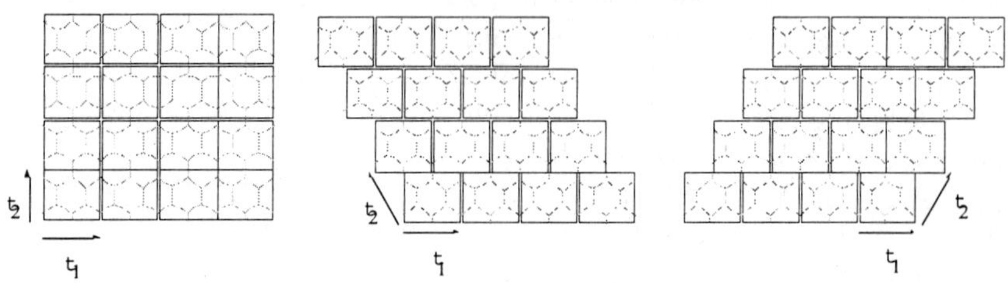

Figure 2-4 Three stacking modes with rectangular blocks with hexagonal arrangement of atoms

arrangement of atoms within the block. Stacking of blocks along these directions produces honeycomb-like crystal structures that are shown as dotted lines in Fig.2-4.

B. Single-, Poly- and Non-crystalline(Amorphous) states

Any solid exists in one of three states : single crystal, polycrystal and non-crystalline (amorphous). All have periodicity ranging from perfect for a single crystal to essentially nonexistential for the amorphous state.

A single crystal in the ideal state has perfect periodicity throughout the whole solid and, consequently, the lowest energy state. In contrast, the amorphous (non-crystalline) state lacks long range order or periodicity and as a result has a higher energy state than the corresponding single crystalline state.

For the crystalline states illustrated in Fig.2-1, there is in discontinuity crossing the boundary into the amorphous state since the energy state is different on both sides of the boundary. This discontinuity typically separates smaller regions in which ordering exists e.g. in glasses, most polymers and liquids. In this respect, the amorphous state is different from the crystalline state because it lacks long range order, which usually corresponds to 100 ~ 1000 unit cells within a crystal.

A polycrystal is different from the amorphous state in that it has long range ordering but there are boundaries that disrupts this long range ordering. Fig.2-5 illustrates the microstructural differences between the amorphous, polycrystal and single crystal states.

Due to the periodic structure, crystals are depicted by simple building blocks and their stacking directions. If the building block is regarded as a point, the resulting array of points, that are arranged by the stacking directions, forms a lattice. Hence, a lattice is a collection of points in space which bear regular and periodic relationship to each other such that environment of each point is

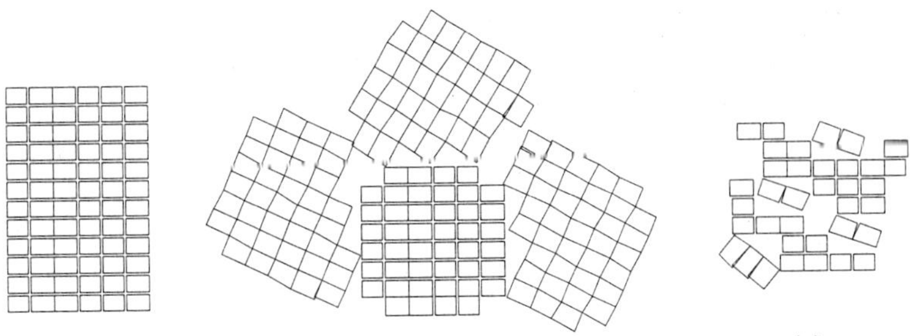

(a) (b) (c)

Figure 2-5 illustration of (a) single crystal, (b) polycrystal and (c) amorphous state

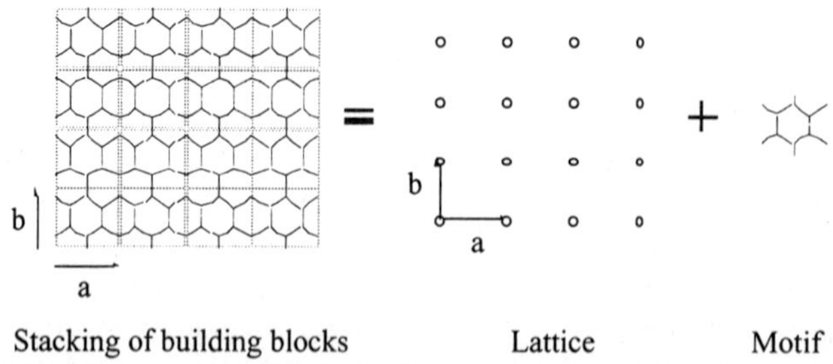

Figure 2-6 Equivalent descriptions of a crystal by blocks, and lattice and motif

identical. Repeating pattern or motif centered at each point still preserves the lattice structure. Since the building block is a spatial identity surrounded by imaginary faces, it can be replaced by the actual atomic configuration, which is called a motif. The motif, therefore, describes the environment or pattern which is centered around each lattice point. The pattern or a crystal may be generated from a motif at a single lattice point by the operation of lattice vectors. Fig.2-6 illustrates how a crystal is described by a lattice and motif.

C. Symmetry

1. Symmetry elements

Symmetry develops from an orderly and predictable arrangement of objects such as to constitute a pattern. Symmetry exists in a crystal because of the natural ordering that takes place when atoms or ions are packed together in a periodic array. This ordering is characterized by the symmetry of the crystal system in terms of symmetry elements and symmetry operations.

A symmetry element is a geometric entity with respect to which one or more symmetry operations may be performed and is invariant with respect to the operation. There are exactly 3 symmetry elements:

a) A point
b) An axis
c) A plane

A point is a dimensionless entity in space with coordinates that specify its position. An axis is a line joining 2 points. A plane is formed by 2 intersecting lines as illustrated in Fig.2-7.

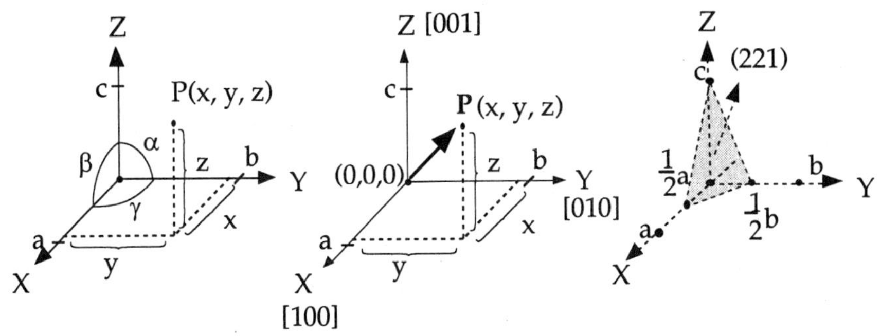

Figure 2-7 Symmetry elements : Point, direction and plane

a. Point : <u>A point</u> has <u>position</u> defined by coordinates (xyz) with regard to unit vectors $\vec{a}\,\vec{b}\,\vec{c}$ along direction axes x,y,z

b. Axis: An axis has direction. The vector $\vec{p}$ connects points (000) and (xyz). $\vec{a}\,\vec{b}\,\vec{c}$

$$\vec{p} = x\vec{a} + y\vec{b} + z\vec{c}$$

i.e. for (x, y, z) = $(\frac{1}{4}, \frac{1}{2}, 1)$,

$$\vec{p} = \frac{1}{4}\vec{a} + \frac{1}{2}\vec{b} + 1\vec{c} \text{ or } \vec{p} = 1\vec{a} + 2\vec{b} + 4\vec{c}$$

i.e. , [uvw] = [124]

where [u v w] defines the direction of $\vec{p}$. The length of the vector $\vec{p}$ is given by $|p| = \sqrt{x^2+y^2+z^2}$ in the orthogonal system

c. Plane : <u>A plane</u> has <u>intercepts</u> on $\vec{a}$, $\vec{b}$, $\vec{c}$. These intercepts define the plane. Refer to Miller indices.

<u>Note</u>: The equation for the plane (m) is given by: $hx + ky + \ell z = 1$. Thus any point lying in the plane must satisfy this equation.

Ex: For (221): $2x + 2y + \ell z = 1$

The points, (1,-1,1), (-1,1,1), (0,0,1) and $(1,-\frac{1}{2},0)$ all satisfy the equation. Thus all these are within the plane (221).

2. Symmetry operations

A symmetry operation is defined as the movement of a body such that after movement is complete every point on the body is coincident with the original orientation, i.e., to an observer the body would not appear to have been moved. Symmetry operations permit the location of each point (atoms, ions, voids, etc.) with regard to a reference point in a given coordinate system.

All operations that do not involve translation leave at least one point in the system unmoved, therefore these are called point symmetry operations.

Operations involving translation move the reference point, although bringing an equivalent point into coincidence with the original reference point, therefore these are called space symmetry operations.

2-1. Basic Symmetry Operations

There are four basic symmetry operations, as follows:

a <u>Rotation</u> - Around an axis [uvw]
Symmetry: 1,2,3,4,6 fold

b. <u>Reflection</u> - Across a plane (hkℓ)
Symmetry: m, 1/m(mirror plane)

c. <u>Inversion</u> - Through a point (xyz)
Symmetry: $\bar{i}$ (inversion π **t**.)

d. <u>Translation</u> - Along an axis [uvw]
Symmetry: translation vector, **t**

Additionally there are compound symmetry operations produced by combinations of the basic symmetry operations as follows:

2-2. Compound Symmetry Operations

a. Rotoinversion - Rotation + Inversion
Symmetry: $\bar{1}$, $\bar{2}$, $\bar{3}$, $\bar{4}$, $\bar{6}$ Rotary Inversion Axes

b. Rotoreflection - Rotation + Reflection
Symmetry: 1m, 2m, 3m, 4m, 6m, 1/m, 2/m, 3/m, 4/m, 6/m

c. Compound Rotation - Rotation + Rotation
Symmetry: 32, 23, 432, 222, 422, etc. $\bar{3}2$, $\bar{4}2$, etc.

d. Screw Axes - Rotation + Translation

Symmetry: $2_1, 3_2, 4_1, 4_2, 4_3, 6_1, 6_3$, etc.

e. Glide Plane - Reflection + Translation
Symmetry: a, b, c, n, d Glide reflection planes

2-3. Basic Symmetry Operations - Illustrations

a. Rotation

Rotation of a point (xyz) must occur around an n-fold axis symmetry, where the angle of rotation $\gamma = \frac{360}{n}$, n=1, 2, 3, 4, 6. Since rotation is a symmetry operation, starting with a given point (xyz), all other points are generated by symmetry. Convention for rotation operation is anti-clockwise

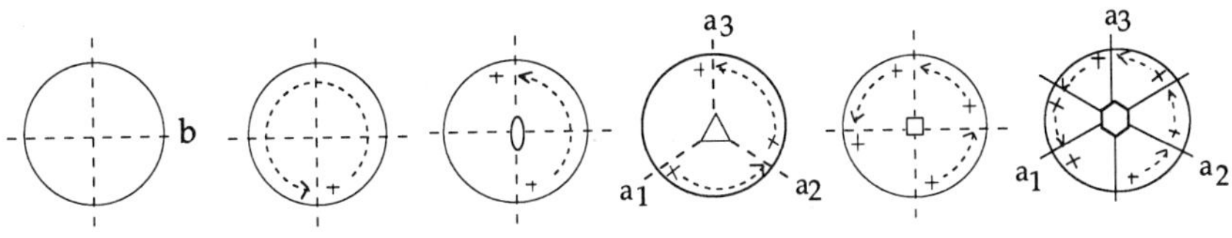

b. Reflection (m, $\frac{1}{m}$)

If the mirror plane is parallel to the rotation axis, reflection is designated by m. If it is perpendicular to the rotational axis, it is designated by n/m where n means n-fold rotation.

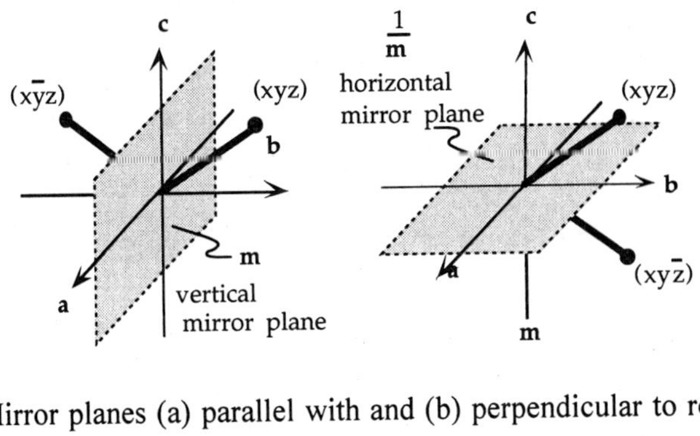

Figure 2-8 Mirror planes (a) parallel with and (b) perpendicular to rotational axis

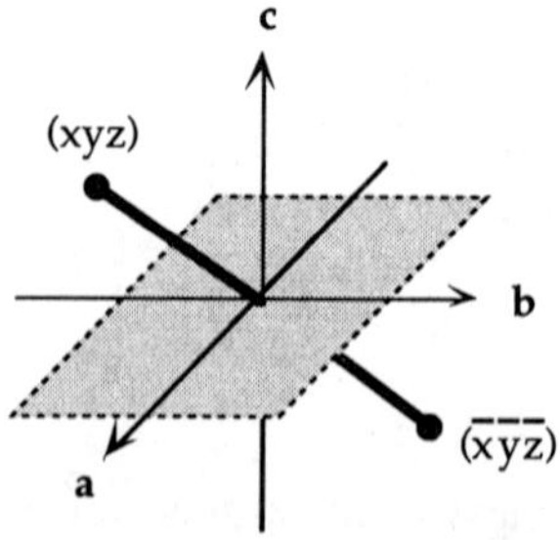

Figure 2-9 Inversion symmetry

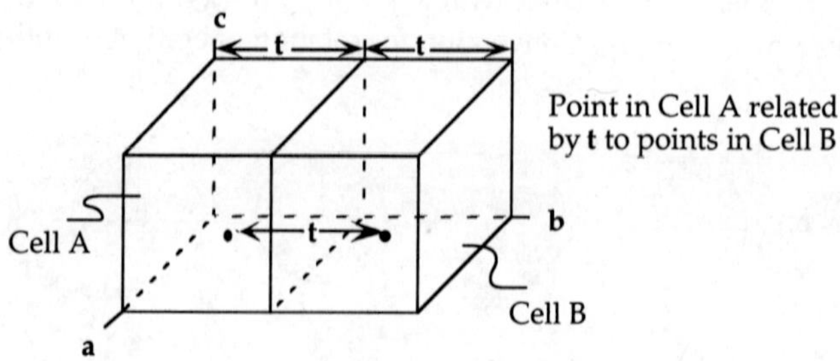

Figure 2-10 Equivalency of cell A and cell B by **t**

c. Inversion (i)

This is the combination of 2-fold rotation and mirror. It changes the coordinate (xyz) of a point to (-x,-y,-z).

d. Translation

Pure translation occurs along lattice vector **a**, **b**, or **c** but does not add any new symmetry. Only fractional translation can contribute to forming new symmetry. The new symmetry operation formed by fractional translation with other symmetry operations are screw axis and glide plane.

D. Lattice

1. Line Lattice (Row)

A line lattice is an array of one dimensional lattice points generated by a vector **a**. Since this generates a lattice, this is called a lattice vector. Fig.2-11 shows a

1-dimensional crystal and its corresponding line lattices with lattice vectors. Depending on the selection of a motif and a lattice vector, the resulting lattice has a different shape. The vector **a** is called a primitive or <u>unit</u> vector; i.e., of smallest magnitude. Among lattice vectors in Fig.2-11, $\mathbf{a}_1$ is a primitive lattice vector. Also, the lattice constructed with the primitive vector is called a primitive lattice.

2. Planar Lattice

A planar lattice is two dimensional array of points in space generated by non-parallel vectors **a** and **b**. In a planar lattice, the primitive lattice is constructed by the two vectors that forms the smallest area. The area surrounded by the lattice vectors is called the unit cell. If it is made by two primitive lattice vectors, it is called a primitive unit cell. Unlike in a linear lattice, many

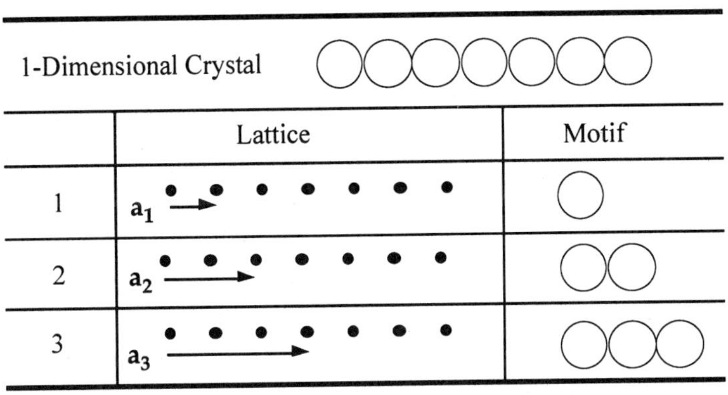

Figure 2-11 Construction of 1-dimensional crystal with various lattices and motifs

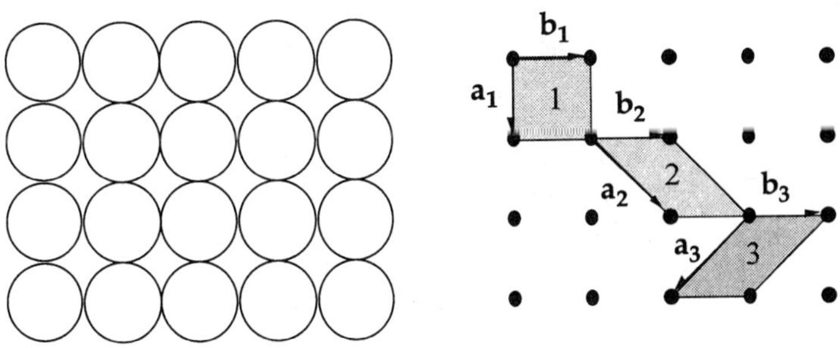

Figure 2-12 2-dimensional crystal and its primitive lattice with different lattice vectors

primitive lattice vectors are possible although they produce the same primitive lattice. This is illustrated in Fig.2-12.

In light of the definition of primitive lattice, there is no preference among primitive unit cells 1, 2 and 3 in Fig.2-12. However, it is conventional to select lattice vectors that best describe the crystal symmetry since the lattice provides the basis for crystal symmetry. Lattice vectors $\mathbf{a}_1$ and $\mathbf{b}_1$ are preferred over other lattice vectors because the primitive unit cell 1 in Fig.2-12 best represents the square-like symmetry of the crystal lattice.

3. Classification of planar lattice

3-1. Geometry of the unit cell

Since a lattice is a collection of unit cells in two dimensional space, the geometry of the unit cell must be related to the symmetry elements of the crystal. If there is a 4 fold rotational symmetry, the geometry of the unit cell must be square. Likewise, if there is six fold rotational symmetry, the unit cell must be a parallelogram whose edges are all the same length with intersecting angles of 60° and 120°. The lattice with a hexagonal unit cell is called a hexanet whereas the lattice constructed by the square unit cell is called a square net. In this respect, 4 fold and 6 fold rotational symmetries cannot simultaneously exist in a lattice although each can exist with 2 fold rotation symmetry. In this regard, lattices can be classified according to the geometry of the unit cell.

Regarding the geometric aspects of symmetry, planar lattices can be classified into 3 major groups : square net group, hexanet group and a group without 3, 4 and 6 fold rotational symmetries. The last group can be further classified into two groups considering the geometry of the unit cell. They are rectangular and parallelogram groups. The parallelogram lattice has a parallelepiped as the unit cell with two edges of unequal length while the rectangular lattice has a parallelepiped with two edges of equal length and intersection angle of 90°. Thus, the symmetry combinations for each group are different. Classification of the planar lattices is given in Table 2-1.

In the parallelogram lattice system, the possible symmetry group is 1 or 2. Hence each group is characterized by the geometry of the unit cell and has unique symmetry groups. The minimum requirement in a symmetry element for a lattice is listed in the last column in Table 2-1.

A planar lattice has only line or plane elements such as a rotational axis and mirror plane, respectively. There is no point symmetry element such as an inversion center since the inversion center is equal to 2 fold rotation in a planar lattice. Hence only the rotational axis and mirror plane are considered as symmetry elements for planar lattices. Also the mirror plane is perpendicular to the planar lattice.

Table 2-1 Classification of planar lattices

Lattice system	Geometric structure (4 unit cells)	Geometry of Lattice	Symmetry element of the lattice system
Parallelogram lattice	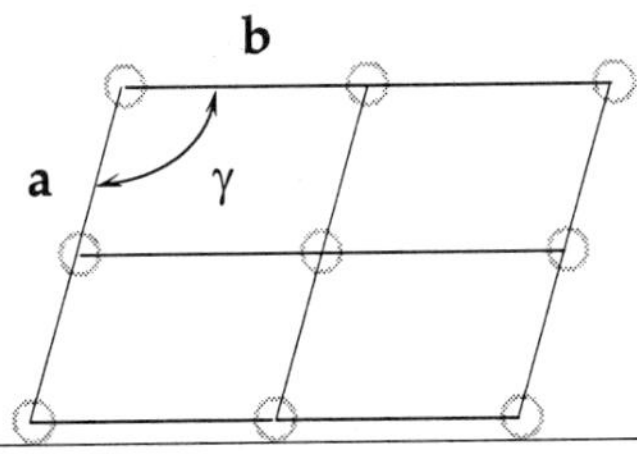	$\lvert \mathbf{a} \rvert \neq \lvert \mathbf{b} \rvert$ $\gamma \neq 90°$	2 fold axis
Rectangular lattice (Orthonet)	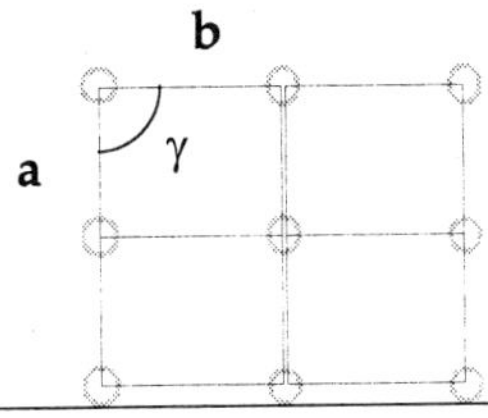	$\lvert \mathbf{a} \rvert \neq \lvert \mathbf{b} \rvert$, $\gamma = 90°$	mirror plane
Square lattice (Tetranet)	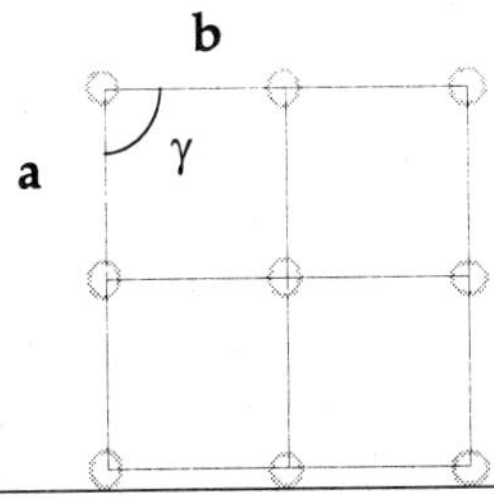	$\lvert \mathbf{a} \rvert = \lvert \mathbf{b} \rvert$, $\gamma = 90°$	4 fold axis
Hexagonal or Trigonal lattice (Hexanet)	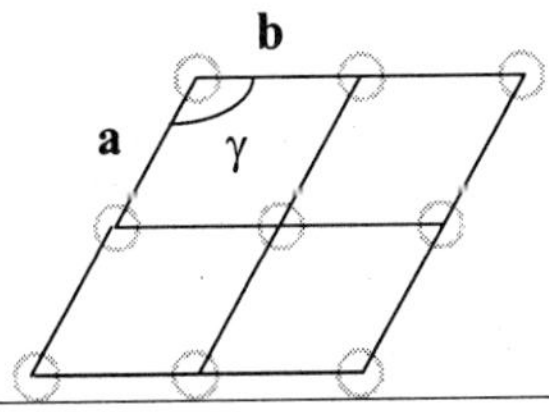	$\lvert \mathbf{a} \rvert = \lvert \mathbf{b} \rvert$ $\gamma = 120°$	3 or 6 fold axis

As mentioned above, planar lattices can have rotation axes: 1, 2, 3, 4, 6. More than one rotational symmetry element can exist in the same lattice along with the mirror plane. It seems that there are many ways to combine the rotational symmetry elements along with mirror planes. However, there are only 10 ways to combine them in a planar lattice if there is no translational symmetry. This small number of ways to combine the symmetry elements is due to the regularity of the lattice points since the positions of a symmetry element are restricted at certain points within the unit cell. The restriction in combining symmetry elements can be illustrated in the relationship between the shape of the unit cell and the symmetry elements that can exist in the lattice.

3-2. Non-primitive unit cell

The lattice types shown in Table 2-1 are called primitive lattice since the unit cell contains only one lattice point. Sometimes a non-primitive unit cell is chosen to match the geometry of the unit cell to that of the corresponding lattice system. The lattice shown in Fig.2-13 has a parallelogram unit cell in that the lattice vectors have same length. Due to the equal length of the lattice vectors, the lattice has a mirror plane in the diagonal direction of the parallelogram unit cell. As illustrated in Table 2-1, the lattice system having a mirror plane as a symmetry element is the rectangular lattice rather than the parallelogram lattice. As a result, this planar lattice is classified as a centered rectangular lattice in an effort to match the geometry of the unit cell with the general shape of the corresponding lattice type having the same symmetry element although its primitive lattice is not a rectangular shape.

The centered rectangular lattice well illustrates the purpose of geometric nomenclature for crystal lattices. When the term "rectangular" is used in the classification of a lattice, it implies a rectangle that has at least a reflection symmetry element. Also when the term "parallelogram" is used, it connotes an only an identity or 2 fold rotational symmetry. Due to this implication, the

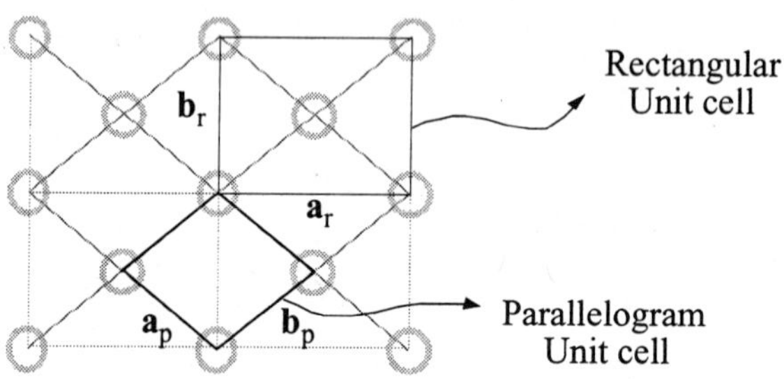

Figure 2-13 Illustration of centered rectangular lattice with circles at lattice points ($\mathbf{a}_r$, $\mathbf{b}_r$ are lattice vectors in rectangular lattice and $\mathbf{a}_p$, $\mathbf{b}_p$ are lattice vectors in primitive parallelogram lattice.)

parallelogram lattice must have a unit cell with the condition of a ≠ b and γ ≠ 90°. Hence the parallelogram unit cell shown in Fig.2-13 is not truly parallelogram unit cell. In order to match the shape of the unit cell with the lattice nomenclature, a non-primitive unit cell is chosen whose shape is rectangular. Hence geometric nomenclature for the crystal lattice is utilized to aid in visualizing the arrangement of lattice points and, therefore, in illustrating the correlation between geometry and the symmetry of lattices.

4. Crystal lattice

A crystal lattice is a 3-dimensional array of points in space, generated by non-coplanar vectors **a**, **b**, and **c** drawn from the same origin. The vectors **a**, **b**, and **c** form a primitive group. The lattice translation vector **t** = u **a**+ v **b** + w **c** locates any point in the lattice. Fig.2-14 illustrates how the crystal lattice is built by translation vector. The volumetric portion defined by these 3 primitive vectors, **a**, **b**, and **c** is called a unit cell within the space lattice.

4-1. Unit Cell

A unit cell is defined as the parallelepiped generated by 3 non-coplanar vectors $\vec{a}$, $\vec{b}$, $\vec{c}$ and included angle α,β,γ all from a common origin. $\vec{a}$, $\vec{b}$ and $\vec{c}$ are defined with respect to a coordinate system appropriate to the unit cell. A unit cell in a crystal is the smallest repeat unit necessary to fill 3-dimensional space.

4-2. Unit Cell/Coordinate System

The geometric elements that characterize a lattice are the lengths of the primitive lattice vectors and the angles between the vectors. These geometric elements (**a**, **b**, **c**, α, β, γ) are called lattice parameters of the unit cell. These are

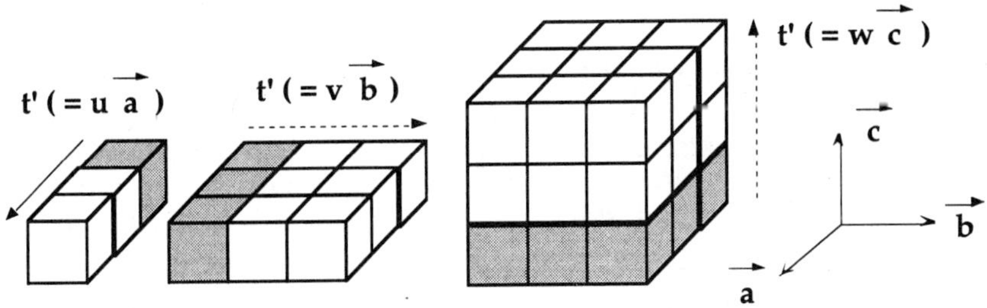

Figure 2-14 Lattice vectors $\vec{a}$, $\vec{b}$, $\vec{c}$ and a space lattice generated by translation vector

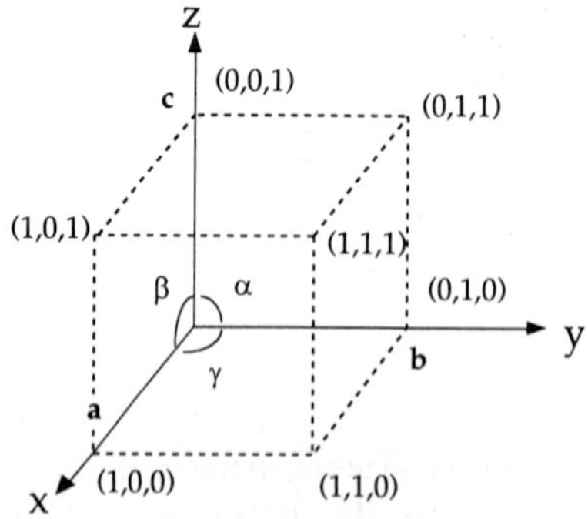

By convention

x, y, z are chosen as shown

a = unit vector along x axis

b = unit vector along y axis

c = unit vector along z axis

α = angle opposite **a** (**b** ^ **c**)

β = angle opposite **b** (**a** ^ **c**)

γ = angle opposite **c** (**a** ^ **b**)

Figure 2-15 Lattice parameters in the unit cell

illustrated in Fig.2-15 in an orthogonal coordinate system. In the orthogonal coordinate system, such as cubic, tetragonal, and orthorhombic lattice, α, β and γ have to be right angles.

5. Classification of crystal lattice

5-1. Geometry of seven lattice systems

Since the crystal lattice extends to all 3-dimensional space, this places a limit on the geometry of the unit cell, which is the repeating unit. As in the planar lattices, the geometries of the crystal lattices are confined to seven lattice types or systems as illustrated in Table 2-2. These geometric aspects contain the symmetric elements required for each lattice system. In this respect, they are not purely geometric factors, but rather symmetry-included geometric factors.

5-2. Non-primitive lattice - Bravais Lattices

With respect to the geometric shape of the unit cell, a non-primitive lattice can always be chosen to so as to match the shape of the unit cell to its name. Hence, a 3-dimensional lattice system can accommodate non-primitive lattices with primitive unit cells that do not have the corresponding geometric shapes. Of the seven lattice systems, three contain an additional point at the center (body-centered, I); two contain three additional points one in each face (face-centered, F); and two contain one additional lattice point in the opposite pair of faces (end-centered, C or A or B) while the basic geometric elements are not lost. These are illustrated in Fig. 2-16. These fourteen types of cell are known as Bravais lattices, after the French crystallographer Bravais.

Table 2-2 Geometric illustration of 7 crystal systems

Lattice system	Triclinic	Monoclinic
Geometric structure	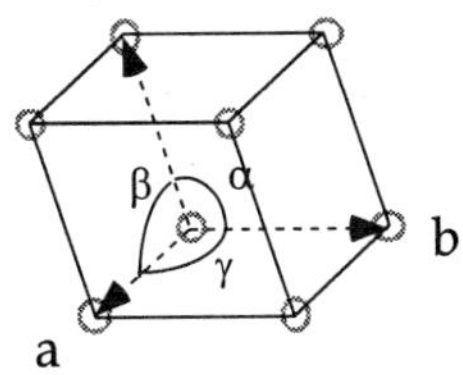	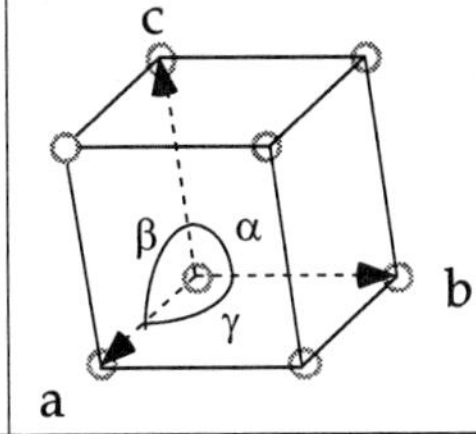
Geometry of Lattice	$\alpha \neq \beta \neq \gamma$ $a \neq b \neq c$	$\alpha \neq \beta \neq \gamma = 90^\circ$ $a \neq b \neq c$
Symmetry element of the lattice system		Mirror plane or 2 fold axis

Lattice system	Trigonal	Hexagonal
Geometric structure	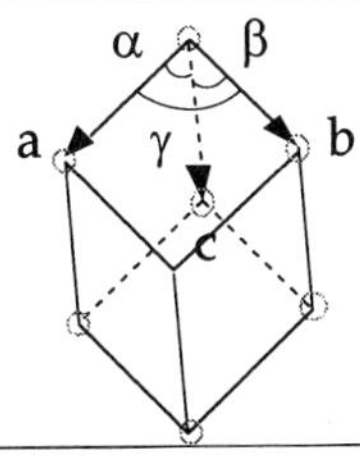	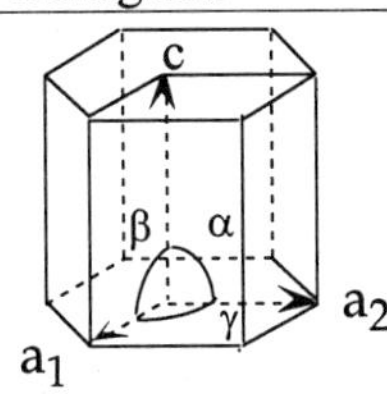
Geometry of Lattice	$\alpha = \beta = \gamma \neq 90^\circ$ $a = b = c$	$\alpha = \beta = 90^\circ$ $\gamma = 120^\circ$ $a = b \neq c$
Symmetry element of the lattice system	3 fold axis	6 fold axis

Lattice system	Orthorhombic	Tetragonal	Cubic
Geometric structure	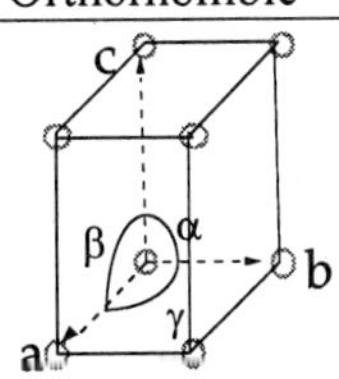	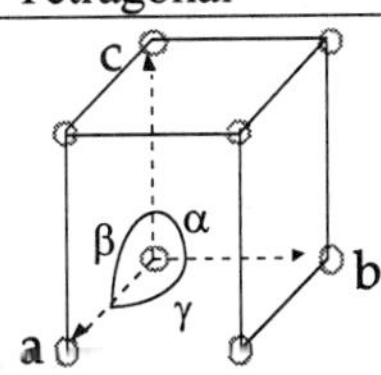	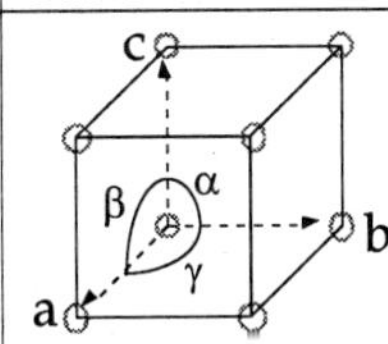
Geometry of Lattice	$\alpha = \beta = \gamma = 90^\circ$ $a \neq b \neq c$	$\alpha = \beta = \gamma = 90^\circ$ $a = b \neq c$	$\alpha = \beta = \gamma = 90^\circ$ $a = b = c$
Symmetry element of the lattice system	2 fold axes or/and mirror planes	4 fold axis	4 fold and 3 fold axis

The fourteen Bravais lattices represent mathematically the only possibilities for uniquely arranging lattice points in space. For example, there is no base centered tetragonal lattice because a base centered tetragonal lattice can be described by a smaller primitive tetragonal lattice as illustrated in Fig.2-17.

In a strict sense, the classification of crystal lattices must be related to the symmetry of crystals. In this regard, the cubic crystal lattice implies 4 fold

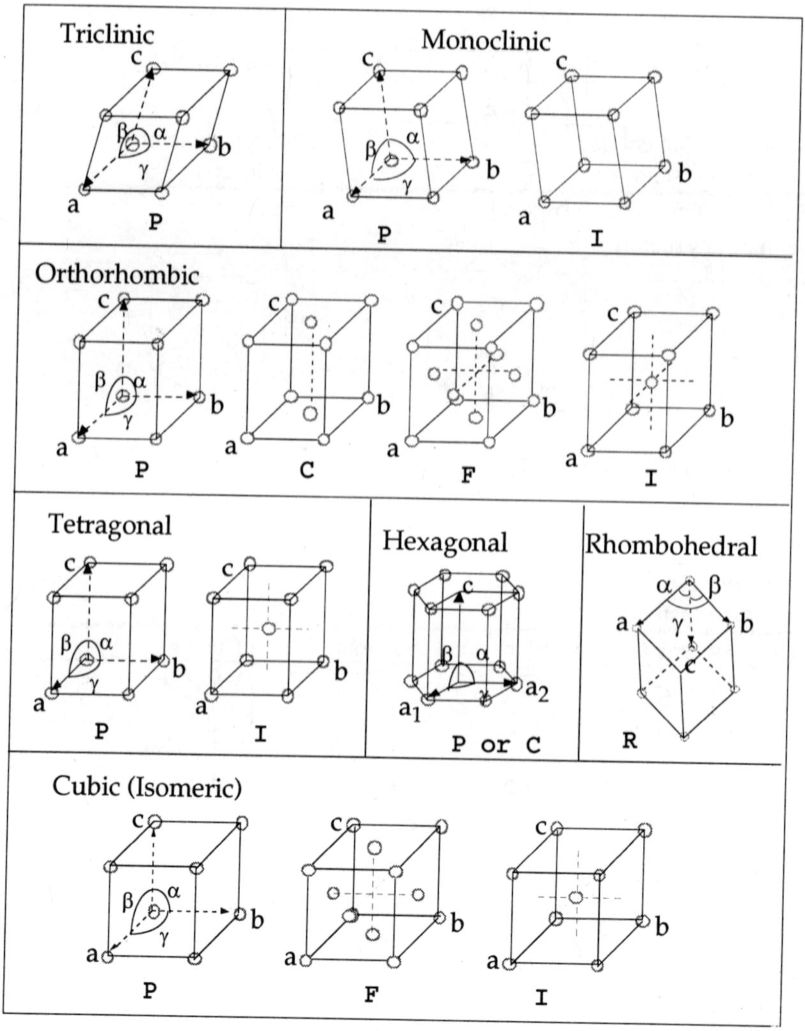

Figure 2-16 Bravais Lattices

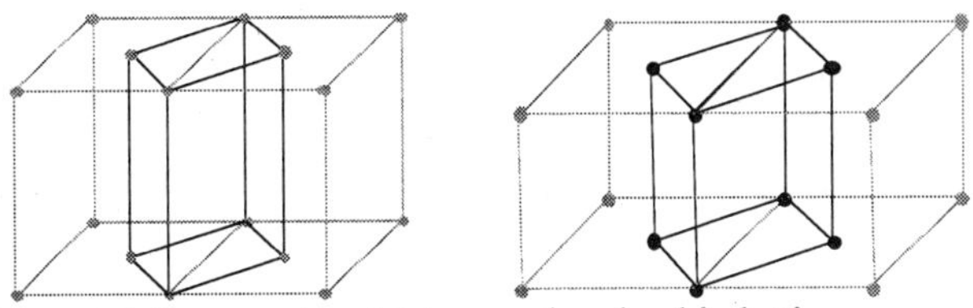

Figure 2-17 Geometric base centered tetragonal and cubic lattices.
(Dotted line indicates non-primitive lattice. Straight line indicates primitive lattice.)

symmetry along all the lattice axes. Due to this implication of additional 4 fold symmetry to the cubic lattice, there is no base centered cubic lattice since an additional lattice point at the center of the base does not satisfy the requirement for 4 fold rotation axis along the other 2 axes. Therefore the base centered cubic is regarded as a tetragonal lattice with a smaller unit cell. These geometric and symmetric aspects for non-primitive lattices for the cubic and tetragonal systems are illustrated in Fig. 2-17.

5-3. Geometric construction of 14 Bravais Lattices from planar lattices

A crystallographic lattice is formed from a repetition of a lattice point by 3 independent vectors. Since two vectors form five unique planar lattices, Bravais lattices can be developed from the five unique planar lattices by translation along a third non-coplanar vector. Fig. 2-18 illustrates the construction of Bravais lattices from planar lattices. The unit cell of the planar lattice is shown by the dark areas. Since the non-coplanar vectors repeat the planar lattice in a unique way for each Bravais lattice, the resulting lattices are unique. The third vector, which extends the planar lattice to 3-dimensional space is represented by an arrow. In some cases this vector is the same as the lattice vector of the Bravais lattice, but in such cases as the F-cubic and trigonal lattices, the translational vector is not the same. Table 2-3 illustrates the constructional relationship between the planar lattices and Bravais lattices.

As shown in Fig.2-18, face centered cubic, body centered cubic and body centered tetragonal are constructed from the square planar lattice in a similar way. The difference arises from the different magnitudes of the translation vector along the "c" axis, which is perpendicular to the square planar lattices. When the magnitude of this translational vector along the c axis is equal to the length of lattice vector of the square lattice, a body-centered cubic lattice is formed. If the length of the translational vector along the c-axis is equal to the length of the diagonal of the square lattice, the translation generates a face-centered cubic lattice. If both of these conditions are not met, a body-centered tetragonal lattice is produced.

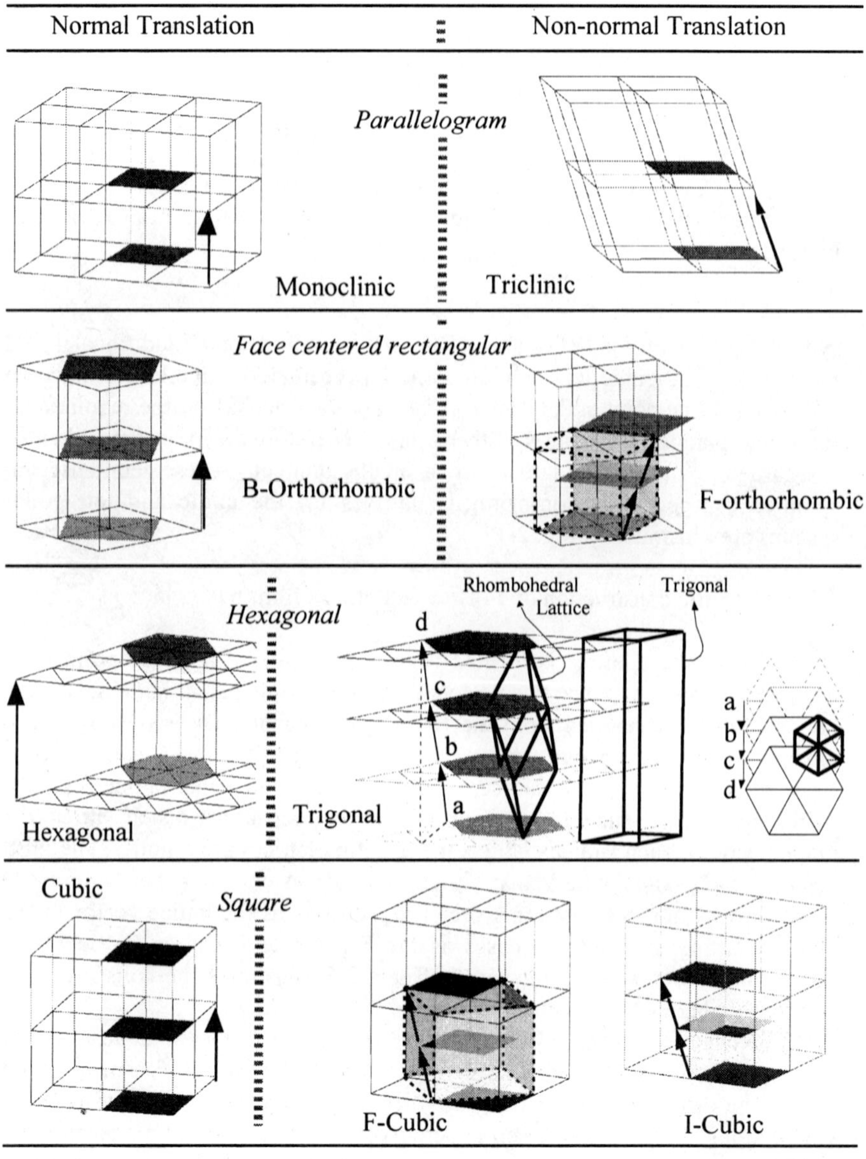

Figure 2-18 Construction of 14 Bravais Lattices by repetition of the planar lattices - dark area - along the third translation vector (P, I - Orthorhombic and P, I-Tetragonal lattices are omitted due to similarity with Cubic and I-cubic lattices.)

Table 2-3 Constructional relationship between planar lattices and Bravais lattices

Planar Lattice	Bravais Lattice
Square lattice	Cubic, I-cubic, F-cubic Tetragonal, I-tetragonal
Rectangular	Orthorhombic, I-Orthorhombic
F-Rectangular	B-orthorhombic, F-orthorhombic
Parallelogram	Monoclinic, I-Monoclinic, Triclinic
Hexagonal	Hexagonal, Trigonal

In Fig.2-18, the top view for the translation of the hexagonal lattice to make the trigonal lattice is shown since it is not clearly shown from the side perspective. The translation vector moves a hexagon in the hexagonal planar lattice by two thirds in a perpendicular direction toward an edge of the hexagon, when viewed from the top. The trigonal lattice is generated by repetition of the translational vector 3 times whereas the hexagonal lattice is generated by one translational vector.

Regardless of the crystal system, the coordinates of lattice points are as follow:

Primitive lattice : $(0,0,0)$

Base centered lattice : $(0,0,0);\ \left(\frac{1}{2},\frac{1}{2},0\right)$

Face centered lattice : $(0,0,0);\ \left(\frac{1}{2},\frac{1}{2},0\right);\ \left(\frac{1}{2},0,\frac{1}{2}\right);\ \left(\frac{1}{2},\frac{1}{2},\frac{1}{2}\right)$

Body centered lattice : $(0,0,0);\ \left(\frac{1}{2},\frac{1}{2},\frac{1}{2}\right)$

6. Volume of Unit Cells

In the analysis of crystal structure, experimental tools such as X-ray diffraction provides unit cell information, such as the length of the primitive vector, rather than on the atomic arrangements. Also it provides information on the symmetry of crystals. The atomic arrangements can be deduced from the geometric and symmetric data.

The first step towards identifying the atomic arrangements in the unit cell is to calculate the number of atoms per unit cell. For this, the volume of the unit cell is calculated as follows.

$$V = (\mathbf{a} \bullet \mathbf{b}) \times \mathbf{c} = (\mathbf{b} \bullet \mathbf{c}) \times \mathbf{a} = (\mathbf{c} \bullet \mathbf{a}) \times \mathbf{b}$$

Here **a**, **b**, and **c** are three lattice vectors. For the unit cell bounded by vectors **a**, **b**, and **c**, the volume is given by in a determinant form

$$V = (\mathbf{a}\bullet\mathbf{b})\times\mathbf{c} = \begin{vmatrix} a^2 & \mathbf{a}\bullet\mathbf{b} & \mathbf{a}\bullet\mathbf{c} \\ \mathbf{b}\bullet\mathbf{a} & b^2 & \mathbf{b}\cdot\mathbf{c} \\ \mathbf{c}\bullet\mathbf{a} & \mathbf{c}\bullet\mathbf{b} & c^2 \end{vmatrix}^{\frac{1}{2}}$$

Where $\mathbf{a}\bullet\mathbf{a} = a^2$ $\mathbf{b}\bullet\mathbf{b} = b^2$ $\mathbf{c}\bullet\mathbf{c} = c^2$. Also: $\mathbf{a}\bullet\mathbf{c} = \mathbf{c}\bullet\mathbf{a} = ac\cos\beta$, $\mathbf{b}\bullet\mathbf{c} = \mathbf{c}\bullet\mathbf{b} = bc\cos\alpha$, $\mathbf{a}\bullet\mathbf{b} = \mathbf{b}\bullet\mathbf{a} = ab\cos\gamma$
Therefore

$$V = abc + \mathbf{a}\cdot\mathbf{b}\ \mathbf{b}\cdot\mathbf{c}\ \mathbf{c}\cdot\mathbf{a} - \mathbf{c}\cdot\mathbf{a}\ b^2\ \mathbf{a}\cdot\mathbf{c} + \mathbf{a}\cdot\mathbf{c}\ \mathbf{b}\cdot\mathbf{a}\ \mathbf{c}\cdot\mathbf{a} - \mathbf{c}\cdot\mathbf{b}\ \mathbf{b}\cdot\mathbf{c}\ a^2 + b^2\ \mathbf{b}\cdot\mathbf{a}\ \mathbf{a}\cdot\mathbf{b}$$

Substituting

$$V = abc\left(1-\cos^2\alpha-\cos^2\beta-\cos^2\gamma+2\cos\alpha\ \cos\beta\ \cos\gamma\right)^{\frac{1}{2}} \quad (1)$$

In relation to the geometry of lattices, the volume of the unit cell expressed by (1) can be simplified for the 7 crystal types. This is listed in Table 2-4.

7. Density of Unit Cell

Density is the mass per unit volume and is expressed as :

$$\text{Density} = \frac{\text{mass}}{\text{volume}} = \frac{g}{cm^3} = \rho$$

Table 2-4 Unit cell volumes of crystal system

Crystal system*	Unit cell volume
Cubic	$V = a^3$
Tetragonal	$V = a^2c$
Orthorhombic	$V = abc$
Monoclinic	$V = abc\sin\beta$
Hexagonal	$V = a^2c\sin60°$
Rhombohedral	$V = a^3(1-3\cos^2\alpha+2\cos^3\alpha)^{1/2}$
Triclinic	$V = abc\ (1 - \cos^2\alpha\cos^2\beta\cos^2\gamma+2\cos\alpha\cos\beta\cos\gamma)^{1/2}$

This density is then converted to atomic density per unit cell according to Gehl's Formula

$$\rho = \frac{\sum(n_i A_i)/N_o}{V} \text{ (g/cm}^3) \qquad (2)$$

Here n_i = # of atoms (ions) of species i/unit cell
A_i = atomic wt. of species i (gm)
V = vol of unit cell (cm^3)
$N_o = 6.023 \times 10^{23}$/mol. (Avogadro's number)

From Eq.(2), the number of atoms in the unit cell can be calculated. From the symmetry aspect of the unit cell, the positions of atoms are specified, i.e., the identification of atomic positions for Na metal.

Ex1. Cu metal

X-ray diffraction indicates that Cu has the face centered cubic lattice type* and the length of the unit cell a_o = 3.615Å. The experimental measurement of density indicates that there are 4 Cu atoms in the unit cell with the above cell length. The position of Cu atoms must be at the face diagonals of the unit cell and at the origin since the number of equipoints for the symmetry space group identified be 4. Here it should be noted how symmetry helps to determine the atomic positions in crystals: This gives 4 Cu atoms/unit cell, 1 atom at (000) and 3 atoms at $\left(\frac{1}{2},\frac{1}{2},0\right)$, $\left(\frac{1}{2},0,\frac{1}{2}\right)$, and $\left(0,\frac{1}{2},\frac{1}{2}\right)$

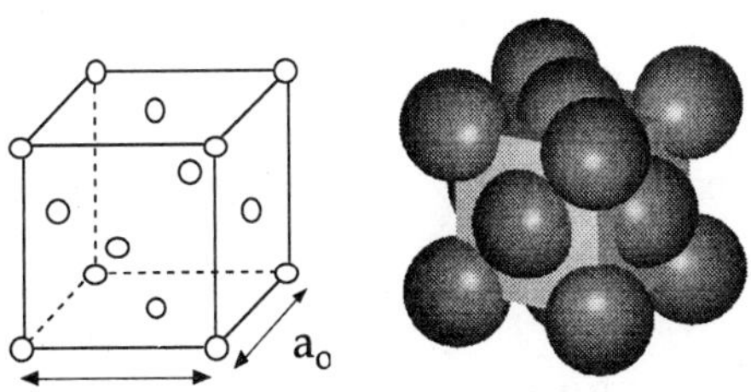

Figure 2-19 Crystal structure of Cu(metal)

* FCC is a non-primitive cubic lattice which has extra lattice points at the face diagonal positions of the unit cell. A more complete description is given in Chapter 3.

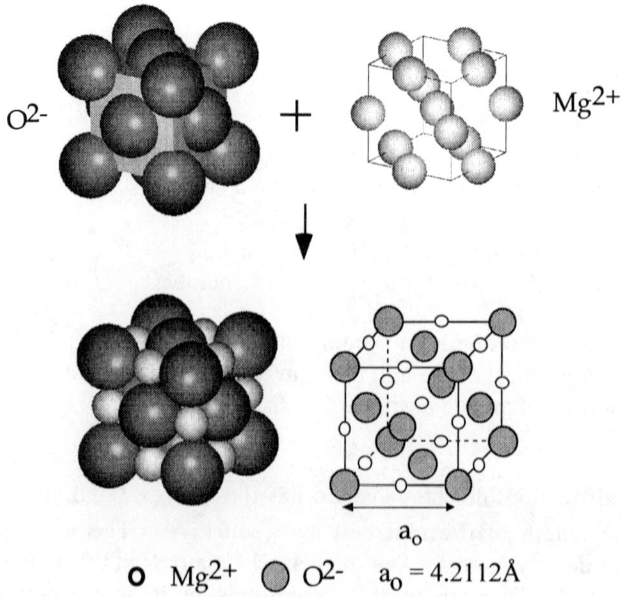

Figure 2-20 Crystal structure of MgO

Ex 2 MgO

X-ray diffraction shows that MgO also has the FCC structure with a_o = 4.2112Å. Density measurements indicate that there are 4 formula units (MgO) per unit cell. With the FCC structure, the possible positions for Mg and O are the special positions of the equipoints which number must be 4. Hence the atomic positions must be those in Fig. 2-20, i.e. 4 O^{2-} ions at (000) and the FCC positions with 4 Mg^{2+} ions at the point of the edges and at the center, $\left(\frac{1}{2},\frac{1}{2},\frac{1}{2}\right)$.

SECTION 2 BASIC CONCEPT IN CRYSTALLOGRAPHY : DIRECTIONS AND PLANES

A. Directions

A direction is specified with respect to a coordinate system (a,b,c) by the vector from the origin to the point (xyz),

$$\mathbf{p} = x\,\mathbf{a} + y\,\mathbf{b} + z\,\mathbf{c}$$

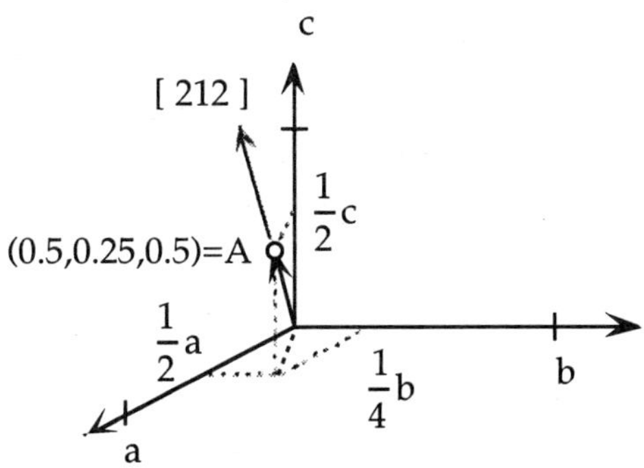

Figure 2-21 Direction [212]

where $x = n_1\,\mathbf{a}$, $y = n_2\,\mathbf{b}$, $z = n_3\,\mathbf{c}$ i.e., n_1, n_2, and n_3 are fractional intercept values along **a**, **b**, and **c** Directions are specified by indices [uvw], where [uvw] = $[n_1, n_2, n_3]$ cleared of fractions and reduced to smallest integers. For example, the direction of [212] as shown in Fig. 2-21 is derived as follow:
The coordinate of Point A is (0.5, 0.25, 0.5). The intercepts of the direction pointing to "A" with respect to a, b and c axes are $n_1 = \frac{1}{2}$, $n_2 = \frac{1}{4}$ and $n_3 = \frac{1}{2}$, respectively, since $x = \frac{1}{2}\mathbf{a}$ $y = \frac{1}{4}\mathbf{b}$, and $z = \frac{1}{2}\mathbf{c}$. Then the notation for the direction is $[uvw] = [\frac{1}{2}\ \frac{1}{4}\ \frac{1}{2}]$. By clearing the fractions, [uvw] = [212]. In other words, [212] means that the direction vector passes through a point whose coordinates is (2,1,2). This is the point closest to the origin having whole integer coordinates. Hence, a direction is represented by the point that the direction vector passes through with integer coordinates and is closest to the origin.

- Procedure for defining direction

a) Given direction coordinates (xyz)	$\left(\frac{3}{8}, \frac{1}{4}, \frac{1}{2}\right)$	$(4, 4, 2)$
b) Express as fractional coordinates (n_1, n_2, n_3)	$\left(\frac{3}{8}, \frac{1}{4}, \frac{1}{2}\right)$	$\left(1, 1, \frac{1}{2}\right)$
c) Reduce to smallest integers	(3, 2, 4)	(2, 2, 1)
d) Enclose in brackets = [uvw]	[324]	[221]

Notes:

a. A family of directions (zone axes) is indicated by:
<324> or <221>

b. A point (xyz) can lie on the line (axis):
$\vec{r} = u\,\vec{a} + v\,\vec{b} + w\,\vec{c}$
only if x, y and z are exact multiples of u, v and w
such that x = mu, y = mv, z = mw

c. For the cubic, tetragonal and orthorhombic systems, planes and directions with like indices are normal to each other.

B. Planes

Like directions, planes are very important in characterizing crystals. The necessity to define planes in crystals is due particularly to the characteristic planar faces of crystals in relation to internal structures. The following method of numerical representation of crystallographic faces is called Miller indices after the who developed them in the present form, although the method was first conceived by Whewell. Like the notation for directions in crystallography, planes are represented by a combination of three integers. While directions are represented by a square bracket, [abc], planes are noted in *parentheses*, (hkℓ). Here a, b, c, h, k, and ℓ are all integers. With this numerical notation for planes characterization of crystal habits and mathematical manipulation of crystallographic indices for crystal structure analysis in schematic and simple ways has been made possible.

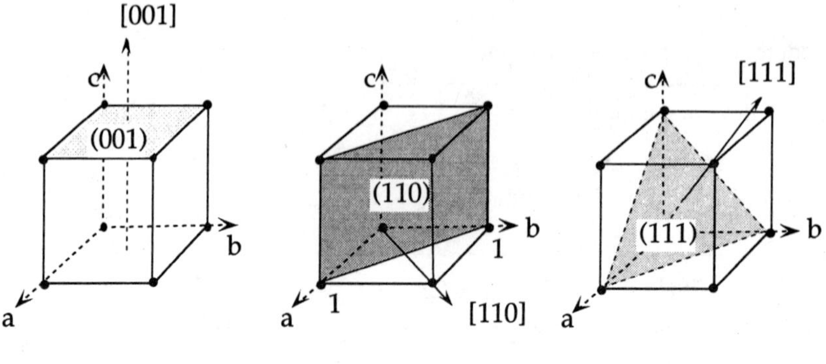

Figure 2-22 Planes (001), (110), and (111)

Miller indices for a crystal plane are defined from the intercepts of the plane with the three crystal axes. Later it will be shown that a crystal plane can be represented by a direction vector that is perpendicular to the plane. The direction of the vector has the same notation as the Miller indices for the plane when the vector perpendicular to the plane is expressed on the basis of the reciprocal lattice. This will be explained in the section on the reciprocal lattice. Hence directions and planes can be handled by vectorial notations, and can be manipulated mathematically, since they have the same dimensional characteristics.

1. Miller Indices

In crystallography planes are described by the Miller Indices. The Miller index for a plane is a set of 3 integers $(hk\ell)$ that describes the orientation of the plane with respect to coordinate axes **a, b, c**(defining a unit cell). The designation of the plane ABC in terms of $(hk\ell)$ is given by $(hk\ell) = (\frac{1}{n_1}, \frac{1}{n_2}, \frac{1}{n_3})$, when the indices are cleared of fractions, reduced to smallest integers and enclosed in parenthesis. Thus, for the plane ABC in Fig.2-23, the intercepts are:

$$A = 1\mathbf{a} = n_1\mathbf{a}$$

$$B = \frac{1}{3}\mathbf{b} = n_2\mathbf{b}$$

$$C = \frac{1}{2}\mathbf{c} = n_3\mathbf{c}$$

where: n_1, n_2, n_3 represent general intercept values along **a, b, c** (or x, y, and z) axes.

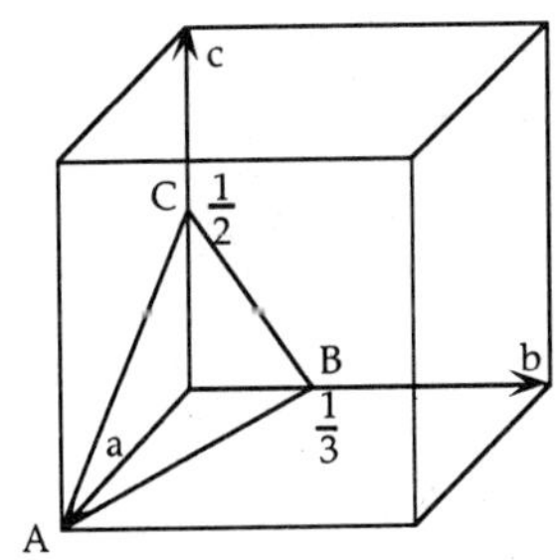

Figure 2-23 Intercepts of plane (132)

Example: Procedure - Plane ABC in Fig. 2-23

a) Given intercepts on **a, b, c** or xyz axes

$$\text{e.g.,} \quad 1, \frac{1}{3}, \frac{1}{2}$$

b) Express in terms of unit coordinate intercepts $n_1n_2n_3$:

$$1, \frac{1}{3}, \frac{1}{2}$$

c) Take reciprocals

$$\frac{1}{1}, \frac{3}{1}, \frac{2}{1}$$

d) Clear fractions and reduce to smallest integers

1 3 2

e) Enclose in parentheses → (hkℓ)

(1 3 2)

2. Planes Parallel to Axes

Planes parallel to one or more coordinate axes in a unit cell do not intercept with those axes. Miller indices for these planes, therefore, cannot be represented as shown in the procedure above. Where this condition exists, the plane is considered to intercept the axis at infinity. Hence the value of the intercept is considered as ∞. The following example illustrates how to calculate (hkℓ) for a plane parallel with one of lattice axes as in Fig. 2-24.

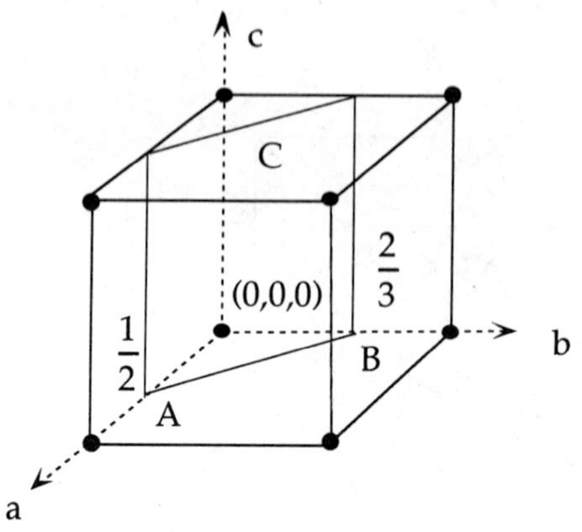

Figure 2-24 Plane parallel with c-axis

For the plane, intercepts are:

$$A = \vec{a} : n_1 = \frac{1}{2}, B = \frac{2}{3}\vec{b} : n_2 = \frac{2}{3}, C = \infty\vec{c} : n_3 = \infty$$

Therefore

$$(n_1, n_2, n_3) = \left(\frac{1}{2}, \frac{2}{3}\ \infty\right) = \left(\frac{1}{n_1}, \frac{1}{n_2}, \frac{1}{n_3}\right) = \left(\frac{1}{\frac{1}{2}}, \frac{1}{\frac{2}{3}}, \frac{1}{\infty}\right) = \left(2, \frac{3}{2}, 0\right)$$

The Miller index for the plane is given as $(hk\ell) = (4\ 3\ 0)$

Note: $(hk\ell)$ represents the <u>first</u> plane out from the origin, i.e., the plane closest to the origin.

3. Drawing planes from Miller Indices

Given $(hk\ell)$, determine the intercepts of the plane by calculating a fraction with an inverse of h, k and ℓ. Recall $(h, k, \ell) = (1/n_1, 1/n_2, 1/n_3)$. Therefore $n_1 = 1/h$, $n_2 = 1/k$, $n_3 = 1/\ell$. i.e., the intercept values on **a, b, c** are the reciprocal of the indices $(hk\ell)$. Draw a plane connecting three intercepting points n_1x, n_2y, and n_3z with x, y and z axes.

<u>Example</u>: $(hk\ell) = (241)$ and (130)

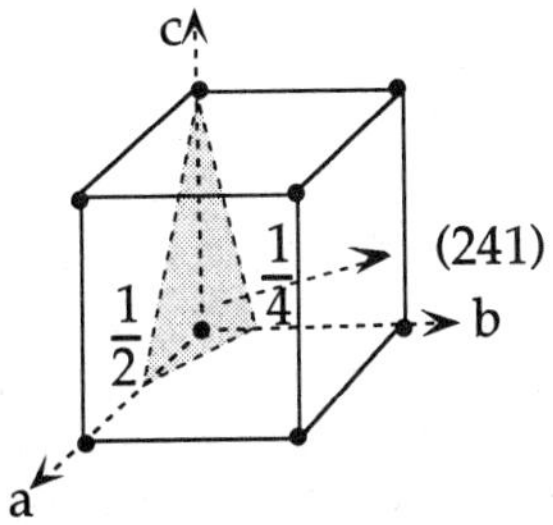

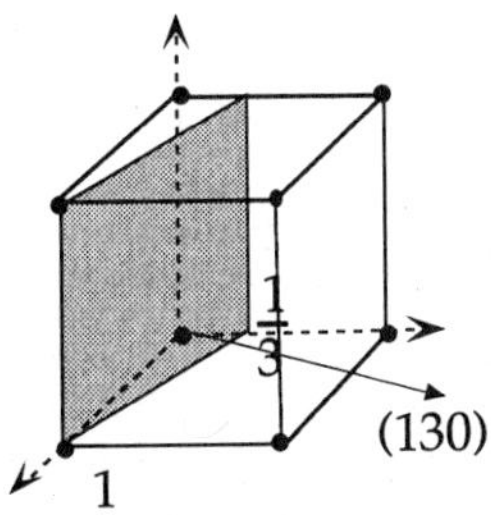

$$n_1 = \frac{1}{h} = \frac{1}{2} \qquad n_1 = \frac{1}{h} = \frac{1}{2}$$

$$n_2 = \frac{1}{k} = \frac{1}{4} \qquad n_2 = \frac{1}{k} = \frac{1}{4}$$

$$n_3 = \frac{1}{\ell} = \frac{1}{1} = 1 \qquad n_3 = \frac{1}{\ell} = \frac{1}{0} = \infty$$

Note that the directions [241] and [130] are perpendicular to planes (241) and (130), respectively.

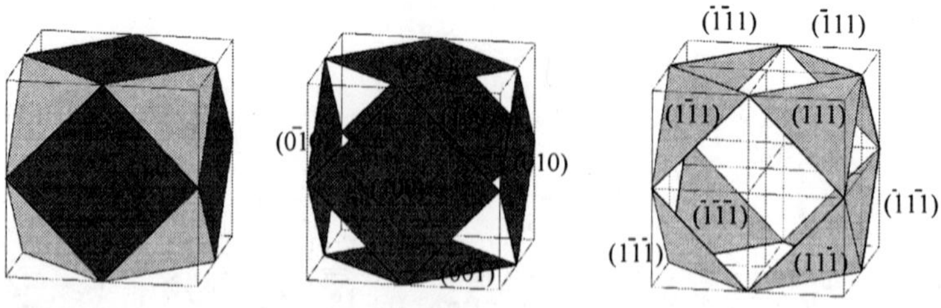

Figure 2-25 Two crystal forms {100} and {111} consisting of a crystal habit in the cubic system

4. Form

A form of a crystal is a set of planes designated by {hkℓ} that includes all the planes of like orientation. In the cubic crystal system, there is no difference among a, b, and c axes. Hence the crystal form {hkℓ} implies that h, k, and ℓ can be permutated in a bracket to generate planes belonging to a crystal form.

Thus planes in the forms of {110}, {111}, and {200} are

$$\{110\} = (110)\,(101)(011)(\bar{1}10)(\bar{1}01)(0\bar{1}1)(1\bar{1}0)(10\bar{1})(01\bar{1})(\bar{1}\bar{1}0)(\bar{1}0\bar{1})(0\bar{1}\bar{1})$$

$$\{111\} = (111)(\bar{1}11)(1\bar{1}1)(11\bar{1})(\bar{1}\bar{1}1)(\bar{1}1\bar{1})(1\bar{1}\bar{1})(\bar{1}\bar{1}\bar{1})$$

$$\{200\} = (200)(020)(002)(\bar{2}00)(0\bar{2}0)(00\bar{2})$$

In relation to crystal habits, outer surfaces of a crystal consist of several forms. Sometimes only planes of the same form are enough to cover all the surfaces. Fig.2-26 illustrates the progress of an NaCl crystal from the octahedral shape to the cubic shape. In the octahedral shape, the outer surfaces consist of the {111} form. In the cubic shape as shown in (e), the outer surfaces consist of the {100} form. The intermediate stage exhibits the coexistence of both {100} and {111} forms. This coexistence is illustrated in Fig.2-25.

The importance of a crystal form is that all the planes of the same form appear on a crystal although each plane of a form sometimes changes in size. In such a case as a crystal grown on a substrate, some of planes may not be developed due to restriction on the development along the direction towards the substrate. However, in a crystal grown under isotropic condition, planes of a form are expected to develop simultaneously as shown in Fig. 2-25. The simplified unit cell arrangements shown in the lower part illustrate planes of a form having the same configuration and equivalent to each other in this respect.

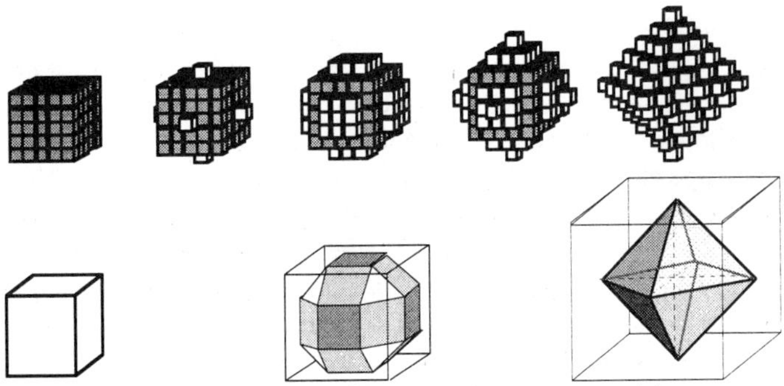

Figure 2-26 Crystal habit development in cubic crystal system from cubic to octahedral shape

5. Crystal habit and planes

As discussed, a crystal is matter with a repeated pattern of atomic arrangements such as the unit cell. Usually this internal structure of repetition of atoms or ions is related to one of the prominent characteristic of crystal such as external shape of a crystal. For example, quartz has the hexagonal crystal system and its crystal geometry takes the form of hexagonal cylindrical shape in that six planar faces along the length of a quartz crystal make 120° between adjacent faces. This relationship, however, does not mean that the crystal habit is the magnified image of the unit cell. In fact, crystal geometry is usually different from that of the unit cell. Fig.2-26 illustrates how different crystal habits can be developed in a cubic crystal unit cell.

Crystals are not formed in the same way as a gas balloon which can expand isotropically along all directions, keeping its shape regardless of its volume. For crystals, growth is by addition of constituent atoms or ions from the outside in sequential movements. As seen in Section 2.E. of Chapter 1, atoms achieve extra stabilization by forming molecules, they achieve the lowest energy state by forming crystals. By the same token, atoms or ions on the surface of a crystal have a higher energy state than those within the crystal. This energy difference is called surface energy of the liquid or solid and is the driving force for tending towards the smallest surface area. In crystal growth, surface energy acts in a similar way although it does not change the shape by moving atoms within crystals. In the other words, crystals grow in directions that generates faces with lowest surface energy. This explains why a crystal develops characteristic planar faces. Since crystals have a lattice structure, a planar face in a crystal must be accompanied by at least one more planar surface in the opposite site. For example, in the crystal shape in Fig.2-26(c) for the cubic crystal system, all the faces have the same atomic arrangement due to the same unit cell structure. Hence each face should have the same surface energy which is developed to the same extent during crystal growth.

The shape change illustrated in Fig.2-26(a) - (c) can be considered as the transition from cubic to octahedral. This occurs when a NaCl crystal is first formed in pure aqueous solution and then grown in a solution with an admixture of urea. The shape of the crystal during the intermediate stage of this transition will be that shown in Fig.2-26(b). These figures illustrate the pairwise development of faces with the same atomic arrangement. The number of pairs and geometric configuration of those faces are a clue to identifying the internal structure. Also this is the reason why crystals have characteristic external shapes. This manifestation of a certain type of shape defined by several planar faces is called "crystal habit."

The relationship between crystal habits, internal structures, planar faces and directions are basic elements in crystallography. Usually a plane is represented by Miller indices derived from the intercepts of the plane with 3 lattice vectors. These indices are the same as the notation for the direction when the direction vector is expressed with respect to the reciprocal lattice vectors instead of (regular) lattice vectors. This relationship of equivalence between the crystal planes and directions will be explained in the section on reciprocal lattices.

6. Important planes in a crystal

Usually important planes are those with low indices such as 0,1 and 2. With respect to the crystal system, important planes are those that exhibit the characteristics of the crystal system, for example in the cubic crystal system, planes of (100), (010), (100), (100), (100), (100), as outer faces of a crystal.

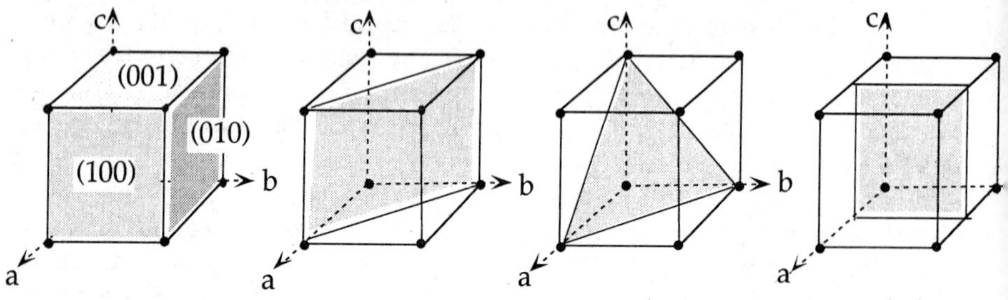

(a) (b) (c) (d)

Figure 2-27 Special planes (a){100} (b){110}, (c){111} and (d) {200} in cubic crystal system

Since these indices are the reciprocal of the intercepts with crystal vectors, low indices mean that the distance of the plane from the origin is large. The distance gets larger with lower indices. (100) has a larger distance than (110) and (200) from the origin regardless of the crystal system. This can be applied to planes having different indices related to only one crystal axis. The distance from the origin to planes (110) and (200) cannot be simply compared since their indices are different for both a and b axes.

The importance of these low-index planes is their distinguished geometry with respect to the crystal vectors or lattice. As shown in Fig. 2-27, (100) is parallel with axes b and c. (100), (010), and (001) are perpendicular to each other. When crystal habits consist of these planes, the crystal takes cubic shape. In the cubic crystal system, (100), (010), and (001) planes are indistinguishable since they have the same crystal configurations and atomic arrangements on these planes. A group of planes that are symmetrically equivalent or have the same atomic arrangements is called a "form." A crystal form of {100} in a cubic crystal system represents a group of crystal planes of (100), (010), and (001). However, in the orthorhombic crystal system, (100), (010), and (001) do not constitute a form since they are not equivalent, having different atomic arrangements and crystal orientations.

C. Miller-Bravais indices in hexagonal system

1. Introduction

The hexagonal system is referenced by 4 axes instead of 3 axes. These four digit indices arise from the inconsistency in the representation of a form by permutation of indices in the hexagonal crystal system. As shown in the cubic crystal system, (khl) as well as (hkl) belong to the same form since a_1 and a_2 are

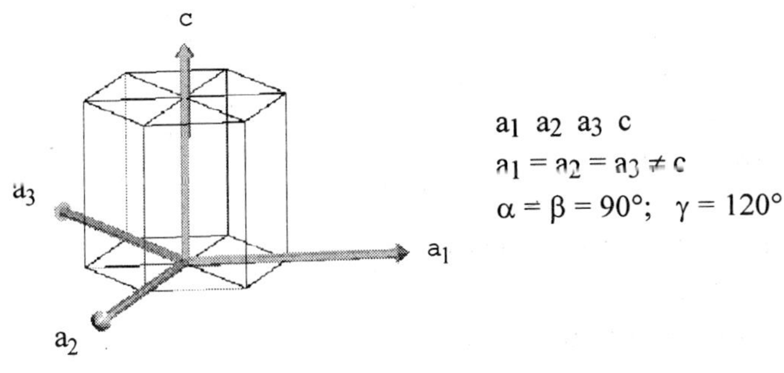

Figure 2-28 Four lattice axes in hexagonal system

equivalent to each other. This permutation of indices between the equivalent axes does not apply to the hexagonal system. For example, planes in the form of {100} are not only (100) and (010) but also (110). The inclusion of $(1\bar{1}0)$, $(\bar{1}10)$ in the form of {100} indicates the inconsistency in representation of a form in the hexagonal lattice. By implementing another index or axis "a_3" shown in Fig.2-28, representation of a form by permutation of indices becomes to be consistent.

Equation for a plane in the Miller Bravais system is given by :

$$h\,\vec{a}_1 + k\,\vec{a}_2 + i\,\vec{a}_3 + \ell\,\vec{c} = 1$$

Thus the indices are: $(hki\ell)$

where $i = -(h+k)$ note: $h + k + i = 0$

Thus "i" is redundant, but this notation ensures that forms of the sort: $\{10\bar{1}0\}$ represent a family of planes. By permutating between h, k and i, planes of a form can be represented. For example, the form {100} denotes $\{10\bar{1}0\}$ with respect to Miller-Bravais indices. By permutating 1,0,-1 for h, k and i, the planes $(10\bar{1}0)$ and $(1\bar{1}00)$ are expected to be planes belonging to the form that cannot be indicated from the permutation of Miller indices, (100). For the form of $\{hki\ell\} = \{10\bar{1}0\}$, the planes in this form are

$$(hki\ell) = (10\bar{1}0),\ (\bar{1}100),\ (01\bar{1}0),\ (0\bar{1}10),\ (\bar{1}010),\ (1\bar{1}00)$$

2. Miller-Bravais Directions

The direction vector from the point (0, 0, 0) to an indicated point xyiz (or uviw) is given by:

$$\vec{t} = u\,\vec{a}_1 + v\,\vec{a}_2 + i_3\,\vec{a}_3 + w\,\vec{c}$$

where $i = -(u+v)\ \therefore\ u + v + i = 0$

3. Procedure - Directions

a. Write the intercept values in terms of all <u>4</u> axes.
b. Clear of fractions.
c. Enclose in square brackets $[hki\ell]$
d. (Do not take reciprocals.)

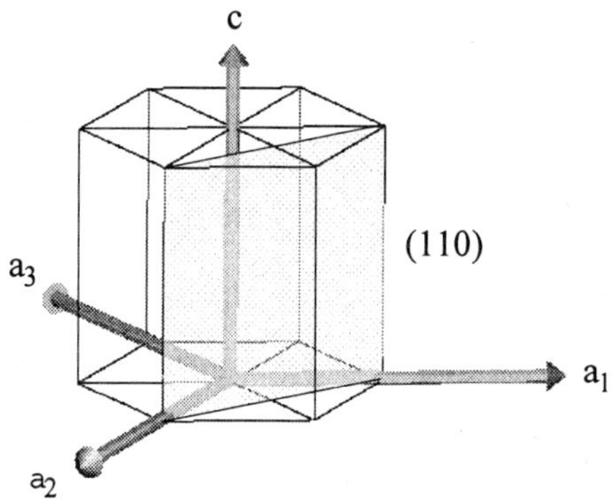

Intercepts for the given plane:

$$A_1 = 1a_1\,; \qquad n_1=1$$
$$A_2 = 1a_2\,; \qquad n_2=1$$
$$A_3 = -\frac{1}{2}a_3\,; \qquad n_3 = -\frac{1}{2}$$
$$C = \infty\,; \qquad n_4 = \infty$$

$$(hki\ell) = \left(\frac{1}{1}, \frac{1}{1}, \frac{1}{-\frac{1}{2}}, \frac{1}{\infty}\right) = (11\bar{2}0)$$

Figure 2-29 Plane $(11\bar{2}0)$ in the hexagonal system

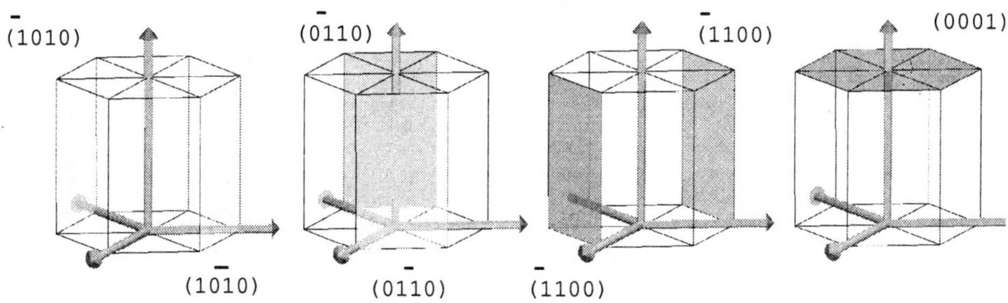

Figure 2-30 Illustration of some important planes with four indices

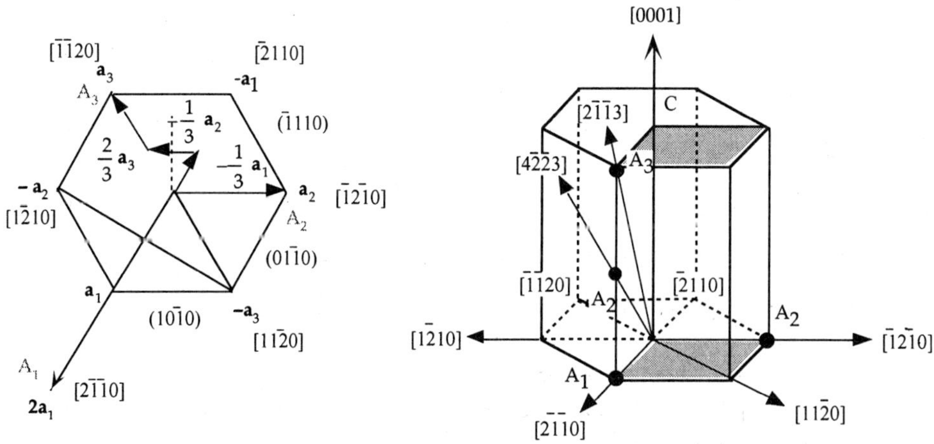

Figure 2-31 Illustration of some important directions in hexagonal lattice

Table 2-5 The Intercepts of the unit vector (A_1,A_2,A_3) of hexagonal lattice

Direction	n_1	n_2	n_3	n_4	[uviw]
A_1	2	-1	-1	0	$[2\bar{1}\bar{1}0]$
	$\frac{2}{3}$	$-\frac{1}{3}$	$-\frac{1}{3}$	0	
A_2	$-\frac{1}{3}$	$\frac{2}{3}$	$-\frac{1}{3}$	0	$[\bar{1}2\bar{1}0]$
A_3	$-\frac{1}{3}$	$-\frac{1}{3}$	$\frac{2}{3}$	0	$[\bar{1}\bar{1}20]$
C	0	0	0	1	[0001]
A'_1	$\frac{1}{3}$	$-\frac{1}{3}$	$-\frac{1}{3}$	1	$[2\bar{1}\bar{1}3]$
A'_2	$-\frac{2}{3}$	$-\frac{1}{3}$	$-\frac{1}{3}$	$\frac{1}{2}$	$[4\bar{2}\bar{2}3]$

4. 3 indices vs 4 indices

(i) Use of 3 indices [UVW] vs 4 indices [uviw] or (HKL) vs (hkiℓ)

a. 3 indices are favored by some authors for the hexagonal system.

b. Directions and planes referenced only to the primitive unit cell.

$$a_1 = a_2 \neq c$$
$$\alpha = \beta = 90°; \quad \gamma = 120°$$

c. Leads to some obscuring of connections between directions and planes.

<u>Example</u> - 3 vs 4 Indices

Planes:	<u>3 Indices</u>		<u>4 Indices</u>
1)	$(010)\ (100)\ (\bar{1}00)\ (0\bar{1}0)\ (\bar{1}10)\ (1\bar{1}0)$	→	$\{10\bar{1}0\}$
2)	$(110)\ (\bar{1}20)\ (\bar{2}10)\ (\bar{1}\bar{1}0)\ (1\bar{2}0)\ (2\bar{1}0)$	→	$\{2\bar{1}\bar{1}0\}$

The planes in 1) and 2) are parallel in the hexagonal system, but appear to belong to different "forms" with three indices. Four indices show the equivalency more clearly.

ii) Three and Four Index Notation and Transformation

a) Directions

In the hexagonal system, a vector can be expressed by a three-index notation [UVW] or a four-index notation [uviw]. The vector [UVW] means that it is the same direction as $U\mathbf{a}_1 + V\mathbf{a}_2 + W\mathbf{c}$. Likewise, the vector [uviw] is in the direction of $u\mathbf{a}_1 + v\mathbf{a}_2 + i\mathbf{a}_3 + w\mathbf{c}$. From the coordinate relationships in a hexagonal system, it is known that $\mathbf{a}_3 = -(\mathbf{a}_1 + \mathbf{a}_2)$.

Thus the vector **A** = [u v i w] can be represented by a three-index notation, since:

$$\begin{aligned}\mathbf{A} &= u\mathbf{a}_1 + v\mathbf{a}_2 + i\,\mathbf{a}_3 + w\mathbf{c} \\ &= u\mathbf{a}_1 + v\mathbf{a}_2 - i\,(\mathbf{a}_1 + \mathbf{a}_2) + w\mathbf{c} \\ &= U\mathbf{a}_1 + V\mathbf{a}_2 + W\,\mathbf{c}\end{aligned} \tag{3}$$

$$\text{Therefore: } \begin{cases} U = u - i \\ V = v - i \\ W = w \end{cases} \tag{4}$$

b) Planes

A plane in the hexagonal system can be expressed by Miller-Bravais indices (hkiℓ) if it follows the equation:

$$h\,\vec{a}_1 + k\,\vec{a}_2 + i\,\vec{a}_3 + \ell\,\vec{c} = 1$$

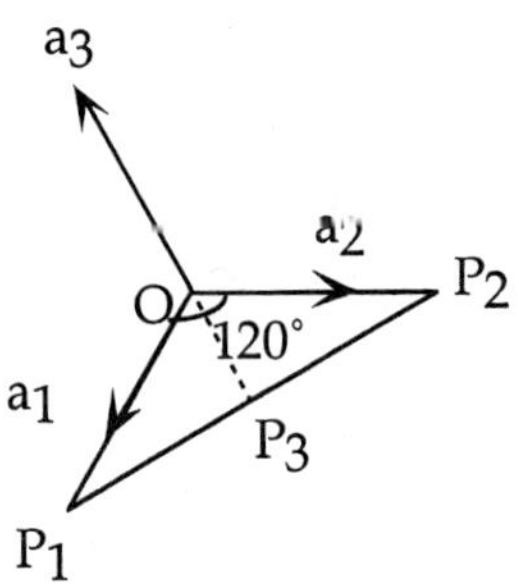

Intercepts

$$OP_1 = \frac{1}{h} \qquad OP_2 = \frac{1}{k} \qquad OP_3 = \frac{1}{i} \tag{5}$$

Figure 2-32 Relationship between h, k and i

For the plane (hkiℓ) the intercepts with axes $\vec{a}_1$, $\vec{a}_2$, $\vec{a}_3$ and $\vec{c}$ are

$$\frac{1}{h}, \frac{1}{k}, \frac{1}{i} \text{ and } \frac{1}{\ell}$$

To plot the intercepts of plane (hkiℓ) and the basal plane, we have:

Since area (OP_1P_2) = area (OP_1P_3) + area (OP_3P_2) (6)

i.e.,

$$\left(\frac{1}{h} \bullet \frac{1}{k}\right)\sin120^\circ = \left(\frac{1}{h} \bullet \frac{-1}{i}\right)\sin60^\circ + \left(\frac{-1}{i} \bullet \frac{1}{k}\right)\sin60^\circ \quad (7)$$

Since $\sin120^\circ = \sin60^\circ = \frac{\sqrt{3}}{2}$, Eq.(7) above can be reduced to:

$$i = -(h+k) \quad (8)$$

It can be seen, therefore, that the third index in the (hkiℓ) notation is somewhat redundant since it is determined by the h and k values. Therefore, (hkiℓ) is sometimes written in shorthand as (hk.ℓ) or in less regular (or strict) notation (hkℓ) that looks like a three-index notation in a cubic system.

c) Transformation Relationships

It is more convenient to use the four-index vector in expressing the normal vector to a plane in a hexagonal system. In general, the normal vector of the plane (hkiℓ) is equal to [hkiℓ]. e.g., vector $[10\bar{1}0]$ is perpendicular to the plane $(10\bar{1}0)$ or (1010). The three-index notation is simply derived by using the **a**, **b**, and **c** vectors to indicate the position of a point relative to the origin without taking into account the a_3 axis. It is easier, therefore, to use a three-index vector notation. However, the vector **B** = (U V W) can be represented by a four-index notation. The relevant expressions are:

$$\begin{aligned} \mathbf{B} &= U\mathbf{a_1} + V\mathbf{a_2} + W\mathbf{c} = u\mathbf{a_1} + v\mathbf{a_2} + t\,\mathbf{a_3} + w\mathbf{c} \\ &= u\mathbf{a_1} + v\mathbf{a_2} - t\,(\mathbf{a_1} + \mathbf{a_2}) + w\mathbf{c} \\ &= (u-t)\,\mathbf{a_1} + (v-t)\,\mathbf{a_2} + w\mathbf{c} \end{aligned} \quad (9)$$

where t can be chosen analogous to i = - (h+k) i.e.,

$$t = -(u+v) \quad (10)$$

from equations (9) and (10), we have

$$U = 2u + v \quad (11\text{-a})$$

$$V = 2v + u \quad (11\text{-b})$$

$$W = w \quad (11\text{-c})$$

Solving for u and v by equations (11-a) and (11-b), gives:

$$2 \times (11\text{-a}) - (11\text{-b})$$

$$2U - V = 3u$$

$$u = \frac{1}{3}(2U-V)$$

Likewise

$$v = \frac{1}{3}(2V-U)$$

Finally, the transformation relationships become:

$$u = \frac{1}{3}(2U-V)$$

$$v = \frac{1}{3}(2V-U)$$

$$t = -(u+v) = -\frac{1}{3}(U+V)$$

$$w = W \qquad (12)$$

An example will demonstrate the use of these notations to express the vectors and planes in the hexagonal system.

In Fig.2-33, vector **ON** is ⊥ to the plane PQRS. To find the Miller-Bravais indices for the plane PQRS one simply notes the intercepts of plane and vector bases, (that is 1, ∞, -1, ∞)

$$\therefore\ h:k:i:\ell = \frac{1}{1}:\frac{1}{\infty}:\frac{1}{-1}:\infty = 1:0:-1:0$$

Therefore the index for the plane PQRS is $(10\bar{1}0)$. Checking the relation i = – (h+k)

$$i = -1 = -(h+k) = -(1+0).$$

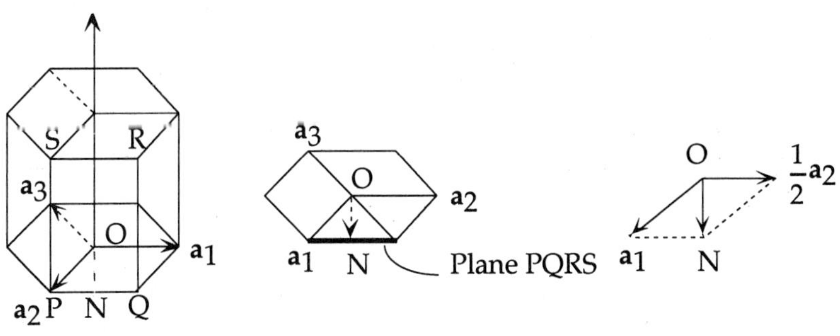

Figure 2-33 Relationship between plane $(10\bar{1}0)$ and direction $[10\bar{1}0]$

It is easy to find a three-index notation for the vector **ON** that is

$$\mathbf{ON} = \mathbf{a_1} + \frac{1}{2}\,\mathbf{a_2} + 0\,\mathbf{c} \qquad \therefore\ \mathbf{ON}\ \text{is}\ [210].$$

By using equations (10) **ON** can be written as

$$u = \frac{1}{3}(2\times 2 + 1) = 1$$

$$v = \frac{1}{3}(2\times 1 - 2) = 0$$

$$t = -(u+v) = -1, \quad w = 0$$

$\therefore$ **ON** is equal to $[10\bar{1}0]$

It is seen in this example that the vector $[10\bar{1}0]$ is $\perp$ to the plane $(10\bar{1}0)$

d. Summary

1) Any plane $(hki\ell)$ in a hexagonal system can be written as $(hk\ell)$ where $i = -(h+k)$.

2) A vector [uvtw] will be $\perp$ to plane $(hki\ell)$ if $u:v:t:w = h:k:i:\ell$.

3) A vector in the hexagonal system can be expressed either as a three-index notation [uvw] or a four-index notation [uvtw]. The transformation relationships are according to eqs. (2) and (9). A three-index notation for a vector is obtained by "decomposing" it to $\vec{a}_1$, $\vec{a}_2$ and $\vec{a}_3$ directions. A four-index vector notation might be easily found if the plane normal to this vector is known. For example vector $\left[21\bar{3}1\right]$ can be readily obtained if we know it to be $\perp$ to the plane $\left(21\bar{3}1\right)$.

4) Note that there is no such thing as three-index notation (HKL) for a plane in the hexagonal system. Hence there is <u>no</u> analog to the transformation relations given in equation (2),

i.e., $H = h\text{-}i$. $K = k\text{-}i$ and $L = \ell$. The plane in the hexagonal system is <u>only</u> expressed in the four-index notation following the Miller-Bravais indices method. The appropriate notation is $(hki\ell)$ or $(hk\ell)$. Use of the abbreviated from $(hk\ell)$ is not recommended as it can lead to misunderstanding.

D. Reciprocal lattices

1. Mathematical manipulation of vectors

A vector is a geometrical entity that has a direction and magnitude. Hence it has a starting and ending point. The magnitude of a vector is the length between starting point and ending point. The direction is defined by the coordinates of the ending point when the starting point is put at the origin of the coordinate system. Since vectors have magnitudes, they can be manipulated mathematically as numbers, but this manipulation is different than with simple numbers since vectors also have directions. In crystallography, the summation, subtraction and product of vectors are important. These vector operations are explained as follows.

1-1. Sum of vectors

$$\mathbf{a} = (x_a, y_a, z_a),\ \ \mathbf{b} = (x_b, y_b, z_b)$$
$$\mathbf{a} + \mathbf{b} = (x_a, y_a, z_a) + (x_b, y_b, z_b) = (x_a + x_b, y_a + y_b, z_a + z_b)$$

1-2. Subtraction of vectors

$$\mathbf{a} - \mathbf{b} = (x_a, y_a, z_a) - (x_b, y_b, z_b) = (x_a - x_b, y_a - y_b, z_a - z_b)$$

1-3 Product of vectors

a. Scalar product of vectors ($\mathbf{a} \bullet \mathbf{b}$)

Scalar product of two vectors is defined as

$$\mathbf{a} \bullet \mathbf{b} = a\, b \cos\theta = x_a x_b + y_a y_b + z_a z_b$$

Here "a" and "b" are the magnitudes of vectors **a** and **b**, respectively and θ is the angle between vectors **a** and **b**.

b. Vectorial products of vectors ($\mathbf{a} \times \mathbf{b}$)

The vectorial product of two vectors (**a, b**) is perpendicular to both vectors. This means that this vector is perpendicular to the plane that includes both **a** and **b**. The magnitude of the vector product, $\mathbf{a} \times \mathbf{b}$, is "ab sin$\alpha$" in that α is the angle between vectors **a** and **b**. The direction of the vector product follows the right-hand rule, which means the thumb indicates the direction of the vector product when the other fingers follow the direction of the arrow as shown in Fig. 2-34. The arrow starts from the first vector in the product. Hence the direction of the vector product of $\mathbf{b} \times \mathbf{a}$ is the reverse of that for $\mathbf{a} \times \mathbf{b}$. (See Fig.2-34)

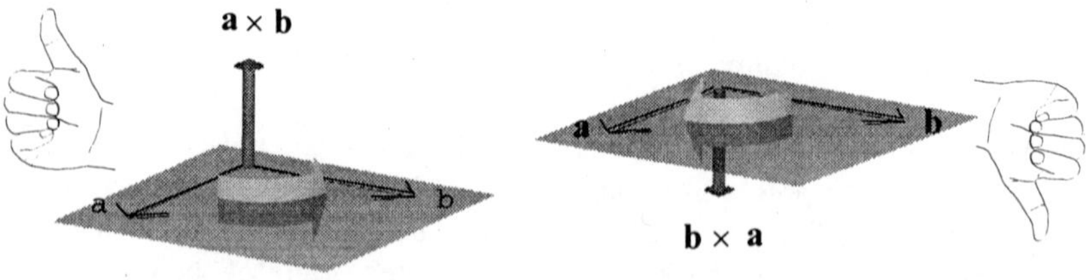

Figure 2-34 Direction of a vector generated by a product of two vectors

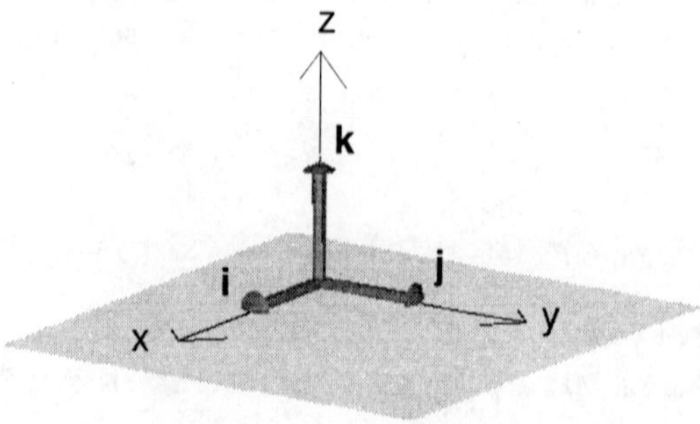

Figure 2-35 Configuration of three axes in the rectangular coordinate system to self-generate between unit vectors by vectorial product as in Eq.(13)

In the rectangular coordinate system a vector can be expressed in terms of three unit vectors that are along x, y and z axes. These unit vectors are named **i, j** and **k** as shown in Fig. 2-35. Among the unit vectors, **i, j** and **k**, the vectorial product of two vectors results in the third. The vector product of the same vector is '0'. Thus :

$$\mathbf{i} \times \mathbf{j} = \mathbf{k} \qquad \mathbf{j} \times \mathbf{k} = \mathbf{i} \qquad \mathbf{k} \times \mathbf{i} = \mathbf{j} \tag{13}$$

$$\mathbf{i} \times \mathbf{i} = \mathbf{0} \qquad \mathbf{j} \times \mathbf{j} = \mathbf{0} \qquad \mathbf{k} \times \mathbf{k} = \mathbf{0} \tag{14}$$

With respect to these unit vectors, vectors **a** and **b** are expressed as:

$$\mathbf{a} = [x_1, y_1, z_1] = x_1\mathbf{i} + y_1\mathbf{j}, + z_1\mathbf{k}\,, \quad \mathbf{b} = [x_2, y_2, z_2] = x_2\mathbf{i} + y_2\mathbf{j}, + z_2\mathbf{k}$$

Using the relationship in Eq.(13), the vector product of two general vectors can be manipulated

$$\begin{aligned}\mathbf{a} \times \mathbf{b} &= (x_1\mathbf{i} + y_1\mathbf{j}, + z_1\mathbf{k}) \times (x_2\mathbf{i} + y_2\mathbf{j}, + z_2\mathbf{k}) \\ &= x_1y_2\mathbf{k} - x_1z_2\mathbf{j} - y_1x_2\mathbf{k} + y_1z_2\mathbf{i} + z_1x_2\mathbf{j} - z_1y_2\mathbf{i}\end{aligned}$$

$$= (y_1z_2 - y_2z_1)\mathbf{i} + (x_2z_1 - x_1z_2)\mathbf{j} + (x_1y_2 - x_2y_1)\mathbf{k} \tag{15}$$

$$\mathbf{a} \times \mathbf{b} = [\, y_1z_2 - y_2z_1,\ x_2z_1 - x_1z_2,\ x_1y_2 - x_2y_1] \tag{16}$$

By using the determinant of a matrix, Eq.(16) is simply expressed as

$$\mathbf{a} \times \mathbf{b} = \begin{vmatrix} \mathbf{i} & \mathbf{j} & \mathbf{k} \\ x_1 & y_1 & z_1 \\ x_2 & y_2 & z_2 \end{vmatrix} \tag{17}$$

$$= \mathbf{i}\begin{vmatrix} y_1 & z_1 \\ y_2 & z_2 \end{vmatrix} - \mathbf{j}\begin{vmatrix} x_1 & z_1 \\ x_2 & z_2 \end{vmatrix} + \mathbf{k}\begin{vmatrix} x_1 & y_1 \\ x_2 & y_2 \end{vmatrix}$$

$$= (y_1z_2 - y_2z_1)\mathbf{i} + (x_2z_1 - x_1z_2)\mathbf{j} + (x_1y_2 - x_2y_1)\mathbf{k}$$

This determinant for a matrix applies only to square matrices which means the number of rows and columns are the same, an $n \times n$ matrix

For a 2×2 matrix

$$\begin{vmatrix} y_1 & z_1 \\ y_2 & z_2 \end{vmatrix} = y_1z_2 - y_2z_1$$

For an $n \times n$ matrix

$$\begin{vmatrix} a_{11} & a_{12} & a_{13} & \cdot & a_{1n} \\ a_{21} & a_{22} & a_{23} & \cdot & \cdot \\ a_{31} & a_{32} & a_{33} & \cdot & \cdot \\ \cdot & \cdot & \cdot & \cdot & \cdot \\ a_{n1} & \cdot & \cdot & \cdot & a_{nn} \end{vmatrix} = a_{11}D_{11} + (-1)^{1+2}a_{12}D_{12} + \ldots + (-1)^{1+n}\, a_{1n}D_{1n}$$

$$= \sum_{j=1}^{n} a_{1j}(-1)^{1+j} D_{1j} \tag{19}$$

In a general way, the determinant is expressed as

$$\sum_{j=1}^{n} a_{ij}(-1)^{i+j} D_{ij}$$

Here D_{ij} is determinant of a submatrix that is formed by removing the i-th row and j-th column of the parent matrix. In Eq.(19),

$$D_{11} = \begin{vmatrix} a_{22} & a_{23} & \cdot & a_{2n} \\ a_{32} & a_{33} & \cdot & \cdot \\ \cdot & \cdot & \cdot & \cdot \\ a_{n2} & \cdot & \cdot & a_{nn} \end{vmatrix}, \quad D_{12} = \begin{vmatrix} a_{21} & a_{23} & \cdot & a_{2n} \\ a_{31} & a_{33} & \cdot & \cdot \\ \cdot & \cdot & \cdot & \cdot \\ a_{n1} & \cdot & \cdot & a_{nn} \end{vmatrix}, \ldots, D_{1n} = \begin{vmatrix} a_{21} & a_{22} & \cdot & a_{2\,n-1} \\ a_{31} & a_{32} & \cdot & \cdot \\ \cdot & \cdot & \cdot & \cdot \\ a_{n1} & \cdot & \cdot & a_{n\,n-1} \end{vmatrix}$$

2. Reciprocal axes and reciprocal lattices

Planes are indexed by the intercepts with the lattice vectors of a crystal system. Alternatively a plane can be represented by a vector perpendicular to the planes since this normal vector is unique to the plane. In a cubic crystal system, the vector perpendicular to a plane, say ($hk\ell$), has the direction of [$hk\ell$]. For example, the vector perpendicular to the (001) plane has the direction [001]. Also the (110) plane has the normal vector [110]. These are shown in Fig. 2-36 in which the normal vectors are indicated by cylindirical arrows and the direction of the lattice vectors are shown by line arrows. A unit cell is illustrated by a cube.

The normal relationship between a plane and the direction vector of like indices in the cubic crystal systems arises from the fact that lattice vectors are perpendicular to each other and their lengths are the same. In other crystal systems, therefore, this normal relationship between like indices for planes and direction vectors is generally invalid. However, some planes such as (100), (010) and (001) planes are normal to the vectors with like indices in the tetragonal and orthorhombic crystal systems, but this is not valid for a plane with general indices ($hk\ell$). To make the indices of a plane coincident with the notation of the normal vector to the plane, another lattice scheme is devised. This is called a reciprocal lattice to the corresponding crystal system. An example is shown in Fig. 2-37 for the orthorhombic crystal system. The new lattice vectors, **a***, **b*** and **c*** are defined with respect to the original lattice vectors. **a*** is perpendicular to a plane formed by lattice vectors **b** and **c**. Likewise **b*** and **c*** are perpendicular to planes formed by a and c and a and b,

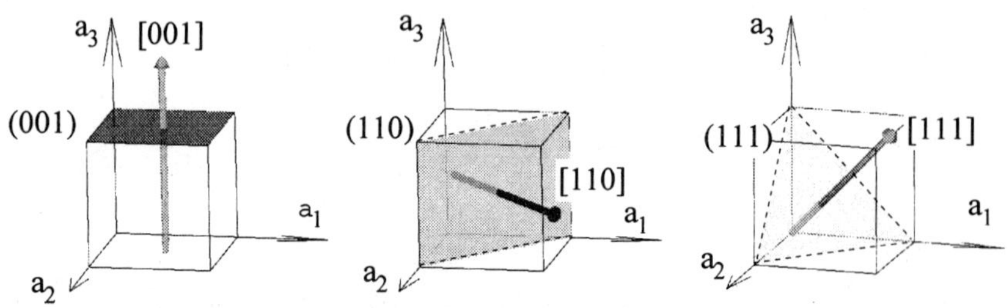

Figure 2-36 (001), (110), and (111) planes and their normal vectors in a cubic crystal system

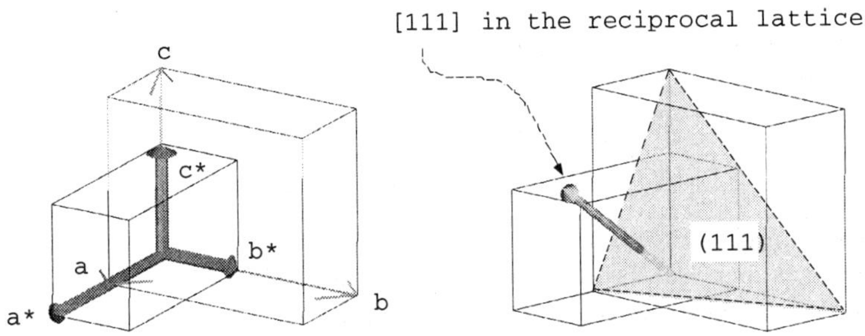

Figure 2-37 Illustration of orthorhombic lattice and its reciprocal lattice and the representation of the normal vector to (111) plane in the reciprocal lattice

respectively. The magnitudes of **a***, **b*** and **c***, they are reciprocal to the magnitudes of lattice vectors, **a**, **b**, and **c**, respectivley. In a mathematical way, the relationships between lattice vectors and reciprocal lattice vectors is given as follows:

$$\mathbf{a}^* = \frac{1}{|\mathbf{a}|}\frac{\mathbf{b}\times\mathbf{c}}{|\mathbf{b}\times\mathbf{c}|},\ \mathbf{b}^* = \frac{1}{|\mathbf{b}|}\frac{\mathbf{c}\times\mathbf{a}}{|\mathbf{c}\times\mathbf{a}|},\ \mathbf{c}^* = \frac{1}{|\mathbf{c}|}\frac{\mathbf{a}\times\mathbf{b}}{|\mathbf{a}\times\mathbf{b}|} \quad (18)$$

$\frac{\mathbf{b}\times\mathbf{c}}{|\mathbf{b}\times\mathbf{c}|}$ in Eq.(18) represents a unit vector that is perpendicular to both vectors **b** and **c**.

The orthorhombic lattice shown in Fig. 2-37 has the ratio of 0.7:1.5:1.3 for a:b:c. (Here plane type font indicates the magnitude of a vector whereas bold type fonts indicate the vectors.) According to Eq.(18), the ratio of a*:b*:c* is 1.43:0.67:0.77. Plane (111) of the orthorhombic lattice has a normal vector that passes through the point whose coordinates on the reciprocal lattice are (1,1,1). Hence the direction of the normal vector to the (111) plane is [111] on the reciprocal lattice. It is easily shown that, unlike in the cubic crystal system, the direction vector passing through the point (1,1,1) in the orthorhombic lattice is not perpendicular to (111) plane.

By using the reciprocal lattice, which can be defined from the crystal lattice, planes can be identified or drawn from the coordinates of the normal vectors. Hence the utilization of normal vectors defined on the basis of the reciprocal lattice makes it convenient for many crystallographic operations and mathematical manipulations to be performed, since planes can be defined by a point rather than by three intercepts. This aspect will be discussed in the following section for zone and zone axis.

SECTION 3 BASIC CONCEPT IN CRYSTALLOGRAPHY : ZONE AND ZONE AXIS

A. Zone

In crystallography, certain planes have a special relationship with a given direction, axes and vectors. Especially if they are parallel to each other, the planes are called zone planes and the axis is called a zone axis. Use of the zone and zone axes illustrate how Miller indices can be used to define the relationship between planes and directions or between planes as shown in the following section on coplanar and tautozonal planes.

A zone is a set of faces $\{hk\ell\}$ that are parallel to a given direction or axis [uvw] with respect to a given crystal. The axis [uvw] is called a zone axis. For the hexagonal crystal, the form $\{10\bar{1}0\}$ belongs to the zone axis [0001]. This is shown in Fig.2-38(a). Sometimes, the zone axis which is typically parallel to the major faces of a crystal coincides with the lattice vectors. As shown in Fig.2-38(b) for staurolite, the zone axis for (100), (110), and $(\bar{1}10)$ is easily recognized as the 'c'-axis for the crystal lattice. Before diffraction theory and the development of the diffractometer for the chracterization of crystal structures, zone and zone axis were key concepts in understanding crystals and their structures. Even with modern instrumental methods such as TEM (transmission electron microscopy) and x-ray diffraction, the zone and zone axis play a important role in the characterization of crystal structures.

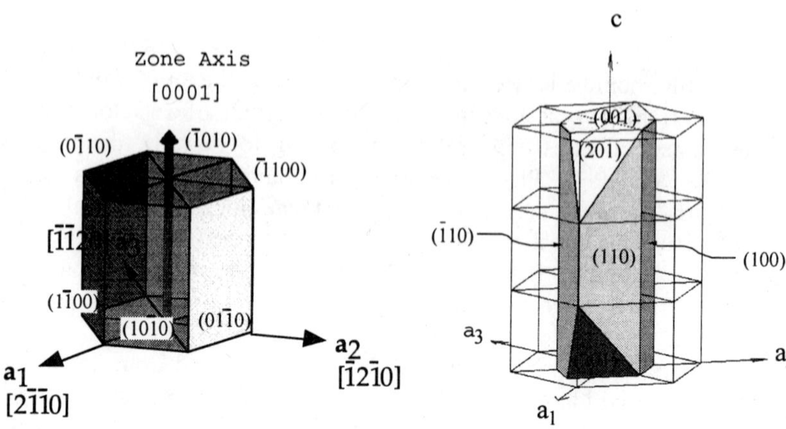

Figure 2-38 (a) Zone and zone axis [0001] in hexagonal lattice and
(b) Zone axes and zones in staurolite having hexagonal lattice
[001] – (110), (100), (110), $[11\bar{2}]$ – (201), (201)
[100] – (001), and [011]. – (100)

Table 2-6 Some important zone axes and zone planes in hexagonal and cubic system

Crystal system	Zone axis	Form or planes
Hexagonal	[0001]	$\{10\bar{1}0\}$*
Cubic	[001]	$(1\bar{1}0)$, (110), (010),$(0\bar{1}0)$
	$[11\bar{2}]$	(201), $(1\bar{1}0)$
	[100]	(001), (010)
	[010]	(201), $(20\bar{1})$, (001)

*Form $\{10\bar{1}0\}$ Parallel to zone axis [0001] in full hexagonal cell.

B. Zone Law

Two intersecting vectors $\mathbf{A}[u_1v_1w_1]$ and $\mathbf{B}[u_2v_2w_2]$ define a plane $(hk\ell)$, and two intersecting planes $(h_1k_1\ell_1)$ and $(h_2k_2\ell_2)$ define a direction or axis $\mathbf{C}[uvw]$.

The normal vector to a plane is perpendicular to any directional vector or axes on the plane. Therefore, the normal vector to a plane $(hk\ell)$ defined by intersecting axes **A**, $[u_1v_1w_1]$, and **B**, $[u_2v_2w_2]$, must be perpendicular to both axes **A**, $[u_1v_1w_1]$, and **B**, $[u_2v_2w_2]$. Since the normal vector to a plane $(hk\ell)$ in the rectangular coordinate system is defined as **H**, $[hk\ell]$, the scalar product of two perpendicular vectors **H** and **A** or **H** and **B** must be zero. The following zone law conditions apply between indices of the plane and the two axes defining the plane as :

$$\mathbf{H}\cdot\mathbf{A} = u_1h + v_1k + w_1\ell = 0, \quad (19a)$$
$$\mathbf{H}\cdot\mathbf{B} = u_2h + v_2k + w_2\ell = 0 \quad (19b)$$

By multiplying by u_2 in Eq.(19a) and by u_1 in Eq.(19b)

$$u_2u_1h + u_2v_1k + u_2w_1\ell = 0,\ \ u_1u_2h + u_1v_2k + u_1w_2\ell = 0$$

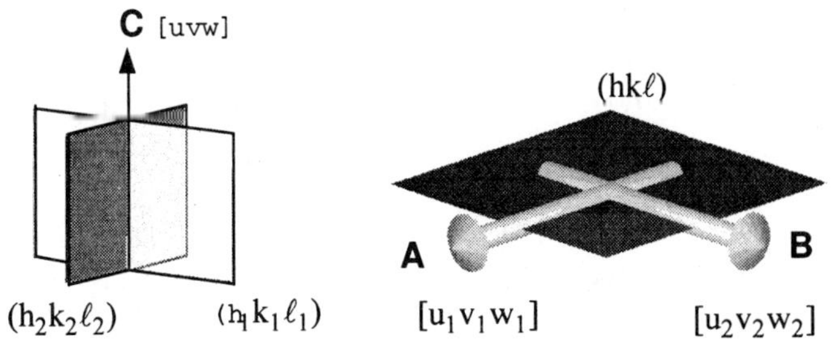

Figure 2-39 Defining of axis,[uvw], and plane, $(hk\ell)$ by two intersecting planes and axes

By subtracting the second from the first

$$k(u_2v_1 - u_1v_2) + \ell(u_2w_1 - u_1w_2) = 0 \text{ and } k = \frac{u_1w_2 - u_2w_1}{u_2v_1 - u_1v_2}\ell$$

Inserting this for k into Eq.(19-a)

$$h = \frac{w_1v_2 - v_1w_2}{u_2v_1 - u_1v_2}\ell \tag{20}$$

Therefore the ratio of $h:k:\ell$ is

$$h:k:\ell = w_1v_2 - v_1w_2 : u_1w_2 - u_2w_1 : u_2v_1 - u_1v_2 \tag{21}$$

The above simultaneous equations can best be solved in a simpler way by the use of determinants for a vector product according to Eq.(17)

$$\mathbf{A} \times \mathbf{B} = \begin{vmatrix} \mathbf{i} & \mathbf{j} & \mathbf{k} \\ u_1 & v_1 & w_1 \\ u_2 & v_2 & w_2 \end{vmatrix}$$

$$= (v_1w_2 - v_2w_1)\,\mathbf{i} + (w_1u_2 - w_2u_1)\,\mathbf{j} + (u_1v_2 - u_2v_1)\,\mathbf{k} \tag{22}$$

Thus $h = v_1w_2 - v_2w_1$; $k = w_1u_2 - w_2u_1$; $\ell = u_1v_2 - u_2v_1$ However, Eq.(22) is different from Eq.(21) by the negative sign. This difference arises from the fact that the derivation of Eq.(19) does not concern how a plane is defined. A plane can be defined by either upward normal vector or downward normal vector to the plane. If the plane formed by two vectors, A and B, is defined by the vector product $\mathbf{B} \times \mathbf{A}$ instead of $\mathbf{A} \times \mathbf{B}$, there should be no difference between Eqs.(21) and (22) for the indices for the plane. There is certainly no difference to the plane itself although its representation by normal vectors is different.

Eqs. (21) and (22) have been derived on the basis of the rectanuglar coordinate system. In general, planes and directions are defined on the basis of lattice vectors that do not form a rectangular coordinate system unless the crystal has the cubic lattice. In those cases, derivation of Eqs.(21) and (22) must be carried out on the basis of reciprocal lattices. In order to make things general, let's assume that unit vectors along the lattice axes are $\mathbf{n}_1$, $\mathbf{n}_2$ and $\mathbf{n}_3$. Then the two vectors are expressed as

$$\mathbf{A} = u_1\,\mathbf{n}_1 + v_1\,\mathbf{n}_2 + w_1\,\mathbf{n}_3 \quad \mathbf{B} = u_2\,\mathbf{n}_1 + v_2\,\mathbf{n}_2 + w_2\,\mathbf{n}_3$$

In contrast to directions, planes are denoted by a normal vector, $\mathbf{N} = [hk\ell]$, to the plane

$$\mathbf{N} = h\,\mathbf{n}_1{}^* + k\,\mathbf{n}_2{}^* + \ell\,\mathbf{n}_3{}^* \tag{23}$$

Here the reciprocal vectors, $\mathbf{n}_1{}^*, \mathbf{n}_2{}^*, \mathbf{n}_3{}^*$, are defined according to Eq.(18). Hence $\mathbf{n}_1 \bullet \mathbf{n}_1{}^* = \mathbf{n}_2 \bullet \mathbf{n}_2{}^* = \mathbf{n}_3 \bullet \mathbf{n}_3{}^* = 1$ and $\mathbf{n}_i \bullet \mathbf{n}_j{}^* = 0$ if $i \neq j$. From these relationships between lattice vectors and reciprocal vectors, Eq.(21) and (22) are derived regardless of the bases for the normal vector to the plane.

$$\mathbf{A} \bullet \mathbf{N} = (\, u_1\,\mathbf{n}_1 + v_1\,\mathbf{n}_2 + w_1\,\mathbf{n}_3\,) \bullet (\, h\,\mathbf{n}_1{}^* + k\,\mathbf{n}_2{}^* + \ell\,\mathbf{n}_3{}^*\,) \bullet$$

$$= h\,u_1\,\mathbf{n}_1 \bullet \mathbf{n}_1{}^* + k\,v_1\,\mathbf{n}_2 \bullet \mathbf{n}_2{}^* + \ell\,w_1\,\mathbf{n}_3 \bullet \mathbf{n}_3{}^* + \ldots$$

$$= u_1 h + v_1 k + w_1 \ell = 0$$

Likewise, $u_2 h + v_2 k + w_2 \ell = 0$

Also for the general lattice, the matrix form for the normal vector to a plane in Eq.(23) is changed to

$$\mathbf{A} \times \mathbf{B} = \begin{vmatrix} \mathbf{n}_1{}^* & \mathbf{n}_2{}^* & \mathbf{n}_3{}^* \\ u_1 & v_1 & w_1 \\ u_2 & v_2 & w_2 \end{vmatrix} = \mathbf{N}$$

$$= (v_1 w_2 - v_2 w_1)\,\mathbf{n}_1{}^* + (w_1 u_2 - w_2 u_1)\,\mathbf{n}_2{}^* + (u_1 v_2 - u_2 v_1)\,\mathbf{n}_3{}^* \tag{24}$$

$$\mathbf{A} \bullet \mathbf{N} = (\, u_1\,\mathbf{n}_1 + v_1\,\mathbf{n}_2 + w_1\,\mathbf{n}_3\,)$$

$$= u_1(v_1 w_2 - v_2 w_1) + v_1(w_1 u_2 - w_2 u_1) + w_1(u_1 v_2 - u_2 v_1)$$

$$= u_1 v_1 w_2 - u_1 v_2 w_1 + v_1 w_1 u_2 - v_1 w_2 u_1 + w_1 u_1 v_2 - w_1 u_2 v_1 = 0 \tag{25}$$

As a conclusion, the zone law expressed in Eqs.(24) and (25) always holds regardless of the lattice or coordinates systems. By using the reciprocal lattice vectors, therefore, the derivation of zone law is simplified and generalization of zone law to any crystal system is easily achieved. The rectangular coordinate system (or the cubic lattice) is a special case in that the lattice vectors and reciprocal lattice vectors are the same.

A given face $(hk\ell)$ in a crystal is parallel to a given zone axis, [uvw], if and only if

$$uh + vk + w\ell = 0 \tag{26}$$

As shown in Fig.2-32(b), for 2 intersecting planes ($h_1\ k_1\ \ell_1$) and ($h_2\ k_2\ \ell_2$), the unique axis [uvw] is defined by:

$$uh_1 + vk_1 + w\ell_1 = 0$$
$$uh_2 + vk_2 + w\ell_2 = 0$$

$$u = k_1\ell_2 - k_2\ell_1 \ ; \ v = \ell_1 h_2 - \ell_2 h_1 \ ; \ w = h_1 k_2 - h_2 k_1$$

Example 1

a) Plane defined by two axes, $[u_1v_1w_1] = [102]$ and $[u_2v_2w_2] = [0\bar{1}0]$

$$h = v_1w_2 - v_2w_1 = 0 \times 1 - \bar{1} \times 2 = 2$$

$$k = w_1u_2 - w_2u_1 = 2 \times 0 - 0 \times 1 = 0$$

$$\ell = u_1v_2 - u_2v_1 = 1 \times \bar{1} - 0 \times 0 = -1$$

Therefore the plane is $(20\bar{1})$.

b) Plane defined by two axes, $[u_1v_1w_1] = [111]$ and $[u_2v_2w_2] = [10\bar{1}]$

$$h = v_1w_2 - v_2w_1 = 1 \times \bar{1} - 0 \times 1 = -1$$

$$k = w_1u_2 - w_2u_1 = 1 \times 1 - \bar{1} \times 1 = 2$$

$$\ell = u_1v_2 - u_2v_1 = 1 \times 0 - 1 \times 1 = -1$$

Therefore the plane is (–1, 2,-1).

Example 2

a) Unique axis, [uvw], defined by two planes, $(h_1k_1\ell_1) = (\bar{1}20)$

and $(h_2k_2\ell_2) = (11\bar{2})$

$$u = k_1\ell_2 - k_2\ell_1 = 2 \times \bar{2} - 1 \times 0 = -4$$

$$v = \ell_1h_2 - \ell_2h_1 = 0 \times 1 - \bar{2} \times \bar{1} = -2$$

$$w = h_1k_2 - h_2k_1 = \bar{1} \times 1 - 1 \times 2 = -3$$

Therefore the axis is $[\bar{4}\bar{2}\bar{3}]$

C. Tautozonal planes and coplanar directions

1. Tautozonal

Three planes, $(h_1k_1\ell_1)$, $(h_2k_2\ell_2)$, $(h_3k_3\ell_3)$, are tautozonal if they satisfy the condition:

$$\begin{vmatrix} h_1 & k_1 & \ell_1 \\ h_2 & k_2 & \ell_2 \\ h_3 & k_3 & \ell_3 \end{vmatrix} = 0 \qquad (27)$$

i.e. : $h_1k_2\ell_3 - h_3k_2\ell_1 + k_1\ell_2h_3 - k_3\ell_2h_1 + \ell_1h_2k_3 - \ell_3h_2k_1 = 0$

Such planes lie in the same zone and, therefore, belong to the same zone axis. Example :

$(111)\,(1\bar{1}1)\,(101) = 0$ are tautozonal,

because $\begin{vmatrix} 1 & 1 & 1 \\ 1 & \bar{1} & 1 \\ 1 & 0 & 1 \end{vmatrix} = -1 + 0 + 1 - (-1) - 0 - 1 = 0$

2. Coplanar

Three directions are coplanar if they all lie in or are parallel to the same plane in a crystal. The vectorial product of two vectors is perpendicular to another vector since they are coplanar. Let say coplanar three vectors be **u**,$[u_1\ u_2\ u_3]$, **v**,$[v_1\ v_2\ v_3]$ and **w**,$[w_1\ w_2\ w_3]$. The vectorial product of v and w is

$$\mathbf{v} \times \mathbf{w} = \begin{vmatrix} \mathbf{i} & \mathbf{j} & \mathbf{k} \\ v_1 & v_2 & v_3 \\ w_1 & w_2 & w_3 \end{vmatrix}$$

Since scalar product of two perpendicular vectors is "0," $\mathbf{u} \cdot \mathbf{v} \times \mathbf{w} = 0$. In matrix form,

$$\mathbf{u} \bullet \mathbf{v} \times \mathbf{w} = \begin{vmatrix} u_1 & u_2 & u_3 \\ v_1 & v_2 & v_3 \\ w_1 & w_2 & w_3 \end{vmatrix} = 0$$

$$u_1v_2w_3 - u_3v_2w_1 + v_1w_2u_3 - v_3w_2u_1 + w_1u_2v_3 - w_3u_2v_1 = 0$$

These satisfy the condition:

Example:

[101], [313] and [111] are coplanar

$$\text{because} \begin{vmatrix} 1 & 0 & 1 \\ 3 & 1 & 3 \\ 1 & 1 & 1 \end{vmatrix} = 1 + 3 + 0 - 1 - 3 - 0 = 0$$

3. Addition of subtraction of indices for zone faces

For the zone axis [uvw], zone planes ($h_1\, k_1\, \ell_1$) and ($h_2\, k_2\, \ell_2$) must satisfy the condition in Eq.(25). Indices of zone planes ($hk\ell$) may be added or subtracted to give other zone faces. The following equation illustrates that new indices obtained by addition or subtraction of indices, ($h_1 + h_2$, $k_1 + k_2$, $\ell_1 + \ell_2$) or ($h_1 - h_2$, $k_1 - k_2$, $\ell_1 - \ell_2$), satisfy the same condition.

In determinant form, the condition for ($h_1 - h_2$, $k_1 - k_2$, $\ell_1 - \ell_2$) to be a zone plane is:

$$\begin{vmatrix} u & v & w \\ h_1 - h_2 & k_1 - k_2 & \ell_1 - \ell_2 \\ h_2 & k_2 & \ell_2 \end{vmatrix} = 0 \tag{28}$$

The left side can be expanded into two separate determinants as

$$\begin{vmatrix} u & v & w \\ h_1 & k_1 & \ell_1 \\ h_2 & k_2 & \ell_2 \end{vmatrix} - \begin{vmatrix} u & v & w \\ h_2 & k_2 & \ell_2 \\ h_2 & k_2 & \ell_2 \end{vmatrix} \tag{29}$$

The second determinant is zero since the second and third rows are the same. The first matrix of Eq.(29) is the same as Eq.(25). Hence Eq.(28) is proved to be true. Therefore addition or subraction of indices of zone planes results in an index of a new zone plane.

SECTION 4 SYMMETRIC CHARACTERISTICS OF CRYSTAL STRUCTURE

A. Symmetry of crystals

1. Symmetry Operations and Constraints in Crystals

The periodic arrangements of points in the planar and space lattices described above were depicted as being generated solely from the operation of a lattice translation vector. While true, this is by no means the only way that lattice points can be generated. Within the unit cell, lattice points are more commonly generated by symmetry operations such as rotation and reflection.

Because of the constraints imposed by the symmetry operations, the number of unique cell configurations that can independently exist are severely restricted. That is to say, these constraints limit the number of truly independent space lattices as in the planar lattices that have only four lattice systems. In the case of the space lattices, there are seven lattice systems and only fourteen unique unit cell configurations exist. The fourteen types of cell are known as Bravais lattices, after the French crystallographer Bravais. The fourteen Bravais lattices represent mathematically the only possibilities for uniquely arranging a collection of points in space. Like lattice systems, Bravais (non-primitive) lattices imply certain symmetry aspects that were partly explained in the planar lattice. In the planar lattice there is only one non-primitive lattice. In contrast there are 7 non-primitive lattices for 3-dimensional lattices. The symmetric aspects of Bravais lattices will be discussed in the section for space groups.

Symmetry is a property of an ordered structure of any object if they can be divided into subsections having the same shape. The subsection can be a point, a set of points or a pattern. The operation that transfers one subsection to the others of identical shape is a called symmetry operation. Rotation and mirror as shown in Fig. 2-40 are fundamental symmetry operations. Other symmetry operations can be constructed by a combination of these two kinds of symmetry operations. A point, line or plane which is not moved during the symmetry operation is called symmetry elements. For example, in Fig.2-40, the rotation axis and mirror plane represent the rotational symmetry and reflection, respectively. Since the symmetry elements may denote symmetry operations, they are used interchangeably. However, if they are expressed mathematically, symmetry elements are represented by a vector whereas symmetry operations are represented by a 3×3 matrix. The matrix representation of a symmetry operation will be described later.

Restriction of rotational symmetry in a planar lattice or crystals

Rotational symmetry operation moves a point to another point on a plane perpendicular to the rotational axis. In crystals, the possible rotational symmetries are 1, 2, 3, 4, and 6 fold rotations. 5, and 7 fold rotational symmetries are not possible as illustrated for a planar lattice in Fig.2-41. The distance between lattice points A and B that are in the direction of lattice vector **a** is 2a. By a rotation symmetry operation around Lattice point A, Lattice point O moves to a new position A' and by a rotation around B it moves to B'. Since A' and B' are also lattice points, their distance must be a multiple of lattice vector length, 'a'. Let's say it is ka.

$$\overline{A'B'} = 2a(\cos\theta - 1) = ka, \qquad \cos\theta = 1 - \frac{k}{2}$$

Since -1 < cosθ < 1, the possible values for k are 0, 1, 2 and 3. The corresponding θ's for these k values are 60°, 90°, 120°(equivalent to –60°). Hence the possible rotational symmetry operations are 6 (60°), 4 (90^{a}) and 3 (120°) fold rotation operations.

Rotation					Reflection
1-fold	2-fold	3-fold	4-fold	6-fold	1m = m

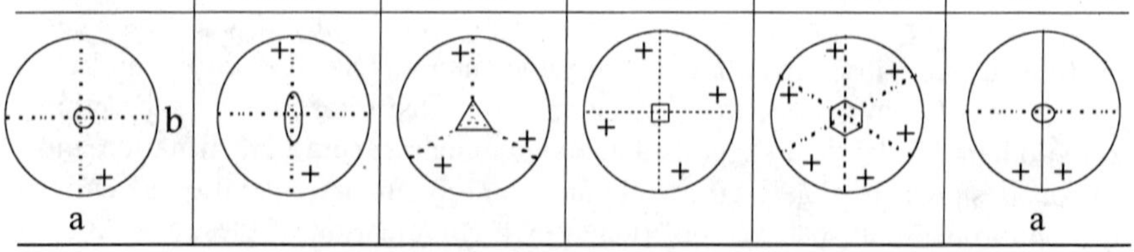

Figure 2-40 Basic symmetry operations and their corresponding elements for crystals

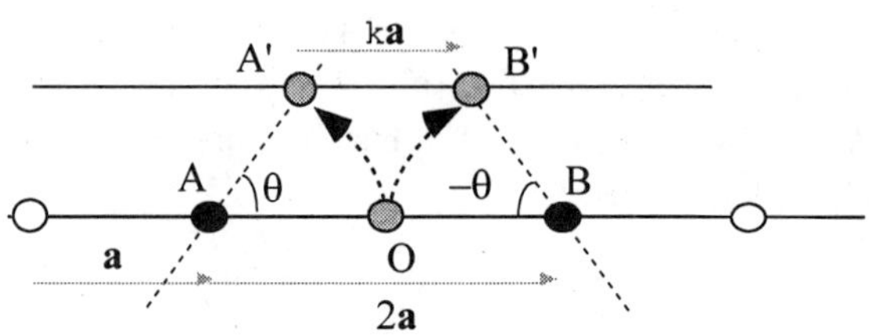

Figure 2-41 Restriction on the rotational symmetry of crystal lattices

2. Conventional rule for symbolic notation for symmetry

2-1 Symmetry element and its representation for a crystal

As indicated, symmetry elements are used interchangeably with symmetry operations. However they are not the same. A rotation axis is a symmetry element while rotation is a symmetry operation.

As described, each lattice is characterized by a combination of symmetry elements. However, this does not mean that the combination must show all the symmetry elements existing in the crystal lattice. In Fig.2-42 (a) and (b), symbolic notations of 4mm and 4 for planar square lattices do not include 2 fold symmetry elements although the lattices have 2 fold symmetry elements along the unit cell edges. In this case, the 2 fold axis is called a secondary symmetry element while the 4 fold axis is a primary symmetry element. In the symbolic notation for the symmetry of a lattice, only primary symmetry elements are listed. Since, in a planar lattice, all the symmetry elements lie perpendicular to the plane, there is no need to specify the direction of the rotation axis and mirror plane. However, in a crystal lattice, the direction of rotation axis and mirror plane must be indicated in the symmetry notation for a crystal. For example, a symbolic notation 4mm for a tetragonal lattice indicates that there is a 4 fold rotational axis along the "a" axis. (Conventionally the position of a 4 fold axis is defined as "c" axis. For purpose of explanation, it is assumed here that the first position represents 'a' axis.) Also there are two mirror planes whose normal directions lie along the other two lattice axes, 'b' and "c". This aspect of symbolic notation of symmetry elements for a crystal will be explained in detail when the point groups and space groups are described.

2-2 Equipoints, primary and secondary symmetry elements

Equipoints (or equivalent points) are positions within a unit cell that are equivalent by virtue of symmetry. As lattice points are related to each other by primitive lattice vectors, equipoints in the unit cell are related to each other by symmetry operations. In Fig. 2-42(a), 4 points are equipoints with respect to the 4 fold axis since they move one another by successive rotation around the axis. Fig. 2-42(b) illustrates the primary and secondary symmetry elements with respect to equipoints. The four equipoints in Fig. 2-42(a) imply a 2 fold rotational axis at the middle of a unit cell edge. However, these are not equipoints for a 2 fold axis since the 2 fold axis cannot move one point to the other 3 points. Rather it moves one point to another point in the opposite direction across the axis. Also the 2 fold rotational axis does not generate new equipoints compared to those for the 4 fold rotational symmetry. The 4 fold rotational axis implies a 2 fold rotational axes. Hence symbolic notation "4" is

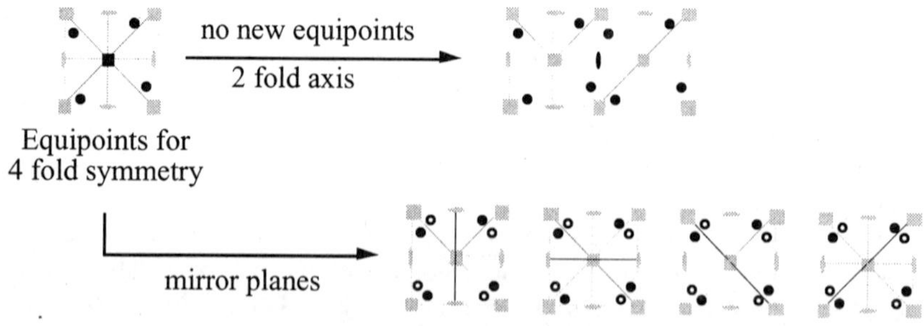

Figure 2-42 Formation of new equipoints from those for 4 fold symmetry

sufficient to represent 2 fold rotational axes on the edges of the unit cell of the square lattice.

The symmetry elements that generate equipoints are called primary symmetry elements while the "implied" symmetry elements are called secondary. In Fig. 2-42, the 4 fold aixs is the primary symmetry element while the 2 fold symmetry element is secondary. With repect to this terminology, the rule for combining the symmetry elements to represent the lattice symmetry can be expressed as listing the primary symmetry elements in such way that the highest rotational symmetry elements are listed first.

As shown in Fig.2-42, mirror planes are not implied symmetry elements of a 4 fold symmetry since they produce a new set of equipoints compared to those for the 4 fold rotational symmetry. Between mirror planes, one mirror plane implies other mirror planes because all mirror planes result in the same equipoints. Hence the symmetry notation for the lattice must be 4m according to the combining rule of symmetry elements. However, the conventional notation for the lattice is 4mm. The reason for this notation comes from the same background that is applied to the geometrical defnition of lattices. As the visualizing effect of the term 'square' in the square lattice, the second 'm' in the notation can visualize the second type mirror plane although it is implied by the other type mirror plane. In this respect, coventional notations of 6mm and 2mm are equal to 6m and 2m, respectively. Compared to even fold rotational symmetry elements, 3m is not equivalent to 3mm since there is only one type of mirror plane in that planar lattice.

The lattice shown in Fig.2-42 is represented by symbolic notation "4mm". The necessary condition for the existence of 4 fold rotational symmetry is that the lattice should be a square lattice. From the geometry, a 4 fold rotational axis is located at the center of the unit cell. Due to the repetition of general equipoints in every unit cell, there must be another 4 fold rotational axis at each lattice point. However, the 4 fold rotational axes at these lattice points do not generate new equipoints that are different from those already generated by the 4 fold axis at the center of the unit cell. Hence, the 4 fold rotational axes at the center of the unit cell implies that the other 4 fold rotational axes exist at each lattice point. Since the reverse is also true, the symmetry notation "4" represents both 4 fold symmetry elements.

3. Internal symmetry

As shown in Fig.2-43, the geometry of the unit cell denotes some symmetry elements. For example, if the unit cell has the square geometry, the lattice is supposed to have a 4 fold rotational symmetry. However, this is true only if the motif that is placed in each lattice point is a circle or a point in a planar lattice.

In real crystals, motifs are not always a circle or its equivalent. The non-circular motifs may remove some symmetry elements that exist with a circular or point motif. This may change not only the symmetry but also the lattice type. The effect of the motif on the lattice is illustrated in Fig.2-43 for a square lattice.

The square lattice is characterized in geometric terms as $|\mathbf{a}| = |\mathbf{b}|$ and $\gamma = 90°$. Although all lattices in Fig.2-43 satisfy this condition, the lattice shown in (c) is not a square lattice. Rather it is a rectangular lattice since it has only a mirror plane. The motif placed on each lattice point in Fig.2-43(c) removes the 4 fold rotational symmetry. In this sense, geometrical representation of a lattice is not precise and it is more accurate to define the square lattice as one that has 4 fold rotational symmetry. However, in actual crystals, it is very rare for

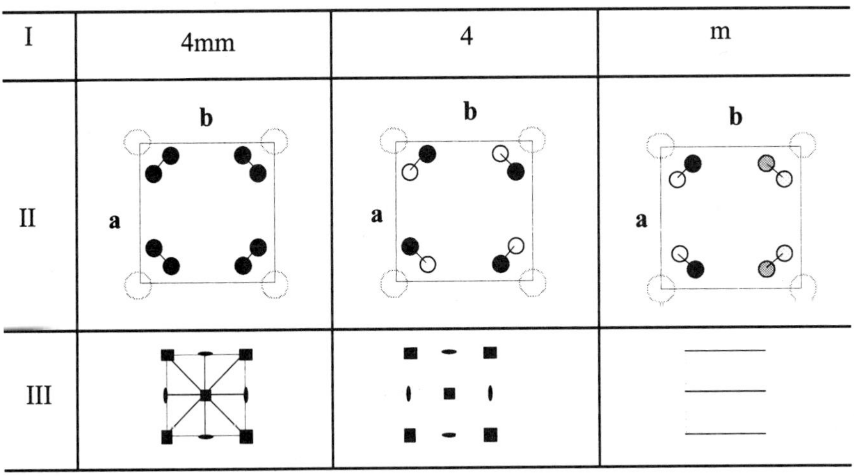

Figure 2-43 The effect of the motif on the symmetry of square-like lattices

a crystal with the atomic arrangement shown in Fig.2-43(c) to have the square geometry. Usually a lattice with that atomic arrangement has a rectangular geometry and consequently is a rectangular lattice. The geometrical representation of a lattice, therefore, is regarded as accurate as the symmetry definition of a lattice. Even with the possiblilty of less accuracy, the use of the geometrical representation in classification of lattices lies in the fact that this type representation more directly visualizes the lattice geometry than does the symmetry representation.

B. Point group and space group

1. Point group

1-1. Point group for planar crystals

The 10 point group notation for planar crystals follows the rules for arranging the symmetry elements that are illustrated below. The first rule is to show the primary rotational symmetry. Next to the rotation axis, other symmetry elements follow. Since, in the point group, all the symmetry elements share a fixed point, the concern in illustrating the symmetry elements is to show relative relationship to each other. For each lattice system, this relative directionality of each symmetry element is determined by their position in the notation of the point group symmetry. The rotation axis is always perpendicular to the planar lattice. Hence the direction of the mirror plane is represented by a normal vector perpendicular to it, and this is implied by where it stands in the notation. The direction of mirror plane in relation to its position in the notation is explained as follows.

1. Parallelogram: 1, 2

   1 - 1 fold (c) rotation axis–No symmetry
   2 - 2 fold (c) axis perpendicular to the plane

2. Rectangular m, 2mm

   m – plane normal parallel with a or b axis

   2 m m

   - 2 — 2 fold axis perpendicular to the plane
   - m — mirror plane parallel with b axis (plane normal $\perp$ a axis)
   - m — mirror plane parallel with a axis (plane normal $\perp$ b axis)

3. Square: 4, 4mm

4 - 4 fold axis
4 m m

- 4 fold perpendicular to the plane
- mirror plane parallel with a_1, or a_2 axis
- mirror plane parallel with face diagonal of unit cell

5. Hexagonal : 3, 3m, 6, 6mm

6(3) m_1 m_2

- 6(3) fold axis perpendicular to the plane
- mirror plane parallel with a, b, or a+b axis
- mirror plane bisecting m_1 mirror planes

1-2. Point group for 3-dimensional crystals

Compared to planar lattices, 3-dimensional crystals have another non-translational symmetry element, inversion, that cannot exist in the planar lattice since it is equivalent to 2 fold rotation. This implies that the mirror plane can exist perpendicular to the rotational axis. This additional element makes the number of point groups increase from 10 for the planar lattice to 32 for the 3-dimensional lattice. Inversion is not itself directly expressed. Rather it is inferred by other symmetry elements. Hence it does not appear in the symmetry group notation as separate from other symmetry elements.

1 Point group illustration

Point group is a representation of an object by a combination of nontranslational symmetry elements. These symmetry elements on a projection circle must share at least one fixed point for the object. Any points that are symmetrically equivalent in an object must be at the same distance from this fixed point. Therefore, equipoints in a 3 dimensional entity locate on the imaginary sphere whose diameter is the distance from the fixed point to the equipoints. In this regard, a point group representation is illustrated by a great circle that is cut through the equator of the sphere. The equipoints are indicated by a projected position on this cross section of the sphere. Usually the great circle is chosen as one that is perpendicular to the highest rotational axis. The illustration of a point group by a circle can be depicted by showing symmetry elements alone or symmetry elements plus general equipoints. As explained for the planar lattice, only one set of general equipoints is enough to represent a point (symmetry)

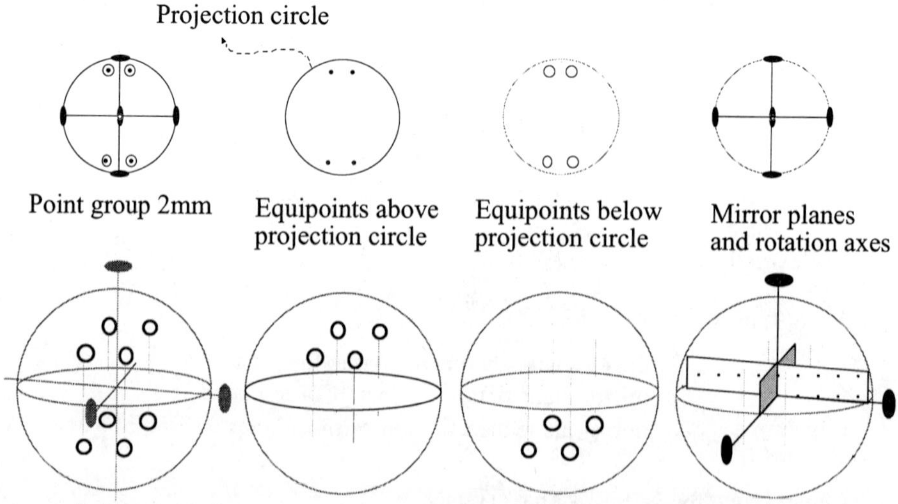

Figure 2-44 Illustration of the representation of the point group, 2mm

group. Fig.2-44 illustrates how a point group is represented by symmetry elements and equipoints on a circular cross section of the imaginary sphere. Table 2-7 lists the graphical representation of 32 point groups with symmetry elements along with equipoints.

2. Point group symbols of crystal system

By virtue of their association with the geometry system of lattices, the point groups themselves can be incorporated into the crystal systems, and further divided into 32 crystal classes. These 32 crystal classes represent a grouping of all the possible symmetry operations for the seven crystal systems that are compatible with cell repetition. As it happens, the natural grouping of real crystals into these 32 crystal classes, was formally documented (by virtue of crystal habit - external form) long before the symmetry connections were understood.

1. <u>Triclinic</u>: 1, $\bar{1}$

1 - 1 fold (c) rotation axis–No symmetry

$\bar{1}$ - 1 fold (c) axis of rotary inversion

2. <u>Monoclinic</u>: 2, m, 2/m

   2 – 2 fold b axis – 2nd setting
   m – mirror plane // to b axis
   $\frac{2}{m}$ – mirror plane ⊥ to 2 fold b axis

3. <u>Orthorhombic</u>: 222, mm2, mmm

   m m 2
   - 2 – 2 fold c-axis
   - m (second) – mirror plane along b axis
   - m (first) – mirror plane along a axis

   $mmm = \frac{2}{m}\frac{2}{m}\frac{2}{m}$

   Note: $\frac{2}{m} \rightarrow m$ in this and other systems except monoclinic

4. <u>Tetragonal</u>: $4,\ \bar{4},\ \frac{4}{m},\ 422,\ 4mm,\ \bar{4}2m,\ \frac{4}{m}mm$

   a) $\bar{4}$ - 4 fold c axis of rotary inversion

   b) $\bar{4}\ 2\ m$
   - m – mirror planes along <110>(bisectors)
   - 2 – 2 fold a_1 a_2 axes
   - $\bar{4}$ – 4 fold rotary inversion c-axis

   c) $\frac{4}{m}\ m\ m$
   - m (second) – mirror planes along a_1 a_2 bisectors
   - m (first) – mirror plane along a_1 a_2 axes
   - $\frac{4}{m}$ – mirror plane ⊥ to 4 fold c-axis

5. <u>Trigonal</u>: $3,\ \bar{3},\ 32,\ 3m,\ \bar{3}m\ (=\bar{3}2)$

   a) System denoted by unique 3 or $\bar{3}$ c-axis of symmetry
   b) Two sublattices occur depending on crystal orientation

   (i) <u>Trigonal</u> (P)

   $a_1 = a_2 \neq c;\quad \alpha = \beta = 90°;\quad \gamma = 120°$
   Same as hexagonal except c-axis is 3 fold
   Distribution of points in space same as hexagonal. Therefore not a new system.

(ii) Rhombohedral (R)

$a_1 = a_2 = a_3$; $\alpha = \beta = \gamma \neq 90°$

Primitive unit cell

3 fold body diagonal taken as c-axis

$3, \bar{3}$ - fold c-axis (both P & R)

3 2

- 2 → 2 fold a_1 a_2 axes (P)
- 3 → 3 fold c-axis (P)

3 2 1

- 1 → third digit refers to bisectors of a_1 a_2 (and a_3)
- 2 → 2 fold a_1 a_2 a_3 axes
- 3 → 3 fold body diagonal (R)

6. Hexagonal: $6, \bar{6}, \frac{6}{m}$, 622, 6mm, $\bar{6}2m$, $\frac{6}{m}$ mm

a) $\bar{6}$ - fold c-axis of rotary inversion

6 2 2

- 2 (third) → 2 fold bisectors of a_1 a_2 axes
- 2 (second) → 2 fold a_1 a_2 axes
- 6 → 6 fold c-axis

$\frac{6}{m}$ mm

- m (third) → mirror planes along bisectors
- m (second) → 2 fold a_1 a_2 axes
- $\frac{6}{m}$ → mirror plane ⊥ to 6 fold axis of rotation

7 Cubic : 23, m3, 432, $\bar{4}3m$, m3m

2 3 → " 3 " in the second position denotes the cubic structure

- 3 → 4 -3 fold diagonal < 111 > axes
- 2 → 2 fold a_1 a_2 a_3 axes

4 3 2

- 2 → 2 fold axes along bisectors of a_1 a_2 a_3 axes
- 3 → 3 fold diagonal < 111 > axes
- 4 → 4 fold a_1 a_2 a_3 axes

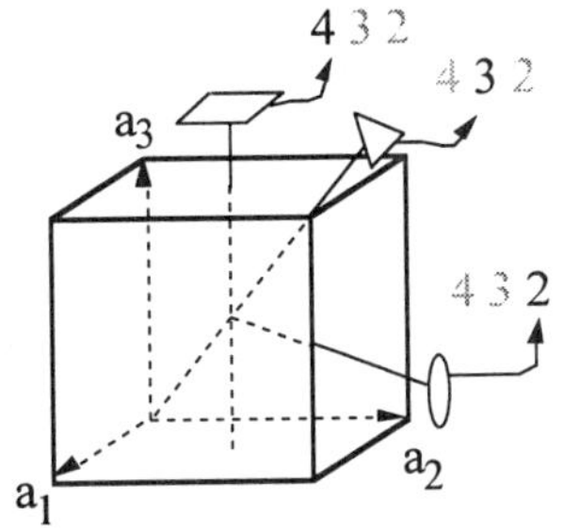

Figure 2-45 Illustration of 2, 3 and 4 fold axes of point group 432

In rotational symmetry operation, axes with higher foldness come before those with lower rotational symmetry. In this respect, the c-axis for a lattice is usually determined to be the direction parallel to the rotational axis with the highest foldness. If there are several rotational axes with the same foldness, the c-axis is identified as the one with the longest lattice vector, as in the orthorhombic lattice. In the cubic system, the point group notation is different from those for other systems and the order of rotational symmetry elements can be reversed as in the point group "23." This reverse ordering in rotational symmetry is due to the 3-fold axis for which the symmetry notation is fixed at the second place, regardless of the foldness of other rotational symmetry elements. Hence the first digit (position) in the notation for a lattice having a cubic system indicates the direction along the edges of the cubic unit cell. The second digit (poisition) indicates existence of the 3-fold axis along the body diagonal direction of the cubic unit cell. In the crystallographic point of view, this is called the <111> direction. Again, the third digit (or position) indicates the direction along the face diagonal of the cubic unit cell. The 3-fold axis in the second digit connotes the cubic system for a crystal while the number "3" in the first digit (position) of the symmetry notation indicates a trigonal system. This unique aspect of the cubic system is shown in Fig.2-45 for point group 432.

2-3. Geometry vs. Symmetry

The geometric concept used in classifying planar lattices can also be applied to the crystal lattices. The difference between planar and crystal lattices is that two lattice vectors are used in the planar lattice while three lattice vectors are used in the crystal lattice. Due to one more lattice vector, the geometric systems are expanded from 4 to 7 systems that are cubic, tetragonal, orthorhombic, hexagonal, rhombohedral, monoclinic, and triclinic. Since the unit cell in a 3 dimensional lattice is a parallelepiped defined by 3 lattice vectors, the

Table 2-7 Classification of point groups by symmetry elements

1. Point group with simple rotation(i)

Notation	1	2	3	4	6

Circular illustration

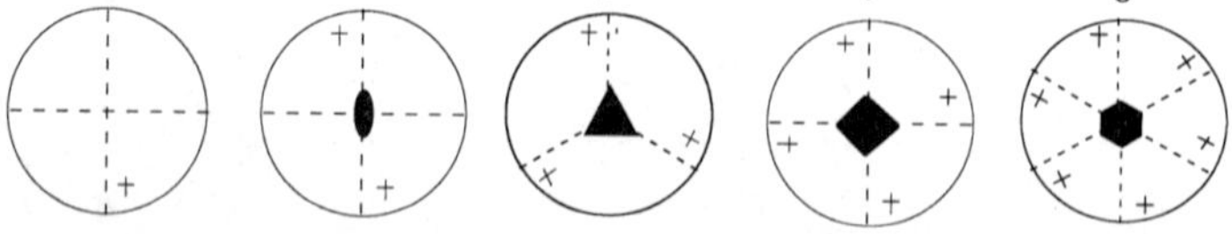

2. Point group with rotation and reflection(m) (Vertical mirror)

Notation	1m = m	2mm	3m	4mm	6mm

Circular illustration

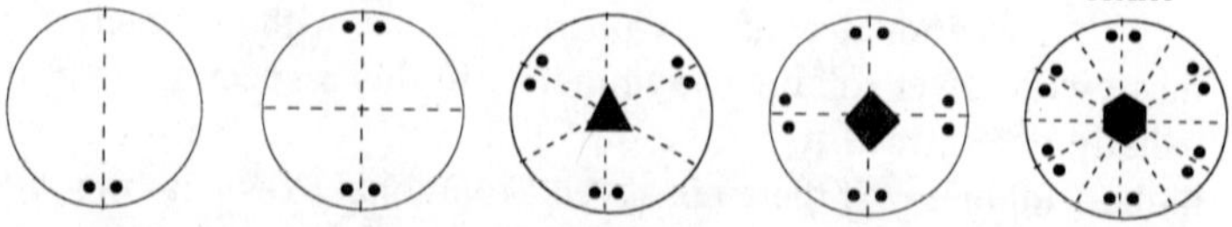

3. Point group with rotation and reflection $\left(\frac{1}{m}\right)$ (Horizontal mirror)

Notation	$\frac{1}{m} = \bar{2}$	$\frac{2}{m}$	$\frac{3}{m}$	$\frac{4}{m}$	$\frac{6}{m}$

Circular illustration

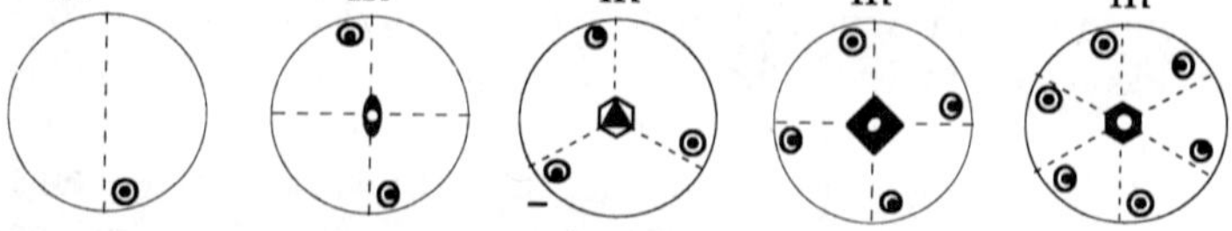

4. Point group with rotary inversion $(\bar{i})$ (through the center)

Notation	$\bar{1}$	$\bar{2}$	$\bar{3}$	$\bar{4}$	$\bar{6}$

Circular illustration

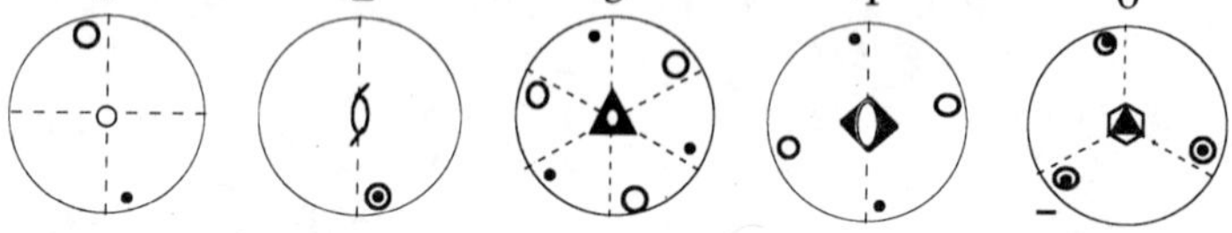

5. Point group with combined rotations

Notation	222	32	422	622

Circular illustration

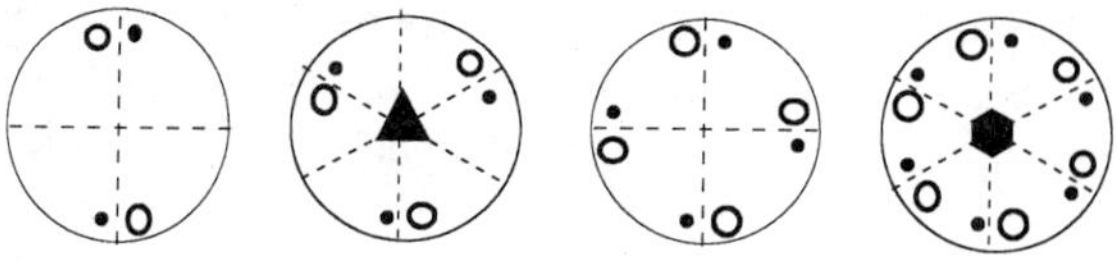

6. Point group by combined rotations and reflections $\left(\frac{1}{m}\right)$ (Horizontal mirror)

Notation: $\frac{2}{m}\frac{2}{m}\frac{2}{m}$ * ; $\frac{3}{m}2 = \bar{6}m2$ * ; $\frac{4}{m}22$ * ; $\frac{6}{m}22$ *

Circular illustration

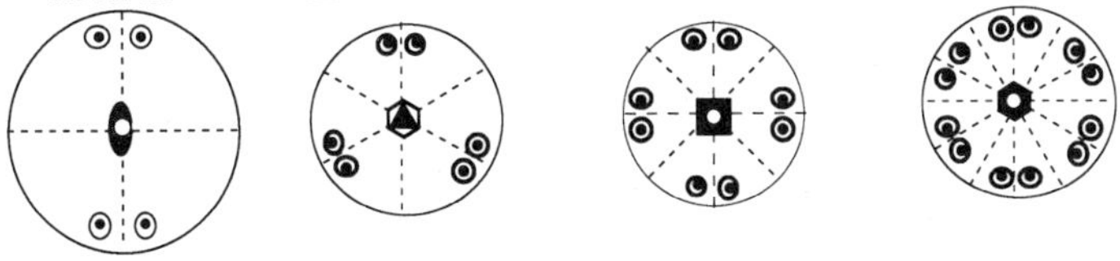

7. Point group by rotoinversion and reflections ($\bar{i}m$ or $\bar{i}im$)

Notation: $\bar{4}2m$; $\bar{3}2 = \bar{3}m$

Circular illustration

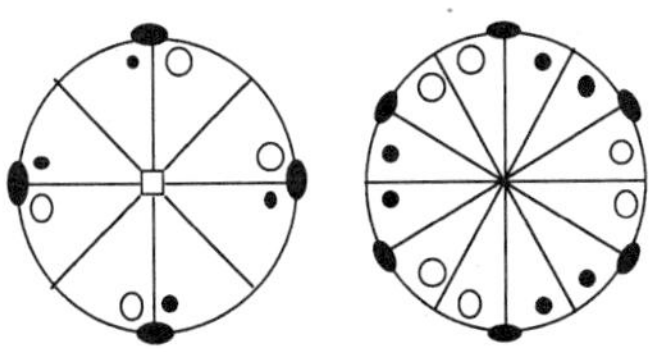

8. Point group with 3 fold rotation** not perpendicular to or parallel with circular projection

Notation*: 23 ; 432 ; $\frac{4}{m}32$ *** ; $\frac{2}{m}3 = m3$; $\bar{4}3m$

Circular illustration

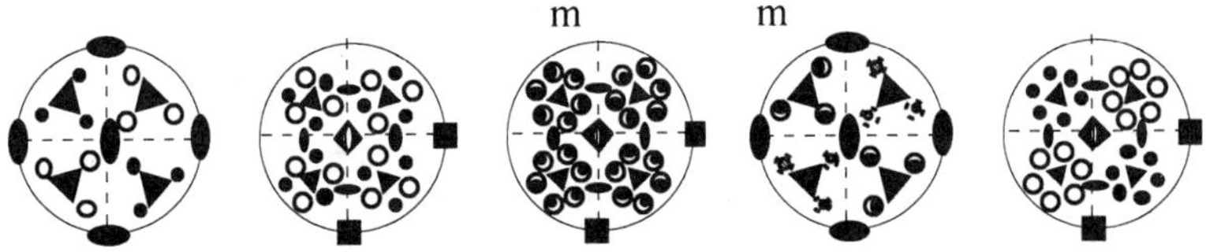

---

* $\frac{2}{m}\frac{2}{m}\frac{2}{m} = mmm$, $\frac{3}{m}2 = \bar{6}m2$, $\frac{4}{m}22 = \frac{4}{m}\frac{2}{m}\frac{2}{m} = \frac{4}{m}mm$, $\frac{6}{m}22 = \frac{6}{m}\frac{2}{m}\frac{2}{m} = \frac{6}{m}mm$

** 3 in the second digit denotes that the point group has cubic structure.

*** $\frac{4}{m}32$ is the same as = $\frac{4}{m}\bar{3}\frac{2}{m}$, $m\bar{3}\frac{2}{m}$, m3m.

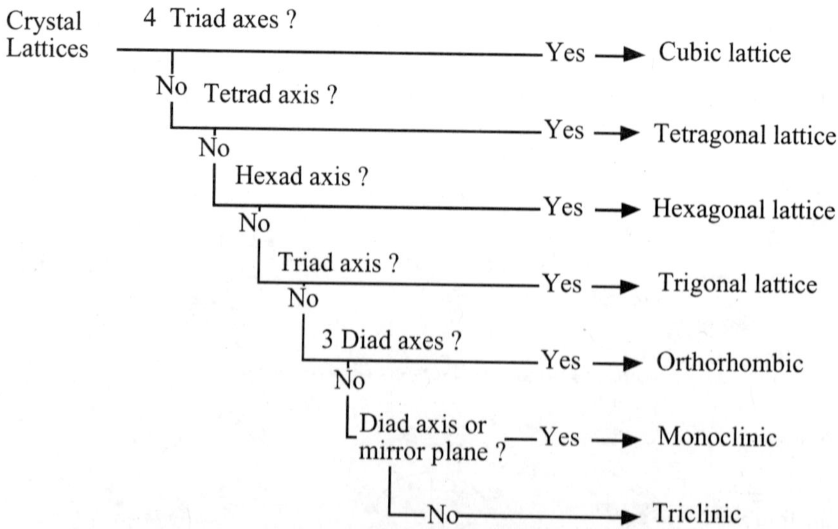

Figure 2-46 Classification of crystal lattices into 7 crystal systems in relation to symmetry with geometry of the lattice

geometrical consideration is consequently based on the symmetry of the parellelpiped unit cell. If all the body diagonals of the parallelepiped are triad(3 fold) axes, the equivalent geometry of the parallelepiped is a cube. If there is only one tetrad(4 fold) axis whose direction is along one of the lattice vectors, the corresponding parallelepiped is an elongated cube in the direction of one edge. This criteria of symmetry for the classification of lattice systems is schematically drawn in Fig.2-46.

The rotational axis for the classification of crystal lattices include not only the pure rotational axis but also the screw axis. Also the mirror plane includes both pure mirror plane and glide plane. The classification of crystal lattices according to symmetry elements, as shown in Fig.2-46, indicates that the classification is not purely geometrical. Rather it implies the underlying symmetry elements associated with the geometry of the crystal systems.

1-2-4 Catagorization of point groups

As shown the total number of point groups are 32. These are sorted out along with the crystal system. These are given in Table 2-8 and illustrated in Fig. 2-47.

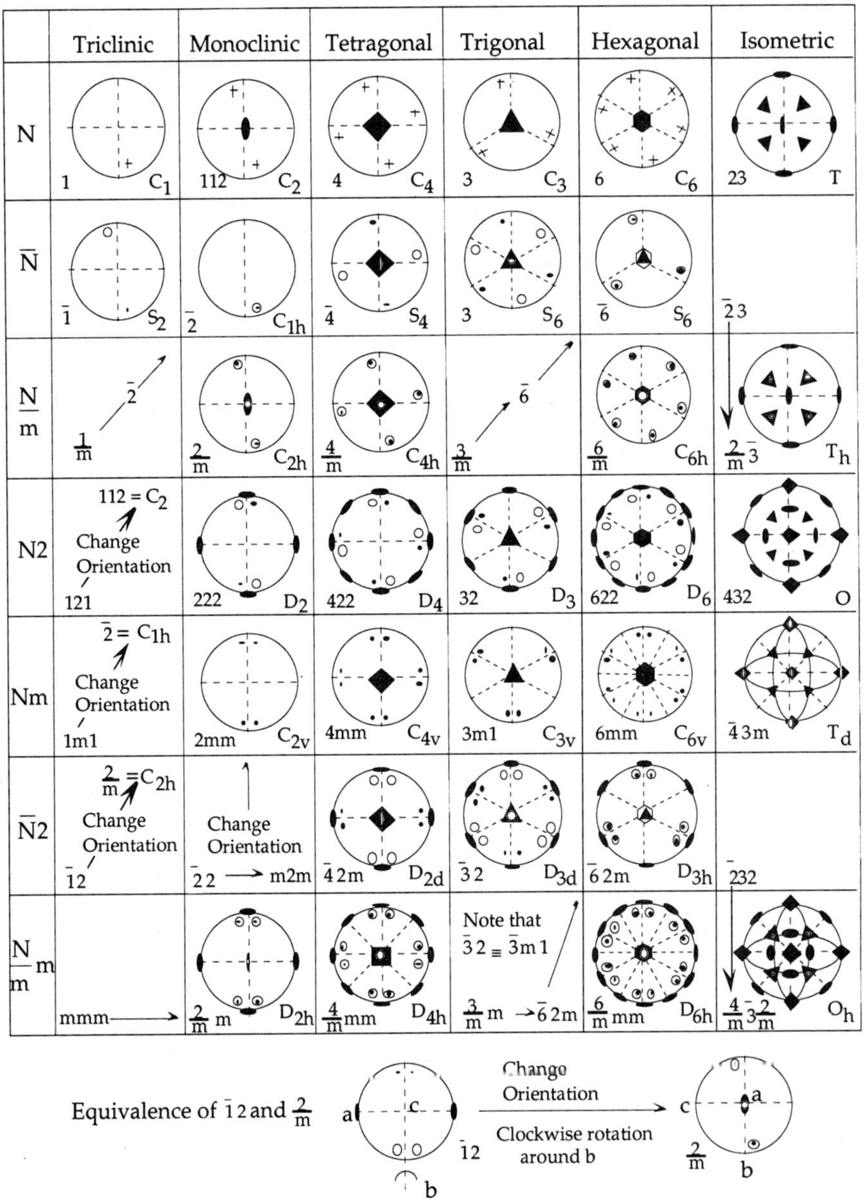

Figure 2-47 Illustration of 32 crystallographic point groups regarding crystal systems

Table 2-8 32 crystallographic points groups

Crystal system	Principal rotation axis	Point groups
Triclinic	$1, \bar{1}$	$1, \bar{1}$
Monoclinic	2 fold, b-axis b,c,a	$2, m, \frac{2}{m}$
Orthorhombic	3 – 2 fold axes (a,b,c)	222, mm2, mmm
Tetragonal	4 fold c-axis	$4, \bar{4}, \frac{4}{m}, 422, 4mm, \bar{4}2m, \frac{4}{m}mm$
Trigonal	3 fold, c-axis	$3, \bar{3}, 32, 3m, \bar{3}m$
Hexagonal	6 fold, c-axis	$6, \bar{6}, \frac{6}{m}, 622, 6mm, \bar{6}m2, \frac{6}{m}mm$
Cubic	4-3 fold <111> axes (diagonal)	$23, m3, 432, \bar{4}3m, m3m$

1-3. Point group operators and equipoints

One characteristic of symmetry operations is that they can be expressed by matrices which change position of the point (xyz) without itself being changed. Hence point groups exhibit the properties of a mathematical group (addition, combination, etc.) since the point symmetry operations can themselves be mathematically combined, as Point Group Operators.

1-3-1. Point group operators in matrix form and their derivations

From Diagram in Fig. 2-48

$$x = r\cos\phi, \quad y = r\sin\phi$$

Also
$$x' = r\cos(\phi + \theta) = r(\cos\phi\cos\theta - \sin\phi\sin\theta)$$
$$y' = r\sin(\phi + \theta) = r(\sin\phi\cos\theta + \cos\phi\sin\theta)$$

Substituting $r\cos\phi$ and $r\sin\phi$ for x and y in above equations:

$$x' = x\cos\theta - y\sin\theta + 0z \qquad (30)$$
$$y' = y\cos\theta + x\sin\theta + 0z \qquad (31)$$
$$z' = 0x + 0y + 1z \qquad (32)$$

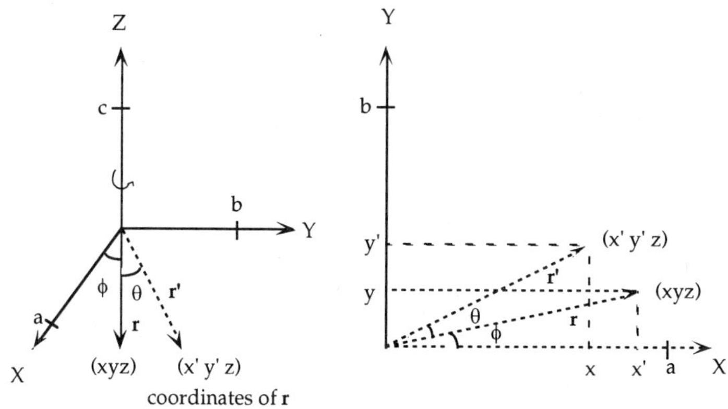

(a) (b)

Figure 2-48(a) Rotation of vector r counterclockwise around c axis and (b) Rotation of vector r shown in x-y plane

Collecting terms and expressing the operation in matrix form

$$\begin{vmatrix} x' \\ y' \\ z' \end{vmatrix} = A_{[001]} \begin{vmatrix} x \\ y \\ z \end{vmatrix} \quad \text{where } A_{[001]} \text{ is } \begin{pmatrix} \cos\theta & -\sin\theta & 0 \\ \sin\theta & \cos\theta & 0 \\ 0 & 0 & 1 \end{pmatrix} \qquad (33)$$

A is the matrix of the coefficients of x, y and z in equation (16d). Use of matrix A will give the coordinates of the new point x'y'z' resulting from rotation of the (original) point x y z around the "C" [001] axis through angle θ. i.e., A transforms xyz → x'y'z'. "A" is not altered in the process and is therefore called a matrix operator. Similarly, rotation around "a" axis and rotation around "b" axis can be expressed by matrices as

$$A_{[100]} = \begin{pmatrix} 1 & 0 & 0 \\ 0 & \cos\theta & -\sin\theta \\ 0 & \sin\theta & \cos\theta \end{pmatrix} \qquad (34)$$

and

$$A_{[010]} = \begin{pmatrix} \cos\theta & 0 & -\sin\theta \\ 0 & 1 & 0 \\ \sin\theta & 0 & \cos\theta \end{pmatrix} \qquad (35)$$

This notation of matrices for coordinates can be applied to directions and planes.

For plane and axes (directions)

$$\begin{vmatrix} h' \\ k' \\ \ell' \end{vmatrix} = A \begin{vmatrix} h \\ k \\ \ell \end{vmatrix}, \quad \begin{vmatrix} u' \\ v' \\ w' \end{vmatrix} = A \begin{vmatrix} u \\ v \\ w \end{vmatrix}, \text{ and } \begin{vmatrix} a' \\ b' \\ c' \end{vmatrix} = A \begin{vmatrix} a \\ b \\ c \end{vmatrix}$$

b. Multiple symmetry operations

$$\begin{vmatrix} x \\ y \\ z \end{vmatrix} \xrightarrow{A'} \begin{vmatrix} x' \\ y' \\ z' \end{vmatrix} \xrightarrow{A''} \begin{vmatrix} x'' \\ y'' \\ z'' \end{vmatrix}$$

$$\begin{vmatrix} x' \\ y' \\ z' \end{vmatrix} = A' \begin{vmatrix} x \\ y \\ z \end{vmatrix}, \quad \begin{vmatrix} x'' \\ y'' \\ z'' \end{vmatrix} = A'' \begin{vmatrix} x' \\ y' \\ z' \end{vmatrix} \rightarrow \begin{vmatrix} x'' \\ y'' \\ z'' \end{vmatrix} = A'A'' \begin{vmatrix} x \\ y \\ z \end{vmatrix}$$

Table 2-9 Operation matrices for n-fold rotations around a, b and c axes

	[100] a-axis	[010] b-axis	[001] c-axis
4 Fold(90°)	$\begin{pmatrix} 1 & 0 & 0 \\ 0 & 0 & -1 \\ 0 & 1 & 0 \end{pmatrix}$	$\begin{pmatrix} 0 & 0 & -1 \\ 0 & 1 & 0 \\ 1 & 0 & 0 \end{pmatrix}$	$\begin{pmatrix} 0 & -1 & 0 \\ 1 & 0 & 0 \\ 0 & 0 & 1 \end{pmatrix}$
Rotation by an angle θ	$\begin{pmatrix} 1 & 0 & 0 \\ 0 & \cos\theta & -\sin\theta \\ 0 & \sin\theta & \cos\theta \end{pmatrix}$	$\begin{pmatrix} \cos\theta & 0 & -\sin\theta \\ 0 & 1 & 0 \\ \sin\theta & 0 & \cos\theta \end{pmatrix}$	$\begin{pmatrix} \cos\theta & -\sin\theta & 0 \\ \sin\theta & \cos\theta & 0 \\ 0 & 0 & 1 \end{pmatrix}$

Table 2-10 Values of θ, cosθ and sinθ for 1, 2, 3, 4 and 6-fold rotations

Fold	0	6	4		3	2	1
θ	0°	60°	90°	270°	120°	180°	360°
cos θ	1	0.5	0	0	-0.5	-1	1
sin θ	0		1	-1		0	0

(i) Cyclic and non-cyclic operators

Among the symmetry operations, cyclic operations are those that can be permutated into multiple symmetry operations. In this sense, simple rotations are cyclic operations since their operators in matrix form permute one another. In matrix form the condition for the cyclic operations is expressed as :

$$A'A'' = A''A' \qquad (36)$$

where A' and A" are matrices representing symmetry operations. Hence the non-cyclic operators in the rectangular coordinate system are inversion, reflection, and combination of these. Table 2-11 illustrate the operators for these non-cyclic operators except for the identity matrix which is cyclic.

Table 2-11 Symmetry operations and their matrices and notations

Sym. Operation	Operation Matrix	Notation
Identity	$\begin{pmatrix} 1 & 0 & 0 \\ 0 & 1 & 0 \\ 0 & 0 & 1 \end{pmatrix}$	I
Inversion	$\begin{pmatrix} -1 & 0 & 0 \\ 0 & -1 & 0 \\ 0 & 0 & -1 \end{pmatrix}$	J
Reflection $\frac{1}{m}$ across c through ab plane	$\begin{pmatrix} 1 & 0 & 0 \\ 0 & 1 & 0 \\ 0 & 0 & -1 \end{pmatrix}$	M
Reflection m across b through ac plane	$\begin{pmatrix} 1 & 0 & 0 \\ 0 & -1 & 0 \\ 0 & 0 & 1 \end{pmatrix}$	D
Reflection m across a through bc plane	$\begin{pmatrix} -1 & 0 & 0 \\ 0 & 1 & 0 \\ 0 & 0 & 1 \end{pmatrix}$	D

1-3-2 Equipoint Representation

Depending on the symmetry operations, there may be more than 2 equipoints. In Fig.2-49 equipoints are illustrated for lattices of 4mm, 4 and m. The lattice represented as 4mm has 8 equipoints in the unit cell whereas the other lattices represented as "4" and "m" have two sets of 4 equipoints and four sets of two equipoints, respectively. In Fig.2-49(a), equipoints i_1, i_3, i_5, and i_7 can move into one another through the 4 fold rotational symmetry operation. With a reflection symmetry operation, these odd number set of equipoints can move to an even number set of equipoints. If there is no reflection symmetry, the odd number set and even number set of equipoints in Fig.2-49(a) are no longer equivalent resulting in two different sets of equipoints as in Fig.2-49(b).

To illustrate the symmetry for a lattice, only one set of general equipoints is enough. For the lattice in Fig.2-50(a), therefore, only one equipoints indicated by "i" or "ii," is enough to represent the symmetry set. Also, in Fig.2-50(b), any one of equipoints sets "i" through "iv" can represent the symmetry of the lattice in Fig.2-50(b). This is illustrated in Fig.2-51.

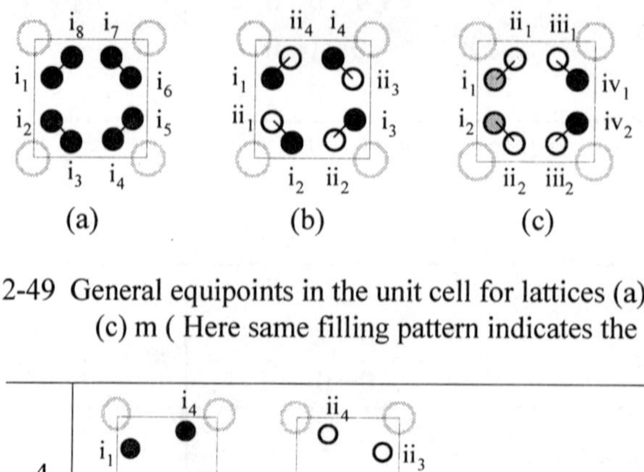

Figure 2-49 General equipoints in the unit cell for lattices (a) 4mm, (b) 4 and (c) m (Here same filling pattern indicates the equipoints.)

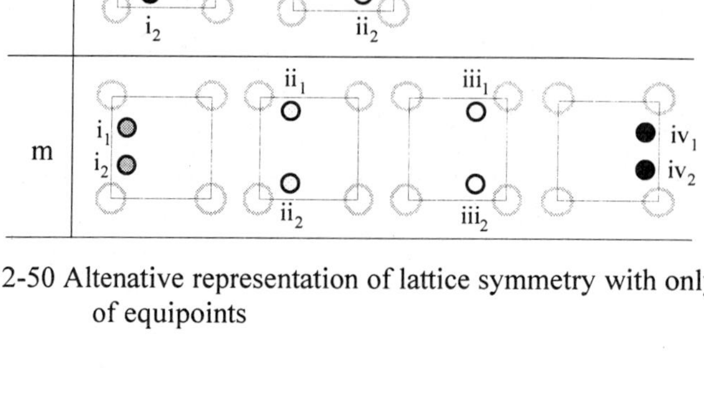

Figure 2-50 Altenative representation of lattice symmetry with only one set of equipoints

As shown in Fig.2-50, there are many ways to represent lattice symmetry with equipoints. Since the purpose of putting equipoints in the unit cell is to make it easy to understand the symmetry operations in the lattice, the position of equipoints do not matter as long as they illustrate all the symmetry elements for a lattice.

1-3-3 Special positions and general positions

Those equipoints shown in Fig.2-50 are said to be in general positions because they move during the symmetry operations. In contrast to general positions, equipoints located on a symmetry element are called special positions since they do not move during the symmetry operation. In a planar latice, a rotation axis and mirror plane are expressed as a point and a line, respectively. Special positions for rotational symmetry fall on those points that are unique positions for the rotation axis. Those points on the mirror plane line are special points for the reflection symmetry. Since equipoints on the special positions do not move during a symmetry operation, their number in the unit cell is less than the number of equipoints in general positions. If the special positions locate at an intersection point of two symmetry elements, the number of equipoints is further reduced. In crystallography, the number of equipoints in the unit cell is called multiplicity. In relation to the symmetry elements that pass through the equipoints, the multiplicity can be replaced by the symmetry element notation that combines symmetry elements common to the equipoints. Another notation was developed by Wyckoff and is adopted in the field of crystallography. He assigns on ascending order of multiplicity for equipoints, using alphabetical notation that gives "a" to the lowest multiplicity and continues with b, c, d and so on. These are illustrated in Table 2-12 for a planar lattice with symmetry of 4mm.

The equipoints in general positions for the lattice have a multiplicity of 8 since no symmetry element falls on the general positions. The symmetry element notation for the equipoints in general positions is "1" that means there is no symmetry element other than identity operation. For equipoints with a multiplicity of 4, there are 3 sets of special positions. For those special positions on the unit cell edges, each position counts as 0.5 since it is shared by two adjacent unit cells. The other special positions are on the diagonals of the unit cell and on the two lines crossing parallel with cell edges at half of the unit cell, respectively. Corresponding symmetry element notation is "m." If the special positions are on the middle of the unit cell edges, two mirror planes cross on these positions. Symmetry element notation for these special positions is "mm." The multiplicity reduces to "2." Each of the special positions at the corner of the unit cell counts as 0.25 since it is shared by four adjacent unit cells. The equipoints with special positions on the unit cell edges have a multiplicity of 1.

Table 2-12 Positions of equipoints for the lattice represented as 4mm

Multiplicity	Symmetry Elements	Wyckoff Notation	Equipoints
8	1	d	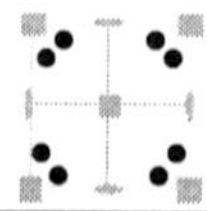
4	m	c	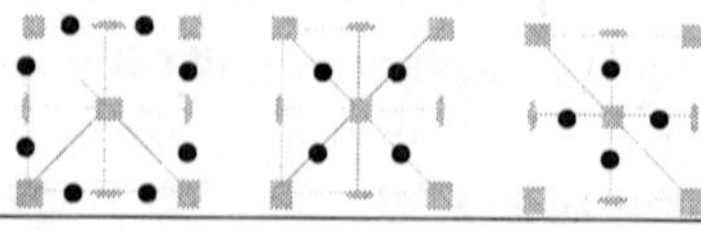
2	mm	b	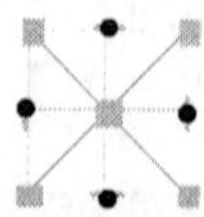
1	4mm	a	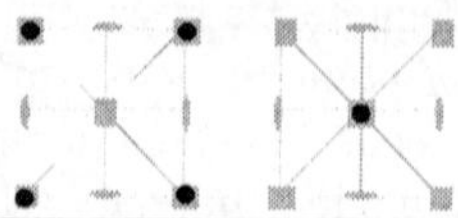

1-3-4 Mathematical illustration of equipoints for point group 222 and $\bar{4}$

As shown above, there is a one-to-one correspondence between symmetry operation and matrix representation. The theoretical aspect of symmetry is handled by these matrices for the symmetry operations. Using matrices for the symmetry operations makes it much easier to characterize physical properties and classification of crystals. Usually the physical properties are vectorial, which means that they have directionality. Mathematically expressed, they are tensorial properties. Hence using a matrix for symmetry is very important in handling those physical properties such as thermal expansion coefficient, mechanical stiffness, and piezoelectric coefficient. Using matrix representation for symmetry operations can provide a simple way to express what seems, otherwise, to be very difficult. Here the mathematical representation of symmetry operation is described for itself and not for any applications.

(i) Point group 222

The circular illustration in Fig. 2-51 represents the point group 222. As shown in Fig. 2-51, atomic position "1" in the unit cell converts to "2," "3" and "4" by

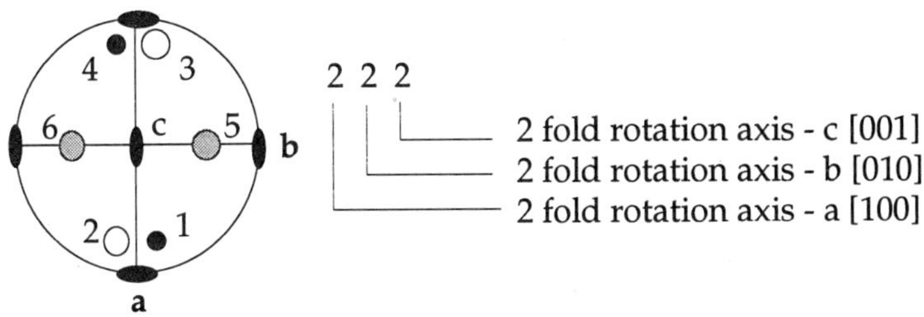

Figure 2-51 A set of general equipoints in 222 point group.

the symmetry operation of rotation around the **a** axis, **b** axis and **c** axis, respectively. Even combinations of these symmetry operations do not place from atomic position "1" to any point other than "2", "3" or "4." Hence the rank of general equipoints in the point group 222 is 4. If an atom is located in the rotation axis as Point 5, its equivalent point is only atom "6" since the rotation around the b axis does not change the positions of "5" and "6." These are the special equipoints. Their rank is 2.

It is conventional to show only general equipoints in a circular illustration of a point group. Generally, therefore, equipoints 5 and 6 are not shown in a circular illustration of point groups. This conventional way is shown in Fig. 2-51.

Illustration of equipoints by symmetry operators can be described by operation matrices. The matrices for rotation around a, b, and c axes in the point group 222 are expressed as :

$$A_{[100]}=\begin{pmatrix}1 & 0 & 0\\ 0 & -1 & 0\\ 0 & 0 & -1\end{pmatrix},\ A_{[010]}=\begin{pmatrix}-1 & 0 & 0\\ 0 & 1 & 0\\ 0 & 0 & -1\end{pmatrix},\ A_{[001]}=\begin{pmatrix}-1 & 0 & 0\\ 0 & -1 & 0\\ 0 & 0 & 1\end{pmatrix}$$

Since $A_{[001]} = A_{[100]}A_{[010]}$, only two matrices are necessary to generate all points in the point group 222. Also matrices commute since symmetry operators are all simple rotations. Therefore they may be taken in any order. The possible matrices for symmetry operations are $A_{[100]}$ $A_{[010]}$ and $A_{[100]}A_{[010]}$. Let us assume the coordinate of atomic position "1" is (xyz). The coordinates of new atomic positions generated by these symmetry operations are

$$\begin{pmatrix}x'\\ y'\\ z'\end{pmatrix} = A_{[100]}\begin{pmatrix}x\\ y\\ z\end{pmatrix} = \begin{pmatrix}1 & 0 & 0\\ 0 & -1 & 0\\ 0 & 0 & -1\end{pmatrix}\begin{pmatrix}x\\ y\\ z\end{pmatrix} = \begin{pmatrix}x\\ -y\\ -z\end{pmatrix}$$

$$\begin{pmatrix} x'' \\ y'' \\ z'' \end{pmatrix} = A_{[010]}\begin{pmatrix} x \\ y \\ z \end{pmatrix} = \begin{pmatrix} -1 & 0 & 0 \\ 0 & 1 & 0 \\ 0 & 0 & -1 \end{pmatrix}\begin{pmatrix} x \\ y \\ z \end{pmatrix} = \begin{pmatrix} -x \\ y \\ -z \end{pmatrix}$$

$$\begin{pmatrix} x''' \\ y''' \\ z''' \end{pmatrix} = A_{[100]}A_{[010]}\begin{pmatrix} x \\ y \\ z \end{pmatrix} = \begin{pmatrix} 1 & 0 & 0 \\ 0 & -1 & 0 \\ 0 & 0 & -1 \end{pmatrix}\begin{pmatrix} -1 & 0 & 0 \\ 0 & 1 & 0 \\ 0 & 0 & -1 \end{pmatrix}\begin{pmatrix} x \\ y \\ z \end{pmatrix} = \begin{pmatrix} -x \\ -y \\ z \end{pmatrix}$$

If an atom locates at the b axis, its coordinates are (0,y,0). The eqivalent point of (0,y.0) is (0,-y,0). There are no equivalent points for (0,0,0). The rank of equipoints of special positions located at b axis is 2. This result agrees with the circular illustration.

The general equipoints are (xyz), (x,-y,-z), (-x,y,-z), (-x,-y,z). Special equipoints are (0,y,0) and (0,-y,0) of rank two and (000) of rank one.

(ii) Point group $\bar{4}$

Rotoinversion(J) is the only symmetry operation in the point group $\bar{4}$ as shown in Fig.2-52. The possible symmetry operations are A_J. $A_J{\cdot}A_J$, and $A_J{\cdot}A_J{\cdot}A_J$. The general equipoints are generated by these matrices as:

$$A_J = \begin{pmatrix} 0 & 1 & 0 \\ -1 & 0 & 0 \\ 0 & 0 & -1 \end{pmatrix},\ A_J{\cdot}A_J = \begin{pmatrix} -1 & 0 & 0 \\ 0 & -1 & 0 \\ 0 & 0 & 1 \end{pmatrix},\ A_J{\cdot}A_J{\cdot}A_J = \begin{pmatrix} 0 & -1 & 0 \\ 1 & 0 & 0 \\ 0 & 0 & -1 \end{pmatrix}$$

$$\begin{pmatrix} x' \\ y' \\ z' \end{pmatrix} = A_J\begin{pmatrix} x \\ y \\ z \end{pmatrix} = \begin{pmatrix} 0 & 1 & 0 \\ -1 & 0 & 0 \\ 0 & 0 & -1 \end{pmatrix}\begin{pmatrix} x \\ y \\ z \end{pmatrix} = \begin{pmatrix} y \\ -x \\ -z \end{pmatrix}$$

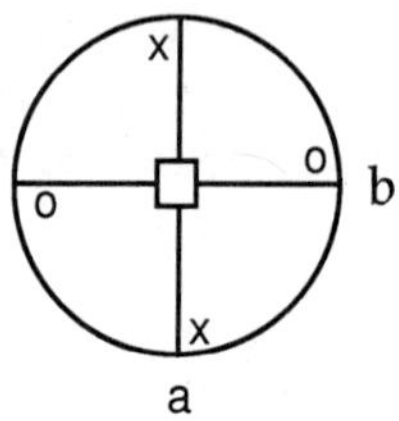

$$A_J = \begin{pmatrix} 0 & 1 & 0 \\ -1 & 0 & 0 \\ 0 & 0 & -1 \end{pmatrix}$$

Figure 2-52 Circular illustration of point group $\bar{4}$ and its symmetry operator.

$$\begin{pmatrix} x'' \\ y'' \\ z'' \end{pmatrix} = A_J \cdot A_J \begin{pmatrix} x \\ y \\ z \end{pmatrix} = \begin{pmatrix} -1 & 0 & 0 \\ 0 & -1 & 0 \\ 0 & 0 & 1 \end{pmatrix} \begin{pmatrix} x \\ y \\ z \end{pmatrix} = \begin{pmatrix} -x \\ -y \\ z \end{pmatrix}$$

$$\begin{pmatrix} x''' \\ y''' \\ z''' \end{pmatrix} = A_J \cdot A_J \cdot A_J \begin{pmatrix} x \\ y \\ z \end{pmatrix} = \begin{pmatrix} 0 & -1 & 0 \\ 1 & 0 & 0 \\ 0 & 0 & -1 \end{pmatrix} \begin{pmatrix} x \\ y \\ z \end{pmatrix} = \begin{pmatrix} -y \\ x \\ -z \end{pmatrix}$$

The general equipoints are (xyz), (y,-x,-z), (-x,-y,z) and (-y,x,-z). Therefore the rank of the general equipoints in the point group $\bar{4}$ is 4. If the matrix form for each symmetry element is known for a space group, it is not necessary to draw a circular projection as in Fig. 2-44 in order to get the equipoints. Using mathematical symbolism with matrices for symmetry operation, a symmetry group is represented as grouping the possible matrices as follows:

$$G \equiv \sum_i W_i = W_1 + W_2 + W_3 + W_4 + \ldots \qquad (37)$$

In this regard, two symmetry groups are said to be the same if they have the same matrices parts as Eq.(37)

2. Space groups

2-1. Translational symmetry operation

If any single point, line or plane remains unmoved during the symmetry, this is described a non-translational symmetry operation. Conversely, if all points are moved during the symmetry operation, this is called a translational symmetry operation. Since the lattice is based on the translational of the lattice points by the primitive lattice vectors, only fractional translations by lattice vectors generate new internal symmetry elements for a crystal. If atomic arrangement is taken into consideration, symmetry operations such as rotation and reflection, plus a fractional translation by lattice vectors, are all needed in characterizing the internal symmetry of the lattice.

2-1-1 Screw axis (rotation + translation)

The notation for the screw axis is n_τ where n describes an n-fold rotation. When the unit cell is projected along the translation vector **t**, the projected image must

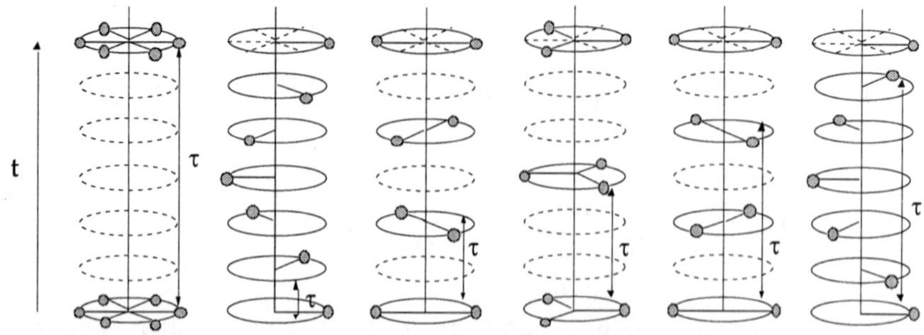

Figure 2-53 Arrangements of atoms in the unit cell for screw axes 6, 6_1, 6_2, 6_3, 6_4, and 6_5 from the left

have n-fold symmetry. π describes the angle of screw operation that is $\frac{360^\circ}{n}$. If the advance of the screw operation is τ, τ is restricted by π. This restriction is expressed as

$$n\tau = \pi t \qquad (38)$$

Fig. 2-35 illustrates the 6-fold screw axis. For 6_1, τ is $\frac{1}{6}$**t** and π is 60°. Hence Eq.(38) argrees with this illustration.

Comparison of similar structures

For easy understanding, similar structures like 3_1 and 6_2 and 6_3 and 2_1 are compared in Fig. 2-54. 3_1 and 6_2 have the same τ and rotation angle of 120°. The difference is in the projected image along translation direction **t**.

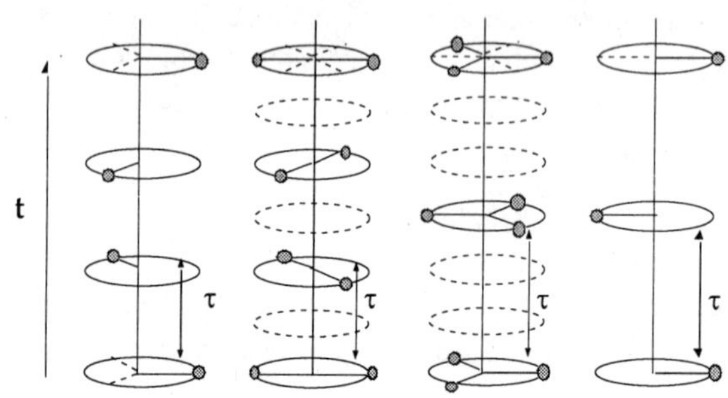

Figure 2-54 Comparison of 3_1 vs. 6_2, and 6_3 vs. 2_1

Symbolic notation for screw axes are shown in the following .

Screw axis	2 fold	3 fold	4 fold	6 fold
Symbolic notation				

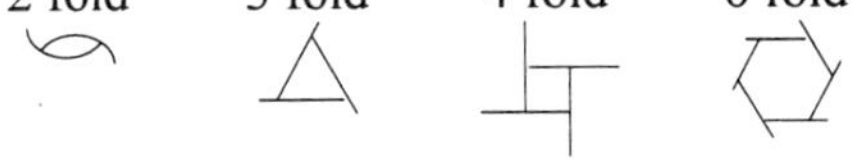

2-1-2 Glide plane (reflection + translation)

(i) Axial glide plane

Reflection across a mirror plane is followed by translation parallel to the plane by $\frac{1}{2}$ a lattice translation vector (**t**). As in screw axis operation, if **t** = 0, it is a simple mirror plane.

If **t** is along the **a** axis, the space symmetry operation is called a-glide. Likewise if it is in **b** or **c** axis direction, it is called b-glide or c-glide according to the direction. For an a-glide, the reflection plane can be "ab" or "ac" plane. Likewise, the reflection plane for c-glide can be either "a" or "b" plane. The notation itself does not distinguish a-glide having "ab" plane from a-glide with 'ac' plane as a reflection plane. The distinction between these two reflection planes for a glide operation is made by the position in a symbolic representation of symmetry for the crystal. For example, a symbolic representation "P2aa" for an orthorhombic crystal system means that there are two a-glide planes. In an orthorhombic crystal system, the first position of the notation indicates "a" axis and the second and third positions indicate "b" and "c" axes, respectively.

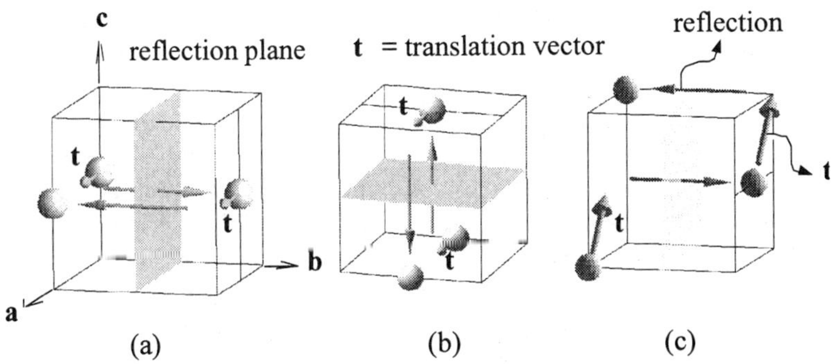

Figure 2-55 Axial glide of a-glide (a) with (010) plane and (b) (001) plane as reflection plane (c) Diagonal glide with (010) as reflection plane

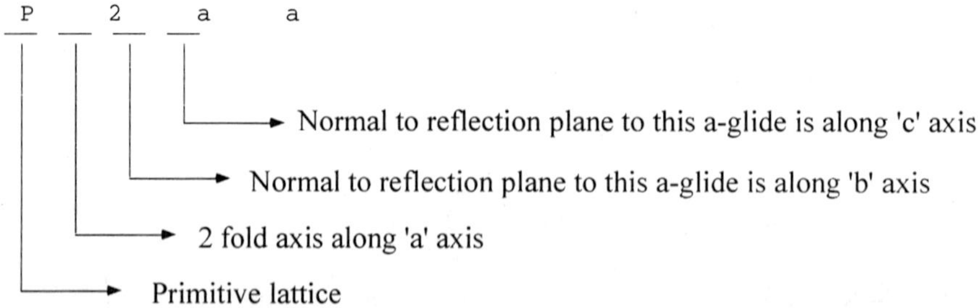

The first 'a-glide' has a reflection plane whose normal direction is along the 'b' axis. By the same token, the second a-glide has a reflection plane whose directionic along the "c" axis. Hence the normal direction of the reflection plane for a glide is identified by the position of the glide in the symbolic notation.

(ii) Diagonal glide and diamond glide plane

Notation for the diagonal glide is n glide. This is the case when **t** shown in Fig. 2-55 is $\frac{1}{2}(\mathbf{a}+\mathbf{b})$, $\frac{1}{2}(\mathbf{a}+\mathbf{c})$, or $\frac{1}{2}(\mathbf{b}+\mathbf{c})$. Since the n glide has components of 2 axes, it is along the diagonal of a face ($\frac{1}{2}$ face translation) Diamond glide is the case when t is $\frac{1}{4}(\mathbf{a}+\mathbf{b})$, $\frac{1}{4}(\mathbf{a}+\mathbf{c})$ $\frac{1}{4}(\mathbf{b}+\mathbf{c})$ or $\frac{1}{4}(\mathbf{a}+\mathbf{b}+\mathbf{c})$. The notations for diagonal and diamond glides are n and d, respectively.

2-2. Symbolic notation crystallographic space groups

A space group, in the crystallographic sense, is a group whose components include both the symmetry elements and the Bravais lattices (lattice translation) of the crystal. The space group notation for a specific crystal consists of two parts: (1) the Bravais lattice and (2) the total symmetry of the array of atoms in the crystal and, therefore, its crystal system. At each point in the Bravais lattice, there exists a collection of symmetry elements that defines the symmetry of the array of atoms (or motifs) about that point. The point group offers information concerning only the symmetry of the array of atoms in the motifs within the crystal. The space group, on the other hand, represents the collection of symmetry elements that are distributed in space according to the Bravais lattices. It, therefore, provides a description of the total symmetry of the array of atoms in the crystal.

The combination of the Bravais lattices and point groups for each crystal class leads to the formation of 73 simple or geometric space groups. Further derivations of all possible space groups must, however, include translational symmetry elements with fractional lattice vectors such as glide planes (reflection + translation) and screw axes (rotation + translation). When these compound symmetry operations are taken into account, a total of 230 space groups are obtained. Properly interpreted, these space groups provide a complete description as to the symmetry and location of each atom and atomic position within the crystal. The 230 space groups consist of 73 symmorphic (simple) space groups and 157 nonsymmorphic space groups. The latter include compound symmetry elements such as glide planes and screw axes. These 230 space groups are used to describe in detail the symmetry of crystals and their structures. These crystallographic space groups are listed in Table 2-15 for crystal systems where the cubic system is arranged with triclinic and monoclinic systems. Other systems follow the numbering sequence for crystallographic space groups in the "International Tables for Crystallography."

In sum, the simple space groups can be derived by combinations of (a) the 14 Bravais lattices, that describe the geometric arrangements of lattice points in space, and (b) the 32 point groups, that contain the symmetry elements associated with each crystal system. The more complex space groups depict the arrangements of atoms (i.e. atomic spacing and bond angle) associated with each lattice point, and must, therefore, incorporate glide planes and screw axes, that describe the detailed translation symmetry operations within the unit cell.

2-3. Equipoints and special points in space groups

The concept of equipoints in a space group is the same as in the point group. For example the space group P23 has 12 general points generated from a point (x,y,z) by successive rotation of the 2-fold and 3-fold symmetry axes. Its special points are as follow:

Special Points	Rank
(0,0,0)	1
$\left(\frac{1}{2},\frac{1}{2},\frac{1}{2}\right)$	1
$\left(\frac{1}{2},0,0\right),\left(0,\frac{1}{2},0\right),\left(0,0,\frac{1}{2}\right)$	3
$\left(\frac{1}{2},\frac{1}{2},0\right),\left(\frac{1}{2},0,\frac{1}{2}\right),\left(0,\frac{1}{2},\frac{1}{2}\right)$	3

<u>Note</u>: If a specific ion or atom occupies a position in a crystal structure then identical atoms must be found to occupy all equivalent positions in the space group of the structure. Special positions can be occupied by only one type of atom or ion.

2-4. Classification of space groups

Space symmetry operations are those symmetry operations that move the central reference point by translation, bringing an equivalent reference point into coincidence. Space groups (230) give complete information as to the location of each atom or ion in a crystal by combination of:

a) The unit cell symmetry i.e., the Bravais lattice, and
b) The point group symmetry and/or
c) Special translation operations, i.e., glide planes, screw axes.

2-4-1 Geometric crystal classes

This classification of crystals is based on the crystal forms that are represented by the point group. Hence crystal classes are nomenclatured by 32 point groups. The (geometric) crystal class is defined as a set of crystals whose point group symmetry are the same. In this sense, a crystal class is represented by the point group symmetry for the crystal, but sometimes point group representation and crystal class are confused since they have the same symmetry representation. The confusion arises from historical reasons, since originally the crystal classes and point groups were developed from a morphological point of view and not from atomic structure or the internal symmetry of the crystals. The difference between the morphological and atomic structucral approaches can be illustrated in the following discussion of point space and vector space for the symmetry operation. A simple way to identify the (geometric) crystal class of a space group is to replace any screw axis and glide plane by its corresponding rotational axis and mirror plane, respectively.

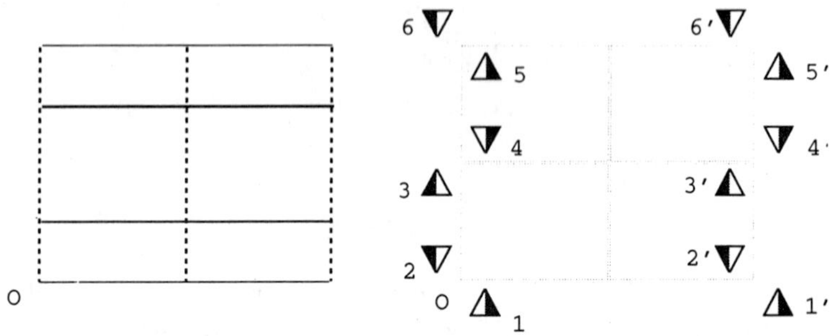

Figure 2-56 Symmetry representation of planar space group p2mg and its equipoints (1, 2, 5 and 6 are occupied by right-handed molecules while left-handed molecules in 3 and 4.)

In Table 2-14, 17 planar space groups are listed along with the corresponding point group symmetry. In a strict sense, this point group stands for the (geometric) crystal class. In some of instances, a point group is assigned although its necessary symmetry element does not exist. For example, in the case of space group p2mg, its corresponding point group or (geometric) crystal class is "2mm" although there is only one kind of mirror plane. This is shown in Fig.2-56.

2-4-2 Arithmetic crystal classes and symmorphic space groups

Symmorphic space groups exist when only a single atom or ion exists at the Bravais lattice points, with each lattice point having the symmetry of the point group. This leads to 73 space groups as shown in Table 2-9. The arithmetic classes and symmorphic space groups have the same symbolic representation.

Table 2-13 73 symmorphic space group

Crystal system	Symmorphic Space group	
Triclinic	P1, $P\bar{1}$	2
Monoclinic	P2, Pm, $P\frac{2}{m}$ C2 Cm $C\frac{2}{m}$	6
Ortho-rhombic	P222, Pmm2, Pmmm C222, Cmm2, Cmmm, Amm2 I222, Imm2, Immm F222, Fmm2, Fmmm	13
Tetragonal	P4, $R\bar{4}$, P422 $P\frac{4}{m}$ P4mm $P\bar{4}2m$, $P\bar{4}m2$, $P\frac{4}{m}mm$ I4, $I\bar{4}$, I422, $I\frac{4}{m}$, I4mm, $I\bar{4}2m$, $I\bar{4}m2$, $I\frac{4}{m}mm$	16
Trigonal	P3, $P\bar{3}$, P312, P321, P3m1, P31m, $P\bar{3}m1$, $P\bar{3}1m$ R3, $R\bar{3}$, R32, R3m, $R\bar{3}m$	13
Hexagonal	P6, $P\bar{6}$, $P\frac{6}{m}$, P622, P6mm, $P\bar{6}m2$, $P\bar{6}2m$, $P\frac{6}{m}mm$	8
Cubic	P23, $Pm\bar{3}$, P432, $P\bar{4}3m$, $Pm\bar{3}m$ I23, $Im\bar{3}$, I432, $I\bar{4}3m$, $Im\bar{3}m$ F23, $Fm\bar{3}$, F432, $F\bar{4}3m$, $Fm\bar{3}m$	15

However, they are represented differently by placing the lattice type in the first for the symmorphic groups and in the last position for the arithmethic classes. For example, P222 stands for the symmorphic space group while 222p represents the crystal class which includes all the space groups having point group symmetry of 222 in the primitive orthorhombic lattice.

With this background for the definition by point group, a geometric crystal class may have a symmetry group that does not have the symmetry element represented in the symbolic notation. For example, the space group "bca" in the orthorhombic system is classified in the crystal class "mmm" although it does not have any mirror plane.

2-4-3 Non-symmorphic space groups

A non-symmorphic space group has translational symmetry elements such as screw axis and glide planes. Since there are 73 symmorphic space groups among the 230 space groups, the number of non-symmorphic space groups is 157. In relation to the arithmetic crystal classes, non-symmorphic space groups are converted by replacing n fold screw axes(n_p) by n fold rotation axes and glide reflection planes a, b, c, n, or d by mirror planes.

2-5. Space group interpretation

Space group interpretation is basically the same as the point groups, except that the Bravais Lattice is included. The interpretation includes the crystal system, Bravais lattice, and Point Symmetry Operation(s).

<u>Examples</u>

1. I$\bar{4}$2m — Tetragonal System
   - Body centered lattice
   - 4-fold c-axis of rotary inversion
   - 2-fold axes along a_1a_2
   - Mirror planes along bisectors of a_1a_2

2. Amm2 — End centered orthorhombic lattice normal to a-axis
   - Mirror plane along a-axis
   - Mirror plane along b-axis
   - 2-fold c-axis

3. P3m1* — Primitive trigonal lattice 3-fold c-axis
   - Mirror planes along a_1a_2 axes

4. P31m* Primitive trigonal lattice 3-fold c-axis
Mirror planes along bisectors of $a_1 a_2$ axis

(* 1-fold axis in 3 and 4 given for orientation purposes only. cf. P312, P321)

2-5-1 Point space and vector space

Historically, crystals were classified accroding to their external appearance which are characterized by crystal form or crystal faces. This classification precedes the classification method using atomic position and symmetry elements of a crystal. If symmetry aspects of a crystal are identified by external faces, translational symmetry elements such as glide plane and screw axis have the same effect as the corresponding non-translation symmetry elements. Fig.2-57 illustrates this point. With respect to coordinates of atomic positions, the crystal shown in Fig.2-57 is moved to the same position by the 6_2 screw operation although mere 6 fold rotation does not move atoms to the equivalent atomic positions. However, as far as planes are concerned, pure rotation and screw operation do move the planes to equivalent planes since the relative position between planes or the angle between planes is important. Since the side faces of the hexagonal cylinder in Fig.2-57 have the same configuration of atomic positions, they are equivalent with respect to physical properties. In this respect they are equivalent to each other although the atoms A, A' and A" are not equivalent to each other with respect to the pure 6 fold axis. Historically, analytical methods for crystal structures such as X-ray diffraction and TEM were developed long after the optical methods used to characterize external appearance and face development in crystals. Before these analytical characterization tools for crystal structure appeared the crystal shown in Fig.2-57 had been classified as a symmetry group that has 6 fold

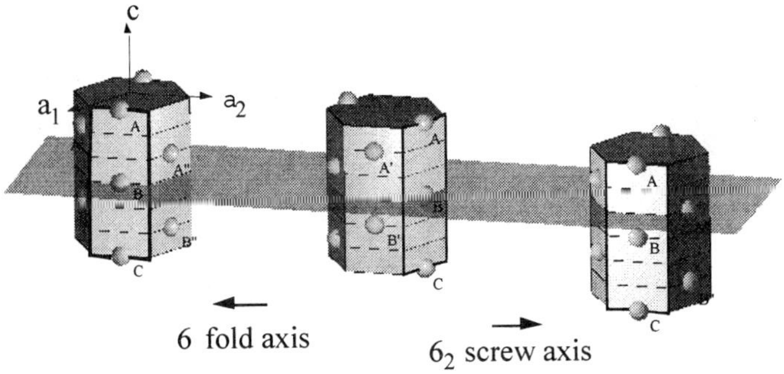

Figure 2-57 Comparison of rotation and screw axis on atomic position and plane

rotational symmetry since the atomic positions cannot be identified without these tools. This has been the convention for the classification of crystals. The reason why the space group "$6_1$22" belongs to the crystal class of "622" is that the crystal is identified as "622" when it is characterized by external appearance of the faces. In a strict sense, '622' is a symmetry representation of the crystal in vector space and not in point space.

Mathematically the difference in classification by atomic coordinates and by planes is explained in terms of point space and vector space. In point space, a line connecting points A and B is said to be same as the line connecting points C and D if the coordinates of A and C and those of B and D are the same. In contrast, a line connecting points A and B in vector space is said to be the same as the line connecting points C and D if their length and directions are the same, regardless of coordinates. This aspect is illustrated by an example of planar space group "p2mg" as shown in Fig.2-58. The glide plane moves atoms from 1 to 3 and from 2 to 4. Also it moves the origin O to the new position O'. Linewise, O1 moves to O3 and O2 moves to O4. In the point space, non-translation symmetry operation exists between O1 and O'3 or O2 and O'4. In vector space, O1 can move to O'3 by a mirror plane since their position does not matter. In Fig.2-58, O and O' are put on the same position to show the mirror plane. (This does not need to be.) In this respect, a glide plane in point space is not equal to a mirror plane although it has the same effect as a mirror plane. In vector space glide plane and mirror planes are regarded as mirror planes. Also rotational axis and screw axis are both regarded as rotational axis. Hence in vector space there is no difference between a point group and a space group for a crystal since translational symmetry operations do not contribute to generating new symmetry elements other than rotation and reflection.

In the morphological approach, crystals are classified according to characteristics of external faces. These faces are defined by face normals that are

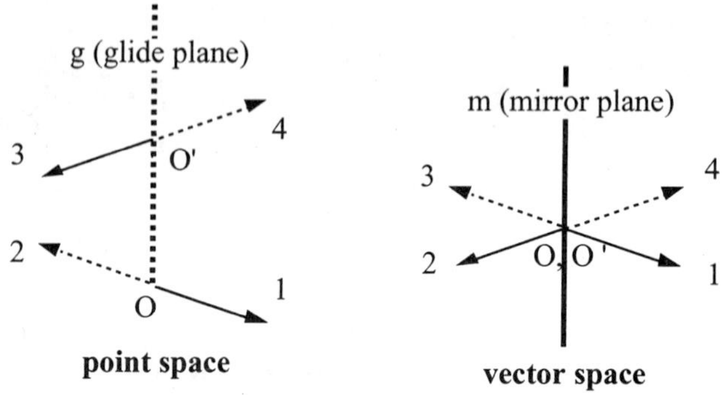

Figure 2-58 Equivalence of a glide plane in point space to a mirror plane in vector space

vectors. Since the symmetry aspects of faces are dependent on the relationship of face normals with one another, the symmetry of crystals must be defined in vector space. This is the reason why a crystal is assigned a point group although the symmetry elements required for the point group do not exist in the space group notation. Historically, the morphological approach had classified crystals into 32 classes before the theoretical approach classified them into the 230 space groups.

In this respect, no symmetry group, regardless of point group or space group, is common to any two of the 4 planar lattice systems. This fact illustrates the effectiveness of four 2-dimensional lattice systems in the classification symmetry groups for planar crystals.

2-5-2 Inclusion of Bravais lattice type in space group notation

Since the geometric shape of lattices makes it easy to classify planar crystals, classification with different symmetry elements can be developed systematically as shown in Fig. 2-46. As indicated in the discussion on equipoints, there are unlimited ways to arrange atoms in a lattice with given symmetry elements. Therefore, there may be many crystals with the same symmetry elements although the atomic arrangements are different from one to another. In this sense, a symmetry notation represents a group of those crystals with a given set of symmetry element and not those with the same atomic arrangements. The symmetric notation for a crystal given by "4" indicates that the crystal belongs to a group that is represented by "4." When crystals are characterized by symmetry elements including translational symmetry elements such as glide plane and screw axis, there are 17 space groups of symmetry. In a 2-dimensional lattice, only the glide plane is a possible translational symmetry operation. Since the screw axis is perpendicular to the planar lattice and it requires a translation parallel to the axis, it is not possible in the planar lattice. As the reflection plane for a glide operation is identified by the position of glide elements in the symbolic notation, so too is a mirror plane identified in the same manner.

By classifying crystals according to their space groups, not only do the number of groups increase but also there is a change in symmetry notation. The latter is inclusion of 'g' for glide plane and putting 'p' and 'c' in front of the symmetry notation that stand for primitive and centered, respectively. In point group symmetry, the centered rectangular lattice does not differ from the primitive rectangular lattice. This is not obeyed in the space group representation. This difference is compared in the case of space group 'cm' that represents the centered rectangular lattice with a mirror plane. Its graphical representation is shown in Fig.2-59(b).

The glide plane in space group "cm" in Fig.2-59 is not an implied symmetry element of the mirror plane. The glide plane is generated by putting an extra motif at the center of the lattice. According to the rule of symbolic notation for a symmetry group, the glide plane must be indicated in the notation. To locate it in the same position of the notation may conflict with the mirror plane since they have the same directions. Hence, instead of indicating explicitly the glide plane in the notation, the glide plane is connoted by putting "c" in the symmetry notation that represents a centered lattice in the corresponding symmetry group. Here "c" does not mean merely existence of an extra motif at the center of the lattice. Rather it is used to designate the generation of a new symmetry element, and not one which is implied by the existence of an extra motif at the center of the lattice. This double-feature of the term "c" in the notation can be well illustrated in the space group "pg." In the case of space group "pg," even with an extra motif at the center of the lattice, a new symmetry element is not generated. This is shown in Fig.2-59(a). Geometrically "cg" is a centered rectangular lattice whereas it is not symmetrically "cg." Hence geometrical "cg" corresponds to the space group "pg" with double motifs of the original primitive lattice. This is similar to the fact that the rectangular lattice in Fig.2-59 is not symmetrically rectangular. Sometimes a geometical centered lattice generates new symmetry elements. However, it may belong to the same symmetry group as the primitive lattice. For example, a geometrical centered lattice of 'p2gm' results in the same symmetry group 'c2mm'. These facts explain why there are only two space groups with a centered rectangular lattice although it is possible to make a centered lattice by putting an extra lattice at the center of the unit cell of any primitive lattice.

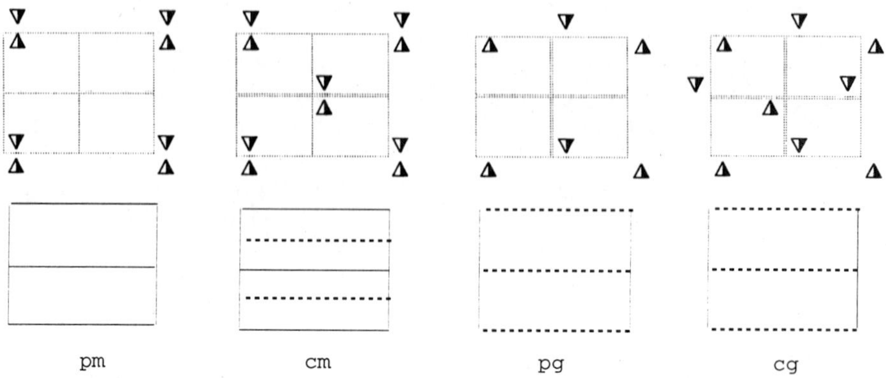

Figure 2-59 Comparison of geometrical and symmetrical centered lattice systems of pm and pg (Half filled triangle shows right or left-handedness of atoms or molecules. Thick line and dotted line indicate mirror plane and glide plane, respectively.)

Table 2-14 Classification of 17 planar space groups with lattice systems

Lattice system	Point group	Space group	Short Notation	Lattice system	Point group	Space group	Short Notation
Parallelogram	1	p1	p1	Square	4	p4	p4
	2	p211	p2		4mm	p4mm	p4m
Rectangular	m	p1m1	pm			p4gm	p4g
		p1g1	pg	Hexagonal	3	p3	p3
		c1m1	cm		3m	p3m1	p3m1
	2mm	p2mm	pmm			p31m	p31m
		p2mg	pmg		6	p6	p6
		p2gg	pgg		6mm	p6mm	p6m
		c2mm	cmm				

For lattices other than rectangular lattices, one centered lattice is equivalent to another primitive lattice with the same geometry. For example, if one extra motif (lattice point) is put at the center of a square lattice, another smaller geometric square lattice that is primitive is produced. Hence a centered square lattice is geometrically and symmetrically regarded as another primitive square lattice. For this reason there is no other centered lattice except the rectangular lattice.

D. Summary: Relationship to Crystal Structure

The goal of crystal chemistry is to identify the crystal structures in solids and to connect them to the physical properties of the solids. Crystallography employs a symbolic language that is based on the lattice and symmetry of a crystal in order to facilitate analysis of the structure. The analytical methods utilized in crystallogrphy provide information about the lattice of a crystal. The actual atomic positions are obtainable from the analysis of the lattice of the crystal.

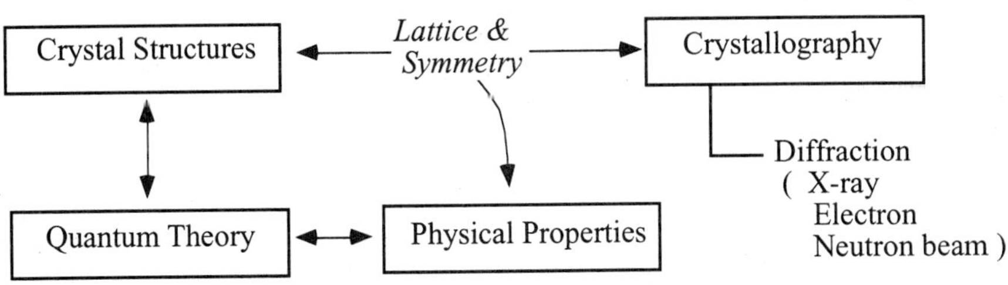

Figure 2-60 The relationship between crystal structures, their properties and crystallography

This symbolic language of crystallography offers an efficient way for the systematic analysis of the relationship between crystal structures and their properties without reference to quantum theory. Even in quantum theory, the lattice concept is utilized in developing the band theory of crystals. The simple correlations between crystal structure, crystallography and quantum theory are shown in Fig.2-60.

Crystals are classified according to the symmetry elements which the crystals possess. Hence, a crystal class means a group of crystals that share common symmetry elements although their atomic arrangements may be different. Within the crystal classes are grouped symmetry operations which produce repetitive planar and 3-dimensional lattices. The combining of symmetry operations to make up the lattices is not arbitray, the number of ways being limited, due to the unique repetition of the lattice points. In the absence of translational, there are only 10 ways to arrange symmetry operations resulting with 10 crystal classes in the planar lattices. Likewise, only 32 crystal classes exist in the 3-dimensional lattices since there are only 32 ways to arrange non-translational symmetry operations. Each class defined in this way is called a point group since non-translational symmetry operations leave at least one point unmoved. In contrast to the point group, a space group is a symmetry group that may contain symmetry operations such as screw axis and glide plane. By including non-translation symmetry operations, the crystal classes increase from 10 to 17 in the planar lattice and from 32 to 230 in the 3-dimensional lattice.

Knowing the space group and equipoints (Tabulated in International Tables -- Wyckoff), one can readily determine the crystal structure and arrangement of the atoms or ions in the unit cell or crystal. This has a direct bearing on such questions as the defect state, substitution and properties of the crystal. The crystal properties then needed to make a complete determination of the structure are:

(i) the crystal system;
(ii) the lattice parameters;
(iii) the Bravais lattice;
(iv) the number and kinds of atoms in the unit cell;
(v) and the locations of the atoms in the unit cell.

Experimentally these determinations are carried out by X-ray or neutron diffraction analysis of the crystal, and from a knowledge of the chemical composition and known accurate density of the crystal.

A full understanding of crystal chemisry requires, therefore, knowledge of the basics of crystallography. Such concepts as lattice systems and classification of space groups are essential not only for identifying crystal structures but also for understanding the chemistry of crystals.

Table 2-15 List of 230 space groups

Triclinic

#	Point group	P	Schoenflies Notation
1	1	1	C_1^1
2	$\bar{1}$	$\bar{1}$	C_i^1

Monoclinic

	Point group	P	C	Schoenflies Notation
3	2	2		C_2^1
4		2_1		C_2^2
5			2	C_2^3
6	m	m		C_s^1
7		c		C_s^2
8			m	C_s^3
9			c	C_s^4
10	2/m	2/m		C_{2h}^1
11		2_1/m		C_{2h}^2
12			2/m	C_{2h}^3
13		2/c		C_{2h}^4
14		2_1/c		C_{2h}^5
15			2/c	C_{2h}^6

Orthorhombic

	Point group	P	C	F	I	A	Schoenflies Notation
16	222	222					D_2^1
17		222_1					D_2^2
18		2_12_12					D_2^3
19		$2_12_12_1$					D_2^4
20			222_1				D_2^5
21			222				D_2^6
22				222			D_2^7
23					222		D_2^8
24					$2_12_12_1$		D_2^9
25	mm2	mm2					C_{2v}^1
26		$mc2_1$					C_{2v}^2
27		cc2					C_{2v}^3
28		ma2					C_{2v}^4
29		$ca2_1$					C_{2v}^5
30		nc2					C_{2v}^6
31		$mn2_1$					C_{2v}^7
32		ba2					C_{2v}^8
33		na2					C_{2v}^9

(continued)

	Point group	P	C	F	I	A	Schoenflies Notation
34		nn2					C_{2v}^{10}
35			mm2				C_{2v}^{11}
36			$mc2_1$				C_{2v}^{12}
37			cc2				C_{2v}^{13}
38						mm2	C_{2v}^{14}
39						bm2	C_{2v}^{15}
40						am2	C_{2v}^{16}
41						ba2	C_{2v}^{17}
42				mm2			C_{2v}^{18}
43				dd2			C_{2v}^{19}
44					mm2		C_{2v}^{20}
45					ba2		C_{2v}^{21}
46					ma2		C_{2v}^{22}
47	mmm	mmm					D_{2h}^{1}
48		nnn					D_{2h}^{2}
49		ccm					D_{2h}^{3}
50		ban					D_{2h}^{4}
51		mma					D_{2h}^{5}
52		nna					D_{2h}^{6}
53		mna					D_{2h}^{7}
54		cca					D_{2h}^{8}
55		bam					D_{2h}^{9}

	Point group	P	C	F	I	A	Schoenflies Notation
56		ccn					D_{2h}^{10}
57		bcm					D_{2h}^{11}
58		nnm					D_{2h}^{12}
59		mmn					D_{2h}^{13}
60		ncn					D_{2h}^{14}
61		bca					D_{2h}^{15}
62		nma					D_{2h}^{16}
63			mcm				D_{2h}^{17}
64			mca				D_{2h}^{18}
65			mmm				D_{2h}^{19}
66			ccm				D_{2h}^{20}
67			mma				D_{2h}^{21}
68			cca				D_{2h}^{22}
69				mmm			D_{2h}^{23}
70				ddd			D_{2h}^{24}
71					mmm		D_{2h}^{25}
72					bam		D_{2h}^{26}
73					bca		D_{2h}^{27}
74					mma		D_{2h}^{28}

Tetragonal

	Point group	P	I	Schoenflies Notation
75	4	4		C_4^1
76		4_1		C_4^2
77		4_2		C_4^3
78		4_3		C_4^4
79			4	C_4^5
80			4_1	C_4^6
81	$\bar{4}$	$\bar{4}$		S_4^1
82			$\bar{4}$	S_4^2
83	4/m	4/m		C_{4h}^1
84		4_2/m		C_{4h}^2
85		4/n		C_{4h}^3
86		4_2/n		C_{4h}^4
87			4/m	C_{4h}^5
88			4_1/a	C_{4h}^6
89	422	422		D_4^1
90		42_12		D_4^2
91		4_122		D_4^3
92		4_12_12		D_4^4
93		4_222		D_4^5
94		4_22_12		D_4^6

	Point group	P	I	Schoenflies Notation
95		4_322		D_4^7
96		4_32_12		D_4^8
97			422	D_4^9
98			4_122	D_4^{10}
99	4mm	4mm		C_{4v}^1
100		4bm		C_{4v}^2
101		4_2cm		C_{4v}^3
102		4_2nm		C_{4v}^4
103		4cc		C_{4v}^5
104		4nc		C_{4v}^6
105		4_2mc		C_{4v}^7
106		4_2bc		C_{4v}^8
107			4mm	C_{4v}^9
108			4cm	C_{4v}^{10}
109			4_1md	C_{4v}^{11}
110			4_1cd	C_{4v}^{12}
111	$\bar{4}$2m	$\bar{4}$2m		D_{2d}^1
112		$\bar{4}$2c		D_{2d}^2
113		$\bar{4}2_1$m		D_{2d}^3
114		$\bar{4}2_1$c		D_{2d}^4

(continued)

	Point group	P	I	Schoenflies Notation
115		$\bar{4}m2$		D_{2d}^{5}
116		$\bar{4}c2$		D_{2d}^{6}
117		$\bar{4}b2$		D_{2d}^{7}
118		$\bar{4}n2$		D_{2d}^{8}
119			$\bar{4}m2$	D_{2d}^{9}
120			$\bar{4}2c$	D_{2d}^{10}
121			$\bar{4}2m$	D_{2d}^{11}
122			$\bar{4}2d$	D_{2d}^{12}
123	4/m mm	4/m mm		D_{4h}^{1}
124		4/m cc		D_{4h}^{2}
125		4/n bm		D_{4h}^{3}
126		4/n nc		D_{4h}^{4}
127		4/m bm		D_{4h}^{5}
128		4/m nc		D_{4h}^{6}
129		4/n mm		D_{4h}^{7}
130		4/n cc		D_{4h}^{8}
131		4_2/m mc		D_{4h}^{9}
132		4_2/m cm		D_{4h}^{10}
133		4_2/n bc		D_{4h}^{11}
134		4_2/m nm		D_{4h}^{12}
135		4_2/m bc		D_{4h}^{13}
136		4_2/m nm		D_{4h}^{14}

	Point group	P	I	Schoenflies Notation
137		4_2/n mc		D_{4h}^{15}
138		4_2/n cm		D_{4h}^{16}
139			4/m mm	D_{4h}^{17}
140			4/m cm	D_{4h}^{18}
141			4_1/a md	D_{4h}^{19}
142			4_1/a cd	D_{4h}^{20}

Trigonal

	Point group	P	R	Schoenflies Notation
143	3	3		C_3^{1}
144		3_1		C_3^{2}
145		3_2		C_3^{3}
146			3	C_3^{4}
147	$\bar{3}$	$\bar{3}$		C_{3i}^{1}
148			$\bar{3}$	C_{3i}^{2}
149	32	312		D_3^{1}
150		321		D_3^{2}
151		$3_1$12		D_3^{3}
152		$3_1$21		D_3^{4}
153		$3_2$12		D_3^{5}

	Point group	P	R	Schoenflies Notation
154		3_221		D_3^6
155			32	D_3^7
156	3m	3m1		C_{3v}^1
157		31m		C_{3v}^2
158		3c1		C_{3v}^3
159		31c		C_{3v}^4
160			3m	C_{3v}^5
161			3c	C_{3v}^6
162	$\bar{3}$m	$\bar{3}$1m		D_{3d}^1
163		$\bar{3}$1c		D_{3d}^2
164		$\bar{3}$m1		D_{3d}^3
165		$\bar{3}$c1		D_{3d}^4
166			$\bar{3}$m	D_{3d}^5
167			$\bar{3}$c	D_{3d}^6

Hexagonal

	Point group	P	Schoenflies Notation
168	6	6	C_6^1
169		6_1	C_6^2
170		6_2	C_6^3
171		6_3	C_6^4
172		6_4	C_6^5
173		6_5	C_6^6
174	$\bar{6}$	$\bar{6}$	C_{3h}^1
175	6/m	6/m	C_{6h}^1
176		6_3/m	C_{6h}^2
177	622	622	D_6^1
178		$6_1$22	D_6^2
179		$6_5$22	D_6^3
180		$6_2$22	D_6^4
181		$6_4$22	D_6^5
182		$6_3$22	D_6^6
183	6mm	6mm	C_{6v}^1
184		6cc	C_{6v}^2
185		6_3cm	C_{6v}^3
186		6_3mc	C_{6v}^4
187	$\bar{6}$m2	$\bar{6}$m2	D_{3h}^1
188		$\bar{6}$c2	D_{3h}^2
189		$\bar{6}$2m	D_{3h}^3
190		$\bar{6}$2c	D_{3h}^4
191	6/m mm	6/m mm	D_{6h}^1
192		6/m cc	D_{6h}^2
193		6_3/m cm	D_{6h}^3
194		6_3/m mc	D_{6h}^4

(continued)

Cubic system

	Point group	P	F	I	Schoenflies Notation
195	23	23			T^1
196			23		T^2
197				23	T^3
198		2_13			T^4
199				2_13	T^5
200	m3	m3			T_h^1
201		n3			T_h^2
202			m3		T_h^3
203			d3		T_h^4
204				m3	T_h^5
205		a3			T_h^6
206				a3	T_h^7
207	432	432			O^1
208		4_232			O^2
209			432		O^3
210			4_132		O^4
211				432	O^5
212		4_332			O^6
213		4_132			O^7
214				4_132	O^8
215	$\bar{4}3m$	$\bar{4}3m$			T_d^1
216			$\bar{4}3m$		T_d^2
217				$\bar{4}3m$	T_d^3
218		$\bar{4}3n$			T_d^4
219			$\bar{4}3c$		T_d^5
220				$\bar{4}3d$	T_d^6
221	m3m	m3m			O_h^1
222		n3n			O_h^2
223		m3n			O_h^3
224		n3m			O_h^4
225			m3m		O_h^s
226			m3c		O_h^6
227			d3m		O_h^7
228			d3c		O_h^8
229				m3m	O_h^9
230				a3d	O_h^{10}

CHAPTER 3

CRYSTAL STRUCTURES

SECTION 1 SPACE FILLING OF ATOMS

As indicated in Chapter 2, the crystal structures formed in a solid must belong to one of the 230 space groups. In terms of thermondymanics the particular structure adopted will be the lowest energy form for the given environmental conditions. This lowest energy state is determined primarilly by geometric factors relating to the packing of constituent atoms or ions within the solid. The concept of space filling or packing, as a determinant of crystal structure development is based on the postulate that: the most probable structures formed are those for which the most economical use of space is made, For crystals, this translates into those structures having the hightst density of packing - consistent with other requirements (bonding characteristics, size effect, charge balance), since these structures experience better shielding and, hence, lower potential energy results. Conditions for densest packing to occur are that the atoms must be:

a) essentially spherical
b) of identical size, and
c) non-directional in bonding

These conditions are, for the most part, met by molecular crystals (Ne, Ar, etc.), and most metals. Also alloys and most ionic crystals, although their components have different sizes, can be depicted by the atomic packing concept since these do not have directionality like covalent crystals. In contrast, covalent structures cannot be developed from this simple packing concept since each atom has directional bonding "arms". Nonetheless some packing concept can also be applied to covalent crystals. The crystal structure of metals will first be described according to the space filling postulate since metals satisfy the above 3 conditions followed by ionic crystals which consist of components with different sizes. Finally characteristics of covalent crystals will be described that have directional bonding with even different sizes.

A. Closest Atomic packing

The concept of closest atomic packing in a plane is illustrated in Fig.3-1, which shows the arrangement of atomic spheres in a square array(a) and in a hexagonal array(b). With the square array, each atom touches or is coordinated to 4 nearest neighbor atoms. For the hexagonal array, this 1st coordination number is 6. If the space filling is considered in a two-dimensional plane, the void area of the square array is $(A_{square}-A_{circle})/A_{square}$ that is $1-\frac{\pi r^2}{4r^2}$. The resulting void area is 21.5%. In the hexagonal array, the void area is $(A_{triangle}-A_{circle})/A_{triangle}$ that is $1-\frac{\pi r^2}{2\sqrt{3}r^2}$. Thus the void area is 9.3%. Hence the hexagonal array of atomic spheres in a plane results in a denser packing and more generally gives the densest structures. In consequence, the densest structures in 3 dimensional space results from the stacking of hexagonal layer arrays as shown in Fig. 3-2. Depending on the stacking mode, two closest packed structures result, which are called hexagonal closest packing(HCP) and cubic closest packing(CCP).

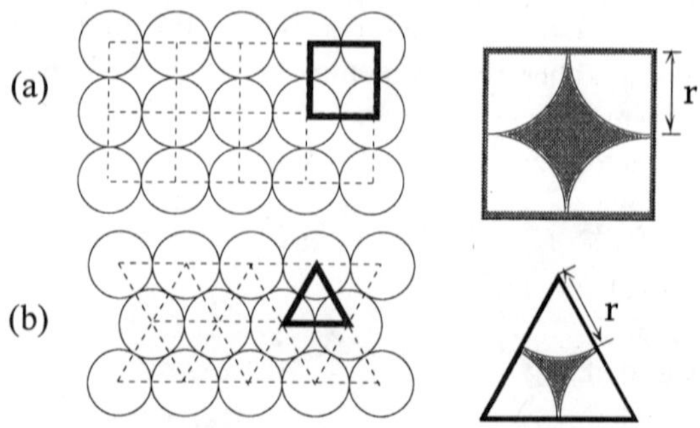

Figure 3-1 Packing of spheres in a plane with (a) 4 and (b) 6 coordination

B. Hexagonal closest packing (HCP) and Cubic closest packing (CCP)

The difference between HCP and CCP in Fig.3-2 results from the relative positions of the stacked layers. To illustrate this concept with respect to the stacked layers, equivalent positions in the close packed hexagonal layer are named as A, B, and C sites. This is shown in Fig. 3-2(a). Let's assume that each sphere in the base layer occupies an A site. Then the spheres in the second layer must occupy B or C sites for the densest packing to occur. If B sites are occupied for the second layer, the possible position for the third layer

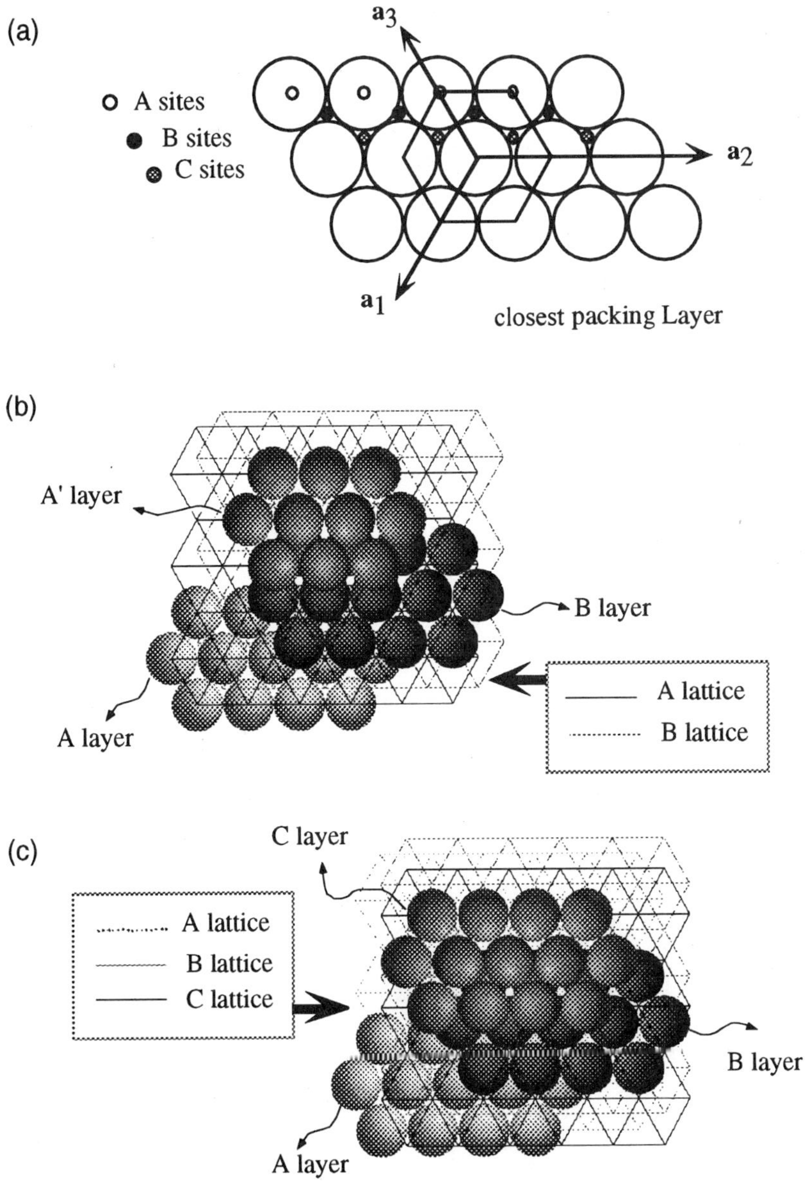

Figure 3-2 (a) Arrangement of atoms in the closest packing
(b) AB stacking and (c) ABC stacking

is on A or C site. If C sites are occupied for the second layer, the possible position for the third layer is a A or B site.

In the stacking of the CCP layers B or C sites may be chosen as the location for the second layer spheres, as indicated. The CN of each A atom in the plane is 6 (nearest neighbors). Each A atom is surrounded by 3 B sites and 3 C sites which are the possible site locations for stacking subsequent layers. B and C sites cannot both be occupied on the same plane by equal sized atoms. If B sites are chosen, these now become tetrahedral void spaces (surrounded by 4 atoms) and the C sites then become octahedral void spaces (surrounded by 6 atoms). A third layer in the CCP stack may be located either on A or C sites. If A, then the stacking sequence becomes ABABAB–, which leads to the hexagonal close packed structure hcp. If C, the sequence becomes ABCABCABC – , that leads to the face centered structure – fcc.

Within the CCP or HCP stack, each atom is surrounded by 12 nearest neighbors, i.e., CN – 12. 6 atoms in the plane A. 3 atoms above the plane +B (or C). 3 atoms below the plane -B (or C).

1. CCP and FCC(Face centered cubic) structure

Cubic closest packing of atoms results in the face centered cubic structure. Their equivalence is illustrated in Fig.3-3. Due to this equivalence, FCC and CCP are interchangeably used. FCC is the symmetry-emphasized term while CCP emphasizes more the packing of atoms resulting from this stacking of atomic layers. In consequence, it is hardly possible to show the four fold rotational symmetry associated with the layer structure of the cubic closest packed structure illustrated in Fig.3-3(a). Fig.3-3(b) illustrates, however, how the FCC unit cell is generated from four stacking layers of atoms. With respect to the stacking sequence of ABCABC in the CCP structure, the length of repeating unit is the distance between two adjacent A, B or C layers. This distance is the length of the body diagonal in the FCC unit cell shown in Fig.3-3(b). Atoms "0" and "10" belong to the same type layer in the stacking sequence.

2. Symmetry elements in HCP and CCP

In Fig.3-3, it was shown how the CCP structure generates the FCC structure. In terms of symmetry, the FCC structure has 3-fold symmetry along with 4-fold rotational symmetry. Like the equivalence of CCP to FCC, HCP is equivalent to the hexagonal lattice in terms of the symmetry although there is no 6-fold rotational symmetry element present in the CCP structure. The term "hexagonal" implies that 6 fold rotational symmetry is present in each layer of atoms. However, when these layers of atoms are stacked, the top layer is shifted parallel to the layer below by a fraction of the distance between the two

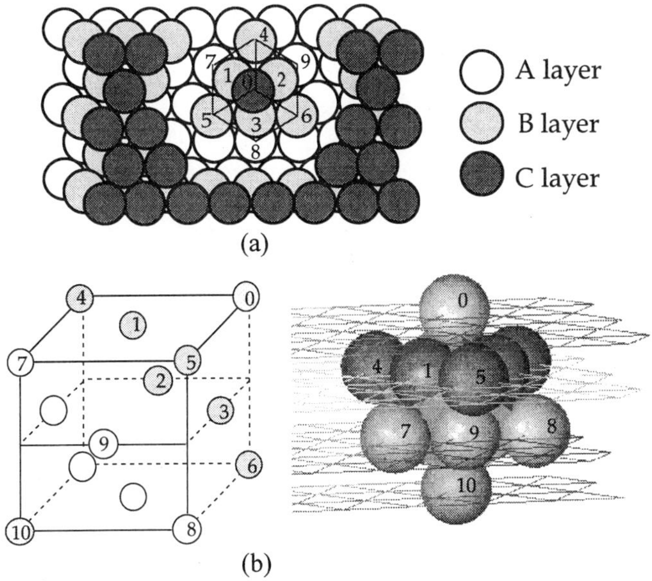

Figure 3-3 Equivalence of FCC structure and CCP structure

touching atoms. As a result of this shift, the 6 fold rotational axis turns into a 6_3 screw axis in the HCP structure while the 6 fold rotational axis becomes a 3 fold rotational axis in the CCP structure. Due to the presence of a 6_3 screw axis, the HCP structure of atoms can be regarded as a hexagonal lattice structure although it does not have true 6 fold rotational symmetry. The FCC structure, in contrast, loses its 6-fold symmetry and acquires both 3 fold and 4 fold rotational symmetries.

C. Packing density

1. CCP structure

Packing density for FCC structure may be estimated from the cell geometry as follows:

The face diagonal of the FCC unit cell is equal to $4r_a$ as shown in Fig.3-4.

Therefore $4r_a = \sqrt{2}a_o$

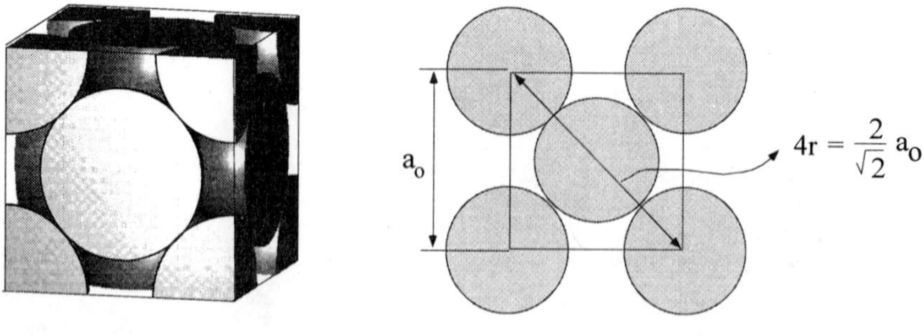

Figure 3-4 View of FCC unit cell and cross section from [100] direction

$$a_o = \frac{4}{\sqrt{2}} r_a = 2\sqrt{2} r_a$$

The volume of unit cell = $(a_o)^3 = \left(2\sqrt{2} r_a\right)^3 = 16\sqrt{2} r_a{}^3$

Since there are 4 atoms in the unit cell

the volume of these atoms = $4 \times \frac{4}{3} \pi r_a{}^3 = \frac{16}{3} \pi r_a{}^3$

The resultant stacking density = $\frac{16}{3} \times \frac{\pi r_a{}^3}{16\sqrt{2} r_a{}^3} = \frac{\pi}{3\sqrt{2}}$

$= 0.741$ or 74.1%

and void volume = 25.9%

2. HCP

Packing density for HCP is estimated according to the geometry illustrated in Fig.3-5. The cross sections of HCP stacking of spheres in Fig.3-5 are shown to illustrate the relationship between the radius of the sphere and the unit cell dimensions. On the left in Fig.3-5 is a hexagonal cylinder whose volume is 3 times that of the unit cell. The actual unit cell is shown in bold lines. The expanded picture in Fig.3-5 shows the cross section of the hexagonal cylinder that is cut at the half of c-axis with the upper half of it cut again along the c-axis. From the Pythagorean relationship for the triangle of PO_2O_3, $\overline{PO_3}$ is $\sqrt{3}$r since $\overline{O_2O_3}$ is 2r and $\overline{O_2P}$ is r. The point T is the center of gravity that is illustrated by drawing a perpendicular line from the vertices of the triangle to the facing lines. Since the triangle $O_1O_2O_3$ is an equilateral triangle, the area of $\Delta\ PTO_2$ is one third of $\Delta\ PO_2O_3$, and, consequently, $\overline{PT}$ is one third of $\overline{PO_3}$. Therefore

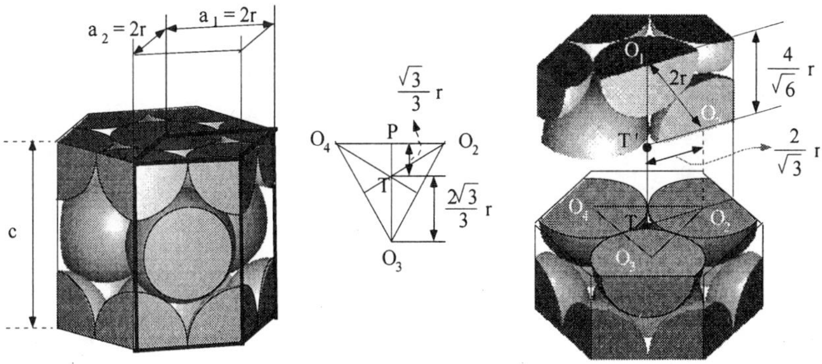

Figure 3-5 View of HCP unit cell and view of cross section from [001] direction

$\overline{PT}$ is equal to $\frac{\sqrt{3}}{3}$ and $\overline{TO_3}$ is $\frac{2\sqrt{3}}{3}$. For the Δ $T'O_1O_2$, $\overline{O_1O_2}$ is 2r and $\overline{T'O_2}$ is equal to $\overline{TO_2}$ that is $\frac{2\sqrt{3}}{3}$. From the Pythagorean relationship for the triangle of $T'O_1O_2$, $\overline{TO_1}$ is $\frac{4}{\sqrt{6}}$ r. From these relationships between the radius of the sphere and unit cell dimensions, the ratio of $\frac{c}{a}$ for the unit cell is calculated as 1.633 where "a" is equal to 2r.

The packing density is consequently calculated from these relationships as follows :

$$\text{Volume of unit cell} = a^2c \sin 60° = a^2c\frac{\sqrt{3}}{2}$$

where c = 1.633a

$$\text{Volume} = 1.633\frac{\sqrt{3}}{2}a^3 = 1.633\frac{\sqrt{3}}{2}(2r)^3$$

since a = 2r

$$\text{Volume of atoms} = 2 \times \frac{4}{3}\pi r_a{}^3 = \frac{8}{3}\pi r_a{}^3$$

Since there are 2 atoms per unit cell

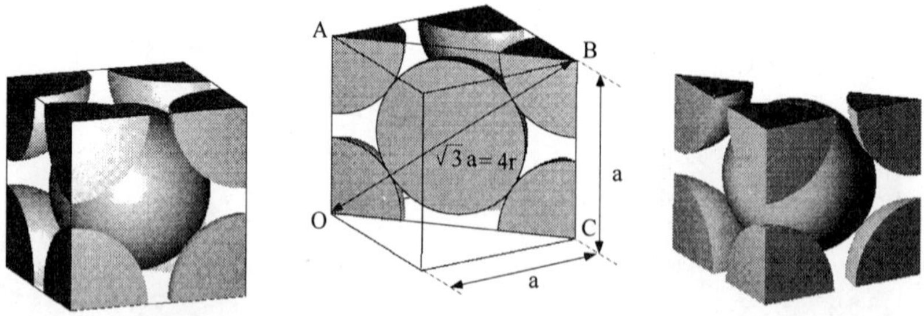

Figure 3-6 Unit cell of BCC packing and view of cross section cut by (110) plane

$$\text{Packing density} = \frac{8}{3} \times \frac{\pi r_a^{\ 3}}{1.633 \times \frac{\sqrt{3}}{2}(2r)^3} = \frac{2\pi}{1.633 \times 3\sqrt{3}}$$

$$= 0.741 \text{ or } 74.1\%$$

Void volume = 25.9%

2. Body centered cubic(BCC)

From the view of the cross section along the (110) plane in Fig.3-6, the body diagonal, $\overline{OB}$, is 4r. According to the Pythagorean theorem, $\overline{AB}$ is $\sqrt{2}a$ and $\overline{OB}$ is $\sqrt{3}a$. Hence, $4r = \sqrt{3}a$. From this result, the relationship between unit cell dimension and radius of the sphere is expressed as $a = \frac{4}{\sqrt{3}}r$.

Packing density for BCC structure

The number of atoms per unit cell is 2.

$$\text{Volume occupied} = \frac{\#\text{of atoms} \times \text{Volume of atom}}{\text{volume of unit cell}}$$

$$= \frac{2 \times \frac{4}{3}\pi r^3}{\frac{64}{3\sqrt{3}}r^3} = \frac{\sqrt{3}\pi}{8} = 0.68$$

Therefore volume occupied = 68%
Void volume = 32%

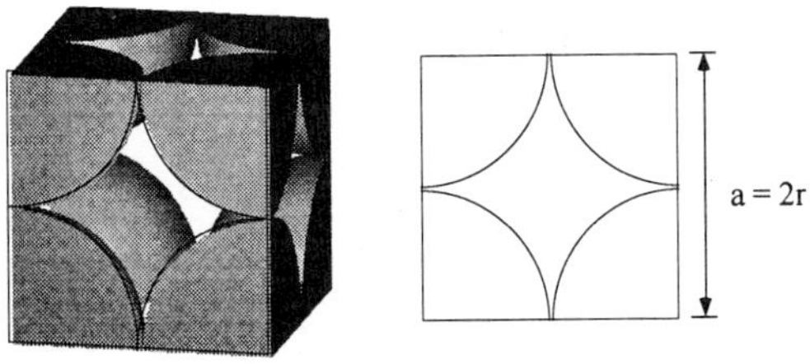

Figure 3-7 Stacking of atoms in simple cubic structure and view of cross section from [100] direction

3. Simple cubic

Packing density of simple cubic structure as shown in Fig.3-7 is calculated as:

Number of atoms/unit cell = 1

$$\text{Volume} = \frac{4}{3} \times \pi r_a^3$$

$$\text{Unit cell volume} = (a_o)^3 = (2r_a)^3 = 8r_a^3$$

$$\text{Volume occupied} = \frac{\frac{4}{3}\pi r_a^3}{8r_a^3} = \frac{\pi}{6} = 0.52$$

Therefore volume occupied = 52%

Void volume = 48%

Note: This is not a common structure due to the low packing densities.

SECTION 2 METALLIC CRYSTALS

With respect to space filling, the simple cubic structure is less efficient than CCP, HCP, or BCC structures since its void volume is close to 50%. Therefore, the simple cubic structure is not a common structure in metals. As for BCC, it is little different from CCP or HCP by 6% in packing density. Hence its energy is expected to be close to those of CCP or HCP structures. Calculations show that energy differences between these structures – HCP, CCP, and BCC – are not very great, hence some metals may change structure with applied temperature or pressure. E.g.. the BCC structure being less dense, tends to be adopted by metals at high temperatures. Nearly all metals adopt the BCC, FCC or HCP basic structures. Structure types of metal elements are listed in Fig.3-8. In this figure, only the crystal structure type at room temperatue is listed for each metal element.

Table 3-1 The relationship between crystal structure of metals and sp electrons of an atom

Crystal structure	sp electrons per atoms in bonding
BCC	< 1.5
HCP	1.7 – 2.1
FCC	2.5 – 3

The crystal structures of metals according to Engel's model is strongly influenced by the sp selectron bonding per atom as summarized in Table 3-1. The sp electrons play a major role while d electrons affect much less the crystal structure. Alkali elements have one valence electron in the s-orbital. Hence all the alkali elements have the BCC structure. For lighter alkali earth elements like Be and Mg, there are two valence electrons in the s-orbital. In the metalic state, the valence electrons are expected to be promoted to the sp hybrid orbital. This promotion results in the two sp electrons for the bonding. Hence Be and Mg are expected to have the HCP structure. As atomic number increases, the energy difference between s, p and d orbitals decreases. The smaller energy difference between s, p, and d orbitals in the heavy elements make it possible to promote valence s electrons of alkali earth elements to the d orbital. The hybrid orbital with the d-orbital for the bonding electrons consequently decrease the number of sp electrons. In this respect, the crystal structure of Ba as BCC can be explained while it is very difficult to explain why Ca and Sr have the FCC structure which requires more sp electrons than the BCC or HCP structure. However, Engel's model agrees with the general trend of crystal structures in metals in that elements on the left side of the periodic table take the BCC structure and the metal elements on the right side of the periodic table take the FCC structure as shown in Fig.3-8, since the number of *sp* electrons increases as the atomic number increases along the same row (period) in the period table.

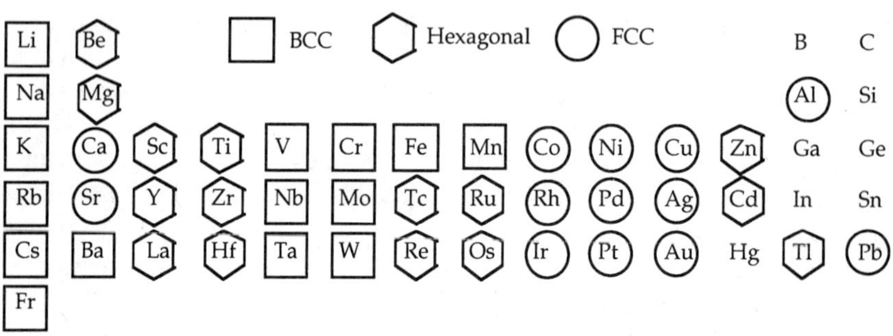

Figure 3-8 Crystal structures of metals at room temperature

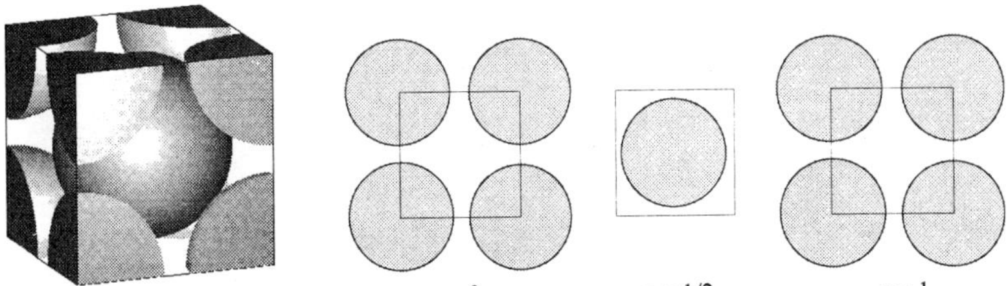

Figure 3-9 Unit cell for BCC structure and illustration of atomic positions along the z-axis of the unit cell.

Although the Engel's model derives from empirical relationships, the electronic configurations that are proposed for the crystal structures in Table 3-1 according to Engel's model were later confirmed by Brewer using thermodynamic data for the bonding energy of metals and the energy states of the metal atoms. Brewer's data indicates that the bonding orbital in V and Cr are d^4s and d^5s, respectively, and according to Engel's model, both V and Cr are expected to have BCC structure. Hence, Engel's model gives a hint that the electronic structure of metals is related to their crystal structure.

BCC Metals (W Type) adopted by

a) Alkali metals

	Li	Na	K	Rb	Cs
a_o(Å)	3.50	4.29	5.27	–	–

b) Other metals

	Ba	Ta	W	α-Fe
a_o(Å)	5.025	3.306	3.160	

FCC METALS Cu Type

a) Most noble metals

	Ag	Au	Pt	Rh
a_o(Å)	4.086	4.078	–	–

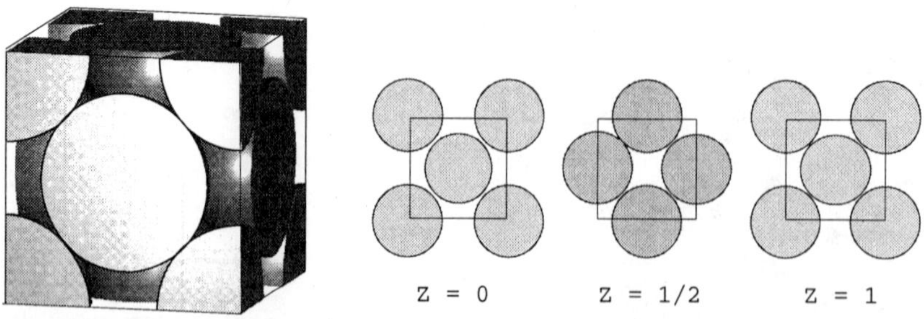

Figure 3-10 Unit cell for FCC structure and illustration of atomic positions along the z-axis of the cell.

b) Other metals

	Cu	Al	Ca	Co	Ni	γ-Fe	Pb
a_o(Å)	3.615	4.05	5.58	3.55	–	3.59	–

3. HCP METALS Mg Type

Many alkali earth and divalent group II metals

	Be	Cd	Mg	Zn	
a_o(Å)	2.287	2.979	3.209	2.605	P6mmc

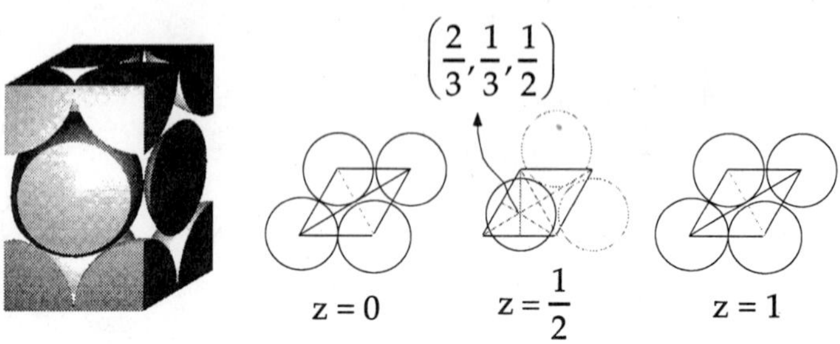

Figure 3-11 Unit cell for HCP structure and illustration of atomic positions along the relative z-axis of the drawing which is the c-axis in the unit cell

SECTION 3 IONIC CRYSTALS

A. Space filling by different size atoms

For metal crystals, a single atomic size is assumed. In contrast, for ionic crystals, the condition of single atomic size is no longer applicable for filling space, since anions are usually larger than cations. However, with elements of different sizes, the efficiency of space filling may be even higher than that achieved with closest packing of single sized atoms if the larger atoms (anions) form a closest packed structure, with the smaller atoms(cations) filling the void spaces.

The possible sites for the smaller atoms(cations) in a closest packed (CCP or HCP) structure are the octahedral or tetrahedral sites. Since these void sites have different sizes, they are selectively occupied by the cations. Also due to the charge neutrality requirement, occupancy of cations in the void sites is limited. Hence the structure of ionic crystals can be developed from the filling or partial filling of voids by cations in closest packed anion structures according to:

a) the ratio the radius of the cations to that of packed anions
b) the ratio of the number of anions to that of the cations in the chemical formula.

B. Interstitial voids in FCC and HCP structure for cations of ionic crystals

1. Void Location in FCC Unit Cell

Octahedral and tetrahedral sites are shown in Fig.3-12. With respect to the closest packing of anions, the number of octahedral sites is equal to the number of anions regardless of FCC or HCP packing. As shown in Fig.3-12, there are 8 tetrahedral sites in each unit cell. Since the number of anions per FCC unit cell is four, this number of tetrahedral sites is twice the number of anions. In the case of octahedral sites, the number per anion is not as easily recognized. There are 12 such sites on the edge of the unit cell, each contributing one quarter and a net total of 3 per unit cell. The other octahedral site is located at the center of the unit cell. Therefore, there are 4 octahedral sites in the FCC unit cell, which means there is one octahedral sites per anion in closest packed structure.

2. Void Location in Hexagonal Structure

Octahedral sites are located at the half height of the interlayer of packing atoms. Since the length of the unit cell in the HCP structure, along the perpendicular

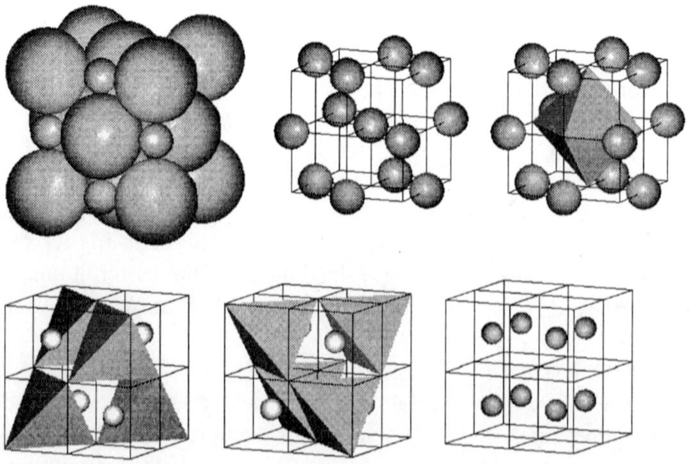

Figure 3-12 (a) Octahedral and (b) tetrahedral void sites in FCC structure

direction to the layer, is twice the interlayer distance, the z-coordinate of octahedral sites is either 1/4 or 3/4 on the scale of the unit cell. Also the tetrahedral sites have z-coordinates of 1/8, 3/8, 5/8 and 7/8. These fractional coordinates of tetrahedral sites are explained by the tetrahedra in Fig.3-12 in which it is shown that the height of the tetrahedral sites from the basal plane consisting of three packed atoms is one quarter of the interlayer distance. A full description of octahedral and tetrahedral sites in the HCP structure is shown in Fig.'s 3-13 and 3-14, respectively. Their coordinates with respect to the unit cell are summarized in Table 3-2.

Table 3-2 Summary of coordinates of octahedral sites and tetrahedral sites in the unit cell of HCP structure

Octahedral site location	Tetrahedral site location
$\left(\frac{2}{3},\frac{1}{3},\frac{1}{4}\right)\left(\frac{2}{3},\frac{1}{3},\frac{3}{4}\right)\left(-\frac{1}{3},\frac{1}{3},\frac{1}{4}\right)$	$\left(0,0,\frac{3}{8}\right)\left(0,0,\frac{5}{8}\right)$
$\left(-\frac{1}{3},\frac{1}{3},\frac{3}{4}\right)\left(-\frac{1}{3},\frac{2}{3},\frac{1}{4}\right)\left(-\frac{1}{3},-\frac{2}{3},\frac{3}{4}\right)$	$\left(\frac{1}{3},\frac{2}{3},\frac{1}{8}\right)\left(\frac{1}{3},\frac{2}{3},\frac{7}{8}\right)\left(-\frac{2}{3},-\frac{1}{3},\frac{1}{8}\right)$
	$\left(-\frac{2}{3},-\frac{1}{3},\frac{7}{8}\right)\left(\frac{1}{3},-\frac{1}{3},\frac{1}{8}\right)\left(\frac{1}{3},-\frac{1}{3},\frac{7}{8}\right)$

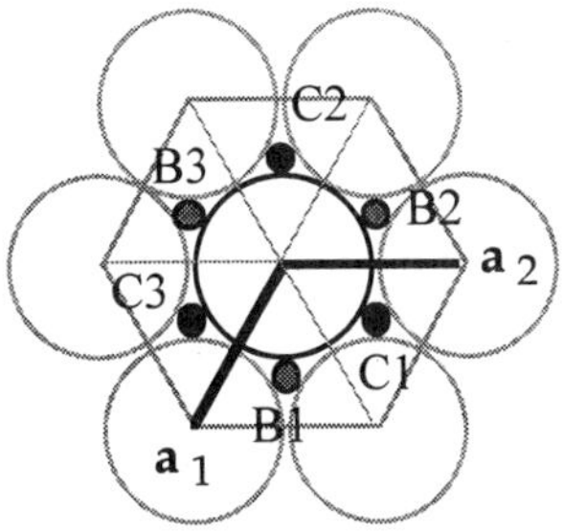

$$B1 = \left(\frac{2}{3}, \frac{1}{3}, z\right) \quad C1 = \left(\frac{1}{3}, \frac{2}{3}, z\right)$$

$$B2 = \left(-\frac{1}{3}, \frac{1}{3}, z\right) \quad C2 = \left(-\frac{2}{3}, -\frac{1}{3}, z\right)$$

$$B3 = \left(-\frac{1}{3}, -\frac{2}{3}, z\right) \quad C3 = \left(\frac{1}{3}, -\frac{1}{3}, z\right)$$

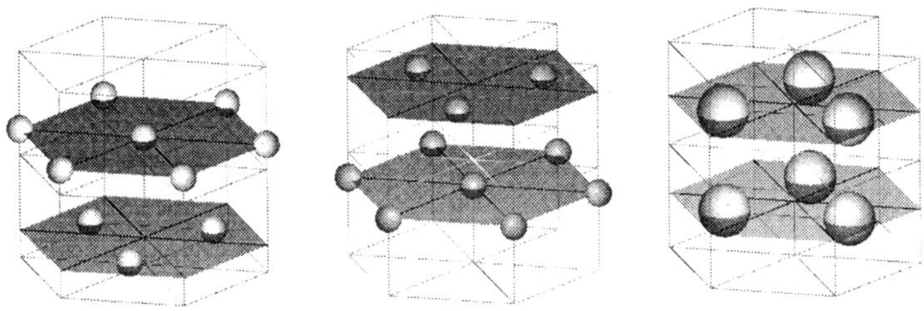

Figure 3-13 Position of tetrahedral sites and octahedral sites in HCP unit cell

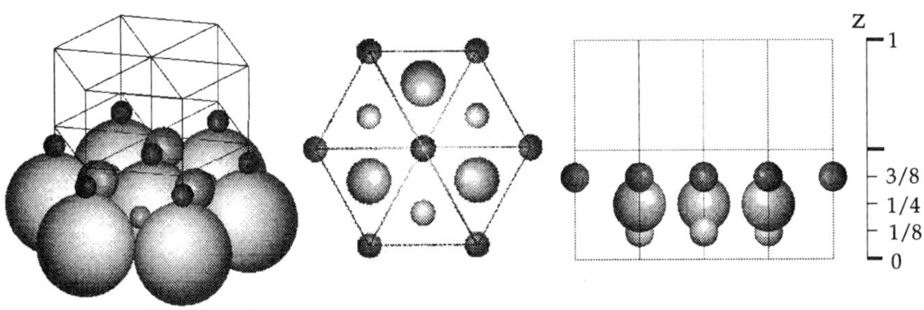

Figure 3-14 View of tetrahedral and octahedral voids in z direction

3. Comparison of HCP with CCP structures

3-1. Configuration of voids around a packed atoms(anions)

The hcp and fcc unit cells result from ABABAB or ABCABC stacking of the layers of packed atoms. The arrangement of tetrahedral and octahedral sites around an atom is easily seen in the CCP structure as shown in Fig.3-12. Each atom is surrounded by 6 octahedral sites and by 8 tetrahedral sites in the CCP structure. The octahedral sites surrounding an atom in the CCP form an

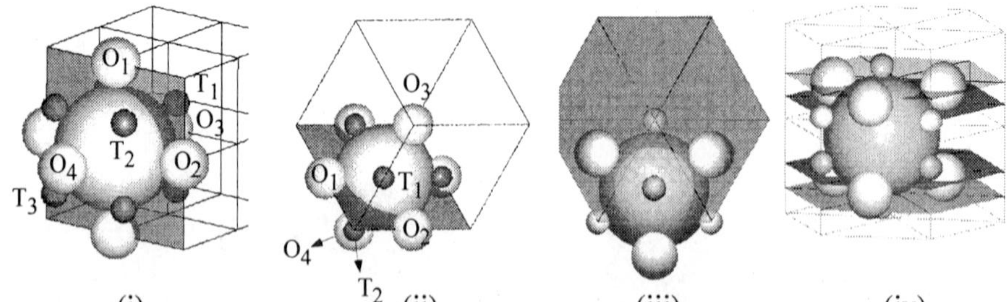

(i) (ii) (iii) (iv)

Figure 3-15 Arrangements of tetrahedral and octahedral sites around an atom in (i) and (ii) FCC and (iii) and (iv) in HCP

octahedral shape and the tetrahedral sites form a simple cube. To make the arrangement in the CCP similar to that in the HCP, the view along the line connecting T_1 and T_4 in Fig.3-15(i) is shown in (ii). The arrangement of tetrahedral and octahedral sites around an atom in HCP structure is shown in (iii) and (iv) of Fig. 3-15. Like in CCP, each atom is surrounded by 8 tetrahedral voids among which 3 are located at +1/8c, and another 3 are positioned at –1/8c. These overlap with the first three when viewed along the c-axis. The other two void positions are just above and below the atom, by 3/8c from the center of the atom. Hence tetrahedral sites do not from a cubic shape. Similarly, in HCP, each atom is surrounded by 6 octahedral voids. Their positions are located at ± 1/4c from the center of the atom. Like tetrahedral sites, these 6 octahedral sites form two groups of 3 octahedral sites that overlap each other when viewed along c-axis. Therefore the octahedral sites in the HCP form a trigonal prism instead of an octahedron, as in CCP. With respect to symmetry, the configuration of octahedral and tetrahedral sites around an atom are arranged to generate an inverse symmetry in CCP and, in HCP, a mirror plane that passes through the center of the atom and is perpendicular to the c-axis.

3-2. Stacking modes of octahedra and tetrahedra in HCP and CCP

3-2-1 Octahedra

With respect to closest packing, three geometries for octahedra have been used, and these are illustrated in Fig.3-16 (i), (ii) and (iii). Shown in Fig.3-16(i) and (ii) are views along the c and a-axes, and in (iii) is the projected view on the ac plane. Sometimes different perspectives of octahedra can be used in order to illustrate a more realistic picture as in the insert in Fig.3-16 (vii) and (viii) that are perspective views from slightly above the octahedra sheet. The advantage of the simple three geometries is, however, that they can describe 3-dimensional structure in a 2-dimensional plane in a very simple and schematic manner.

The octahedra sheet in Fig.3-16(iv) implies two layers of closest packed atoms. Also it indicates the presence of smaller atoms inside the octahedra. This feature will be further illustrated in the later discussion of TiO_2 (Rutile) and $BaTiO_3$ in this chapter. The layer of octahedra in Fig.3-16(vi) shows that adjacent octahedra are connected by edge sharing in any direction. That means that the adjacent two atoms (or cations) in the octahedral void shares the two larger packed atoms (or anions).

With this simplified representation of octahedra, the difference between CCP and HCP is illustrated in Fig.3-17(I). There is no difference between HCP an CCP regarding the arrangement within an octahedra sheet. The difference arises from the stacking modes of the octahedra sheets. In CCP when viewed from the top, the octahedral sheet is shifted in a parallel and opposite direction layer. In contrast, in HCP, there is no shift of a layer from the one above or below that layer. To show this overlapping between adjacent octahedral sheets in HCP, one and two rows of octahedra are removed from the second and third

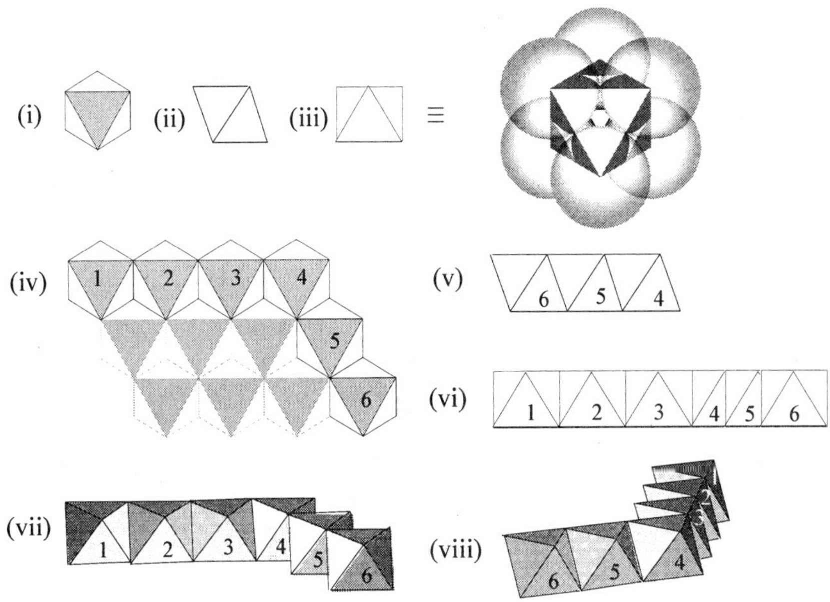

Figure 3-16 Simple representation of octahedron by (i) top, (ii) front and (iii) right side views and schematic illustration of a layer of octahedra by (iv) top, (v) right side and (vi) front views. (vii) and (viii) are front and right side perspectives for 1-6 octahedra

to the second layer which in turn is shifted from atomic positions in the first layer as illustrated in Fig.3-17(i). This shift makes the position of the octahedra sheet return to the same position in every three layers with respect to the base layers, respectively, in Fig.3-17(ii). Octahedra 6 in Fig.3-17(ii) is in the first layer and Octahedra 5 and 4 are in the second and third layers, respectively. The location of these octahedra 4-6 in different layers are more apparent in Fig.3-17(iv).

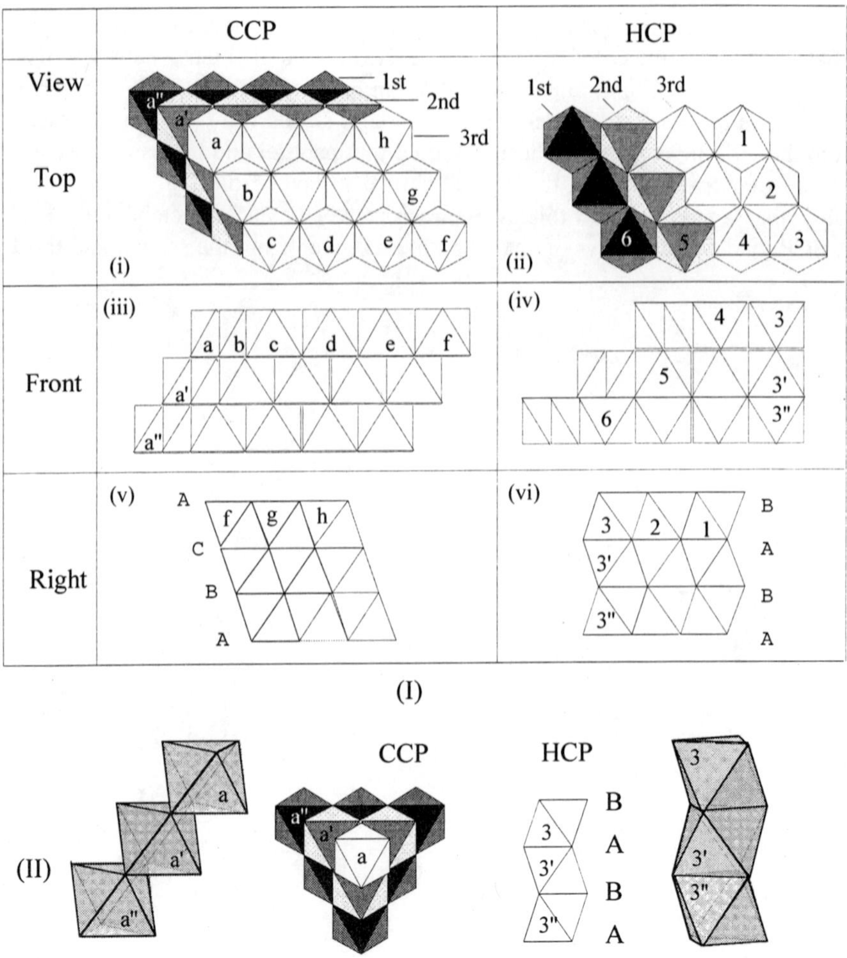

Figure 3-17 (I) Stacking mode of octahedra sheets in HCP and CCP structures and (II) Schematic illustration of the difference by top, front and side(right) views

It is already mentioned that the HCP structure has a mirror plane while the CCP structure has 3 fold rotational symmetry. Fig.3-17(II) is the portion of those in Fig.3-17(I). This arrangement of octahedra illustrates why CCP has 3 fold rotational symmetry and HCP has a mirror plane. Another characteristic of CCP and HCP shown in Fig.3-17(II) is that the octahedra between two layers are connected by edge sharing and by face sharing in HCP as shown by octahedra a, a' and a" in CCP and 3, 3', and 3" in HCP. Regarding stacking of octahedra the difference of CCP and HCP is that in CCP the octahedra are connected by edge sharing in any direction while in HCP connection is by edge sharing in the transverse direction of the layer and face sharing in the perpendicular direction of the octahedra sheet. This combination of edge sharing and face sharing of octahedra in the HCP structure is to be expected in that the interaction between octahedra in adjacent layers like 3 and 3' is different from that between adjacent octahedra within a layer. Two different distances exist between the octahedra. This is the reason why ionic crystals of the AX type take the FCC structure rather than the HCP, since ionic structures do not have directionality. This means that distances between octahedra must be the same regardless of direction. This difference is well illustrated in the structures of NaCl and NiAs that take the FCC and HCP structures, respectively.

3-2-2 Tetrahedra

There are two types of tetrahedral sites in an interlayer of closest packed atoms. When these sites are illustrated by tetrahedra, one type is the dark tetrahedra in Fig.3-18 (iv) that point downward and the other type is Fig.3-18(iii) pointing upward. For simplicity these are called a D-type tetrahedra sheet and a U-type tetrahedra sheet which imply downward and upward, respectively. Here the term "sheet" is used to make a distinction from the tetrahedra layer that is

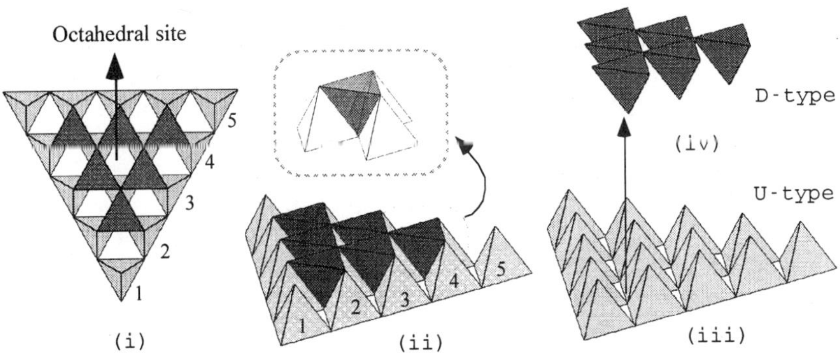

Figure 3-18 Configuration of connection of D and U type tetrahedra sheet

formed within an interlayer of closest packed atoms and the combination of two types of tetrahedral *sheets*. Within the same type of sheet, tetrahedra are connected to adjacent ones by corner sharing. Between D and U type tetrahedra in the same layer, the connection is by edge sharing and this is well illustrated in the insert of Fig.3-18(ii). Fig.3-18(i) is the top view of a tetrahedral layer. As shown in Fig.3-18(i), the octahedral site is surrounded by two sets of three tetrahedra. Six faces from the surrounding six tetrahedra plus two extra faces that are shown as empty triangles at the top and bottom of the octahedral sites in Fig.3-18(i) form eight (octa) faces (hedra).

The HCP and CCP, both have the same configuration of tetrahedral layer as shown in Fig.3-18(i). The difference arises from the stacking mode of the tetrahedral layers.

In HCP there is a mirror plane between interlayers. This means that any layer of the closest packed atoms is a mirror plane and across this mirror plane, atoms are arranged in a mirrored image. This point was also illustrated in the discussion of the stacking mode of octahedra in HCP. The tetrahedra layers are also stacked in a mirrored image with the next tetrahedra layer. When two adjacent tetrahedra layers are viewed in a direction perpendicular to the layer, the U-type sheet of one tetrahedra layer overlaps the D-type sheet of the other. If the two tetrahedra layers are distinguished from each other by the subscripts of "ℓ" and "h" for lower and higher, respectively, the tetrahedra in the U_ℓ sheet is connected to those in the D_h by corner sharing while D_ℓ-U_h connection is the face sharing of tetrahedra. These connection modes of tetrahedra in HCP are summarized as :

Edge sharing (D-U) and Corner sharing (D-D or U-U)
in the transverse direction within the same tetrahedral layer
Corner sharing (D_ℓ-U_h) and Face sharing (U_ℓ-D_h)
in the vertical direction to the tetrahedra layer

The difference in the connection modes of tetrahedra in the transverse and vertical directions to the layer in the HCP structure is expected to give rise to the distortion of the tetrahedra but with the same length for all edges. In contrast to HCP, CCP structures that have both edge sharing and corner sharing between adjacent tetrahedra regardless of transverse or vertical direction to the layer. Also there is no face sharing in the CCP structure. These aspects in connection modes of tetrahedra in HCP and CCP are illustrated in Fig.3-19.

Face sharing (D_ℓ–U_h) in HCP makes the distance between cations closer and produce greater repulsive interaction between cations than edge sharing. On the other hand, face sharing accompanies the more stable configuration of tetrahedra in the corner sharing (D_ℓ–D_h , U_ℓ–U_h and D_h–U_ℓ). This extra destabilization energy due to face sharing is more than the extra stabilization

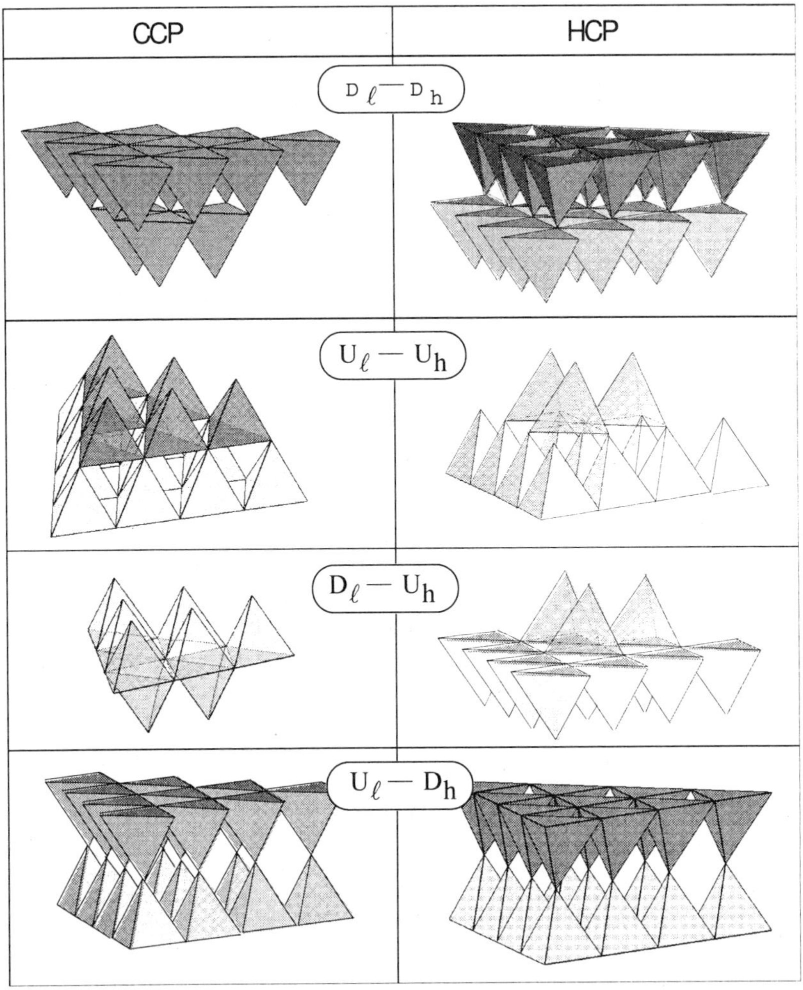

Figure 3-19 Comparison of stacking modes of tetrahedra in CCP and HCP

energy from the configurational change in corner sharing and the overall lattice energy of HCP is lower than that of CCP. This is the reason why ionic crystals such as Li_2O, in which all the tetrahedral sites are occupied take the CCP instead of the HCP structure.

C. Site selection of cation in ionic crystals

1. Site selection criteria

As mentioned in Chap. 1, the driving force for formation of a certain structure in ionic crystals is the lattice energy. The lattice energy is significantly dependent on the Madelung constant which is determined from the crystal structure. The Madelung constant is calculated by including the long range charge interaction between cations and anions. Estimation of the minimum value of the Madelung constant for an ionic crystal is the object of structure determination in these type materials. The determination of the maximum value for the Madelung constant for the structure of an ionic material can be done by comparison of the many possible structures for the material. As a rough estimation of the Madelung constant, Verwey-Heilmann's "principle of maximal charge neutralization"Verwey-Heilmann's principle; and Pauling's coordination theory have been utilized. The maximal charge neutralization principle emphasizes that the major contributing factor to the stability of an ionic crystal is the interaction of a cation with the nearest anions. However, there is a reverse effect by putting many anions around a cation although this will increase the electrostatic energy between the cation and anions. When anions are packed such that the cation rattles inside the void space formed by the anions, this action destabilizes the ionic crystal structure. To compromise between these two opposite effects, anions will assume the closest packing in order to generate maximal charges around the cation. The cations in turn will occupy the void space formed by the closely packed anions which permits no rattling. Therefore, the structures developed for ionic crystals are mainly in accordance with the coordination number (CN) and compactness predicted by Pauling's rulesas follows:

1-1. Pauling's Rule 1

A coordination polyhedra of anions is formed around each cation. The cation-anion distance is determined by the radius sum(r_c+r_a) and the coordination number by the radius ratio(r_c/r_a))

Radius Sum : $r_o = r_a + r_c$

r_o = interionic distance

r_a = anion radius

r_c = cation radius

The coordination number is related to the <u>radius ratio</u> (r_c/r_a). This ratio is the single most important concept regarding the ionic crystal structure. Here, the coordination polyhedra is defined as a group of nearest neighbor anions surrounding a central cation, while the number of such anions is called the coordination number (CN). The relationship between the coordination polyhedra and radius ratio is summarized in Table 3-3. This relationship represents Pauling's Rule 1. Also the coordination polyhedra are shown in Fig.3-20 for coordination numbers 3, 4, 6 and 8. In Fig.3-20, cations are drawn by dark circles and anions are shown by gray circles. The size of cations in Fig.3-20 is seen to increase progressively with coordination number although the sizes of anions remain the same. As shown in Table 3-3, the ratio of the cation to anion size is 0.115: 0.225: 0.414: 0.727.

Table 3-3 Summary of Radius Ratio/CN Ranges

r_c/r_a	CN	Type	Example
-0.155	2	Linear	CO_2
0.155-0.225	3	Triangular	B_2O_3
0.225-0.414	4	Tetrahedral	SiO_2
0.414-0.732	6	Octahedral	TiO_2
0.732-1.0	8	Cubic	CsCl

Note: 1. CNs given for cations only. Anion CN determined by rule 2.
2. The ranges given for r_c/r_a represent stability range for the CN.

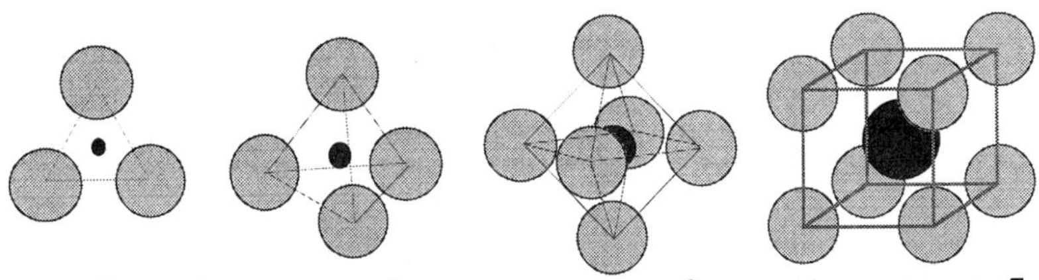

$[BO_3]^{3-}$ $[SiO_4]^{4-}$ $[TiO_6]^{8-}$: $[CsCl_8]^{7-}$

Figure 3-20 Anion Polyhedral Structures with CN of 3, 4, 6 and 8

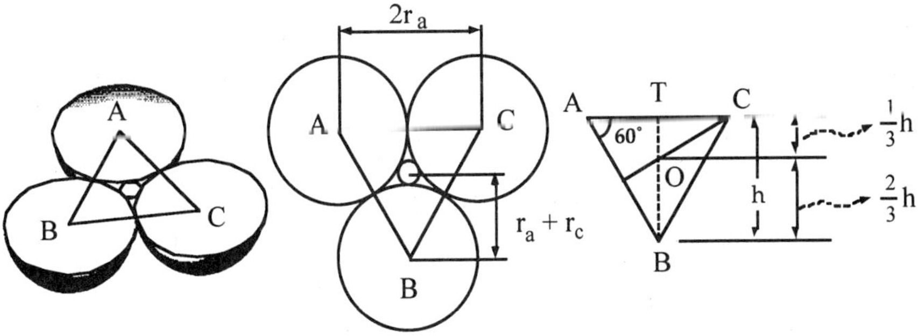

Figure 3-21 View of cross section cut through the centers of spheres in triangular stacking

1-2. Size of void site in HCP and CCP packing

a) Triangular void site

The triangle ABC in Fig.3-21 is an equilateral triangle. The anion is located at the center of gravity for this triangle. From the geometry, $\overline{AC}$ is $2r_a$ and $\overline{TB}$ is $\sqrt{3}\,r_a$. Since $\overline{OB}$ is $\frac{2}{3}\,\overline{TB}$ and is equal to $r_a + r_c$.

$$\text{Therefore } \frac{2}{3}\,\overline{TB} = \frac{2\sqrt{3}}{3}\,r_a = (r_c + r_a)$$

$$r_c = r_a\left(1 - \frac{2}{\sqrt{3}}\right) = r_a(1.155 - 1) = 0.155r_a$$

$$\text{Therefore, } \frac{r_a}{r_c} = 0.155$$

b) Tetrahedral Void

Triangle ABC in Fig.3-22 is an isosceles triangle with $\overline{AB} = \overline{BC} = \sqrt{3}\,r_a$ and $\overline{AC} = 2r_a$. For the triangle ABM and CBM, $\overline{BM}$ is $\sqrt{2}\,r_a$. $\overline{OM}$ is $\frac{\sqrt{2}}{2}r_a$ since "O" is the center of the tetrahedron APQC and, therefore, $\overline{BM} = \overline{OM}$. From

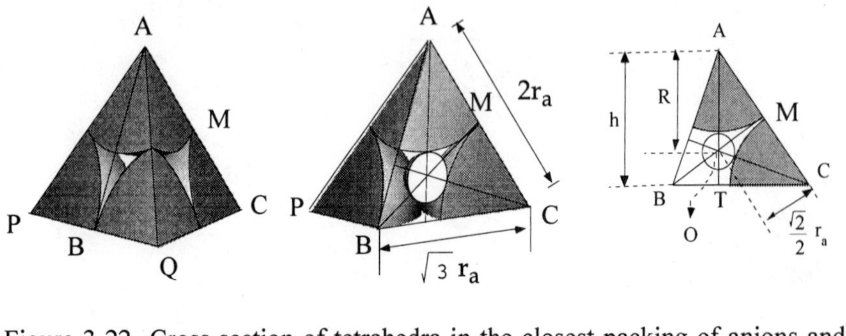

Figure 3-22 Cross section of tetrahedra in the closest packing of anions and void space for a cation

Triangle AOM, $\overline{AO}$ (= R) is $\sqrt{\frac{3}{2}}r_a$ according to the Pythagorean theorem. Since R is $r_a + r_c$, ,

$$\sqrt{\frac{3}{2}}r_a = r_c + r_a$$

Solving for this,

$$r_c = \left(\sqrt{\frac{3}{2}} - 1\right)r_a \text{ and } \frac{r_c}{r_a} = \left(\sqrt{\frac{3}{2}} - 1\right) = 0.2247$$

Geometric relationships that are important in the characterization of the closest packing and tetrahedral sites are the ratio of h to $2r_a$ (= $\overline{AC}$ = "a" of the corresponding hexagonal or FCC lattice) and the ratio of $\overline{AO}$ to $\overline{TO}$ or $\overline{BT}$ to $\overline{CT}$. These geometric characteristics in closest packing are helpful in characterization of crystal structures such as ZnS, NiAs, CdI_2.

For Triangles BOT, the length of $\overline{OT}$ can be expressed according to the Pythagorean theorem as :

$$\overline{OT} = \sqrt{\overline{OB}^2 - x^2} = \sqrt{\frac{1}{2}r_a^2 - x^2}$$

Here x is the length of $\overline{BT}$. Also the length of $\overline{CT}$, therefore, is expressed as $\sqrt{3}r_a - x$. For Triangle COT, the length of $\overline{OT}$ can be expressed according to the Pythagorean theorem as:

$$\overline{OT} = \sqrt{\overline{OC}^2 - \left(\sqrt{3}r_a - x\right)^2} = \sqrt{\frac{3}{2}r_a^2 - \left(\sqrt{3}r_a - x\right)^2}$$

Solving for x

$$x = \frac{1}{\sqrt{3}}r_a \text{ and, therefore, } \overline{OT} = \frac{1}{\sqrt{6}}r_a$$

Also $\overline{CT}$ can be calculated as

$$\overline{CT} = \overline{BC} - x = \sqrt{3}r_a - x = \frac{2}{\sqrt{3}}r_a$$

From these results, the ratios of $\frac{\overline{OT}}{\overline{OA}}$ and $\frac{\overline{BT}}{\overline{CT}}$ are $\frac{1}{3}$ and $\frac{1}{2}$, respectively. And the ratio of h to r_a is 1.633 since h is equal to the sum of $\overline{OA}$ and $\overline{OT}$ that is $\sqrt{\frac{3}{2}}r_a + \sqrt{\frac{1}{6}}r_a = \frac{4}{\sqrt{6}}r_a$. This ratio is equivalent to the ratio of $\frac{c}{a}$ in the hexagonal lattice.

c) Octahedral Void

In Fig.3-23, $\overline{AC}$ is the face diagonal of a square whose edge length is $2r_a$. Hence

$$\overline{AC} = 2\sqrt{2}\ r_a = 2\,(r_a + r_c)$$

$$(r_a + r_c) = \sqrt{2}r_a$$

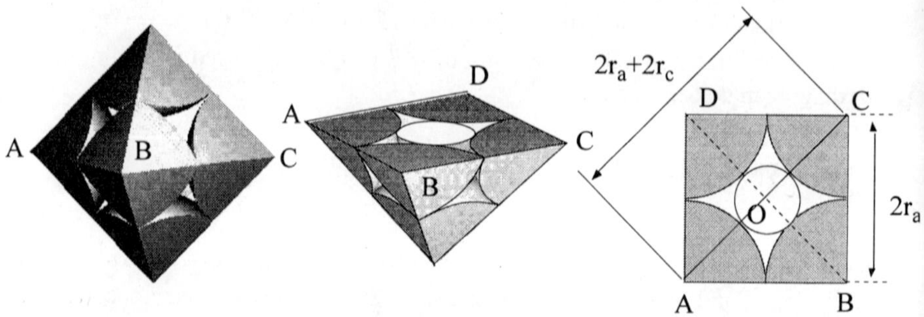

Figure 3-23 Cross section of octahedra in the closest packing of anions and void space for the cation

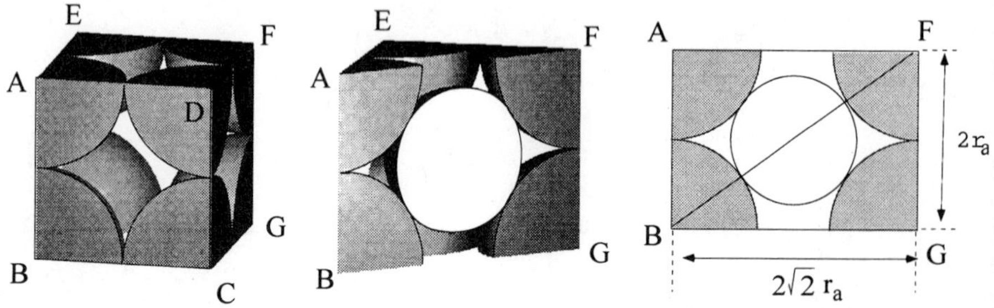

Figure 3-24 Cross section of simple cubic structure of anions and void space for the cation

$$r_c = r_a\left(\sqrt{2}-1\right) = 0.414r_a$$

$$\frac{r_c}{r_a} = 0.414$$

d) Cubic Void

A cubic void site is not generated within the scheme of HCP or CCP structure. Pauling's rule, however, predicts this void structure for the non-rattling condition with maximum charge in relation to the ratio.

In Fig.3-24, $\overline{BF}$ is the body diagonal of a cube whose edge length is a. Also $\overline{BF}$ is a line connecting centers of anions and cations.

Therefore $$2(r_a + r_c) = \sqrt{3}a_o = \sqrt{3} \times 2r_a$$

$$(r_a + r_c) = \sqrt{3}r_a$$

$$r_c = r_a\left(\sqrt{3}-1\right) = 0.732r_a$$

$$\frac{r_c}{r_a} = 0.732$$

1-3. Pauling's Rule 2

Pauling's Rule 2 governs how coordination polyhedra are connected to one another while Rule 1 predicts how cation and anion configurations are connected. The rule states that the strength of the electrostatic bond that reaches an <u>anion</u> from neighboring <u>cations</u> is equal to the charge on the anion, i.e., an ionic structure can have no net charge. This electrostatic bond strength(EBS) can be defined as:

$$EBS = \frac{Z}{CN} \quad (1)$$

Where: Z = Charge on the cation (or anion)
CN = Coordination number

It should be noted that the EBS in only a <u>rough approximation</u> of the bond strength, since bond length is not considered. In spite of this limitation, Rules 1 and 2 are the main criteria determining ionic structures.

2. Application of Pauling's rule to ionic crystal structure

Since the triangular anion polyhedra is rare in ionic crystals, only tetrahedral, octahedral and cubic anion polyhedra are considered in the following discussion. According to Pauling's first rule, the selection of these polyhedra depends solely on the radius ratio. Also to satisfy charge neutrality requirements, the ratio of cation to anion must be the same as the coordination ratio of cations and anions, defined as $\frac{x}{y}$ in A_xY_y

For example

A_xX_y : A - cation, X - anion

and $$\frac{x}{y} = \frac{\text{cation coordination \#}}{\text{anion coordination \#}} = \text{CN ratio}$$

This CN ratio is also related to the number of void sites per anions for particular packing. In CCP or HCP structure, there are two tetrahedral sites and one octahedral site per each atom. With respect to the CN ratio, the possible structure for Li_2O whose CN ratio is 2, is occupancy of 2 tetrahedral sites by Li in closest packing of O. Also this radius ratio favors the tetrahedral site. Conventionally the crystal structure of Li_2O is described as anti-fluorite structure since the fluorite structure is more prominent than Li_2O with many crystals having this crystal structure. Fluorite is the name for the mineral CaF_2. As 'anti-' indicates, fluorite takes the Li_2O structure with switching of the positions for cation and anion in Li_2O. Detailed structure will be given later when AX_2 compounds are described. Actual crystal structures of Li_2O and CaF_2 are shown in Fig.3-30.

As in Li_2O, the binary ionic compounds (A_nX_m type) can take the crystal structure whose interstitial void spaces are occupied according to the coordination ratio and radius ratio as shown in Table 3-4. There are two concerns for the crystal structures developed according to Pauling's rule. They are: (1) the selection of HCP or CCP packing structure for anions that are

Table 3-4 Void occupancies of binary ionic crystals in the closest packing

Compound	AX	AX_2	AX_3	A_2X_3	A_2X
Octahedral	all	$\frac{1}{2}$	$\frac{1}{3}$	$\frac{2}{3}$	
Tetrahedral	$\frac{1}{2}$	$\left[\frac{1}{4}\right]^*$	$\left[\frac{1}{6}\right]^*$	$\frac{1}{3}$	all

*These are rarely observed due to large repulsion between anions.

usually bigger than cations and (2) distribution of cations among the interstitial void spaces if only a fraction of the voids are occupied by cations. For an AX compound whose radius ratio prefers octahedral occupancy, all the octahedral sites are occupied by cations. Hence the only concern is the selection of CCP or HCP for anions (X). For other ionic crystals, however, concern for the description of crystal structures is first to select the closest packing, CCP or HCP for anions, and second to select sites for the cations.

D. Illustration of binary ionic crystals with HCP and FCC structures

1. AX compound

1-1 Occupancy of octahedral sites

Previously CCP and HCP were compared regarding the stacking modes of octahedra. NaCl and NiAs are typical examples of CCP and HCP structures, respectively. Based on electronegativity considerations, NaCl is expected to exhibit high ionic bonding compared to NiAs. Their respective crystal structures illustrate the differences between CCP and HCP regarding the bonding characteristics.

1-1-1 NaCl

The electronegativity value on the Pauling's scale (Table 1-17) for Na and Cl are 0.93 and 3.16Å, respectively. This difference in electronegativity ensures that the bonding between Na and Cl is ionic since atoms having an electronegativity difference of 1.5 ~ 1.8 take on ionic bonding. On the basis of ionic bonding, the radii for Na^+ and Cl^- are 0.98 and 1.81, respectively. The ratio of cation to anion radius is, therefore,

$$\frac{r_c}{r_a} = \frac{r_{Na^+}}{r_{Cl^-}} = \frac{0.95}{1.81} = 0.54$$

According to Pauling's Rule 1, Na and Cl form an octahedral coordination as shown in Fig.3-25. But from Pauling's rule 2, EBS = $\frac{Z}{CN} = \frac{1}{6}$, i.e., the charge reaching each Cl^- = 1/6. Therefore, to realize charge neutrality, 6 Na^+ ions must surround each Cl^- ion. Since $\frac{y}{x} = \frac{6}{6}$, the CNs for both Cl^- and Na^+ must be 6. The configurations for coordination number 6 for both Cl^- and Na^+, are illustrated in Fig.3-25.

The dimension of the unit cell of NaCl is a_o = 5.63Å. Assuming a hard sphere model for the ions, the sum of r(Na+) and r(Cl-) is very close to half of the unit cell length. The sum of the covalent radii for Na and Cl is much smaller than the half of the unit cell length. Hence cell dimension and the adoption of FCC structure indicate a close relationship to the bonding characteristics.

$$r(Na+) + r(Cl-) = 0.95 + 1.81 = 2.76 \approx \frac{a_o}{2} = 2.82 \text{ for ionic bonding and}$$

$$r(Na) \ + r(Cl) \ = 1.54 + 0.99 = 2.53 < \frac{a_o}{2} \text{ for covalent bonding of Na} - \text{Cl}$$

Crystals that take the NaCl structure are halogen compounds of alkali elements and oxide compounds of alkali earth elements.

alkali halide : LiF, LiCl, LiBr, LiI, NaF, NaBr, NaI, KF, KCl, KBr, KI
alkali earth oxide : MgO, CaO, SrO, BaO

The <u>only</u> arrangement compatible with this 6:6 CN is the FCC structure. Hence, <u>ionic</u> CN 6:6 structure must adopt the FCC structure. Cl resides at the

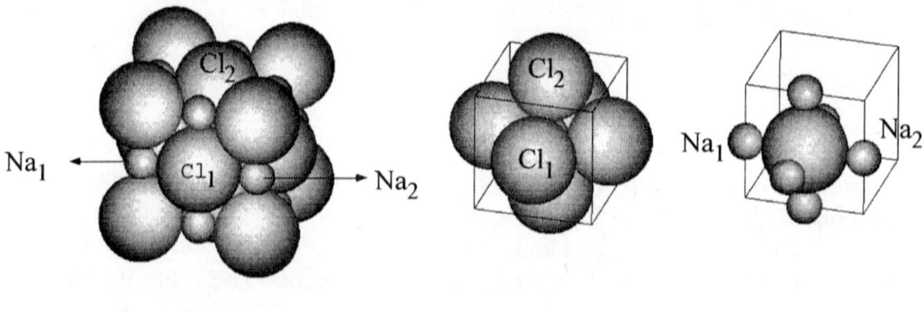

Figure 3-25 Crystal structure of NaCl and coordinations of Na and Cl

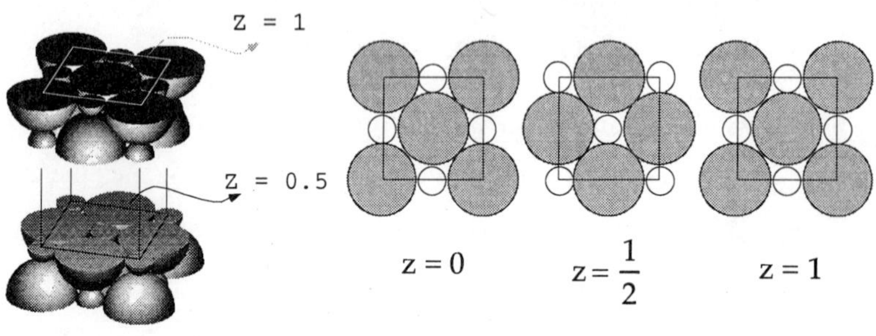

Figure 3-26 Crystal structure of NaCl with atomic positions along z-axis

closest packing positions and Na occupies the octahedral sites in the FCC structure. In terms of the FCC lattice, NaCl structure is a combination of two FCC lattices with one lattice shifted to the center of the other lattice. Each FCC lattice is occupied by Na^+ or Cl^- ion. Two dimensional illustration of the NaCl crystal structure is given in Fig.3-26.

1-1-2 NiAs

The crystal structure of NiAs is shown in Fig.3-27 and Fig.3-28. Less ionic CN 6:6 structures adopt the NiAs arrangement. In this structure, As ion forms the HCP packing and Ni occupies the octahedral sites. The cell dimension of NiAs is a = 3.60Åand c = 5.01Å. The aspect ratio of c/a is 1.392, much lower than the standard value of 1.633 for the HCP packing of anions. From the unit cell dimension, the bond distance between Ni and As is 2.44Å.

ro = 2.44Å	
r(Ni+3) + r(As-3) = 0.62 + 2.22 = 2.84	for ionic bonding
r(Ni) + r(As) = 1.15 + 1.21 = 2.36	for covalent bonding

The comparison between bond distance and radii illustrates the character of the bond as covalent bonding. This bonding character between Ni and As is the expected result from the electronegativity difference of 0.27 between Ni (1.91) and As (2.18), since covalent bonding is formed between two elements whose electronegativities differ by less than 0.7. Another noticeable point for the bonding between Ni–As is the distance between Ni ions. The cell dimension of NiAs shows that the distance of Ni(1)– Ni(1') is different from that of Ni(1)–Ni(2). The former has distance of 2.51Åand the latter has distance of 3.60°A. In terms of the octahedra stacking, the octahedra formed around Ni(1) and Ni(1') are connected by face sharing while those around Ni(1) and Ni(2) are connected by a edge sharing. If NiAs was mainly ionic bonded, the distance of Ni(1)–Ni(1') would be elongated from the standard distance (= half of 1.633 a) due to the repulsive interaction between cations along the c-axis. The actual data shows the reverse interaction. This implies that there is a non-ionic attractive interaction along the c-axis.

In the NiAs structure, charge balance is preserved in accordance with Pauling's second rule, as each Ni atom is surrounded by 6 As atoms and also, each As atom is surrounded by 6 Ni atoms.

Although it seems improper to list NiAs as an ionic crystal, the reason is that a number of important compounds adopt this crystal structure. Typical compounds with NiAs structure are sulfide compounds of transition metal elements in the fourth period such as TiS, CrS, FeS, CoS and NiS. The cell dimensions for the sulfide compounds of these transition elements are listed in Table 3-5. The aspect ratio of c/a ranges from 1.53 ~ 1.95. The observed

Table 3-5 Cell dimensions and bond distances for the sulfides of transition metal elements in the fourth period of the periodic table

	a	c	c/a	Observed	Calculated	
				r_o	r_o(ionic)	r_o(cov.)
Sc*	–	–				
Ti	3.30	6.44	1.95	2.49	2.74	2.36
V*	3.36	5.81	1.73	2.42	2.72	2.26
Cr	3.45	5.75	1.67	2.46	2.68	2.21
Mn*						
Fe	3.44	5.88	1.71	2.47	2.60	2.21
Co	3.37	5.16	1.53	2.33	2.58	2.20
Ni	3.44	5.35	1.56	2.39	2.56	2.19
Cu	–	–				
Zn*	–	–				

* Sc, Mn - NaCl type, V - distorted NiAs, Zn - ZnS (Zinc Blende or Wurtzite - tetrahedra coordination)

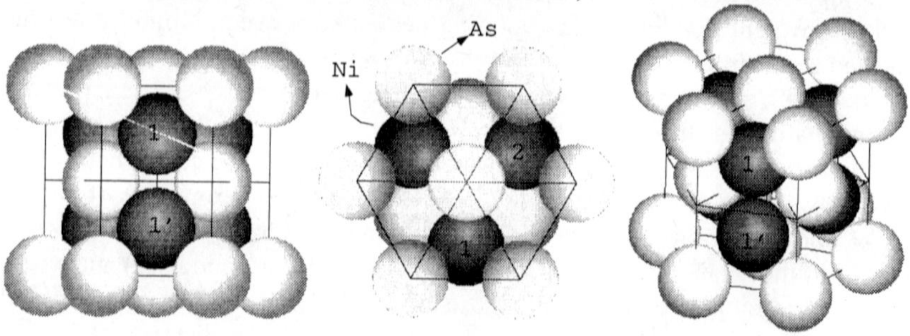

Figure 3-27 Crystal structure of NiAs (a) front view, (b) top view and (c) perspective view

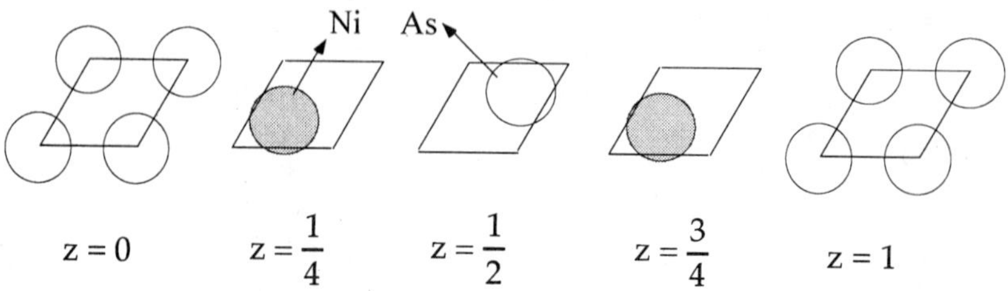

Figure 3-28 Crystal structure of NiAs - HCP structure

bonding distance of M–S is between the sum of ionic radii and the covalent radii for the sulfur and metal elements. These intermediate ranges of bond distances illustrate that sulfides of these transition metal elements have more or less covalent character. Also the sulfide compounds can be compared to the corresponding oxides of the transition elements (TiO, CrO, FeO, CoO) that have the NaCl structure. This distinction between the oxides and sulfides of the transition metal elements in selecting the NaCl and NiAs structures respectively indicates that 1:1 ionic compounds with less ionic bonding prefer the NiAs structure for octahedra coordination. The sulfides clearly have less ionic character than the corresponding oxides due to lower electronegativity of sulfur. Consequently sulfides of transition metal elements such as TiS, CrS, FeS, CoS take the NiAs structure while oxides of these elements take the NaCl structure.

1-2 Occupancy of tetrahedral sites - ZnS

In Table 3-5, the trend in structure change in the first transition elements from NaCl (octahedral coordination in FCC) to NiAs (octahedral coordination in HCP) to ZnS (Tetrahedral coordination) is noteworthy although there is a break at Mn. Among the sulfides of the first transition metal elements, Zn is the only element which occupies a tetrahedral site. According to Table 1-32(a), Co^{+2} (0.74Å), Ni^{+2} (0.72Å), and Zn^{+2} (0.74Å) have very similar ionic radii. Therefore, radius ratios of r_c/r_a for CoS, NiS and ZnS are around 0.4, which is the marginal value between octahedral and tetrahedral coordination formation. The ratio of r_c/r_a seems, therefore, not to be solely responsible for the transition from octahedral to tetrahedral coordination of zinc sulfides. Like the other transition metal sulfides, this relates to the covalent character of the bond. A descriptive explanation for the tetrahedral coordination in ZnS is derived by utilizing the similarity in crystal structure between ZnS and diamond.

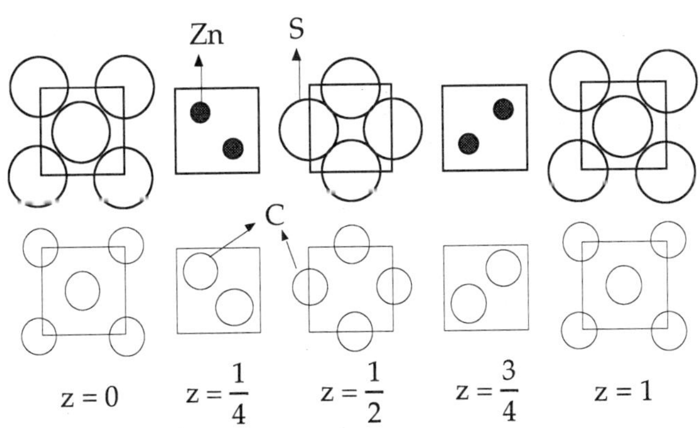

Figure 3-29 Atomic positions in ZnS(Zinc Blende) with z-notation

1-2-1 ZnS (Sphaelerite or Zinc Blende)

There are two crystal structures for ZnS. One form, called Zinc Blende or Sphaelerite, has the CCP structure while the other, called Wurtzite, has the HCP structure. The sphaelerite structure based on CCP structure is first described.

Since Zn is located on the tetrahedral site, its coordination number is 4. Also each S atom is surrounded by 4 Zn atoms because only half of eight tetrahedral sites around each S atom are occupied. The crystal structure of ZnS in the CCP structure is the same as the diamond structure except for the substitution of Zn for carbon in tetrahedral sites.

In sphaelerite, S forms closest packing and Zn is located at the tetrahedral sites. The atomic position within the unit cell is shown in Fig.3-29. Only half of the tetrahedral sites of S^{-2} cubic closest packing are occupied by Zn^{+2} ions. In relation to the D and U notation in Fig.3-18, only D type or U type tetrahedra sites are occupied. By taking only one type of tetrahedral site, the sharing mode between tetrahedra becomes only corner sharing. The corner sharing results in an even distance between adjacent cation inside tetrahedra. In contrast, if part of both D and U types of tetrahedra are occupied, it inevitably results in an uneven distance between adjacent cations inside each tetrahedra. Hence the Zn ions in ZnS have a shorter distance along the direction of the c-axis than in the direction perpendicular to the c-axis. The real reason for the occupancy of only the D or U type by Zn, therefore, is due to the geometric restriction imposed by bonding orbitals which results from the covalent bonding character in ZnS.

The dimension of the unit cell of ZnS is a_o = 5.41Å. From this, the distance of Zn–S is calculated as 2.34Å. This distance is very close to the calculated value for the bond distance using the covalent radii of Zn and S in

Table 3-6 Unit cell dimensions of ZnS, CdS, HgS and BeS and comparison of observed bond distance of M–S with calculated bond distances from ionic radii sum and covalent radii sum

	a_o	Observed	Calculated	
		r_o	r_o(ionic)	r_o(cov.)
Zn	5.41	2.34	2.58	2.29
Cd	5.82	2.52	2.81	2.45
Hg	5.85	2.53	2.94	2.48
Be	4.85	2.1	2.15	1.93

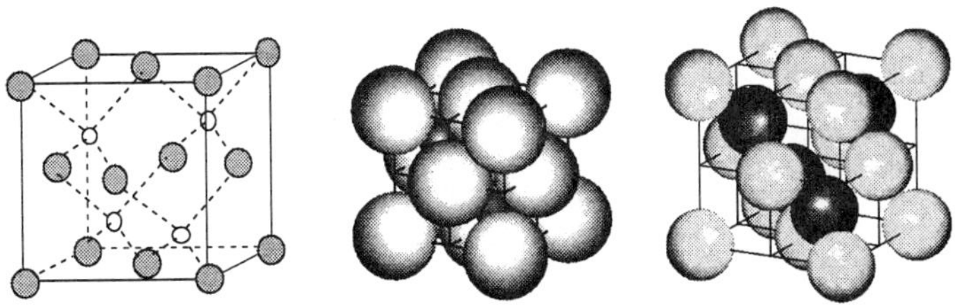

Figure 3-30 ZnS structure $F\bar{4}3m$ in CCP(Sphaelerite, or Zinc Blende) and Diamond Cubic Fd3m (For the purpose of comparison carbon atoms corresponding to Zn position are shown in dark color.)

Table 1-32(a). If the agreement between calculated value and observed value indicates that ZnS is partially ionic and partially covalent, the real atomic packing resembles that of the diamond crystal structure in Fig.3-30, since covalent bond character makes the sizes of Zn and S very similar.

The sulfides of Cd and Hg which belong to the class as Zn in the periodic table have the same crystal structure. Their unit cell dimensions are listed in Table 3-6. Based solely on the bond distance, the data indicate that all have covalent bond character. According to Pauling's first rule, Hg is expected to take the NaCl or NiAs since the ratio of r_c/r_a is 0.61, which is far reduced from the maximum value for occupancy of the tetrahedra sites. This indicates that significant covalent bonding characteristics are involved in the preference of Hg for the sphaelerite structure with tetrahedral coordination of the cation. The covalent bond character can be identified by the similarity between the sphaelerite and diamond crystal structures as shown in Fig.3-30. To show the correspondence of diamond structure to ZnS (sphaelerite), carbon atoms at the Zn positions are shown in a dark color in Fig.3-30. In this structure, each carbon atom forms tetrahedral coordination with the nearest carbon atoms. This is because of the directional covalent sp^3 hybrid bonding between the carbon atoms. The configuration of the valence electrons of the carbon atom in the neutral state is $2s^22p^2$. Although energy is needed to make the transition from the ground state ($1s^22s^22p^2$) to the activated state ($1s^22s^12p^3$), carbon usually takes the hybrid orbital bonding since the energy gained through $sp^3 - sp^3$ bonding exceeds the bond energy obtained through non-hybrid bond orbitals even after compensating for the activation energy of the transition. This is the basis for the crystal structures of elements in the same group as carbon such as Si, Ge and Sn, that have the diamond structure. Si, Ge and Sn have the same valence electron configuration as [Ne]–$3s^23p^2$, [Ar]$3d^{10}$–$4s^24p^2$, and [Kr]$4d^{10}$–

$5s^25p^2$. As atomic number increases, the energy difference between "ns" and "np" orbitals becomes smaller. Therefore the activation energy needed for the formation of sp^3 hybrid orbitals from "ns" and "np" orbitals becomes lower as the atomic number increases in the carbon group. However Pb, in the same group as carbon with electron configuration of $[Xe]4d^{10}–5s^25p^2$ has the pure FCC or HCP structure instead of the diamond structure. As atomic number increases, the size of the atoms become so large that the directionality of the sp^3 hybrid bonding and its effectiveness as a strong bond is loosened. Therefore, the bonding becomes more or less metallic instead of covalent. This aspect of metallic bonding in the crystal structure can be seen in the β-form of tin which does not have the diamond structure. This trend of increasing metallic character in relation to crystal structure is shown in Table 3-7 which gives data on the unit cell dimensions for elements of the carbon group in the periodic table, along with band gap(Eγ), resistivity and melting points.

In the carbon group, the effect of sp_3 hybrid bonding orbital on the crystal structure is illustrated. By the same token, this sp_3 hybrid bonding orbital can be applied to the crystal structure of ZnS (sphaelerite). The valence electron configuration of Zn is $[Ar]4s^2$ and for S, $[Ne]3s^23p^4$. By sharing a valence electron, both Zn and S may have the configuration of s^1p^3, which is necessary for the sp^3 hybrid bonding.

$$\text{Zn} : [\text{Ar}]4s^2 \rightarrow \text{Zn}^{-2} : [\text{Ar}]\ 4s^2\ 4p^2 \rightarrow 4sp_3$$
$$\text{S} : [\text{Ne}]3s^23p^4 \rightarrow \text{S}^{+2} : [\text{Ne}]\ 3s^2\ 3p^2 \rightarrow 3sp_3$$

In contrast to Zn, other first transition elements like Ni, Cu do not have rare gas configurations. Their anionic state is in a higher energy state than Zn^{-2} and sp^3 hybrid bonding is less possible than in Zn. Therefore they do not assume tetrahedral coordination. Rather they take the NiAs structure with octahedral coordination.

Table 3-7 The unit cell dimensions and physical properties of elements in the carbon group in the periodic table

	a_o	Eγ(eV)	σ (ohm-cm)	MP
C	3.567Å	6.03	$10^{14} \sim 10^{16}$	4100°C
Si	5.43	1.11	48	1420°
Ge	5.66	0.668	47	945
α-Sn*	6.48	0.080	$1 \times 10^{3*}$	232
Pb		0	20×10^{-6}	327

*The β-form of tin is metallic and has a tetragonal lattice. Its resistivity is 11×10^{-6}.

Table 3-8 Compounds of elements of II-VI and III-V groups and typical applications for these materials

Groups	Examples	Properties
II-VI	ZnS, ZnSe, ZnTe CdS, CdSe, CdTe	Semiconductor device application
III-V	GaAs, GaP, GaSb AlP, BP, AlSb, AlAs InP, InAs, InSb	Phosphor materials

The compounds between elements in Group II and VI take the same crystal structure as ZnS. Similarly, compounds between elements in Group III and V take the ZnS (sphaelerite) structure. Table 3-8 lists the compounds formed between II-VI and III-V groups. They all exhibit similar physical properties. Regarding the physical property of compounds having ZnS crystal structure, strong sp^3 hybrid bonding is expected between the smaller atomic sizes of elements such as in BN, BP, and SiC. These three compounds, therefore, exhibit great hardness.

1-2-2 ZnO

Another crystal structure of ZnS is shown in Fig.s' 3-31 and 3-32. The mineralogical name for it is wurtzite. The crystal structure wurtzite has long represented the group of compounds with the same crystal structure. In this book, this structure is called ZnO structure instead of ZnS(wurtzite) structure in order to avoid the confusion between the two crystal forms of ZnS. Therefore, ZnS and ZnO will represent the sphaelerite and wurtzite structures, respectively.

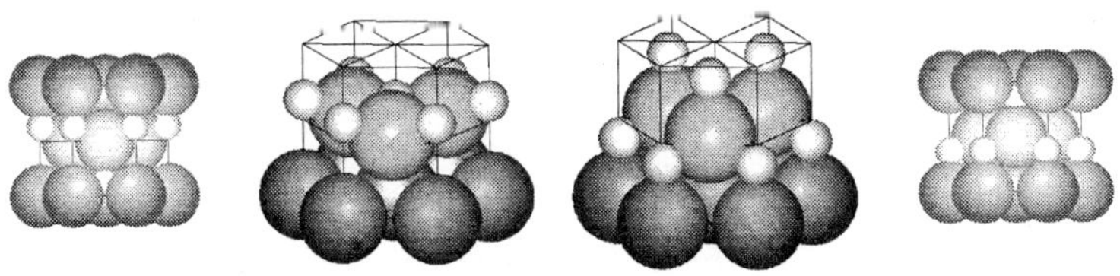

Figure 3-31 ZnS structure in HCP (Wurtzite structure)

In the ZnO structure, Zn occupies the tetrahedral sites within HCP oxygen. The coordination number of Zn is, therefore, four. Also the coordination of oxygen with Zn is four. All the tetrahedra are connected by corner sharing. Only one among U or D-types is chosen. Although they are distinguished by D and U-type, the actual crystal structure is the same since one becomes the other by rotating around the "a" axis of the hexagonal lattice. This is illustrated in Fig. 3-31 by the front view.

The unit cell dimension of ZnO is $a_o = 3.25$ and $c_o = 5.21$. The aspect ratio of c/a is 1.603. The bond distance calculated from this unit cell dimension is 1.99 which is closer to the bond distance of Zn–O calculated using covalent radii. Like ZnS, therefore, the geometrical data indicate that ZnO has a large component of covalent bonding.

	a	c	r_o	$r_{o,ionic}$	$r_{o,coval}$
ZnO	3.25	5.21	1.99	2.19	2.04

The difference of the ZnO structure from that of ZnS is uneven distances between Zn atoms. The distance between the nearest Zn atoms in the lateral direction of the tetrahedra sheet is longer than the distance between the nearest Zn atoms along the vertical direction to the tetrahedra sheet. This difference in the tetrahedral stacking modes in Fig.3-19 for FCC and HCP illustrate why the uneven distances arise in HCP and not in FCC. ZnS has a U_ℓ–U_h connection mode for tetrahedra while ZnO has a U_ℓ–D_h mode.

Mathematical calculation for the Madelung constant indicates that the ZnO structure (4.375) has a higher value than ZnS structure(4.368). This extra lattice energy with higher Madelung constant gives rise to extra stabilization of

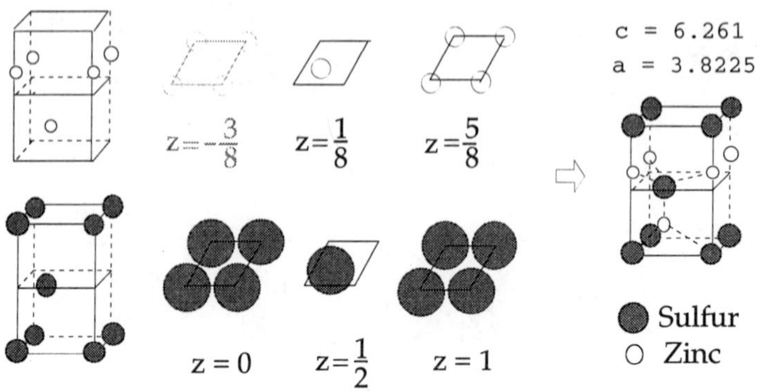

Figure 3-32 Atomic positions in ZnS (Wurtzite) with z-notation

ZnO in the wurtzite structure than in the sphaelerite structure. The reason why there is no ZnO with sphaelerite structure is probably due to this extra lattice energy which is enough to override other factors that might lead to the formation of the sphaelerite structure. This preference for wurtzite structure by ZnO is also revealed in the crystal structures of BeO, BeS, AlN and AlP. BeO and AlN both have the wurtzite structure while BeS and AlP have the sphaelerite structure. BeO and AlN also have greater ionicity than BeS and AlP since the electronegativity differences between Be–O and Al–N are larger than Be–S and Al-P, respectively. Similar to the relationship between ZnO(wurtzite) and ZnS(sphaelerite), the wurtzite structures of BeO and AlN can be justified over the sphaelerite structures of BeS and AlP.

2. AX_2 compound

2-1. Tetrahedral site occupancy

The typical structure types for AX_2 compounds on the basis of closest packing are the $CdCl_2$, CdI_2 and TiO_2 (rutile) structures. All have hexagonal packing of the larger anions with the cations filling half of the octahedral sites. Both structures satisfy Pauling's second rule regarding the ratio of r_c/r_a in the range of 0.414 – 0.722 for octahedral occupancy by cation. However, the arrangement of cations among octahedral sites is different. Every other octahedral site in every other layer along the c-axis is missed in the $CdCl_2$ and CdI_2 structure while every other octahedral site along the direction of the a or b axis is vacant in the TiO_2 structure. If the bond character is ionic, the crystal structure of $CdCl_2$ and CdI_2 is not possible since there is a huge repulsive interaction between anion layers and the empty octahedral sites.

If the ratio goes beyond this optimum range for octahedral site occupancy, cations will occupy tetrahedral or cubic sites. Since there are no cubic sites in HCP or CCP packing, anions are packed in a simple cubic array according to Pauling's Rule 1 and cations occupy every other vacant site in the cubic packing of anions. This has become known as the fluorite structure, in which , three quarters of the tetrahedral cation sites must be empty for the AX_2 compound to form since there are four tetrahedral sites per atom (or anion in ionic crystals). According to Pauling's second rule, the coordination number of A is twice that of X. Since A occupies the tetrahedral sites, its coordination number is 4 and the coordination number of X is two. To satisfy this requirement, tetrahedra must be connected by corner sharing because face sharing or edge sharing cannot satisfy the coordination number of X

In an ionic crystal, the binding forces for the crystal is the net attractive force overall between cations and anions and this depends most importantly on the interaction between nearest neighbor cations and anions. Therefore, a postulate can be made concerning the stability of an ionic crystal.

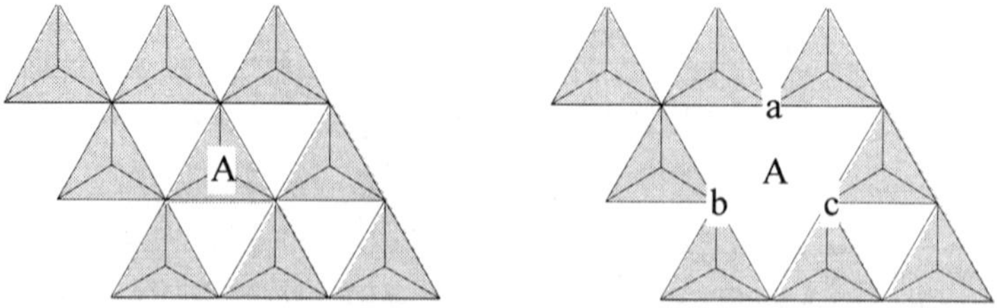

Figure 3-33 Removal of one tetrahedral site in a U-type tetrahedra layer

In the closest packing of anions, the repulsive force between contacting or nearest neighbor anions must be compensated by the attractive force between the anions and nearest neighbor cations. Otherwise the anions will be separated from each other.

Let's suppose there is a crystalline compound of formula AX_2 in which crystal has one quarter of the tetrahedral sites are filled with cations. On the basis of the above postulate, the tetrahedral sites must be filled in such a way as to avoid having anions that do not have direct sharing of a cation being in contact with each other.

If it is not possible to fill one quarter of the tetrahedral sites in each layer, it will be also impossible to fulfill the condition in the postulate for the whole crystal.

In a closest packed structure of anions (X), therefore, AX_2 compounds can be considered as a structure consisting of one quarter of the tetrahedra connected by corner sharing. In fact, there is no crystal of formula AX_2 having such a configuration since on a geometrical basis one quarter of tetrahedra in closest packing cannot be connected. The proof that AX_2 ionic compounds cannot exist with tetrahedral coordination in a closest packed structure of anions will be done by showing that the condition of the postulate cannot be satisfied by filling one quarter of tetrahedral sites in a tetrahedral layer of CCP or HCP.

In Fig.3-33, a top view of U-type tetrahedra layer is shown. One layer of U or D-type implies that one half of tetrahedral sites are filled. In order to make one quarter of tetrahedral sites filled with cation, one tetrahedra is removed in Fig.3-33(a). Removal of tetrahedra "A" results in three anions ("a", "b", and "c") on the basal plane of the tetrahedra layer and one anion on the apex of the tetrahedron "A," that do not share a common cation. The three anions on the basal plane can share a common cation by making a D-type tetrahedron in the next lower tetrahedra layer. However, the anion on the apex of tetrahedron "A" still does not share a common cation with "a', "b" and "c." If cations are removed from the half of the tetrahedra in Fig.3-33(a), the overall fraction of filled tetrahedral sites becomes one quarter. The removal of cations make one half of the anions on the apices of tetrahedra that are removed like "A" in Fig.3-33(a). This illustrates that there is no way in which all the contacting anions can share a common cation with a quarter of tetrahedral sites filled. In other words, at least one half of tetrahedral sites must be filled by cations in order to satisfy the condition in the postulate. Real crystals such as SiO_2 that has tetrahedral coordination, have a framework of closest packing of anions (O^{-2}). However, half of the oxygen and one quarter of the Si are removed instead of filling only one half of the tetrahedral sites, in order to avoid the repulsive interaction between cations that do not share a common cation. This will be discussed in Chap. 4. As a result, there are no ionic compounds in the AX_2 formula with tetrahedral coordination of cations.

2-2. Occupancy of octahedral sites

2-2-1 CdI_2

In the CdI_2 structure, ctahedral sites in every other interlayer of the larger anions are missing. A layered structure results, as shown in Fig.3-34. If CdI_2 has the true hexagonal packing of I, the location of I in the direction of the c-axis

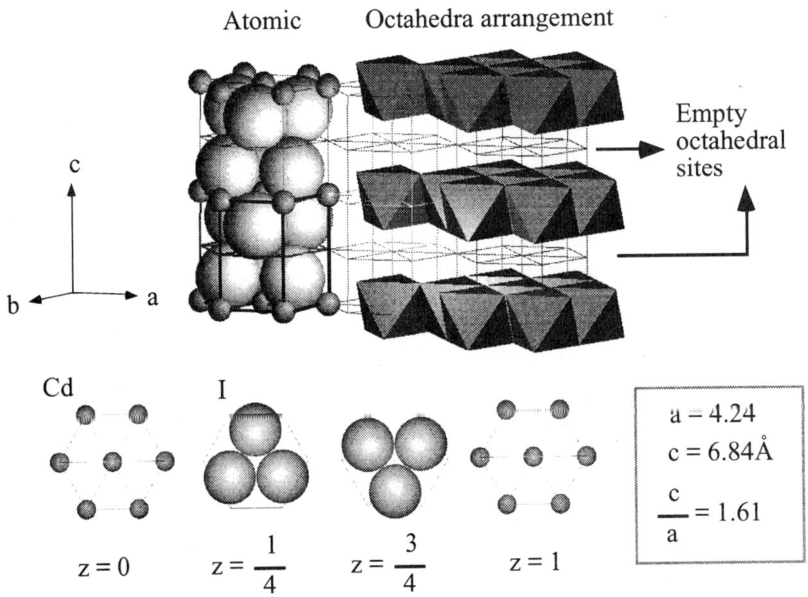

Figure 3-34 Crystal structure of CdI_2 showing the empty octahedral sites in every other interlayer of I

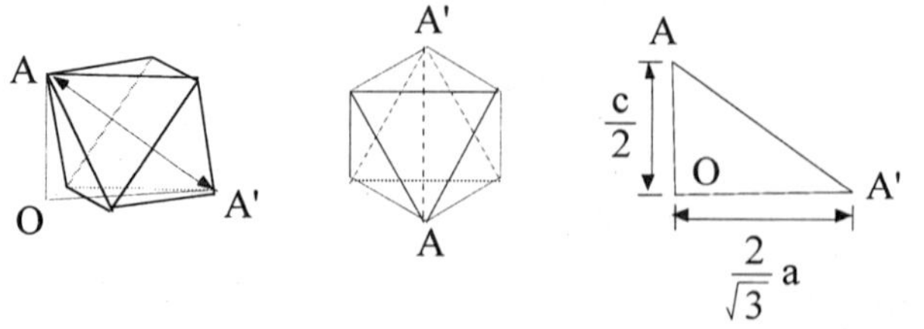

Figure 3-35 Geometry for the calculation of the bond distance Cd-I in the flattened octahedron for CdI_2

would be 0.25 in fractional coordinates with respect to "c." Due to a missing Cd atom in every other interlayer of I, the length between the first and second layers of Cd in Fig. 3-34 is different from that between the second and third layers of I. Usually in the isostructure of CdI_2, the position of the anions (atoms) deviates from 0.25 which is the value for the true hexagonal packing. In the case of CdI_2 the coordinate of I along the c-axis is about 0.25.

The unit cell dimension for CdI_2 is a = 4.24 and c = 6.84. The aspect ratio of c/a is 1.61 which is less than the 1.633 for a regular hexagonal lattice. Hence the octahedron around Cd is a flattened octahedron along the c-axis. From this unit cell dimension, the bond distance between Cd and I is 2.99 from the relationship in Fig.3-35. The ionic radii of Cd^{+2} and I^{-} are 0.97 and 2.16Å, respectively. The covalent radii of Cd and I are 1.41 and 1.33. Since the ionicradii sum and covalent radii sum are 3.13 and 2.74Å, respectively, the bond in Cd–I has both ionic and covalent character. Hence, the condition for the stability of anions (I^{-}) is violated in CdI_2 since cations in every interlayer are missing. The physical property of CdI_2 indicates the existence of van der Waals bonding between octahedral sheets. This also supports the non-ionic bonding character in CdI_2.

2-2-2 TiO_2 (Rutile)

The actual crystal structure of TiO_2 (rutile) is not represented by a closest packing of anions. The unit cell has a tetragonal lattice with dimensions of a = 4.59 and c = 2.96Å. The crystal structure is shown in Fig.3-36 with octahedra stacking (chain structure), viewed along the c-axis. Also, the coordinates of Ti and O are shown in Fig.3-37. The ionic and covalent radii are 0.68 and 1.32 for Ti and 1.40 and 0.66 for O. The ratio of r_c/r_a is 0.49, which is in the range of octahedral site preference. Since the sums for ionic radii and covalent radii are close, it is not easy to guess the ionicity or covalency of the bond from the bond distance alone since there two different Ti–O distances, one 1.98 and the other is 1.95Å. The longer represents the distance between Ti and O lying on the

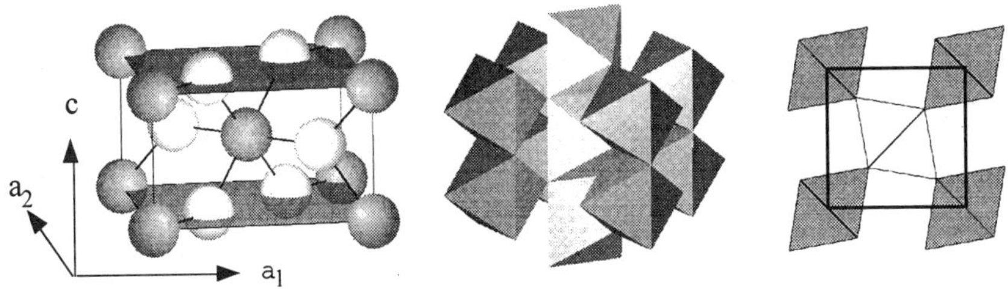

Figure 3-36 (a) Crystal structure of TiO_2(rutile) and (b) illustration by octahedra and (c) view along the c-axis

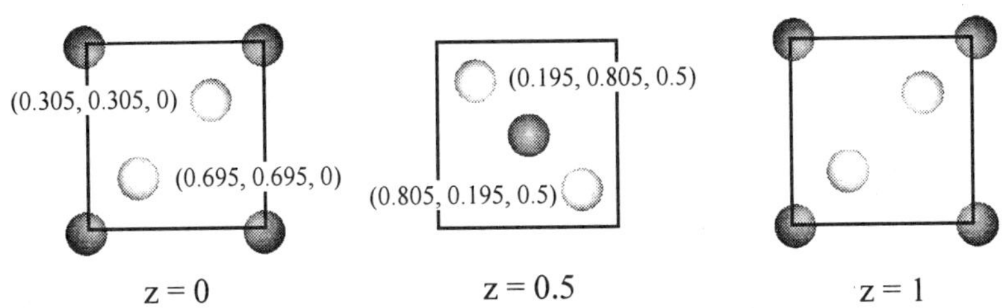

Figure 3-37 Coordinates of Ti and O along the c-axis of TiO_2
(Ti locates at the origin, (0,0,0) and the body center, (0.5, 0.5, 0.5))

same plane that is parallel with ab-plane. Also there are two different O–O distances. The shorter is 2.55 and the longer is 2.96Å. The shorter distance for O–O is less than the ionic radii sum of O^{-2}. This implies that the bond has definitely covalent character.

Although the crystal structure of TiO_2 is based on closest packing of oxygen, it can be described with the HCP structure. Since the ionic radii ratio indicates octahedral site preference for Ti+4 and the concentration of Ti is half of the oxygen concentration, half of the octahedral sites must be empty. One structure for which half of the octahedral sites are empty is that of CdI_2, but this structure is not suitable for ionic crystals since there are two entire contacting oxygen layers without common cations. For TiO_2, which is expected to be more ionic than CdI_2 half the octahedral sites are missing in each octahedral layer, rather than all octahedral sites in alternate octahedral layers. This configuration of half octahedral site occupancy is shown in Fig. 3-38(a). The removal of octahedral sites in a row results in a configuration of octahedra in a chain formation. The octahedral stacking in a chain shown in Fig. 3-38(b) is more illustrative than in Fig.3-38(a). Along the chain, octahedra are connected to the next octahedra by edge sharing (connection mode of dark octahedra in

Fig.3-38(b)) while they are connected to other tetrahedra in other chains by the corner sharing (connection mode between dark octahedra and white in Fig.3-38(b)). The direction of the octahedral chain corresponds to the c-axis of the tetragonal lattice.

By removing cations from half of the octahedral sites, some anions do not share a common cation. According to the postulate in Section 2-1, this structure should be avoided for stabilization of the crystal. For example, oxygen atoms "A" and "C" make contact with each other while there is no common cation between them with half the octahedral sites vacant. This configuration around oxygen atoms "A" and "C" in the standard HCP structure are shown in Fig.3-39(I). In order to force them apart so as to reduce the repulsive interaction, oxygen atoms are shifted from the closest packing. With respect to the octahedral connection, the shifting of oxygen is equivalent to the rotation of octahedra around the axis of the octahedral chain in clockwise and counterclockwise directions, which alternates every other octahedra sheet as shown in Fig.3-39. This rotation takes apart oxygen "A" from "C" without breaking other coordinations. Therefore, the TiO_2 (rutile) structure can be

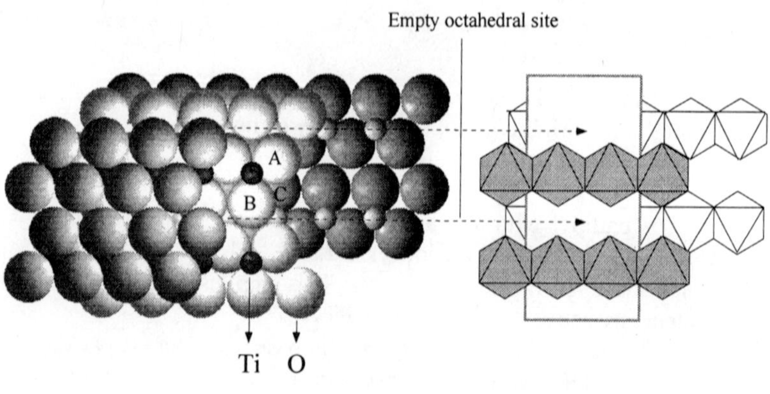

Figure 3-38 Occupancy of half of octahedral sites by Ti in HCP of oxygen

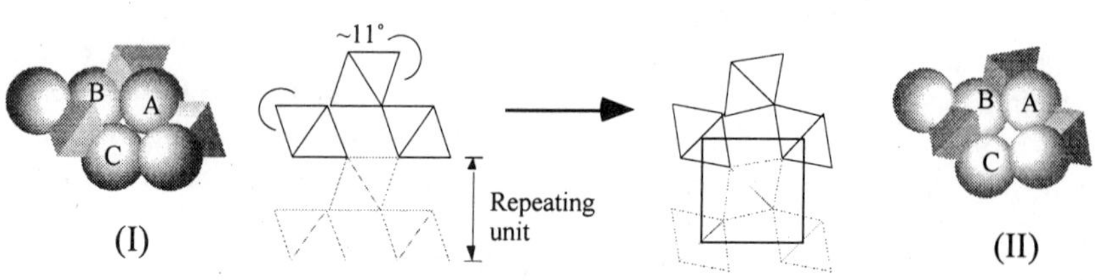

Figure 3-39 Rotation of octahedra by about 11° to separate oxygen "A" from "C" and converts the standard HCP structure(I) to real TiO_2(II)

regarded as a modified HCP structure with half of the octahedral sites empty. This modification is rationalized according to the postulate in Section 2-1.

Due to the HCP structure of oxygen, the length of the repeating unit along the direction perpendicular to the octahedral sheet is twice the height of the octahedra along that direction. Also the rotation of octahedra around the c-axis or the chain direction must alternate in direction with every other octahedra sheet because the anions suffering the same condition as "A" and "C" in the next octahedra sheet need the opposite rotational direction in order to avoid the contacts between those anions. The pattern of octahedral vacancy in each octahedra sheet plus the rotational repetition in every other layer results in the tetragonal symmetry in the actual TiO_2 crystal. This is one example where other crystal lattices can be generated by arrangement of cations within the framework of closest packing which usually is associated with the FCC or hexagonal lattice.

2-3. Occupancy in cubic site: CaF_2

The typical compound in this category is CaF_2, also known as fluorite. Its crystal structure is shown in Fig.'s 3-40 and 3-41. In the fluorite structure, cations are located at the face center position and anions are located at the tetrahedral position of the corresponding FCC lattice formed by the cations. For similarity, the crystal structure of Li_2O is shown on the right in Fig.3-40.

The unit cell dimension of fluorite (CaF_2) is a = 5.45Å. The bond distance of Ca-F is $\sqrt{3}\,a/4$ which gives 2.39Å. The ionic radius of Ca^{+2} is 0.99 in Table 1-32. This radius is that of 6-coordinated Ca+2. If this is corrected for the 8-coordinated Ca^{+2}, it is 1.03. The radius of O^{-2} is 1.36. Hence the ionic sum is equal to the bond distance of Ca–F. The covalent radii sum also gives

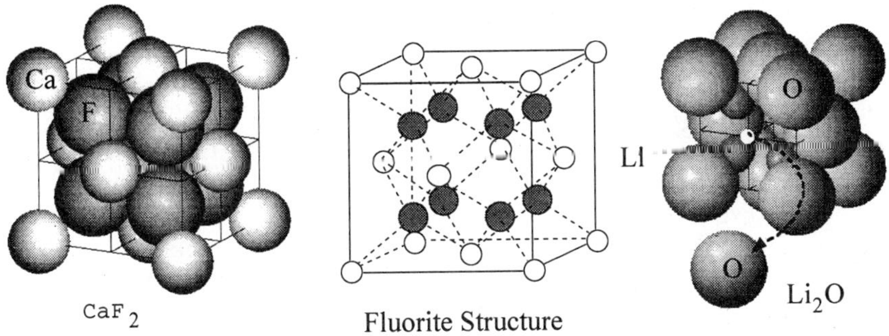

Figure 3-40 Crystal structure of fluorite structure and Li_2O in which the position of cation and anions are switched with respect to the fluorite structure

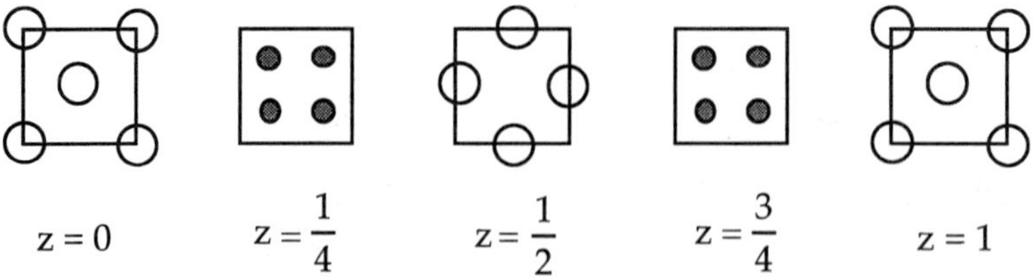

Figure 3-41 Representation of CaF_2 (Fluorite) structure in the z notation

Table 3-9 Comparison observed structures of AX_2 oxides (MO_2) to those predicted by the radii ratio according to Pauling's Rules

MO_2	(r_c/r_a)	Predict	Observed	Structure Type
CO_2	0.11	2:1	2:1	Linear
SiO_2	0.29	4:2	4:2	Quartz (Tetragonal)
SnO_2	0.51	6:3	6:3	Rutile (TiO_2)
PbO_2	0.60	6:3	6:3	Rutile
ThO_2	0.76	8:4	8:4	Fluorite (CaF_2)

2.38Å which is almost the same as the bond distance. From the radii sum alone, the bond character is difficult to determine.

According to Pauling's first rule, the appropriate coordination of Ca with F is 8 since the ratio r_c/r_a is 0.73 which is just over the minimum value for 8 coordination in the cubic structure of anion.

In terms of Pauling's second rule, EBS for Ca is 1/4 since the electric charge on Ca is +2 and its coordination is 8. Therefore the charge reaching each $F^- = 1/4$. From this consideration of charge balance, 4 Ca^{+2} ions must be bonded to F^- to compensate the –1 on F^-. Hence CN (Ca:F) = 8:4. A structure compatible with this CN is the cubic fluorite structure.

Compounds with the fluorites structure are oxides of heavy metals such as UO_2,(rc/ra = 0.71) ThO_2(rc/ra = 0.73) and ZrO_2(rc/ra = 0.80). Lead oxide takes the rutile structure while PbF_2(rc/ra = 0.86) has the fluorites structure. This structural difference between oxide and fluoride of lead agrees with the Pauling's first rule since the ratio or rc/ra for PbO_2 is 0.618 and it is 0.86 for PbF_2. Some compounds with the AX_2 formula are listed in Table 3-9 in order to show the agreement between the prediction of coordination according to the Pauling's first rule and the coordination number in the actual crystal structures.

3. AX_3 compound - Octahedral site occupancy

3-1. CCP – $CrCl_3$

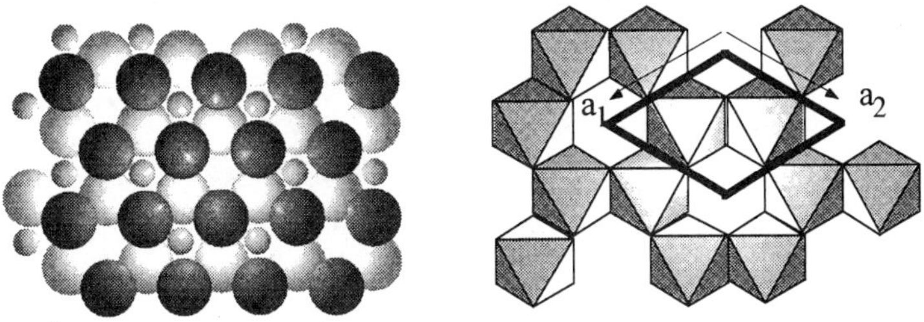

Figure 3-42 Top view of $CrCl_3$ structure showing one third octahedral sites empty

On the basis of the closest packing, two-thirds of the octahedral sites should be vacant in order to satisfy the formula AX_3. If two-thirds of the octahedral sites are vacant in each interlayer of packed atoms, strong repulsive interaction between adjacent packed anions would be expected, according to the the discussion of AX_2 structures, in that cations might take all the tetrahedral sites in the closest packing of anions. By the same token, the occupancy of only one-third of the octahedral sites would be improbable. In fact, two-thirds of the octahedral sites in every other interlayer are filled and the others are empty. Due to the empty octahedral sites in every other interlayer, $CrCl_3$ has a sandwich type structure that is the same as the CdI_2 structure except for the fact that two-thirds of the octahedral sites are filled. Fig.3-42 illustrates the occupancy of octahedral sites by cations and the resulting octahedra sheet.

The repeating pattern of vacant octahedral sites is an equilateral triangle. Two sides of the equilateral triangle as shown in Fig.3-42 correspond to two lattice vectors of $CrCl_3$. Two vectors "a_1" and "a_2" in Fig.3-42 and the other one "c" along the direction perpendicular to the layer form the hexagonal lattice.

If octahedral sites are named according to the method used for the naming of packed atoms in the lateral direction, "c" is assigned to a octahedral site that is formed between "A" and "B" type packing atoms. Capital letters are used for the packing atoms (anions) and lower case letters are used for the cations that occupy the interstitial sites within the packing atoms. "c" and "C" designate the same lateral position on the plane of the layer as shown in Fig.3-2. The relationships between the types of octahedral sites and the corresponding packing layers are illustrated in Fig.3-43. With this system of nomenclautre for the lateral position of cations and anions on the layer, the sequence of octahedra sheet is –cc'cc'– in CdI_2 while it is –a(b)c(a)b(c)a(b)c(a)b(c)– in $CrCl_3$. Here the parenthesis indicates that the octahedra on that layer are vacant. From this repeating sequence of octahedral sheets, the unit cell length along the c-axis that is perpendicular to the octahedra sheets can be estimated. The length of the c-axis in CdI_2 is twice the height of the octahedra whereas the length of the c-axis

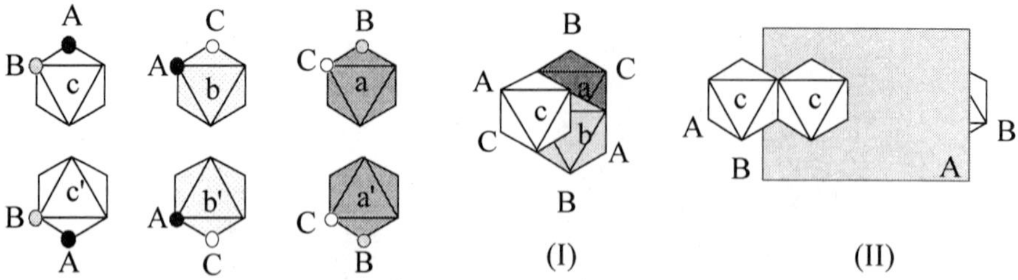

Figure 3-43 3 types of octahedra ('a', 'b' and 'c') in the closest packing and the stacking mode of the octahedra in (I) $CrCl_3$ and (II) CdI_2

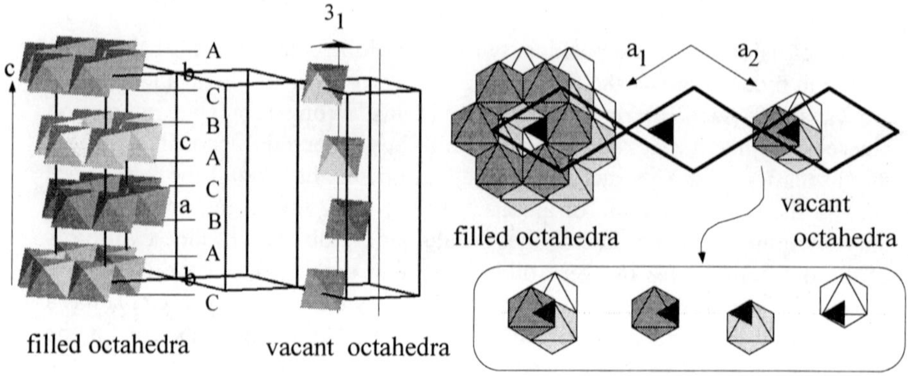

Figure 3-44 Sandwich structure of $CrCl_3$ and illustration of 3 fold screw axis in it by top view of filled octahedra and vacant octahedra in the perspective view on the left

in $CrCl_3$ is six times the height of the octahedra. From a perspective view, the sandwich structure of $CrCl_3$ is shown in Fig.3-44. The unit cell dimension of CrCl3 is given as a = 6.00Åand c = 17.3Å.

Although the pattern of vacant site in each octahedra layer is an equilateral triangle, the actual lattice is trigonal rather than hexagonal, due to the stacking mode of octahedral sheets which reduces 6-fold symmetry to 3-fold symmetry. In Fig.3-44, the location of the 3 fold screw axis is shown with vacant octahedra since only one octahedra is needed for the illustration of the 3-fold screw axis. With filled octahedra, the triangular hole at the center of stacking corresponds to the position of the 3-fold screw axis. The bold line in the perspective view of $CrCl_3$ in Fig.3-44 represents the unit cell. Unlike other crystals previously described in this chapter, lattice points do not reside at the center of any atom. For the octahedra sheet shown in Fig.3-42, identification of atomic positions in the unit cell structure becomes simpler if the lattice is defined with two vectors which connect two adjacent vacant octahedral sites. With respect to the cell dimensions or shape of the unit cell, both unit cells based on Fig.3-42 or on Fig.3-44 are equivalent. However, as far as systematic representation of the crystal with symmetry elements or atomic positions are concerned, there is a difference between them. For example, the atomic positions with reference to the unit cell shown in Fig.3-44 are given as follows:

Atom	Position	x	y	z
Cr(1)	3b	2/9	4/9	0
Cr(2)	3b	5/9	1/9	0
Cl(1)	6c	2/9	4/9	0.26
Cl(2)	6c	5/9	1/9	0.26
Cl(3)	6c	8/9	7/9	0.26

Here "3b" in the "Position" column indicates that there should be two other Cr atoms whose coordinates can be derived from the coordinates given. Those coordinates are given by (u,-u, 2/3) and (-2u, -u, 1/2). Here u is the x-coordinate of the first Cr atom. Likewise, "6c" indicates that there are five other atoms of the same kind totaling 6 atoms in the unit cell. If the coordinate of the first atom is (x, y, z), the other five atoms are given as (y-x, -x, z+2/3), (-y, x-y, z+1/3), (y-x, y, -z), (-y, -x, 1/3-z) and (x, x-y, 2/3-z). The "Position" column number indicates the total number of atoms of the same kind. The letter represents a special relationship between these atoms and is called Wyckoff notation. The Wyckoff notation implies a total number of equivalent points or rank of equipoints. Hence 'a' in the Wyckoff notation is assigned to the equipoint having the least number of equivalent points. The number in front of the alphabet in the Wyckoff notation denotes the total number of equivalent points.

When atomic coordinates are given on the basis of the unit cell in Fig.3-44, only five coordinates are given from which other atomic positions are derived. In contrast, if the unit cell is defined with two vectors in Fig.3-42, atomic positions must be given for each atom for all 6 Cr atoms and 18 Cl atoms since there is no special relationship between atoms as the letter of the notation in the "Position" column implies. In the following, the position of lattice points in a crystal and the determination of the position of the unit cell is further examined using TiO_2 (rutile) as an example.

In this chapter, crystal lattices are drawn by connecting the centers of two atoms or ions that are symmetrically equivalent in such a way as to visualize the atomic positions within the lattice system. This illustrates the lattice system and the correlations between atoms within the unit cell. Since

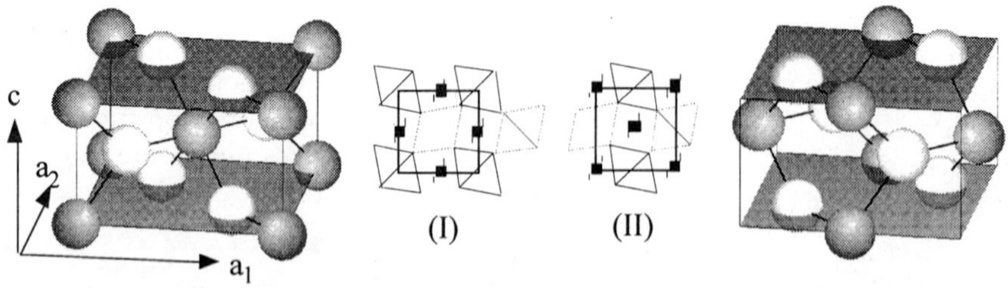

Figure 3-45 Location of the unit cell of TiO_2(rutile) (I) in the conventional way and (II) in a way to show 4 fold symmetry

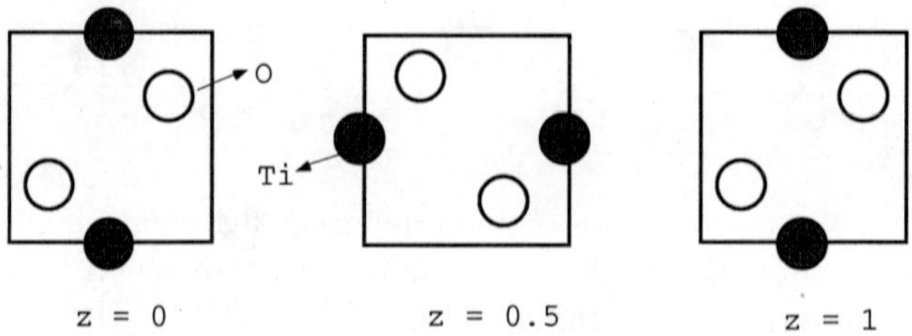

Figure 3-46 Representation of TiO_2 for the lattice on the right in Fig.3-45(II) in the z-notation

some atoms reside at special positions: corner, edge and center of the unit cell, they serve as reference points for other atoms. For example, in Fig.3-45(I), lattice points are located at the center of Ti atoms. It is easily seen that the distances between four octahedra are the same and TiO_2 has a tetragonal lattice. On the other hand, if it is necessary that the major primary symmetry elements must pass through the lattice points, then the resulting lattice points do not reside on the center of atoms. In this regard, the lattice shown in Fig.3-45(II) was generated by shifting the lattice on the left by half of "a_1" in order to make 4 fold screw axis(4_2) pass through the lattice points. The advantage of the lattice for TiO_2 shown in Fig.3-45(II) over that of Fig.3-45(I) is that four fold symmetry can be seen more easily while the positions of atoms within the unit cell and especially the equal distance between Ti atoms along "a_1" and "a_2" axes are also identified more easily as in Fig.3-45(I). When atomic positions are depicted by z-notation as in Fig.3-46, four fold symmetry (in a strict sense, four fold screw axis) is more easily recognized than the unit cell structure in Fig.3-45(I). In this respect, Fig.3-45 illustrates that the lattice of a crystal can be drawn in several ways depending upon what is to be emphasized.

3-2. HCP – $FeCl_3$

$FeCl_3$ has the similar layered structure of $CrCl_3$. While $CrCl_3$ is based on the CCP structure of Cl, $FeCl_3$ is based on the HCP structure of Cl. For the AX_3 formula, two-thirds of total octahedra sites are empty. The vacant octahedral sites are distributed in the same way as $CrCl_3$. Every other interlayer in HCP structure of I has no cations at all in the octahedral interstitials. In the partially empty interlayers, one-third of the octahedral sites are empty in order to match the molar ratio between A and X of AX_3.

Due to the presence of vacant octahedral sites in the octahedral sheets, these sheets degenerate into three types that are shown in Fig.3-47. The bold line denotes the two probable lattice vectors and serves as a reference line to distinguish the relative positions of vacant octahedra in the lateral direction within the sheet. If the notation scheme in Fig.3-47 is followed, the sequence of the octahedral sheet is $-c_1cc_2cc_3c-$ assuming chlorine atoms form layer types A and B. Here "c" is inserted to designate an empty octahedra sheet. From this sequence, the unit cell length along the c-axis that is perpendicular to the layer is six times the height of octahedra sheet. The hexagonal lattice is depicted by the bold line in Fig.3-48. The cell dimension of this hexagonal lattice is given as a = 6.06Å, and c = 17.38Å

Due to the presence of three fold rotational symmetry, the lattice also can be described as a rhombohedral lattice. The rhombohedral lattice is that which is generated by connecting vacant octahedra on the left of Fig.3-48. This relationship between the hexagonal lattice and the corresponding rhombohedral lattice will be illustrated in the discussion of Al_2O_3 following. The hexagonal (trigonal) lattice is the non-primitive lattice of the rhombohedral lattice. The hexagonal lattice has 3 lattice points while the rhombohedral lattice has only one lattice point. If the lattice parameters are given on the basis of the rhombohedral lattice, they are a = 6.758Åand $\alpha = 53°11'$. Since all the lattice vectors have the same length and the same angle between any two lattice vectors, the lattice parameters for a rhombohedral lattice are given by the length of the lattice vectors and the angle (α) between two lattice vectors. If lattice parameters are given by two lengths, therefore, they represent the hexagonal

Table 3-10 Lattice parameters of compounds with the $FeCl_3$ structure

	$CrBr_3$	BiI_3	SbI_3
a	6.26	7.50	7.47
c	18.20	20.68	20.89

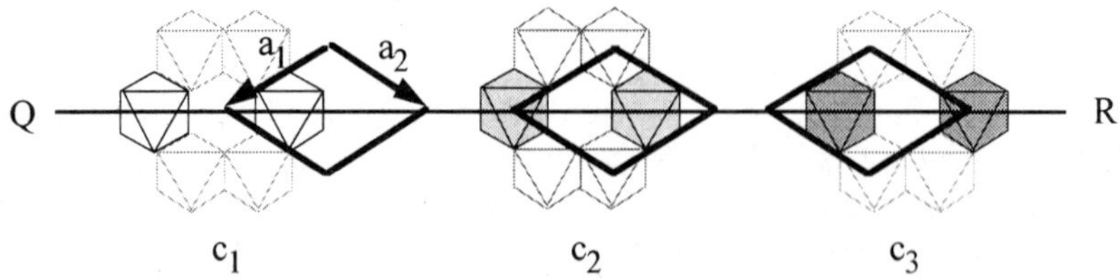

Figure 3-47 Three possible configuration of octahedra sheet having one third of octahedra is empty

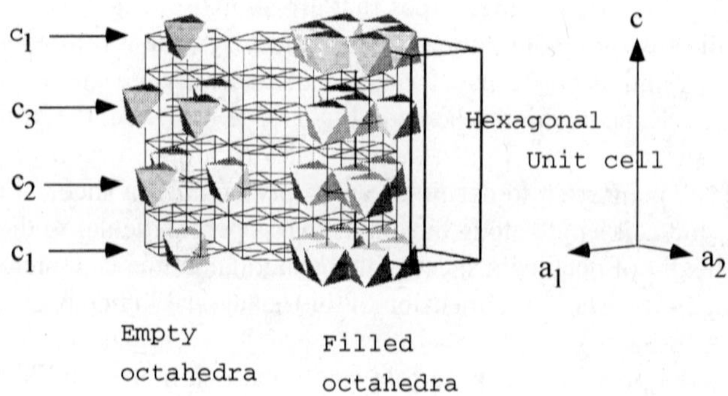

Figure 3-48 Crystal structure of $FeCl_3$ in a hexagonal (trigonal) lattice along with rhombohedral lattice shown by empty octahedra

lattice, and the corresponding rhombohedral lattice if they are given by one length plus an angle.

Compounds with the same crystal structure of a $FeCl_3$ are $CrBr_3$, BiI_3, and SbI_3. Their lattice parameters are listed in Table 3-10.

4 A_2X_3 compound

4-1. Octahedral site occupancy – Al_2O_3

In the closest packing of anions, one-third of the octahedral sites should be empty to satisfy the formulation A_2X_3. The typical example in this group is Al_2O_3. Each octahedra sheet has the configuration in Fig.3-47. Hence "a_1" and "a_2" in Fig.3-47 are also expected to be the two lattice vectors of the crystal lattice for Al_2O_3. Building the crystal of Al_2O_3 reflects the stacking of these octahedral sheets along the c-axis which is perpendicular to the layer. Since O in Al_2O_3 takes the HCP structure, the sequence of octahedra should be –cc'cc'cc'– according to the notation in Fig.3-43, if oxygen atoms are assumed to be located at A (lower) and B (upper) layers. Due to the vacant sites, octahedra sheets degenerate into three types, c_1, c_2 and c_3 as shown in Fig.3-47. By combining these two notations for the octahedral sheets for Al_2O_3, the possible stacking modes for Al_2O_3 are (i) $-c_1c_1'-$, (ii) $-c_1c_2'-$ and (iii) $-c_1c_2'c_3c_1'c_2c_3'-$. The actual crystal structure of Al_2O_3 takes the sequence of $-c_1c_2'c_3c_1'c_2c_3'-$. The crystal structure of this stacking sequence of octahedra sheet is shown in Fig.3-49. The explanation for taking this stacking sequence among the three can be explained by Pauling's Rule 3. This rule, concerning the stability effect of sharing modes between polyhedra can be stated as :

> The presence of shared edges, and particularly of shared faces, in a coordinated structure decreases its stability; this effect is large for cations with large valence and small coordination number, and is especially large in case the radius ratio approaches the lower limit of stability of the polyhedron.

Pauling compared Al_2O_3 with TiO_2 (rutile) regarding the sharing modes between adjoining octahedra. In TiO_2 each octahedra is connected to the adjoning octahedra by edge sharing and corner sharing while octahedra in Al_2O_3 are connected to adjoining octahedra by face and edge sharing. This is explained by the different effects of sharing modes on the stability of the crystal structures. Before examining the stability of the three possible crystal structures of Al_2O_3 with respect to Pauling's Rule 3, their shape will be examined. The crystal structure having three sequences (i) $-c_1c_1'-$, (ii) $-c_1c_2'-$ and (iii) $-c_1c_2'c_3c_1'c_2c_3'-$ are illustrated in Fig.3-49 by a front view of octahedra whose positions lie on the line, QR, of Fig.3-47.

Regardless of the arrangement of vacant octahedra within an octahedral sheet, any octahedra is connected to the neighboring three octahedra by edge sharing in the lateral direction of the sheet. In the vertical direction to the sheet, the different sequence generates different combinations of sharing modes between adjoining octahedra. With the sequence of $-c_1c_1'-$, vacant octahedra are connected to other vacant octahedra by face sharing in the vertical direction while filled octahedra are connected to other filled octahedra by face sharing also. Hence each octahedra shares two faces. With the sequence of $-c_1c_2'$, one half of the octahedra have two shared faces in the vertical direction toward the octahedral sheet. With the sequence of $-c_1c_2'c_3c_1'c_2c_3'-$, all the octahedra have one face sharing with adjoining octahedra. According to Pauling's Rule 3, therefore, the presence of double shared faces in the first and second case is expected to make those cases less stable than the one having the sequence of $-c_1c_2'c_3c_1'c_2c_3'-$ since octahedra in that sequential configuration have only one shared face.

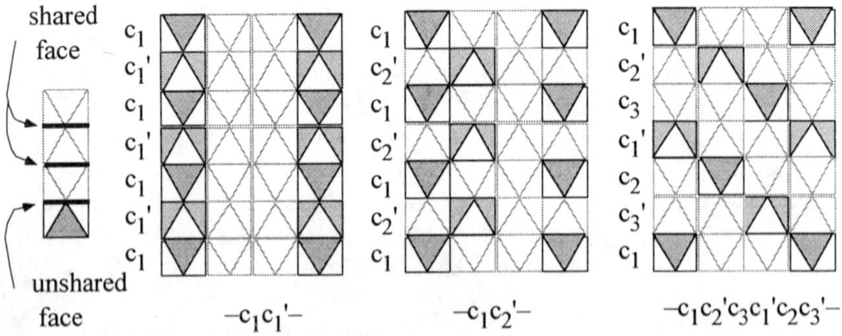

Figure 3-49 Arrangement of empty octahedra along the perpendicular direction to the octahedra sheet. (Only octahedra on the line, QR, of Fig.3-47 are shown. Dark line represents empty octahedra.)

Regarding the unit cell shape, the first two sequences in Fig.3-49 make the length of the unit cell vector along the vertical direction to the octahedral sheet twice the height of the octahedra. The last sequence generates the unit cell length along that direction which is six times the height of the octahedron. The actual lattice parameter for Al_2O_3 is given as

$$a_o = 4.76\text{Å} \text{ and } c_o = 13.00\text{Å}$$

since it is a hexagonal lattice. (In a strict sense, it is not a hexagonal lattice but a trigonal since it does not have six fold rotational symmetry.) For the trigonal lattice, there are two ways in which to describe it. One is a description based on the hexagonal lattice and the other is based on a rhombohedral lattice. When described by a hexagonal lattice, that lattice is a non-primitive lattice. If the lattice parameters for Al_2O_3 are expressed in terms of the corresponding rhombohedral lattice, they are given as

$$a_o = 5.13\text{Å} \text{ and } \alpha = 55°20'$$

Here α is the angle between two lattice vectors. The relationship between the hexagonal (trigonal) lattice and rhombohedral lattice is shown for Al_2O_3 in Fig.3-50. To illustrate the correspondence between them lattice points are drawn on the empty octahedra. In the sequence of octahedra as $c_1c_2'c_3c_1'c_2c_3'$, octahedra in c_1, c_3 and c_2 are mirror images of those in c_2', c_1', and c_3'. In terms of symmetry, they are not equivalent. Hence if a lattice point is located at

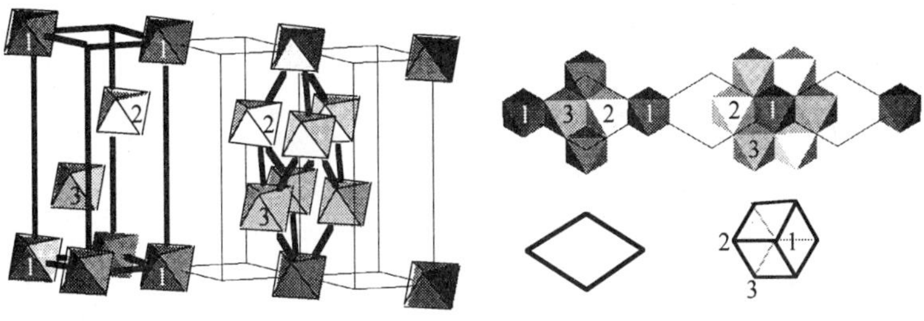

Figure 3-50 Illustration of hexagonal lattice and rhombohedral lattice in Al_2O_3

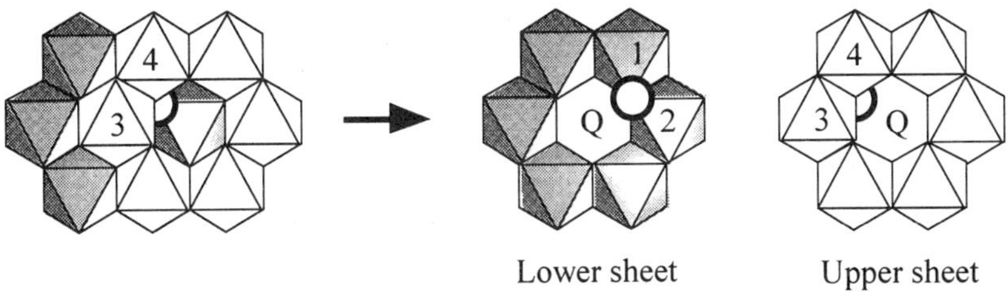

Figure 3-51 Configuration of coordination of oxygen "Q" with 4 neighboring Al

c_1, other lattice points must be at c_2, and c_3 not at either one of c_1', c_2' and c_3'. Only vacant octahedra in the same type group, (c_1, c_2, c_3), within the hexagonal lattice are shown.

Al forms an octahedral coordination with 6 oxygen anions. The electrostatic bond strength of Al–O is 3/6 = 0.5. Each oxygen is bound to 4 neighboring Al cations shown in Fig.3-51. According to Pauling's Rule 2, the charge on the oxygen must be equal to the sum of the electrostatic bond strength in order to balance the electric charges. The EBS for Al–O bond = 0.5 and there are 4 Al atoms around an oxygen anion. The summation of EBS for the oxygen anion is calculated according to Eq.(1) as follows:

$$0.5 \text{ (EBS)} \times 4 \text{ (number Al atoms of surrounding the oxygen)} = 2$$

The electric charge balanced with the coordination of oxygen is 4 while Al has the coordination number of 6. Compounds with the same crystal structure of Al_2O_3 are Cr_2O_3 and α–Fe_2O_3 (hematite). As metallic oxide compounds, $FeTiO_3$ (ilmenite) and $LiNbO_3$ have the same structure as Al_2O_3 except for the arrangement of Fe and Ti in ilmenite and Li and Nb in $LiNbO_3$ among the filled

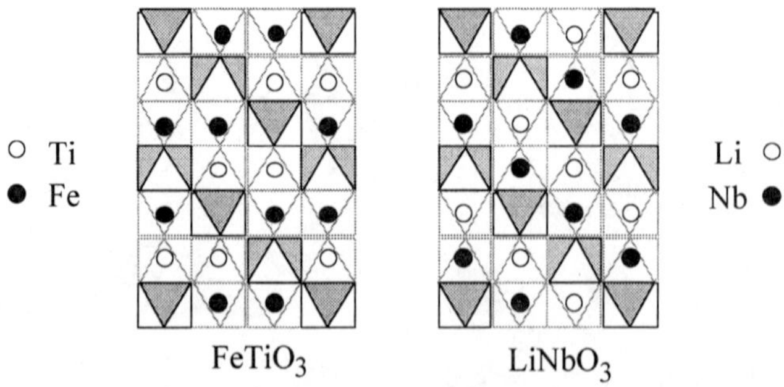

Figure 3-52 Distribution of Fe and Ti in ilmenite and Li and Nb in $LiNbO_3$

octahedral sites. In ilmenite, Fe and Ti are distributed in alternating octahedra sheets while Li and Nb in $LiNbO_3$ are distributed in the same octahedra sheet in such a way that Li and Nb octahedra do not make edge or face sharing with the same species octahedra. The distribution of these metallic elements within the Al_2O_3 structure is shown in Fig.3-52. Accordingly, the space group of $R\bar{3}c$ for Al_2O_3 is changed to $R\bar{3}$ for ilmenite while it is changed to R3c for $LiNbO_3$. Hence, a glide plane is lost in ilmenite and an inversion symmetry is lost in $LiNbO_3$.

4-2. La_2O_3 – quasi-octahedral site

The ionic radii ratio of La^{+3} and O^{-2} is 0.77, which is slightly over the maximum value for the stable octahedra coordination of a cation with anions. Still it can be described by the octahedral coordination of La with closest packing of oxygen atoms although its coordination is seven instead of six. Like Al_2O_3, for La_2O_3, the molar ratio of La to "O" is 0.67. In order to satisfy the molar ratio, one-third of the octahedral sites are empty. On the scheme of the cubic closest packing of O^{-2} anions, every third interlayer is vacant. According to the notation for relative positions of anions and interstitial sites, the stacking sequence of anions and cations is described and compared with the regular FCC structure with all sites filled as

$$La_2O_3 \quad —A_1–c—B_1——C_1–b–A_2–c–B_2——C_2–b–A_3—$$

$$FCC \quad —A_1–c—B_1–a—C_1–b—A_2–c—B_2–a—C_2–b—A_3—$$

Emptying all the octahedral sites in every third interlayer of FCC packing of oxygen results not in the cubic structure but in a hexagonal lattice.

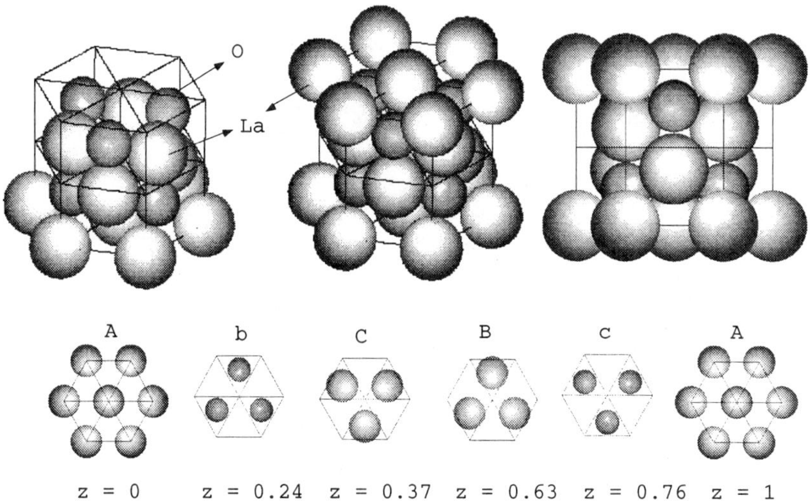

Figure 3-53 Crystal structure of La_2O_3 and atomic positions along c-axis

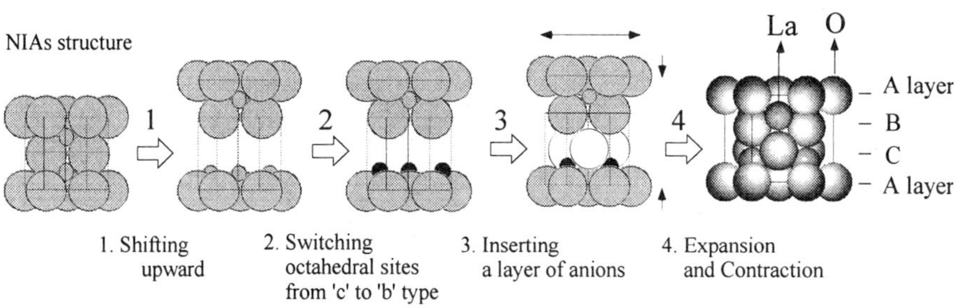

Figure 3-54 Schematic illustration of transition from NiAs to La_2O_3 structure

The resulting hexagonality of La_2O_3 from emptying one-third of the octahedral sites is explained in Fig.3-54 as a modified structure of NiAs whose lattice is hexagonal. The transformation from NiAs structure to La_2O_3 is illustrated in Fig.3-54. Since there are no La atoms between the inserted oxygen layer and one layer of oxygen at z = 0.5 in the NiAs structure, the distance of these two oxygen layers is shorter than the other two oxygen layers. This contraction between the two oxygen layers is related to the discontinuity in the sequence of layers of anions and interstitials from that of the regular FCC as shown above. This contraction destroys the necessary requirement for the CCP structure of oxygen anions to be the FCC lattice. As a result, insertion of one oxygen layer into the NiAs structure keeps the hexagonality of the lattice. The lattice parameters for La_2O_3 are a_o = 3.94 Åand c_o = 6.13Å. Regarding the lattice type, unlike Al_2O_3, La_2O_3 is a trigonal and not a rhombohedral lattice.

In La_2O_3, 'a' type cations are missing between 'B' and 'C' type anions. Due to this shorter distance, cations in 'c_n' position are in contact with oxygen anions in 'C_n' position, that is, to the right of the cation in the sequence above. Likewise, cations in 'b_n' position are in contact with oxygen anions in 'B_n' position that is to the left of the cation in the sequence. In this sequence, left side indicates lower layer and right side indicates upper layer. Considering this configuration of atomic positions, the seven coordination of La in La_2O_3 can be seen in Fig.3-53. Each cation, say 'c' type, makes contact with 3 'a' type, 3 'b' type and 1 'c' type oxygen anion. In relation to the ionic radii ratio, seven coordination is possible with longer distance (3.94Å= a_o) between oxygen anions than that (2.80Å) in closest packing. This is made by putting such a large cation as La^{+3} (1.04Å) in interstitial sites within the oxygen packing.

Although cations make seven coordination with anions, oxygen anions make different coordination numbers. Type 'A' oxygen anions make six coordination while type 'B' and 'C anions make 4 coordination. Due to these different coordination numbers of oxygen anions, their EBS varies. For charge balance, the sum of EBS for a cation or anion must equal to the charge on the ion. EBS for O_B and O_C is 0.5 (2/4) while that for O_A is 0.33(2/6). The sum of EBS for La^{+3} is given as follows :

$$EBS(La^{+3}) = 0.5 \times 1 + 0.5 \times 3 + 0.33 \times 3 = 3$$

Therefore, charge balance is preserved even with seven coordination of La^{+3}.

4-3. Tetrahedral site occupancy – Ga_2S_3

The ionic radii ratio of Ga and S predicts tetrahedra coordination for Ga in a closest packing of S. Ga_2S_3 is similar to ZnS in that it has two types of crystal structures. It has a cubic form (low temperature form) and a hexagonal form (high temperature form). They correspond to zinc blende and wurtzite structures, respectively. Here only the wurtzite type structure of Ga_2S_3 is constructed. The difference of Ga_2S_3 from ZnS is that one-third of the

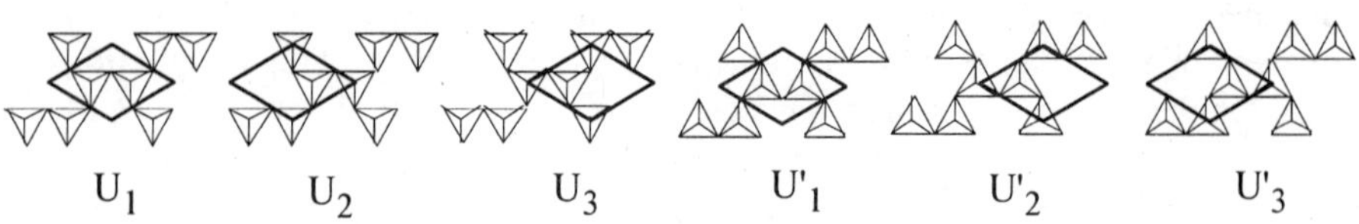

Figure 3-55 Tetrahedra sheet of U and U'-type with one third empty and three possible configurations for U and U' like in Fig.3-46 for $FeCl_3$

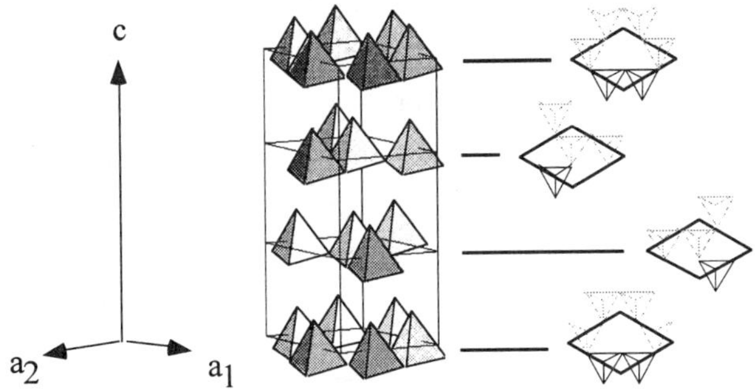

Figure 3-56 Crystal structure of with only U-type tetrahedra

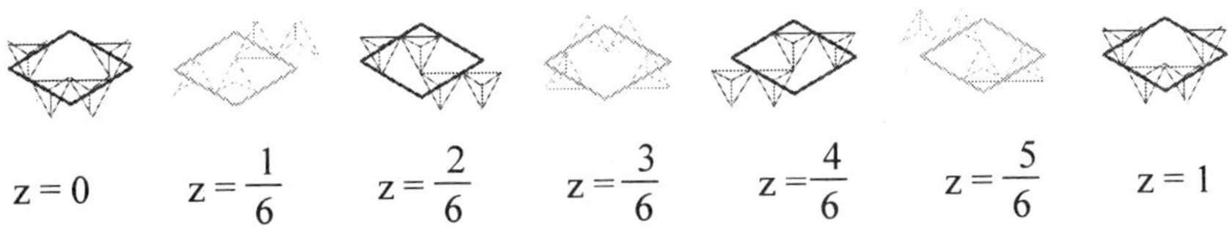

Figure 3-57 Representation of crystal structure of Ga_2S_3 with z-notation

tetrahedral sites that are occupied by Zn are empty so to satisfy the molar ratio of Ga and S. Tetrahedral sheets having one-third tetrahedra empty are constructed as shown in Fig.3-55. Due to the pattern of vacant site, the bold lines in Fig.3-56 and 3-57 define lattice vectors '$\mathbf{a}_1$' and '$\mathbf{a}_2$'.

Building the crystal of Ga_2S_3 is a question of how to stack these tetrahedra sheets along the c-axis that is perpendicular to the layer. Since S in Ga_2S_3 takes the HCP structure, the sequence of tetrahedra should be –UU'UU'– according to the notation in Fig.3-47 if the tetrahedra sheet takes only an upward direction. (If it takes a downward direction, the result is the same except that they become mirror images across the sheet.) Here U and U' indicates that the tetrahedra in the sheet U' is the rotated image of one in the sheet U. Like in Al_2O_3, all three configurations are used resulting in the sequence of – $U_1U'_3U_2U'_1U_3U'_2U_1U'_3$–. Therefore, the length of the c-axis in Ga_2S_3 is six times the height of the tetrahedra. Regarding the lattice type and symmetry, both Al_2O_3 and Ga_2S_3 are very similar except for the different interstitial occupancies. The lattice parameters for Ga_2S_3 are given as a_o = 6.39Åand c_o = 18.09Å. Its crystal structure is shown in Fig.3-56. Its representation with z-notation is illustrated in Fig.3-57.

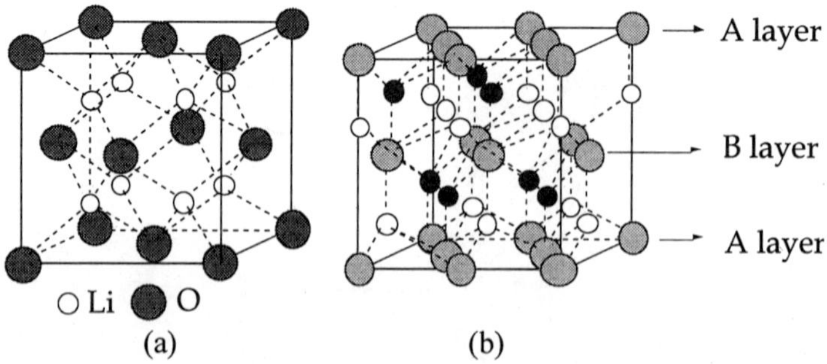

Figure 3-58 Structure of (a) Li_2O and (b) its equivalency with HCP structure

5. A_2X compound - Tetrahedral site occupancy

The only possible structure is the closest packing of "X" with all tetrahedral sites occupied by "A'. There is no equivalent structure by filling octahedral sites since the ratio of octahedral sites and packing atoms is one.

In Li_2O, all of the tetrahedral sites are occupied by Li. The HCP structure equivalent to Li_2O is not known. If there was, it must be the structure shown in Fig. 3-58 (b). Li has a coordination number of 4 and each oxygen is surrounded by eight Li. The EBS of Li is, therefore, 1/4 and the summation of EBS for oxygen is $1/4 \times 8 = 2$. According to Pauling's Rule 2, the charge is balanced. Both Na_2O, and K_2O have the same structure as Li_2O. However, their ionic radii ratio is over the maximum limit of stable conditions for tetrahedral coordination of those cations since there is no possible structure with octahedral coordination for compounds with the A_2X formula.

6. Summary of crystal structures of binary ionic compounds

Atomic packing models for the characterization of ionic crystals provide a schematic way for predicting their structures with the aid of the coordination theories of Pauling. As shown for ZnS and ZnO, bonding in crystals cannot be solely characterized by purely ionic or purely covalent bonding since, in general, they exist together. Sometimes, ionicity of the bond is so dominant over covalency that coordination theory gives a plausible explanation for the crystal structures although it does not provide a quantitative way for predicting the crystal structure of a compound. The summary of crystal structures of binary compounds in Table 3-10 illustrates this aspect of coordination theory and Pauling's rules.

7. Summary of compounds with HCP structures

(i). WC (Tungsten Carbide) – Covalent.

The tungsten carbide(WC) structure is formed by many carbides and nitrides. It is derived directly from HCP basic structure where "W" replaces atoms at the (000) and C atom at the (2/3 1/3 1/2) positions. Tunsgten carbide is an extremely hard material.

(ii). CdI_2 (Cadmium Iodide) Structure.

In the CdI_2 (Cadmium Iodide) structure, HCP iodine atoms occupy interior positions, with cadmium atoms at the corners. It is formed by the more covalent AB_2 phases. Thus the larger (polarizable) atoms, iodides and bromides, form this structure.
e.g., MgI_2, $MgBr_2$–covalent but $MgF_2 \rightarrow$ Rutile (TiO_2–ionic) structures)

(iii). ZnS (Wurtzite) Structure.

ZnS (Wurtzite) structure is formed from the basic HCP structure with S and Zn atoms in separate unit cells, displaced vertically by 3/8. It is formed by

Table 3-10 Summery of binary ionic crystal structures based on CCP and HCP

	Type	CN	(r_c/r_a)	Basic Structure	
				HCP	CCP
1.	A_2X	4:8	> 0.225		Li_2O
2.	AX	4:4	< 0.414	ZnS (Hex)	ZnS (Cubic)
		6:6	> 0.414	NiAs	NaCl
		8:8	> 0.73		CsCl
3.	AX_2	6:3	> 0.414	CdI_2	$CdCl_2$
		6:3	> 0.414	Rutile(TiO_2)	
		8:4	-		CaF_2*
4.	A_2X_3	4:8/3	< 0.414	Ga_2S_3	Ga_2S_3
		6:4	> 0.414	Al_2O_3	
		6:4	> 0.414		Y_2O_3
		7:14/3	> 0.7		La_2O_3
5.	AX_3	6:2		$FeCl_3$	$CrCl_3$

* This is the reverse of the anti-fluorite structure where cations occupy all tetrahedral sites of the closest packing of anions.

Table 3-11 Summary of compounds with HCP structure and their bonding characteristics

Basic HCP (Mg) Structures	WC Structure	CdI_2 Structure	ZnS (Wurtzite) (Hexagonal)
P6mc	P6m2	P3ml	$P6_3mc$
Atoms at	Atoms at	Atoms at	
Mg – (000) (2/3 1/3 1/2)	W– (0 0 0) C–(2/3 1/3 1/2)	Cd– (0 0 0) I– (2/3 1/3 /14) (1/3 2/3 3/4)	Atoms at Zn–(0 0 0) (2/3 1/3 /12) S– (0 0 3/8) (2/3 1/3 7/8)
HCP Metals	WC, RuC Covalent	Covalent AX_2 Composition W_2C, CdI_2, ZnI_2	Covalent Compounds BeO, ZnO, α–SiC AgI, CaI, CuBr

compounds of the (AB) type that are intermediate in covalency between ZnS (cubic) and NaCl ($r_c/r_a = 0.414$); CN = 4:4
eg –SiC, NH_4F, ZnO, BeO, CdSe, etc.

(iv). NiAs (Nickel Arsenide) Structure.

Compounds with the NiAs (Nickel Arsenide) structure are formed by more covalent phases with CN = 6:6 i.e., (r_c/r_a = 0.414-0.732). More ionic phases form the NaCl structure
Ni atoms are at the (0 0 0) and (0 0 1/2) positions with As at (1/3, 2/3,1/4) and (2/3, 1/3, 3/4). These coordinates are are different from those based on the Fig.3-28, in which As atoms are at lattice points.
As atoms at (2/3 1/3 1/4) and (1/3 2/3 3/4).
This structure is adopted by the sulfides, selenides, tellurides and arsenides of the transition metals (Ni, Fe, Co, Mn, Ti).

(v). A rare earth structure.

The A rare earth structure which is mainly ionic, is formed by the larger (and more covalent) rare earth oxides. Smaller ones e.g., Sn_2O_3 form the cubic c-rare earth structure. The unique 7 CN of La+3 in La_2O_3 A-rare earth structure was described in section 4-2.

8. Derivative structures of FCC lattice

CaC_2 and FeS_2 have the NaCl structure. However, they do not have the FCC lattice since their anions are not spherical which removes the 3 fold rotational

Table 3-12 Compounds having similar structures with FCC lattice and their space groups

CaC_2	FeS_2	NaCl	CaF_2
I 4/m mm	Pa3	Fm3m	Fm3m
CaC_2 BaC_2 MgC_2 $MoSi_2$ BaO_2 CaO_2 SrO_2 CsO_2	FeS_2 (Pyrite) MnS_2 CuS_2 ZnS_2 ZnO_2	TaC, AgBr MO oxides NaCl, KCl etc.	ZrO_2 UO_2 ThO_2 SrF_2, BaF_2 Li_2O, Na_2O,K_2O

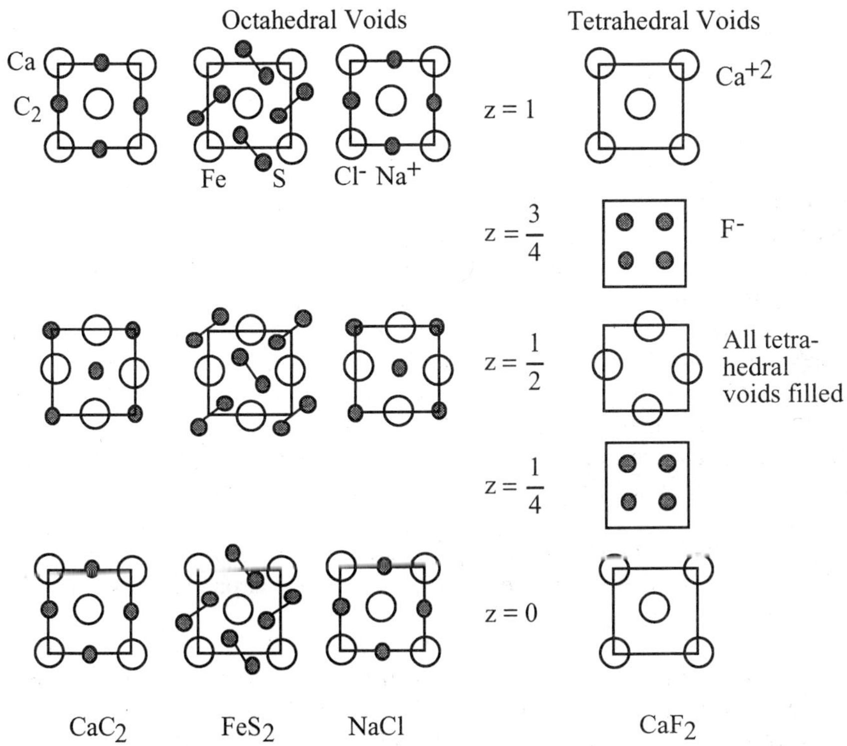

Figure 3-59 Derivative structures of FCC

symmetry that is required for the FCC lattice. In CaC_2, C_2 has an oval shape aligned along the c-axis in one direction, which becomes the c-axis in the resulting tetragonal lattice. In the case of FeS_2, S_2 are aligned in a such way, shown in Fig.3-59, that it keeps its 3-fold symmetry but the FCC structure is broken. Its lattice is a simple cubic instead of FCC. In contrast to these compounds, CaF_2 which has the AX_2 formula keeps the FCC structure by distributing F_2 on the tetrahedral sites. Hence CaF_2 does not lose the symmetry requirement for the FCC structure. These compounds illustrate that the tetrahedral sites of the FCC structure are the only positions for X in the FCC lattice formed by A in compounds with the AX_2 formula. Distribution of X among positions other than the tetrahedral sites destroys the FCC lattice of A. Likewise, for AX compounds, the octahedral sites are the only possible positions for X in the FCC lattice formed by A. AX_2 compounds having derivative structures of the FCC structure are listed in Table 3-12 along with compounds having the NaCl structure.

E. Other complex structures with the closest packing structure

1. AB_nX_m compound

1-1. Perovskite ($BaTiO_3$)

If A has a size similar to X, A and X form closest packing and B locates at the interstitial tetrahedral or octahedral sites according to Pauling's Rule 1. If A is a cation and X is an anion, the possible arrangement of A and X in closest packing is the one that is met by the condition that A must be surrounded by X only. Closest packing is not possible for AX or AX_2 compounds which satisfies this condition. Closest packing satisfying this condition is possible only for AX_3 compounds. This is shown in Fig. 3-60. Fig. 3-60(i) and (ii) show that A is surrounded by only X within a 2-dimension plane and in 3-dimension space, respectively. This arrangement of cations agree with the Pauling's Rule 4 concerning ionic compounds with more than one cation:

> In a crystal containing different cations those with a large valence and small coordination number tend not to share polyhedron elements with each other.

This rule has the same basis as the postulate in Section 2-1 in the sense that the same kind ions tend to be set apart from each other. With the configuration in Fig.3-60, cation A can avoid contacting other A cations. The characteristic of stacking in Fig.3-60(ii) is that no tetrahedra but only octahedra of X can be developed. The number of octahedral sites formed by X only is equivalent to one-third of the number of octahedral sites in the regular closest packing of X. According to the postulate in Section 2-1, anions (X) can avoid direct contact with each other by sharing cation located in the octahedral sites that are indicated by B in Fig.3-60(ii). In this structure, the molar ratio of A, B, and X should be 1:1:3. The requirement for this structure is that one cation (represented as "A") should have similar size to that of the anion and the electric charges must be balanced.

For $BaTiO_3$, the size of Ba^{+2}(1.35Å) is very close to the size of O^{-2} (1.36Å) and electric charges are balanced. In terms of the molar ratio, Al_2AlO_3 ($AlAlO_3$) has the same formula as ABX_3. The size of Al^{+3} is not compatible with O^{-2}. Likewise, ilmenite ($FeTiO_3$) cannot have this structure due to the size incompatibility of Fe^{+2} with O^{-2}, whereas $BaTiO_3$ has this structure. Conventionally this type structure is called the perovskite ($CaTiO_3$) structure for historical reasons as is the case with the fluorite structure. An anti-fluorite structure (Li_2O) is more regular than fluorite (CaF_2) with respect to the closest packing of anions. Similarly, compared to $BaTiO_3$, $CaTiO_3$ is more distorted from the closest packing of anions(O^{-2}). This closeness to regularity with respect to closest packing has nothing to do with the nomenclature for the structures. Hence the perovskite structure will be illustrated with $BaTiO_3$ instead of $CaTiO_3$. Even $BaTiO_3$ undergoes a phase transition from cubic to tetragonal at 125°C and from tetragonal to orthorhombic around –5°C as temperature decreases. Here the purpose of illustrating the crystal structure is not just to give detailed atomic positions but rather to show

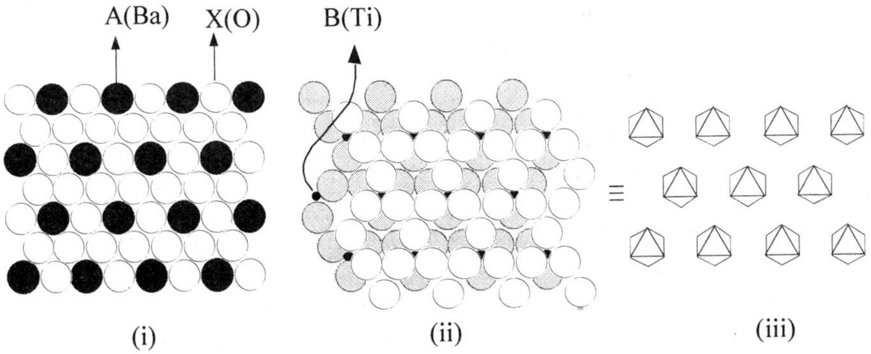

Figure 3-60 Possible stacking of AX_3 compound avoiding direct A-A contact illustrated with $BaTiO_3$ structure.
(i) Arrangement of A(Ba) in a layer of the closest packing of X(O)
(ii) Stacking of two layers of AX_3 with direct contact between A atoms and
(iii) Schematic drawing of the stacking of two layers with octahedra formed with A atoms omitted

how the structure is developed on the basis of coordination theory within the scheme of the closest packing of anions in ionic crystals.

For anlaysis of the crystal structure of $BaTiO_3$ the configuration of the octahedral sheet is shown in Fig.3-60(iii) along the direction vertical to the sheet. Stacking of octahedra sheets by edge sharing of octahedra results in direct contact of A atoms. There are two ways to stack the octahedra sheet in Fig.3-60(iii) by corner sharing or face sharing. Following Pauling's Rule 3, face sharing is not probable as a stacking mode. Therefore, these octahedral sheets must be stacked by corner sharing. This is shown in Fig.3-61. Only octahedra are shown in order to simplify the structure. To make this configuration, the anions must pack in the CCP structure. Hence the structure developed has the cubic structure, not hexagonal. A 3-dimensional perspective of octahedral stacking is illustrated in Fig.s 3-61 and 3-62. Fig.3-62(iii) illustrates especially well the repeating unit as a cubic lattice. To correlate with each other, the octahedral units are numbered.

In Fig.3-63, the detailed atomic positions including all the cations are developed from the octahedral structure in Fig.3-62(iii). As shown in the case of TiO_2, $BaTiO_3$ can be described by two types of unit cell. Type A has "Ti" at the center of the unit cell while Type B has Ba at the center. From Type B, it is easily seen that the coordination number of Ba is 12. The advantage of Type A over Type B is that the movement of Ti accompanying the transition from the cubic to tetragonal structure can be more easily visualized than in Type B. Hence the unit cell among two types can be selected depending on the purpose of the illustration. Atomic positions with respect to these two type unit cells are listed in Table 3-13.

Compounds with the perovskite structure are classified into 3 categories according to the ionic charges on the cations. For oxides, they are (i) $A^{+2}B^{+4}O_3$, (ii) $A^{+3}B^{+3}O_3$ (iii) $A^{+1}B^{+5}O_3$. Typical examples for each category are $BaTiO_3$, $YAlO_3$ and $LiNiO_3$, respectively. For fluorides, only the first category is possible as $A^{+1}B^{+2}F_3$. One of the characteristics of the perovskite

(i) (ii) (iii)

Figure 3-61 Possible stacking mode of octahedra sheet in Fig.3-60(iii) along the perpendicular direction to the sheet avoiding the direct contact of A-A

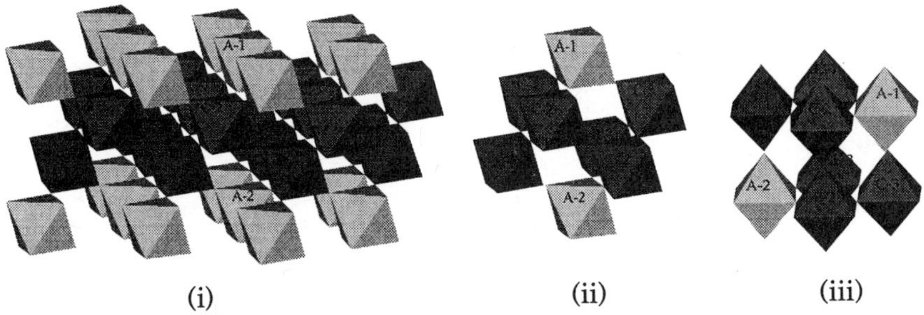

Figure 3-62 3-dimensional perspective of the stackings of octahedra sheet in Fig.3-61(iii). The numbering is the same in Fig.3-61(iii)
(i) Complete stacking of octahedra,
(ii) unit cell and
(iii) rotated image of the unit cell of (ii) around the perpendicular axis to the page by 60° in order to show cubic structure

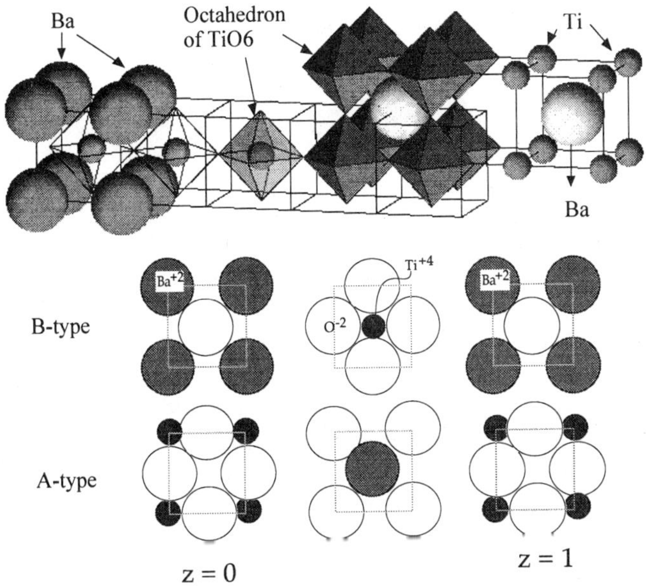

Figure 3-63 $BaTiO_3$ in the cubic form with two unit cell types

structure is that compounds of the perovskite structure easily form solid solutions. This miscibility between perovskite compounds is utilized for customization of physical properties. Also, compounds with the perovskite

Table 3-13 Atomic positions in A and B type cells in $BaTiO_3$

	B Type	A - Type
A ions	(000)	(0.5, 0.5, 0.5)
B ions	(0.5, 0.5, 0.5)	(000)
X ions	(0.5, 0.5, 0) (0, 0.5, 0.5) (0.5, 0, 0.5)	(0.5, 0, 0) (0, 0.5, 0) (0, 0, 0.5)

structure can usually modify the structure by deforming the ideal cubic lattice in order to accommodate the misfit between cations and packing anions. This aspect will be illustrated in Chap. 4.

1-2. Spinel ($MgAl_2O_3$)

Spinel is the name for $MgAl_2O_4$. More generally, this implies that the group that has the chemical formula of AB_2O_4. It has closest packing of O^{-2} ions with the A and B cations distributed among tetrahedral and octahedral sites. The possible arrangement of A and B is that A occupies one-half of the octahedral sites and B occupies one-eighth of the tetrahedral sites. The structure resulting from this distribution is called normal spinel. Conversely, B ions can occupy one-half of the octahedral sites with A and the rest of B ions distributed among the tetrahedral sites. This arrangement gives the inverse spinel structure. With respect to the CCP structure, the distribution of A and B cations in tetrahedral and octahedral sites are shown in Fig. 3-64(a) and (b). The occupancy of cations on the sites follows these two patterns. In one interlayer, three-quarters of the octahedral sites are occupied and none of the tetrahedral sites. In the following interlayer, one-quarter of each tetrahedral site and three-quarters of the octahedral sites are occupied. Hence overall occupancy of A and B cations is one-eighth of the tetrahedral sites and half of the octahedral sites. This distribution of cations form the two types of polyhedral sheets shown in Fig.3-64. Type A consists of only octahedra and Type B consists of octahedra and tetrahedra. The distribution of empty octahedral sites within Type A and the distribution of tetrahedra in Type B is in accordance with Pauling's Rule 4 which separates cations of the same kind far as possible.

Assuming Type A and Type B distribution of cations among the interstitial sites in packed anions, the stacking modes in Fig.3-65 follows Pauling's Rule 3 which indicates the destabilization effect of face sharing between polyhedra. The stacking mode in Fig.3-65 has only edge sharing between octahedra within and across the polyhedral sheet. Otherwise, face sharing is inevitable. Therefore, the stability of the stacking configuration in Fig.3-65 is understood. Since the octahedra are connected by edge sharing, the anion packing must take the CCP structure instead of HCP. The sheet repeats in sequence –ABAB– and the unit cell is expected to extend six polyhedral sheets. This length corresponds to the body diagonal of the spinel unit cell. Therefore, the lattice parameter of the face centered cubic lattice for spinel is given as a_o = 8.07Å. Its space group is Fd3m. The unit cell is shown in Fig.3-66(i). Since this unit cell in Fig.3-66(i) does not show all the details, octahedral cations and tetrahedral cations are indicated within the unit cell in Fig.3-66(ii) and (iii). An even more schematic diagram for the unit cell is drawn in Fig.3-67 with two types of octants for the spinel. As shown in Fig.3-67, spinel is a mosaic of two types of cubic structures. The equivalence between the stacking layers and the actual unit cell of spinel in the FCC structure is illustrated in Fig.3-68.

As indicated, one of the characteristics in the spinel structure is the switching of cations between octahedral sites and tetrahedral sites although octahedral sites are normallyoccupied by cations with a higher oxidation number. With respect to the electric charge balance, these type cations prefer to occupy the octahedral sites in order to make contact with more anions for greater stability. If the formula for a compound with the spinel structure is expressed as $A^{+2}B^{+3}{}_2O_4$, octahedral sites are occupied by B^{+3} and tetrahedral sites by A^{+2} ions.

Spinel compounds are called a normal if a cation with a higher oxidation state occupies the octahedral sites. It is said to have an inverse spinel structure if tetrahedral sites are occupied by cations in higher oxidation states. In Fe_3O_4 one-half of the Fe^{+3} ions takes tetrahedral sites and one-half of Fe^{+3} and all Fe^{+2} occupy the octahedral sites. Although Fe^{+3} is expected to occupy the octahedral sites in order to have lower electrostatic energy, it actually takes the tetrahedral sites. The reason is the extra stabilization energy derived from interaction of the d-orbital electrons of the metal cation with the surrounding anions as explained in the crystal field theory in Chapter 1. By this scheme, Fe^{+3} provides greater stabilization energy than Fe^{+2} to the structure, from the interaction of d-electrons with the anions when it is in the tetrahedral sites. Table 3-14 lists the typical compounds having spinel structure and their switching states.

Binary compounds of transition metal elements having the spinel structure are typically magnetic materials. As magnetic materials, the spin interaction between neighboring cations through oxygen as a bridging element has a major role in determining the magnitude of the magnetic moment. This spin interaction is most effective when two adjacent cations and bridging oxygens are in a straight line. This means that the spin interaction between cations is greatest when the angle between two lines connecting each cation to the bridging oxygen is 180°. In this respect, A (tetrahedral cation) – B(octahedral cation) interaction is most effective since the angle between the two lines connecting A and B to the bridging oxygen is 125°, while the angles for

AA and BB interaction are 79° and 90°, respectively. These angular relationships between A and B sites can be seen from the polyhedra sheet in Fig.3-64. This sheet shows that there is only one connection mode for each pair of nearest A–A, A–B, and B–B cations.

Since the A–B interaction is most effective, the magnetic properties of spinel structures can be controlled by controlling the normal and inverse structures through compositional changes.

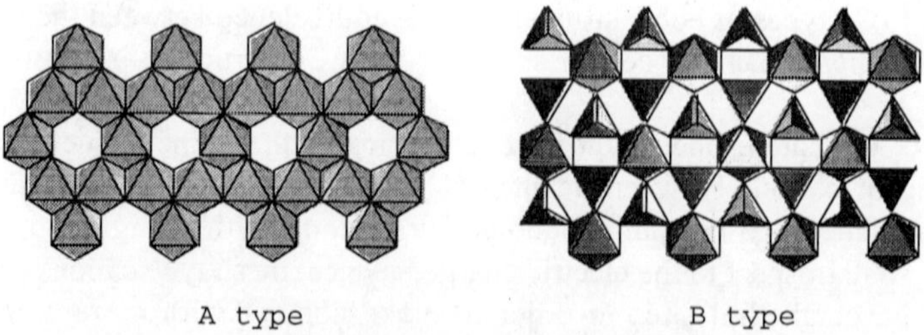

Figure 3-64 Two types of polyhedra sheets in spinel

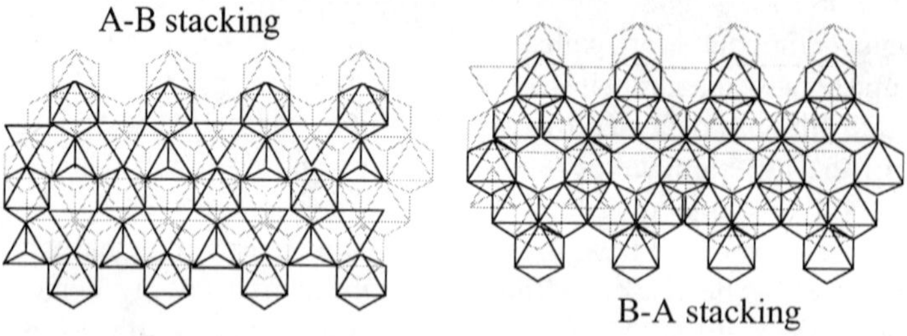

Figure 3-65 Stacking mode of two types polyhedra sheets

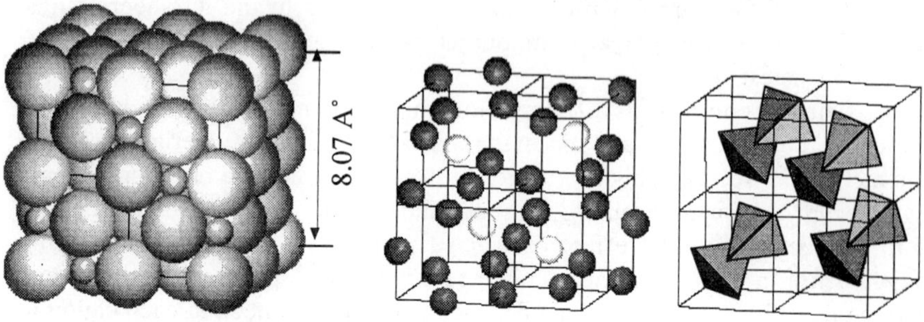

Figure 3-66 (i) Crystal structure of spinel in that anions are on the FCC position (ii) position of cations on the octahedral sites (iii) position of cations on tetrahedral sites shown by tetrahedra

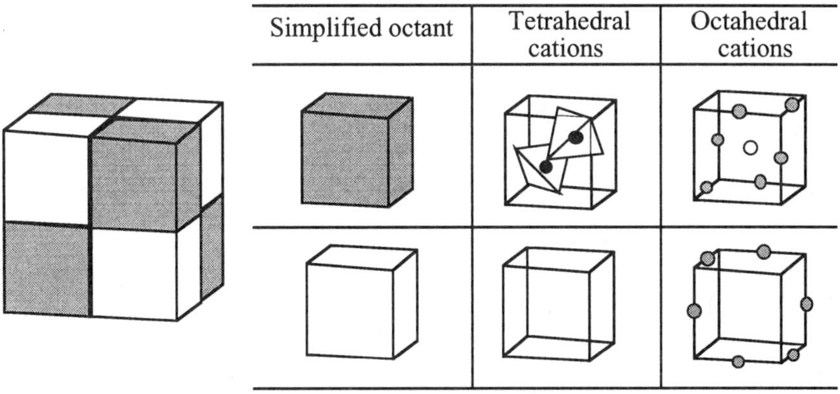

Figure 3-67 Unit cell of the spinel structure illustrated with two types of octant and atomic arrangements within the two types of octants

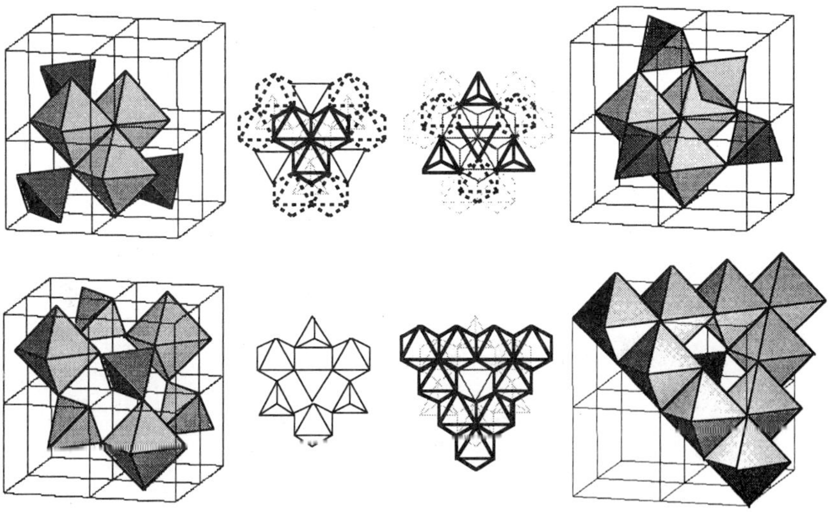

Figure 3-68 Equivalence of stacking mode of layers to the configuration of octahedra and tetrahedra within the unit cell of spinel

Table 3-14 List of compounds of spinel structure with their lattice parameters and normality

Compound	a_o(Å)	normality	Tetra	Octa
$MgAl_2O_4$	8.07	normal	Mg^{+2}	Al^{+3}
Fe_3O_4	8.40	inverse	Fe^{+3}	Fe^{+2}, Fe^{+3}
$ZnFe_2O_4$	8.44	normal	Zn^{+2}	Fe^{+3}
$NiFe_2O_4$	8.35	inverse	Fe^{+3}	Ni^{+2}, Fe^{+3}

2. Compounds with complex anions

In standard crystal structures for both binary and complex ionic crystals, anions are in the closest packing and cations are located at tetrahedral or octahedral sites within the packed structures.

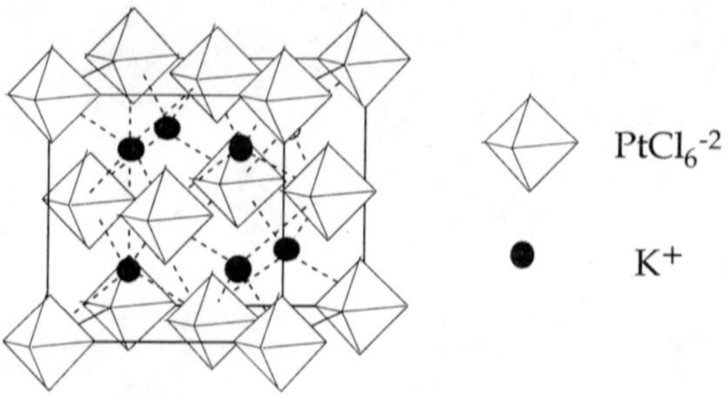

Figure 3-69 K_2PtCl_6 structure

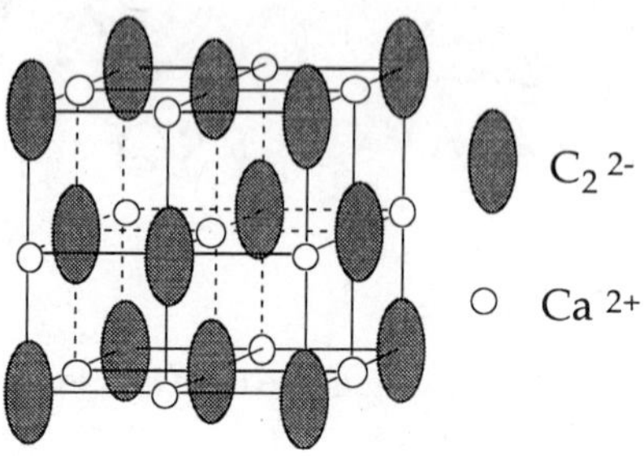

Figure 3-70 CaC_2 structure

2-1. K_2PtCl_6

This compound can be described as a structure of type A_2X. Here X is $PtCl_6$ complex anions that are spherically symmetric. According to Pauling's rule, the radius ratio indicates tetrahedral site occupancy by K ions which results in the anti-fluorite structure in which $PtCl_6$ forms the CCP structure and K locates at tetrahedral sites. Fig. 3-69 shows the crystal structure of $PtCl_6$.

2-2. CaC_2

Here C_2^{-2} are complex anions. This is regarded as an AX compound that takes the NaCl structure as shown in Fig.3-70. C_2^{-2} are the packing anions and Ca occupies the octahedral sites. In contrast to Cl^- of NaCl, C_2^{-2} is not spherically symmetric but rather rod-like. Hence, as illustrated in Table 3-12, its crystal lattice lacks 3-fold rotational symmetry and the lattice symmetry is reduced to the tetragonal lattice. By the radius ratio, Ca^{+2} prefers the octahedral site.

F. Simple cubic and body centered cubic (BCC) structure

1. Simple Cubic - CsCl

The simple cubic structure is not as effective as CCP or HCP in terms of the space filling efficiency. Hence it is not observed in single element crystals such as metals. In ionic crystals, however, it can be chosen if the radius ratio, r_c/r_a, is larger than 0.73 according to Pauling's rule. (r_{Cs}/r_{Cl} = 1.69/1.81= 0.93) As shown in Fig.3-71, the crystal structure can be depicted as consiting of two interpenetrating simple cubic lattices of Cs and Cl. Compounds with this crystal structure are:

CsBr, CsI, CsCl, CsCl, NH_4Br, NH_4I, NH_4Cl.

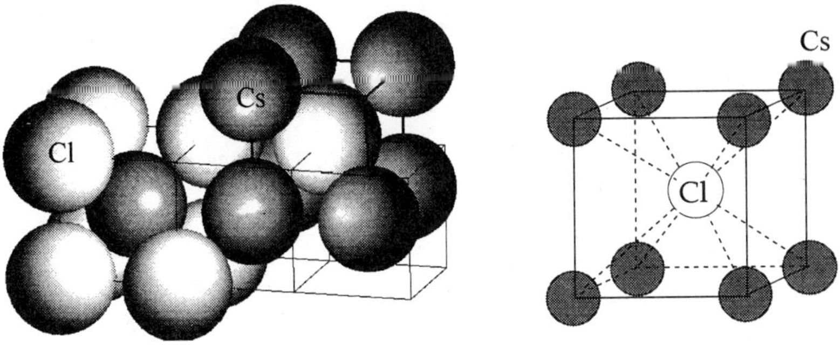

Figure 3-71 CsCl structure

2. Body centered cubic

The efficiency for space filling for a BCC structure is lower than for an FCC or HCP structure by 6%. Other effects such as direction of bonding orbitals can, however, overcome somewhat this lower efficiency. Among metals, all the alkali elements take the BCC structure. Also Fe takes the BCC stricture.

One of the characteristics of the BCC structure is a reversal in the relative sizes of octahedra and tetrahedral sites, as described below.

2-1. Tetrahedral site and Octahedral site in BCC structure

1. Octahedral voids in BCC structure

There are 6 octahedral sites per lattice point. Also there are two atoms per lattice point. Therefore there are 3 octahedral sites per atom in BCC structure. This is calculated as follows:

As shown in Fig.3-72, there is one octahedral site at the center of each face of the cube. Therefore, 6 x 1/2 = 3 octahedral sites belong to each unit cell. With one octahedral site at each of 12 edges, therefore 12 x 1/4 = 3 belong to the unit cell. There are a total of 6 octahedral voids/unit cell with 2 atoms/unit cell. Hence, there are 3 octahedral voids/atom

In Fig.3-73, the body diagonal (a) of the BCC unit cell is equal to $4r_a$.

$$\sqrt{3}a_o = 4r_a, \text{ and } a_o = \frac{4}{\sqrt{3}}r_a = 2.31r_a$$

$$\overline{PQ} = a_o = 2.31r_a = 2(r_a+r_c)$$

$$0.31r_a = 2\,r_c$$

Therefore,

Relative void size: $r_c/r_a = 0.154$

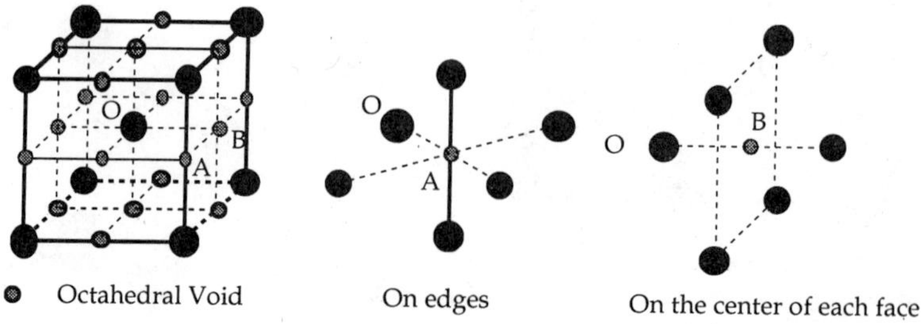

Figure 3-72 Location of octahedral sites in BCC structure

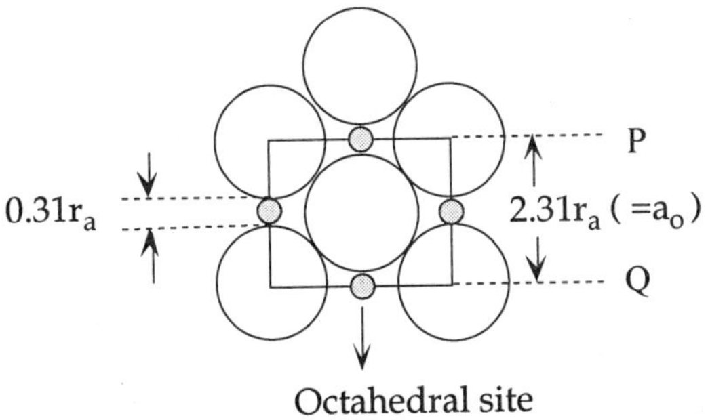

Figure 3-73 Cross sectional view along (110) plane in BCC structure

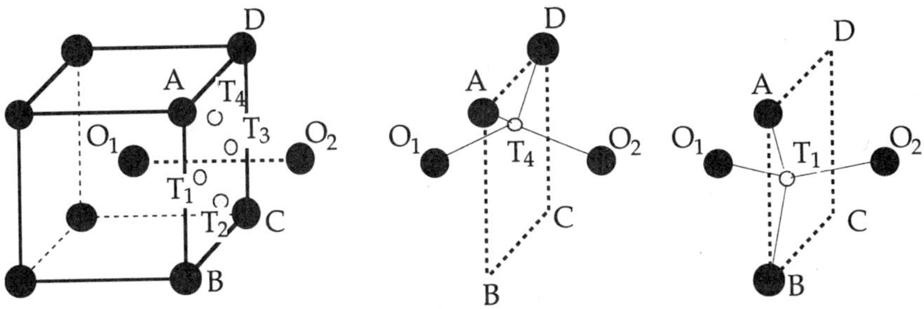

Figure 3-74 Locations of tetrahedral sites in BCC structure and their coordination structures for T_1 and T_4

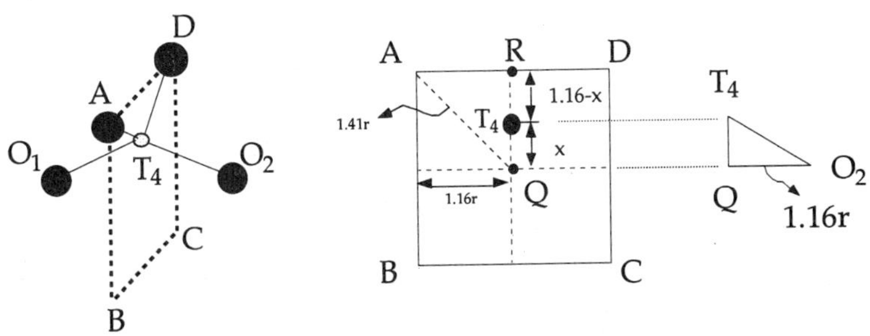

Figure 3-75 Geometric relationship around T_4 in BCC structure

2-2. Tetrahedral Voids in BCC structure

The locations of tetrahedral sites in BCC structure are on each face of the cubic unit cell as shown in Fig.3-74. Also the coordinations of each tetrahedral cation are shown in Fig.3-74. In each face 4 tetrahedra voids are located midway between the center of face and the nearest edge of the unit cell. Since there are 6 faces, 12 tetrahedra voids exist in the unit cell. This means that there are 6 tetrahedra voids per atom in BCC structure.

The center of the tetrahedral site, T_4, is located on the plane ABCD in Fig.3-75 and must satisfy the condition :

$$\overline{T_4 A} = \overline{T_4 D} = \overline{T_4 O_1} = \overline{T_4 O_2}$$

Let's say the distance of T_4 from the line AD is x. Then, from the trigonometric relationship,

$$\overline{T_4 O_2} = \sqrt{(1.16r)^2 + x^2},$$

$$\overline{T_4 D} = \sqrt{(1.16r - x)^2 + (1.16r)^2}$$

$$\sqrt{(1.16r)^2 + x^2} = \sqrt{(1.16r - x)^2 + (1.16r)^2}$$

Therefore,

$x = 0.58r$ and $\overline{T_4 O_2} = 1.291r$.

Since $\overline{T_4 O_2} = r_a + r_c$, $r_a + r_c = 1.291 r_a$.

The relative size, r_c/r_a for the tetrahedral site in BCC is 0.291. Thus the tetrahedral site in the BCC structure is larger than in the CCP structures. It is also larger than the octahedral voids in BCC (the reverse in CCP structures). This fact (tetrahedral > octahedral void site) strongly limits substitution in BCC structures since the octahedral site is normally favored for its lower energy, due to higher CN.

Thus:

α – Fe (BCC Structure)	γ – Fe (FCC Structure)
solid solution of C ~ 0.1 atom %	solid solution of C ~ 3.6 atom %

2-2. Compounds similar to BCC structure

Table 3-15 list compounds whose crystal structures are similar to the BCC lattice. However, due to the atomic arrangement within the unit cell, they have different lattice systems. Their crystal structures are shown in Fig.3-76. For CsCl, the center atom(Cs) is not the same as the atom(Cl) on the lattice point. For rutile and cuprite, the center atom is the same as the atoms at the lattice

Table 3-15 List of compounds having similar structure to BCC

Prototype	BCC	CsCl	TiO_2(Rutile)	Cu_2O(Cuprite)
Space group	Im3m	Pm3m	$P4\frac{2}{m}m$	Pn3m
Compounds	W, Nb, Ta Na, K Mo, V (metals)	CsCl CsBr CsI (Ionic)	GeO_2 SnO_2 PbO_2	Pb_2O Ag_2O

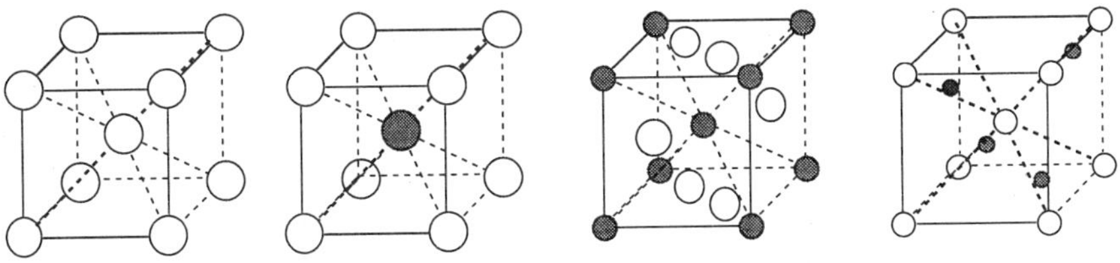

Figure 3-76 Crystal structures of W, CsCl, TiO_2, and Cu_2O

points, but adjacent atoms have a different configuration. Hence the center position is not a lattice point and the lattice type becomes a primitive rather than a non-primitive body centered lattice. This is illustrated in the space group notation for CsCl, TiO_2, and Cu_2O. All have the initial "P" in the space group notation that implies a primitive lattice.

SECTION 4 COVALENT CRYSTALS

A. Introduction

In the last two sections, crystal structures of metal and ionic crystals are described on the basis of space filling. The space filling model is one of the methods that explains actual structures. Although it describes in general metal and ionic crystals, it has certain limitations such as in explaining the fact that some metals have BCC structure and others the FCC structure. This limitation is due to the assumption of non-directionality of metal bonds. Even in pure metal crystals, the structures are somewhat dependent on the bonding orbitals of each atom. Table 3-5 illustrates the dependency of the crystal structure of metals on the bonding orbitals of each metal atom.

In metals, however, bonding orbitals affect the crystal structure only to the extent that their structures, BCC, FCC, and HCP can all be considered within the scope of closest packing. In contrast, covalent bonded crystals can not be described in terms of closest packing. The effect of bonding orbitals is so predominant that the crystal structure must be considered from the shape or directionality of the bonding orbitals.

B. Covalent bonds

1. Coordination

As explained in Chapter 1, covalent bonding has directionality. Also the coordination number which can be considered as the number of the nearest atoms in crystal, is related to the valence state of each atom.

In VSEPR theory, the valence state of each atom determines the shape of the bonding orbitals. Therefore, the crystal structure is determined by the valence state of the atoms since coordination number (the number of the closest atoms) and shape of bonding of each atom are determined by the valence state.

If the coordination number of an atom is one, it cannot form crystals. Rather it forms a molecule instead. If the coordination number is two, the structure is in the form of a ring or chain. For 3 dimensional crystals to form, the necessary coordination number is at least 3. Therefore there must be at least one element among the components of crystals that has a 3 or higher coordination number. For atoms to form covalent bonds, they must have similar electronegativities and be non-metallic. Also, the octet rule predicts that elements in Group III, IV, or V in the periodic chart of elements are suitable for 3 dimensional covalent crystals since those elements can have more than 3 coordination. Therefore elements in these groups will be mainly discussed.

2. Covalent crystals with 3 coordination

2-1. C (graphite)

The electron configuration of atomic carbon is $1s^2\,2s^2\,2p^2$. In graphite, the electron configuration of each carbon atom is $1s^2\,2(sp_2)^3\,2p_z^{\,1}$. The shape of these hybrid orbitals is shown in Fig.3-78. These sp_2 orbitals lie in the same plane and a p_z orbital is perpendicular to this plane. Through the σ bonding between sp_2 orbitals and p_z - π bonding between p_z orbitals, a planar C_6 ring is formed. A schematic drawing of the C6 ring is shown in Fig. 3-77(c). However, actual experimental results indicate that there is only one C – C distance. This result implies that the p_z - π bonding is not localized between 1-6, 2-3 or 4-5 atoms in Fig. 3-77(c). Hence the C6 ring is similar to the benzene structure where p_z electrons are delocalized as shown in Fig. 3-77(c).

Since the electrons in $2p_z$ orbitals are delocalized throughout the crystal, molecular orbital bands can be constructed with the $2p_z$ orbital. If the molecular orbital band is constructed from 1N p_z orbital, electrons fill π_{2pz}. The electron distribution in the molecular orbital band in Fig. 3-79(b) is the typical one for a metal. Since there is no band energy gap between π_{2pz} and π^*_{2pz}, electrons in valence band orbital(π_{2pz}) are easily excited and move readily between the delocalized p_z orbitals leading to high electrical conduction within the layer, and similar high thermal conduction.

Normal to the plane, the weak Van der Waal's bonding permits relatively low electrical and thermal conduction. However, the weak bonds permit easy rupture and sliding of the layers over each other. Graphite is, therefore, an excellent solid lubricant, especially at high temperature. Table 3-16 illustrates the characteristics which result from the layered structure of graphite.

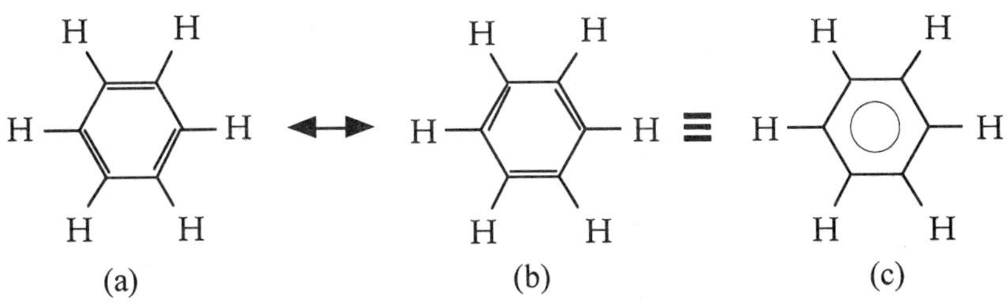

Figure 3-77 Electron configuration of benzene (a) and (b) with localized double bonds and (c) with delocalized double bonds in resonance form

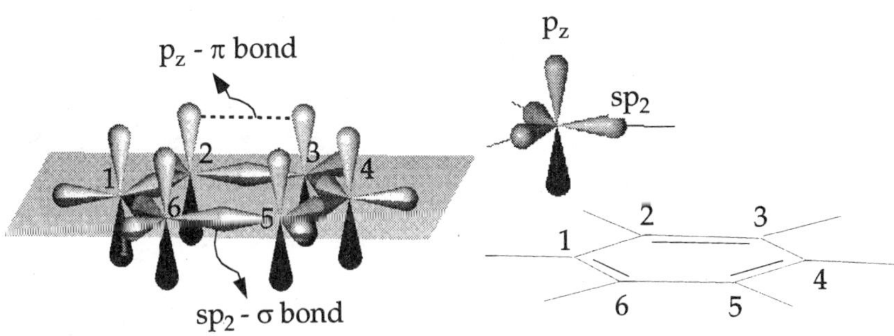

Figure 3-78 Electron configuration of carbon atom in graphite and bonding scheme

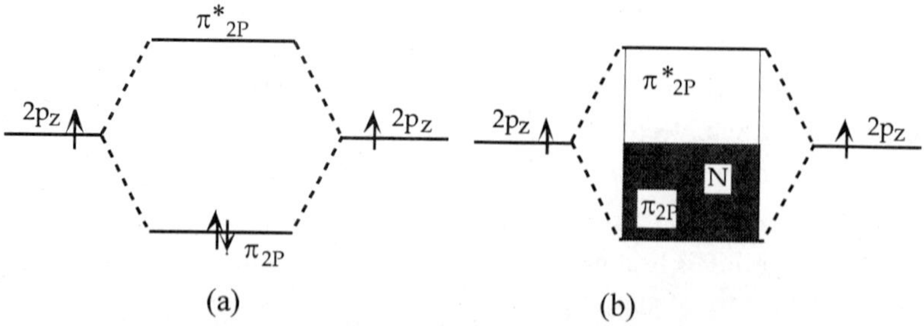

Figure 3-79 Electronic configuration of (a) carbon's atomic orbital and (b) graphite in terms of orbitals band

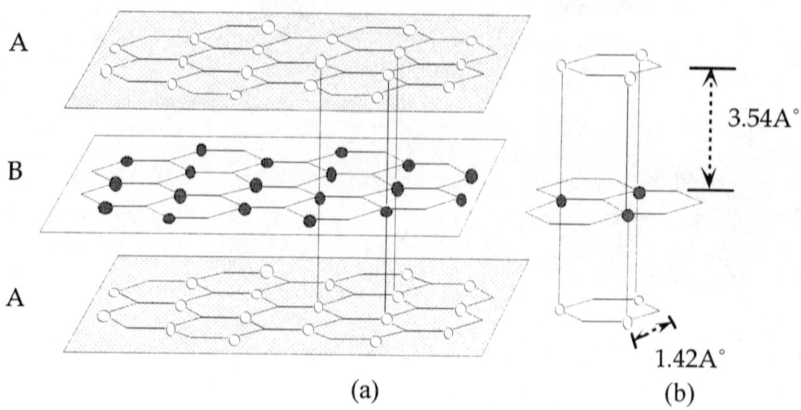

Figure 3-80 (a) ABAB Stacking of Graphite Layers and (b) Relative positions of carbon atoms in layers A and B

Table 3-16 Differences in bond length and physical properties along two distinct crystal directions in graphite

	Along Layer	Across Layer
C-C bond length	1.42Å	3.54Å
Electrical conduction	1	10^4
Thermal expansion coefficient	1×10^{-6}/°C	27×10^{-6}/°C

* Note: The high $\alpha \perp$ to the planes reflects the Van der Waals bonding.

The layers are bonded by Van der Waal's forces. Graphite has an ABAB stacking of layers. Fig.3-80 illustrates the stacking sequence and relative positions of carbons in each layer. This stacking of layers can be visualized in such a way that, in A layer, sp_2 orbitals of circled carbons are toward the left,

, and, in B layer, this is reversed, 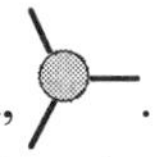. The circled atoms in both A or B layers are located in the same positions in the layer plane so that they can be connected by a vertical line. In another way, the structure can be visualized as one in which there is one atom below and above the center of each hexagonal C6 ring.

2-2. BN (Boron nitride)

BN has the same structure as graphite (and the same stacking sequence) except that B and N atoms alternate in the planes. For the bonding, the electron configuration change of boron and nitrogen can be illustrated as :

$$B : 1s^2\ 2s^2\ 2p^3 \rightarrow 1s^2\ 2s^1\ 2p^1\ 2p^1\ 2p^0 \rightarrow 2sp_2{}^3\ 2p^0$$
$$N : 1s^2\ 2s^2\ 2p^1 \rightarrow 1s^2\ 2s^1\ 2p^1\ 2p^1\ 2p^2 \rightarrow 2sp_2{}^3\ 2p^2$$

Both boron and nitrogen develop sp^2 hybrid orbitals. Bonding in the plane is therefore σ bonding between sp^2 hybrid orbitals as in graphite. Therefore a hexagonal ring of alternating boron and nitrogen is formed as shown in Fig.3-81. However, there are no π - p_z bonds between nitrogen and boron since the $2p_z$ orbitals of B are empty and $2p_z$ orbitals of N are filled. Also, since electrons of the p_z orbitals are localized on nitrogen, p_z orbitals do not form molecular orbital bands as in graphite which shows a typical metallic orbital band structure. Hence boron nitride is an insulator.

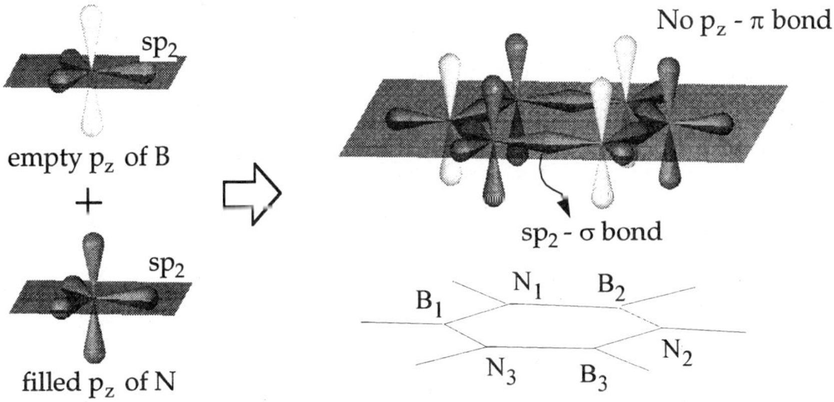

Figure 3-81 Electronic configuration of B and N in boron nitride

As shown in Fig.3-82, the stacking sequence for each layer in boron nitride is different from that of graphite. In graphite, half of the carbon atoms locate above or below one another. In boron nitride, the boron and nitrogen atoms lie above or below one another.

The interlayer distance in boron nitride is 3.33Å. This indicates that the interlayer bonding is van der Waals. The smaller interlayer distance than graphite is due to polar interaction between B and N since the electronegativity difference($|\Delta X|$) between N and B atoms is greater than between carbon atoms. Hence BN is a poorer lubricant than graphite owing to stronger polar bonds between layers.

2-3. P (Black phosphorous)

This is the most stable form of phosphorous crystals. The electronic configuration of phosphorous is illustrated as :

$$P : [Ne]\ 3s^2\ 3p^3 \rightarrow\ [Ne]\ sp_3{}^2\ sp_3{}^1\ sp_3{}^1\ sp_3{}^1$$

Three half-filled sp_3 orbitals participate in bondings. One filled sp_3 orbital is a nonbonding orbital. The shape of these orbitals is shown in Fig. 3-83. In graphite and boron nitride, each atom has 3 coordination in the plane since carbon, boron and nitrogen are bonded through sp_2 hybrid orbitals. In black phosphorous, the hybrid orbital developed is not sp_2 but sp_3 which has the tetrahedral shape. Phosphorous atoms are, therefore, bonded together by σ-sp_3 bonding which does not form a plane. However σ-sp_3 bonding still develops a layer. The layer structure from sp_3 hybrid orbitals in black phosphorous is illustrated in Fig.s 3-84 and 3-85.

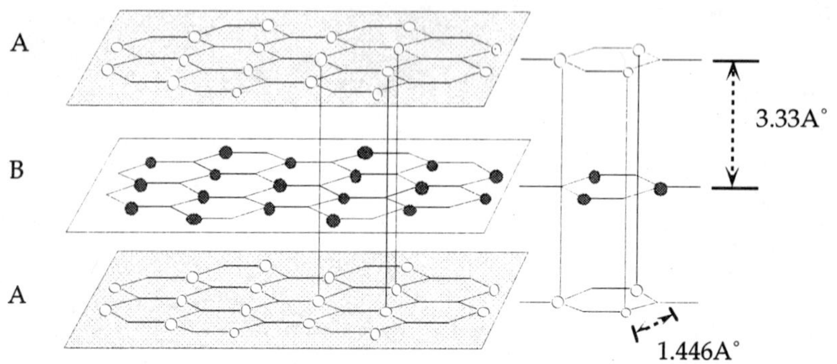

Figure 3-82 Stacking sequence in boron nitride and its interlayer distance

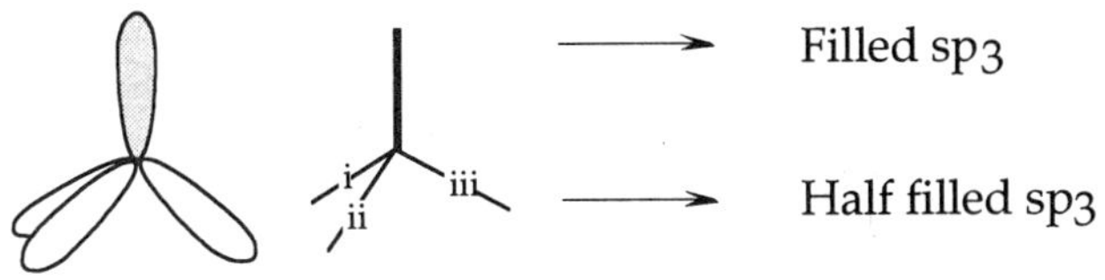

Figure 3-83 Shape of $4sp^3$ orbitals of phosphorous with filled one and three half filled. [Half filled orbitals are numbered for illustration of bonding in black phosphorous in Fig. 3-84(c).]

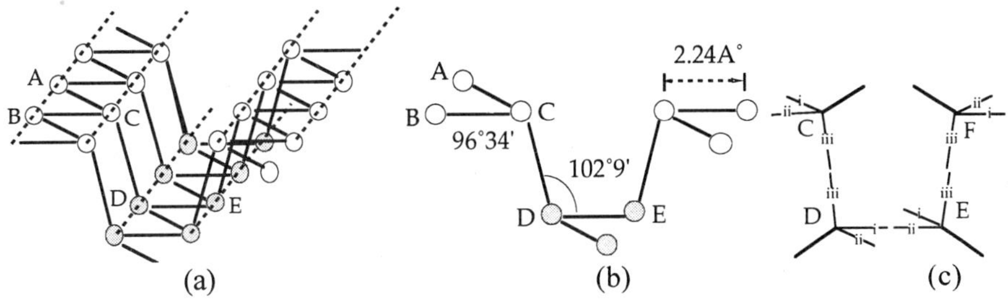

Figure 3-84 (a) Atomic arrangement in a layer of black phosphorous, (b) bonding angle and dimension, and (c) orbital arrangement in the layer structure

Figure 3-85 (a) Schematic drawing of a layer of black phosphorous and (b) the stacking sequence of layers

3. Covalent crystals with 4 coordination

3-1. Si, C, Ge (diamond structure)

These are elements in Group IV in the periodic table of elements. These have the same valence state as [Noble gas] $ns^2\ np^2$.

$$C : [He]\ 2s^2\ 2p^2, \quad Si : [Ne]\ 3s^2\ 3p^2, \quad Ge : [Ar]\ 4s^2\ 4p^2$$

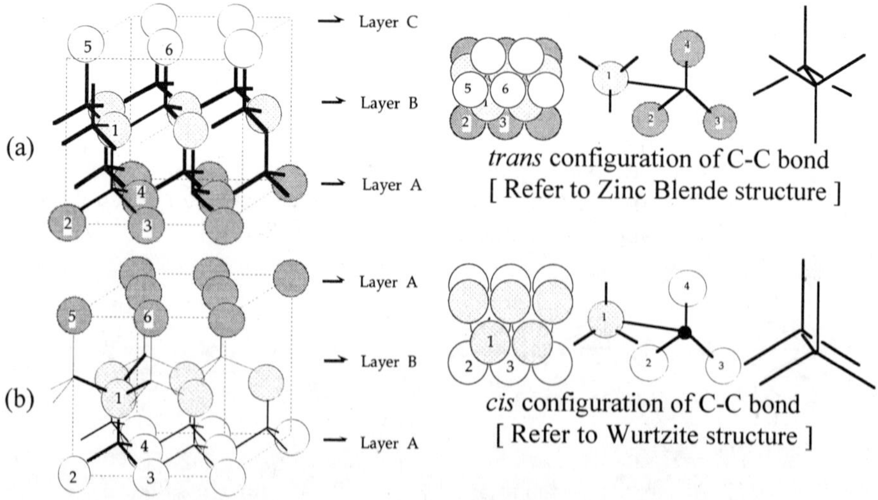

Figure 3-86 (a) Cubic diamond structure and (b) Hexagonal diamond structure and their spatial arrangements of carbon atoms around C-C bond

Hence they form four sp3 orbitals in order to make covalent bonds. The shape of sp3 orbitals is tetrahedral. There are two ways to form a 3-dimensional crystal structure from sp3 orbitals although other structures are possible by distorting the shape of sp3 orbitals from its typical tetrahedral shape. They are cubic and hexagonal diamond structures. The cubic diamond structure is equivalent to a zinc blende structure except that tetrahedral sites are filled by carbon atoms instead of small cations. Also, the hexagonal diamond structure is the same as wurtzite structure except for the substitution of carbon atoms for zinc in tetrahedral sites. These two types of diamond structures are shown in Fig.3-86. In this figure, carbon atoms are omitted in positions corresponding to the tetrahedral sites of zinc blende and wurtzites structures, in order to show the equivalence of the diamond structures to these structures.

With respect to topology of the carbon atoms in these structures, the difference between cubic and hexagonal diamond structures is the spatial arrangement of 6 atoms attached to the C-C bond. This spatial arrangement is shown on the right in Fig.3-86 (a) and (b). In cubic diamond structure, each pair of 3 atoms bonded to C-C bond takes a *trans*-form in that the skeleton of C-C bond has an inversion center at the middle of the C-C bond. In hexagonal diamond structure, the skeleton of C-C bond has *cis*-form that has a mirror plane perpendicular to the C-C bond. In organic chemistry, it is known that the trans-form is more stable than the *cis*-form since the C-C bond acquires the

CHAPTER 4

STRUCTURE CHANGE OF CRYSTALS

SECTION 1 POLYMORPHISM

Polymorphs are solids with the same chemical composition which can exist in different crystal structure forms. Polymorphic transformation describes, therefore, a condition whereby a solid undergoes a structural or phase change as a result of temperature and/or pressure changes. While chemical composition remains the same, changes do occur in both the physical and chemical properties of the solid. The driving force for the phase change is the free energy change that accompanies the structural change. The free energy change with temperature and pressure is given by:

$$\Delta G = \Delta H - T\Delta S$$
$$= \Delta E + P\Delta V - T\Delta S \qquad (1)$$

where ΔH, ΔS and ΔV are the enthalpy, entropy and volume changes of the system during the structural change. ΔE increases with temperature due to a change in bonding (bond angle, bond distance, etc.) between the atoms or ions. Also ΔS increases with temperature due to increased disorder in the solid, which is manifested as vibrational entropy and configurational entropy. The latter is due to generation of point defects (vacancies, interstitials) with temperature. Compared to ΔE and ΔS, ΔV changes little with temperature or pressure for solids. Therefore the structural change in solids (crystals) is determined mainly by changes in the internal energy and entropy. On the assumption that ΔE and ΔS change with temperature at the same rate, ΔS has only a small effect on ΔG at low temperature since the effect is multiplied by T (temperature). However, at high temperature, TΔS may surpass ΔE, although ΔS itself is less than ΔE.

Hence a structure even with higher ΔE is favored at high temperature due to this entropy effect. For this reason, high temperature structures are usually more open and of higher symmetry compared to the lower temperature form.

In Fig.4-1, the changes in free energy(G), entropy(S) and energy(E) are plotted with temperature change for two phases of a system. To distinguish the plots for G from those for S and E, G is shown by a thick line while E and S are drawn by thin lines. Also, Phases 1 and 2 are distinguished by dotted and straight lines, respectively. From a thermodynamic point of view, any change will follow a direction that makes the free energy lower. With respect to phase transformation in crystals, this means that the free energy of the *second* phase must be lower than that of the *initial* phase for the transformation to occur. As shown in Fig.4-1, the order in G is reversed at Tc. Hence phase transformation occurs at T_c from Phase 1 to Phase 2 with temperature increase or from Phase 2 to Phase 1 in the cooling process.

In terms of the mode of the free energy change, the phase transformation occurring in Fig.4-1(a) is called a first order transformation while that in Fig.4-1(b) is called a second order transformation. This terminology is developed from the fact that, at the transformation temperature(T_c), the first order transformation is discontinuous in the first derivative of the free energy(G), such that the heat capacity that may be expressed as $\left(\frac{\partial H}{\partial T}\right)_P = \left(\frac{\partial G}{\partial T}\right)_P$. In the second order transformation, the first deriative of the free energy is continuous although its second derivative is discontinuous at the transformation temperature. Hence first order transformation inplies a sudden change in the internal energy(E) and entropy of the system while the internal energy and entropy change in a continuous manner. The free energy changes smoothly in second order transformation.

In relation to the crystal structure change during the transformation, first order transformation usually involves rearrangement of bonds (reconstructive transformation) and second order transformation is accompanied by the displacement of atoms (displacive transformation) without breaking of bonds. Due to this bond breaking, first order transformations require high activation energy and usually occur at higher temperature than the displacive transformations. This aspect of reconstructive and displacive transformations is illustrated in Fig.4-2 which shows the free energy change for the polymorphs of SiO_2. Reconstructive transformations typically occur above 600°C while displacive transformations occur below 600°C.

It is well known that silica has many polymorphs. They include quartz, tridymite, cristobalite, coesite and stishovite. Si has four coordination to form a tetrahedron except for stishovite which shows six coordination for Si. Different linkages between silica tetraheda result in the many polymorphs for silica. The

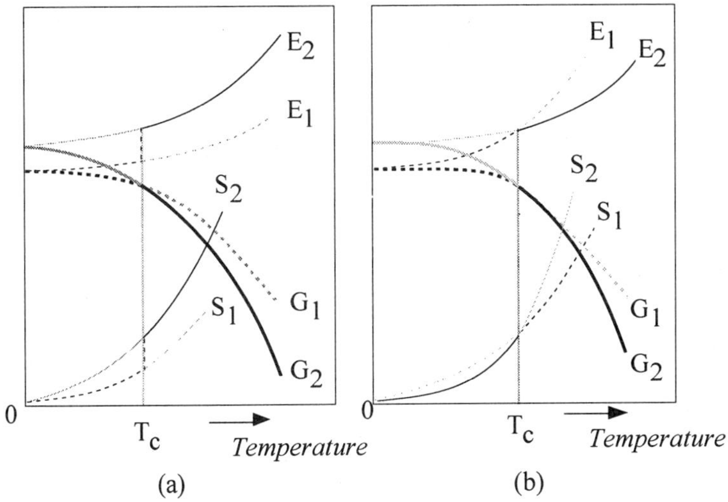

Figure 4-1 Change of enthalpy, entropy and Gibb's free energy with temperature for (a) first order and (b) second order transitions

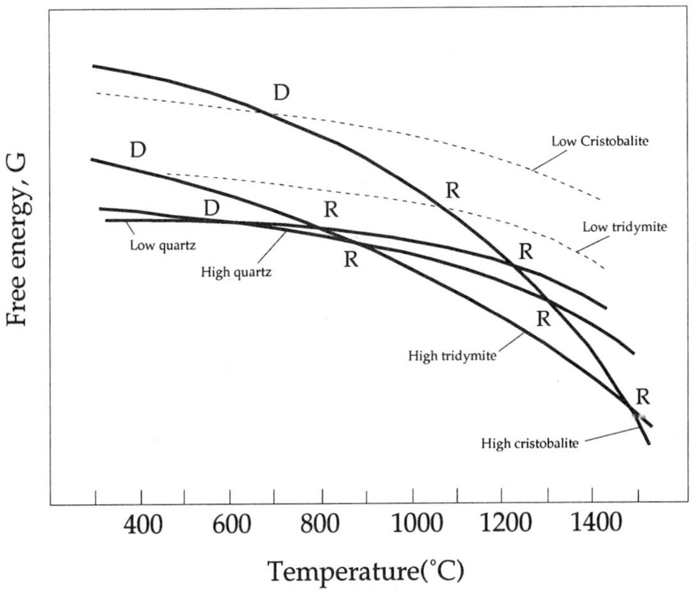

Figure 4-2 Free energy diagram of silica polymorphs with temperature[1]

different linkage exhibits different internal energy and entropy. The Free energy changes of these polymorphs are schematically illustrated in Fig.4-2.

If the free energy is the only criteria for the stability of a phase, liquid silica will follow the path that keeps the free energy lowest when it is cooled to room temperature. Therefore the expected structural change with decrease of temperature is as following :

Liquid silica → β-cristobalite → β-tridymite → β-quartz → α-quartz

Except for the β-quartz → α-quartz transformation, higher activation energy is needed to effect the other transformations. Hence, unless the cooling rate is very slow, these transformations do not occur. This is the reason why cristobalite and tridymite exist with quartz in the earth's crust at ambient temperature although quartz is thermodynamically more stable than cristobalite and tridymite. Transformations such as the β-tridymite → β-quartz where bonds do break are called reconstructive transformations. In contrast, the conversion β-quartz → α-quartz is rapid since no bond breaking is involved. Such transformations without bond breaking are called displacive transformations since atoms are displaced from the original positions by rotation. In Fig.4-2, transition points designated by R denote reconstructive transformations. Due to the reconstructive characteristics of these transition points, the polymorphic transformation of silica can be illustrated as in Fig.4-3

Polymorphs interconverted by displacive transformation are denoted by β and α, representing high and low temperature forms, respectively. Also, with respect to pressure, structural change can occur when pressure is high enough for PΔV to dominate ΔE in Eq(1). This structural change occurs in the direction

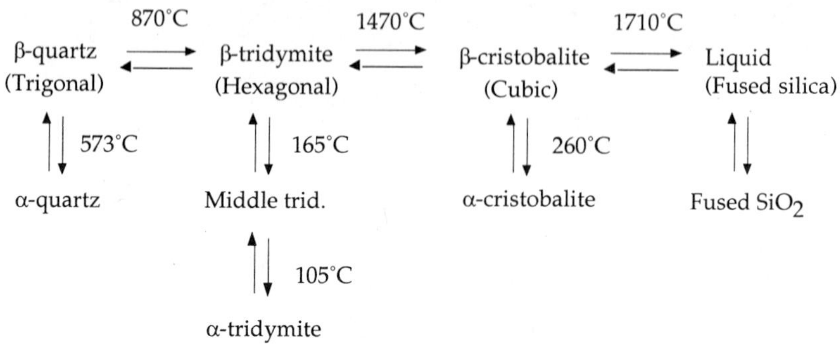

Figure 4-3 Polymorphic forms of SiO_2
(Vertical line : displacive transformation
Horizontal line : reconstructive transformation)

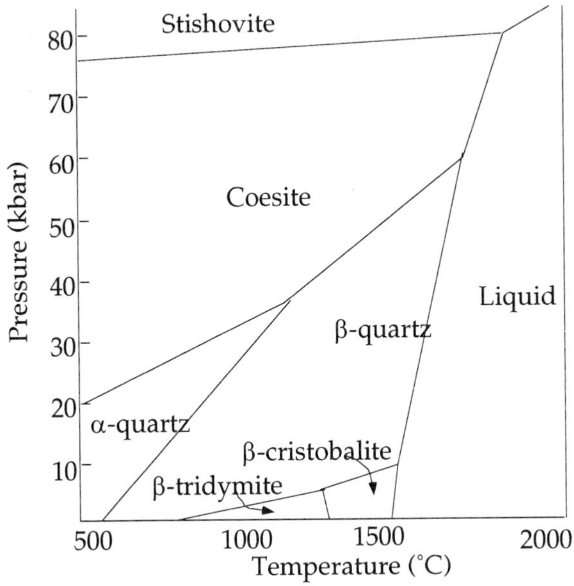

Figure 4-4 Phase diagram of silica [2]

Table 4-1 Summary of polymorphs of silica and their properties

Name	Crystal System	Space group	Density (g/cm3)	Refractive Index
Stishovite	Tetragonal		4.35	1.81
Coestite	Monoclinic	B2/b	3.09	1.59
α-Quartz	Hexagonal	$P3_12$	2.65	1.55
β-Quartz	Hexagonal	$P6_22$	2.53	1.54
α-Tridymite	Monoclinic	-	2.26	1.47
β-Tridymite	Hexagonal	P6/mmc	2.22	1.47
α-Cristobalite	Tetragonal	$P4_12_1$	2.32	1.48
β-Cristobalite	Cubic	Fd3m	2.20	1.48

making ΔV negative, which means a negative change in ΔG. The phase diagram of silica polymorphs with pressure and temperature is shown in Fig.4-4. The structural change with pressure can be explained simply with respect to the densities, which for some polymorphs of SiO_2. are listed in Table 4-1.

Structural changes with pressure increase trend from lower density to higher density as follows :

Cristobalite, Tridymite → β-Quartz → Coesite → Stishovite

These structural changes may occur slowly or rapidly depending on their characteristics. By this criteria, the changes can be divided into two main groups, rapid transformations and slow transformations.

A. Rapid Transformation (Displacive transformation)

This occurs over a narrow range of temperature and/or pressure. It is called dilational transformation when the changing factor is external pressure. If the transformation occurs with temperature, it is a displacive transformation.

1. Dilational displacive transformation

This transformation occurs reversibly under the influence of pressure and temperature change. The structure generally reverts to its original form on release of the pressure or on cooling. This is common in the alkali halides. The typical examples are alkali halides under external pressure and temperature increase.

Examples

Rb Cl (NaCl structure) $\underset{}{\overset{\text{5K-bar}}{\rightleftarrows}}$ Rb Cl (CsCl structure) | ΔV = -6.95cc at –190°C

CsCl $\overset{\text{479ßC}}{\rightleftarrows}$ CsCl (NaCl structure) | ΔV = 10.3 cc

Usually the compressive effect of external pressure is greater on anions than cations. Under external pressure, anions contract more rapidly and this results in an increase in the ratio r_c/r_a with increase of external pressure. As pressure increases, the ratio r_c/r_a reaches the preferential range for the simple cubic structure according to Pauling's rule 1. Then RbCl takes the CsCl (simple cubic) structure. Hence the change in the ratio r_c/r_a caused by the external pressure gives rise to a reversal of lattice energies for RbCl structures with the FCC and simple cubic structures. The energy stability diagram for AX ionic compounds shown in Fig.4-5 illustrates this structure change with pressure.

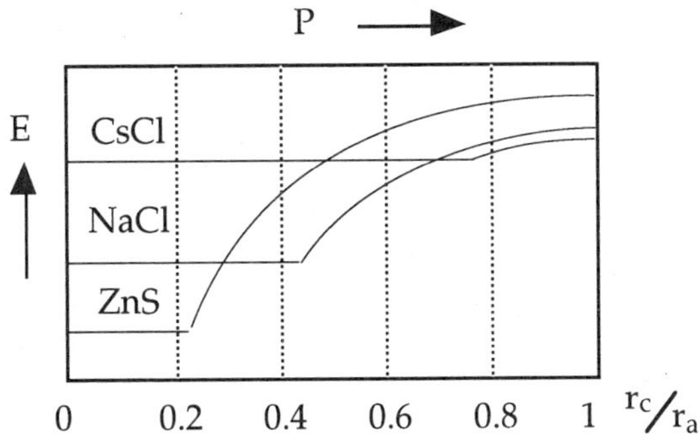

Figure 4-5 Crystal energy change with the ratio, r_c/r_a, under external pressure

With respect to Eq.(1), the phase transition of RbCl from the NaCl to CsCl structure under the influence of external pressure is favored, since the internal energy is lowered and the volume change is negative. In contrast, the phase transition of CsCl with temperature at 479°C is not favored with respect to internal energy and volume changes since both result in an increase in the Gibbs free energy. Therefore the main factor in the transtion of CsCl from simple cubic to the FCC structure is the entropy effect. This can be explained simply in terms of atom mobility, in that the cations acquire greater freedom in being released from the tightly bound 8 coordination to the less tightly bound 6 coordination state. At low temperature this entropy effect is small, compared to the internal energy and volume. But at high temperature, its negative effect on ΔG is enhanced, and the temperature term can overcome the effects of volume and internal energy which favor the simple cubic structure.

By same token, the crystal structure of RbCl acquires the freedom of 6 coordination by taking the NaCl structure even at room temperature. This freedom disappears if very high pressure(10^{+5} times atomospheric pressure) is applied since the effect of the volume change on ΔG is amplified by the pressure.

The examples of RbCl and C3Cl illustrate the effects of internal energy, volume and entropy on ΔG. The dilational phase transition of the alkali halides is usually reversible and rapid since the internal energy difference between the NaCl(FCC structure) and CsCl(simple cubic) structure is slight, as shown in Fig.4-5, and the activation energy for the transtion is expected to be so small.

The steps in the dilational phase transition of RbCl between the FCC and simple cubic structures, are illustrated in Fig.4-6. The direction of dilation is along the body diagonal of the face centered cubic and simple cubic lattices,

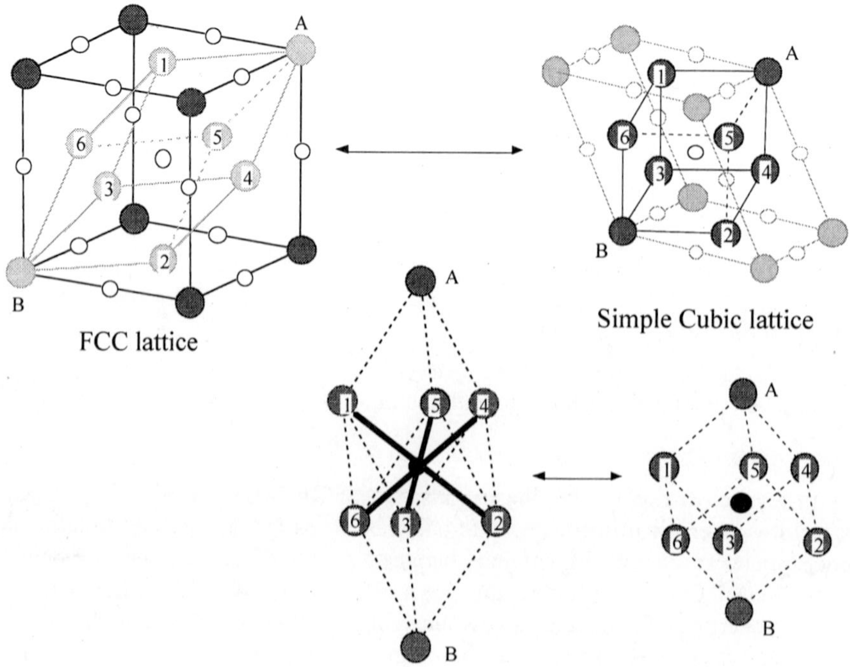

Figure 4-6 Dilational change in crystal RbCl from NaCl(FCC) to CsCl(S.C.)

which are shown in Fig.4-6 by the dark lines on the left side. The dilational phase transition can be compared to elastic compression and expansion as shown in the lower of Fig.4-6.

2. Thermal Displacive Transformation

Displacive transformations typically occur at a fixed temperature (± a few degrees) on heating or cooling and is rapid, spontaneous and reversible. Structure change is associated with bond angle changes or changes in second CNs (not first CN). Volume or thermal expansion changes are also associated with this transformation. The higher temperature phases generally have a higher symmetry and a more open structure (lower density and thermal expansion). The energy difference is typically small between the high and low temperature polymorphs.

2-1 Transformation of β-quartz to α-quartz

2-1-1 β-Quartz

The high temperature form of silica is known as β while the low temperature form is called the α form. Quartz can be described as consisting of double helical chains of silica tetrahedra as shown in Fig.4-7. This helical chain of silica tetrahedra runs along a c-axis that has 6 fold symmetry (6 fold screw axis) for β-quartz and 3 fold symmetry (3 fold screw axis) for α-quartz. Since the helical chain can turn the c-axis in left-handed or right-handed directions, there are two enantiomers whose mirror images are identical to the other. The mirror images of right-handed and left-handed quartz are shown in Fig.4-7. This handedness of quartz does not change during the displacive transformation from β-quartz to α-quartz or vice versa because the transformation occurs by displacive rotations not by bond breaking. Hence α-quartz also has two mirror images. They are defined as α_1 and α_2 depending on the handedness.

The image of β-quartz projected onto the plane perpendicular to the c-axis is shown in Fig.4-8. Each tetrahedron of silica exhibits a square shape which means that two edges are perpendicular to the c-axis. The circles represent the columns of double helical chains of silica tetrahedra.

This representation of β-quartz illutrates the hexagonal symmetry of the c-axis. Although this illustration reveals the 6 fold symmetry, the height of each tetrahedron is different. Hence, in order to show the overall structure of quartz, a schematic illustration of β-quartz is presented in Fig.4-9 with only silicons shown in the helical chains.

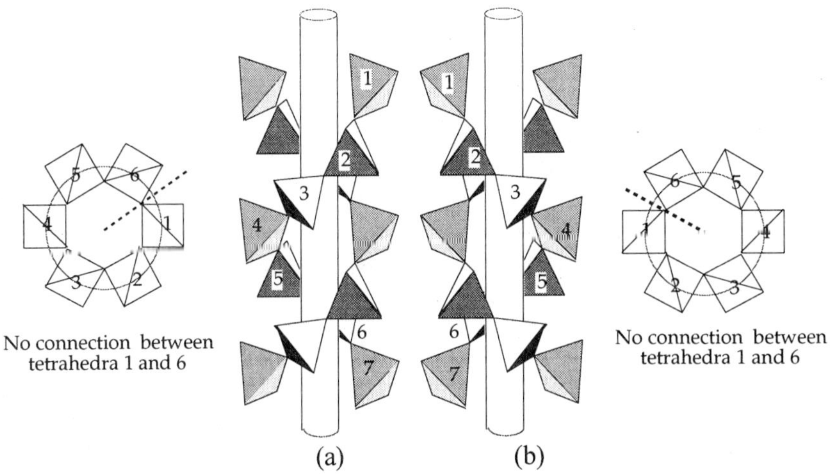

Figure 4-7 β-quartz with (a) right-handedness and (b) left-handedness

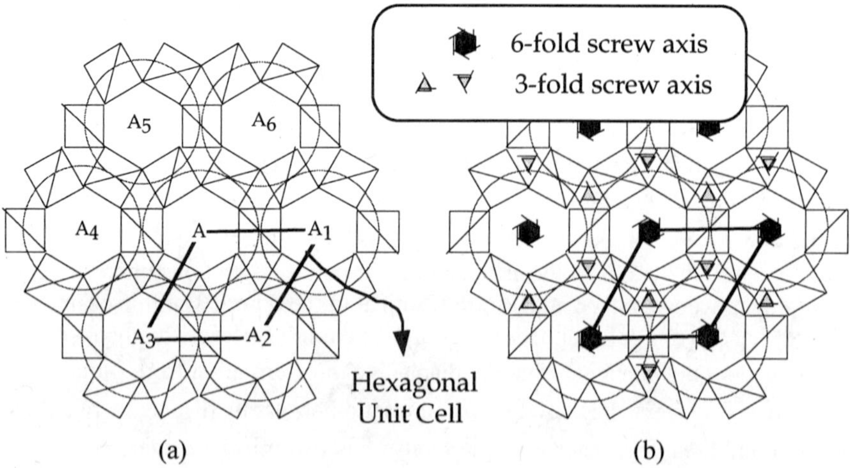

Figure 4-8 Projected image of (a) right-handed and (b) left-handed quartz

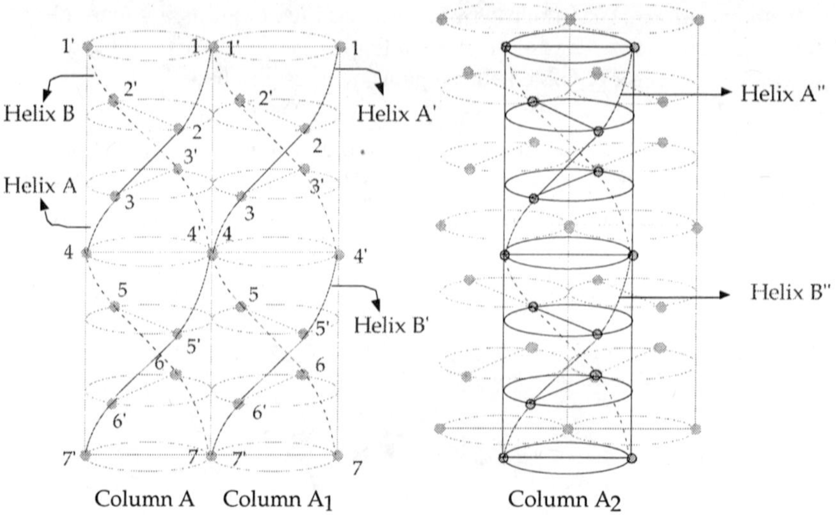

Figure 4-9 Schematic drawing of crystal structure of quartz

Let's call the silica chain connecting tetrahedra 1, 2, 3,... helix A and the chain connecting tetrahedra 1', 2', 3'... helix B. The helices A and B in the same column have no bonding between them. Rather they are bound together by other helical chains in adjacent columns. In order to distinguish between helices of column A_1 and A_2, their helices are represented as A' and B' helices for

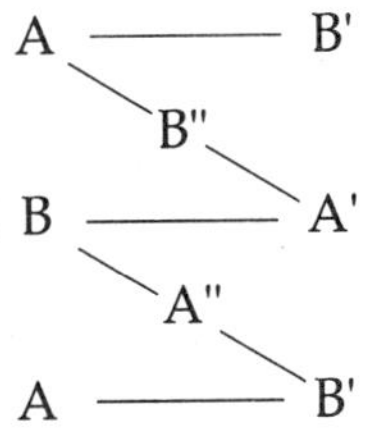

Figure 4-10 Bonding scheme between helical chains of silica in quartz. (Line represents the bonding between helices.)

column A1 and A'' and B'' helices for column A_2. The bonding between helices can then be described as in Fig.4-10.

Without helices A'' and B'', bonding does not form between A and B or A' and B'. This illustration of the bonding scheme between helical chains of adjacent columns show that at least three columns are required to stabilize the double helical structure of quartz.

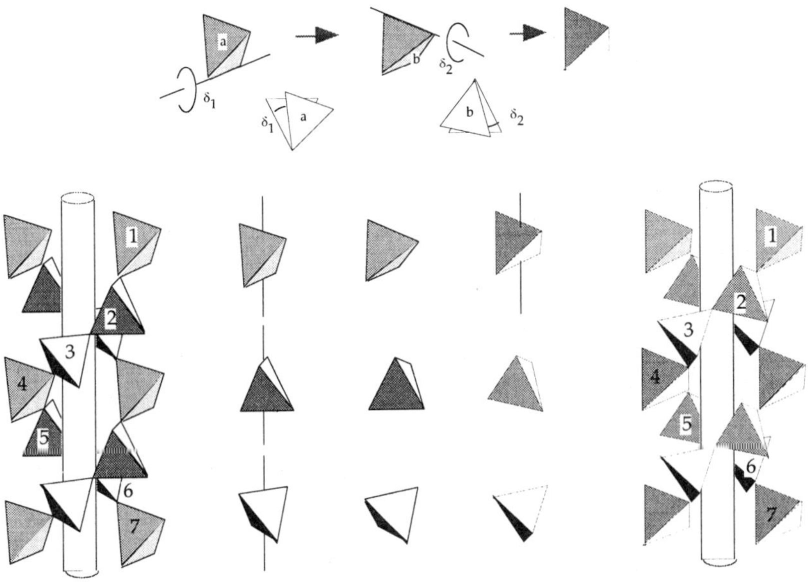

Figure 4-11 Displacive transformation of β-quartz to α-quartz showing rotations of silica tetrahedra

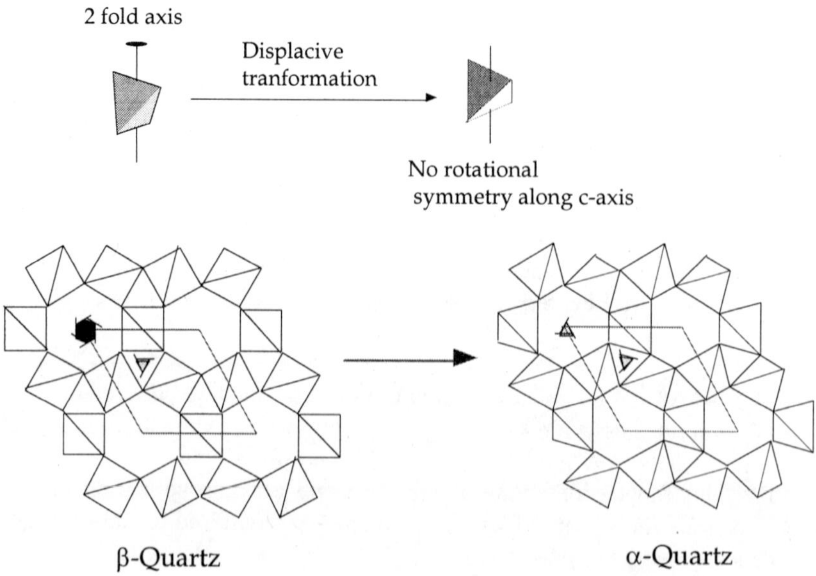

Figure 4-12 Change of images projected on the plane perpendicular to c-axis during β to α quartz transformation

2-1-2 a-Quartz

α-quartz is transformed from β-quartz by displacing tetrahedra of silica. The displacement is purely rotational resulting in no bond breaking. The transformation can be simply depicted as consecutive rotations around two edges that are perpendicular to the c-axis as shown in Fig.4-11. Let's assume δ_1 and δ_2 are rotational angles about these two edges. After this rotation each tetrahedron loses the 2-fold rotational symmetry. Hence tetrahedral 1 and 4 in Fig.4-11 are not equivalent positions in α-quartz since rotation around a previously 6-fold screw axis does not place tetrahedron 1 on tetrahedron 4. This can be seen more easily in Fig.4-12 where images of β and α-quartz projected on a plane perpendicular to the c-axis are depicted.

B. Slow Transformations

1. Reconstructive Transformation

Reconstructive changes require significant rearrangement of the structure. It is accompanied by breaking of bonds and changes in 1st CN or 2nd CN. Sometimes it occurs by diffusion and rearrangement of atoms or ions in the new structure. Also, nucleation of new phase and bond angle changes may be required. Hence reconstructive transformations occur slowly and over a fairly wide temperature range.

1-1. Trydymite, Cristobalite and Quartz

In Chap.3, it was explained why silica tetrahedra are connected by corner sharing only. Since tridymite, cristobalite and quartz have the same composition, SiO_2, they all consist of the same basic units of SiO_4 tetrahedra. The differences between these silica polymorphs are in the connection modes of the tetrahedra. The detailed crystal structures of these polymorphs are illustrated in Section 4-3-7. The reason why the transformations between them are reconstructive can be explained from the crystal structures in that section.

1-2. Graphite and Diamond

Although graphite and diamond are both carbon based materials, their physical properties are very different. The hardness of diamond is 10 on the Mohs' scale whereas that of graphite is just one. The density of diamond is also 1.5 times that of graphite. These differences can be explained with respect to crystal structures. The basic difference between diamond and graphite is in the atomic orbital structrue of carbon. The crystal structure of diamond can be described as FCC lattice with half of the tetrahedral sites filled by carbon atoms. In terms of atomic orbitals, diamond has sp^3 hybrid orbital while graphite has sp^2 hybrid orbital. The length of the C — C bond in diamond is 1.54Å whereas the carbon-carbon bond between atoms in the same layer of graphite is 1.42Å. Although the carbon-carbon bond length is smaller in graphite, diamond has a higher density than graphite since the distance between layers of graphite is 3.35Å. Fig.4-13 depicts in a schematic form the transition from graphite to diamond. This figure illustrates that diamond is formed at high pressure and high temperature. Here, for convenience, only 4 carbon layers in graphite are shown. The first step is a transformation of atomic orbitals of atoms in Layer 1A from sp^2 to sp^3. During this transformation, atoms 1, 2, 3 and 4 form a pyramidal structure. This pyramidal structure can accommodate the increase in bond length between atoms 2 – 1, 2 – 3, and 2 – 4 and retain the horizontal positions of atoms 1, 3 and 4 relative to the atoms "a," "c," and "e" in the next layer. The second step is the transformation of atoms in Layer 1B from sp_2 to sp_3. For this tranformation, bonds between atoms a – 5, c – 8 and e – 7 must be broken. In the third step, the transformed part is shifted to form a bond between b – 6. Steps 4 and step 5 can be considered the same as step 1 and

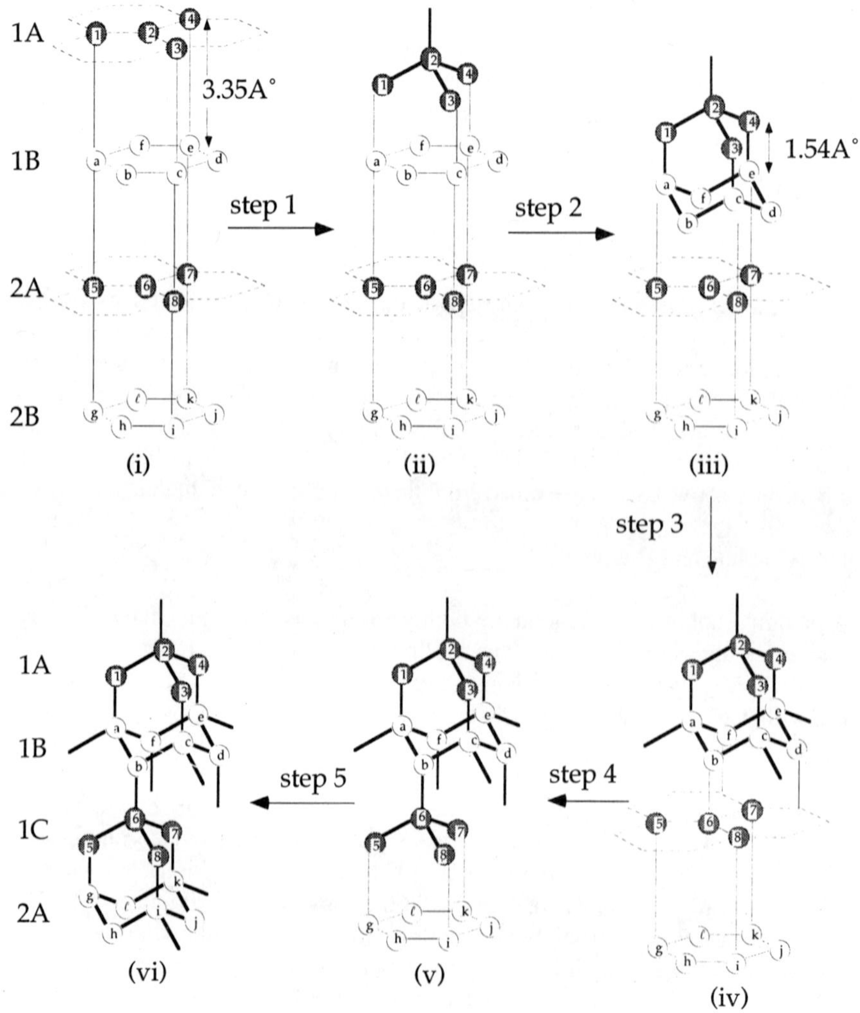

Figure 4-13 Hypothetical description of bonding changes in the transformation from graphite to diamond struture

step 2, respectively. The step following step 5 must be the same as step 3 that requires bond breaking. During step 1 and step 2, there is a large contraction along c direction of graphite. This contraction is related to the fact that diamond is formed at high pressure. Also, bond breaking is inevitable as shown in step 3, which requires a high activation energy and can be accomplished only at high temperature. This breaking of bonds needs more energy if pressure is increased.

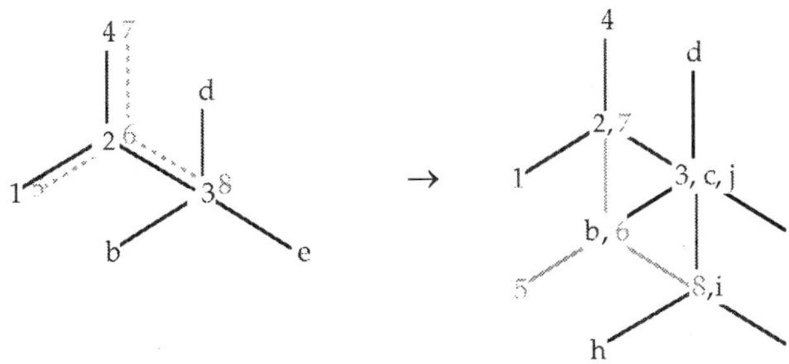

Figure 4-14 Projection of diamond structure of Fig.4-13(vi) along c-axis of graphite

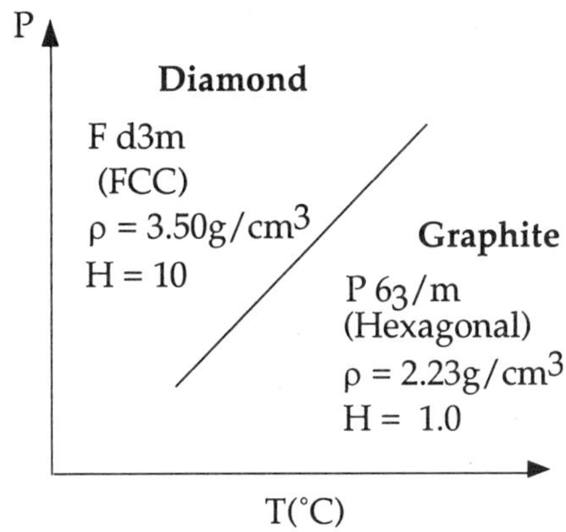

Figure 4-15 Phase diagram of carbon with temperature and pressure

The linear increase of pressure with the transformation temperature shown in Fig.4-15 can be attributed to this fact.

As shown in Fig.4-13, the transformation of graphite to diamond is a reconstructive transformation since graphite cannot be transformed to diamond without bond breaking. The bond breaking during the transition from graphite to diamond can also be considered in terms of the crystal structure. Diamond has the FCC structure and graphite the HCP structure. Therefore, the stacking sequences of carbon layers are ABCABC and ABABAB for diamond and graphite, respectively. If atoms are assigned numbers and letters in alternating layers of graphite as shown in Fig.4-13(i), the atoms designated by numbers overlap with other atoms assigned numbers when viewed along the c-axis of graphite. Also, atoms with letters overlap with other atoms with letters. The transformation of graphite to diamond destroys this overlap pattern. The projected view along c-axis of graphite is illustrated in Fig.4-14.

1-3. Calcite and Aragonite: $CaCO_3$

$CaCO_3$ has two polymorphs which are called calcite and aragonite. Basically calcite has the same crystal structure as NaCl in that Cl atoms are located at FCC sites and Na atoms occupy the octahedral sites of the FCC lattice. In calcite, Ca forms an FCC lattice structure and carbonate (CO_3) groups occupy the octahedral sites. The difference between NaCl and calcite is in the way the carbonates reside in octahedral sites of the FCC lattice. The carbonate is a planar triangle and resides in octahedral sites parallel to the (111) plane of the FCC lattice. In every other interlayer, the carbonates rotate 180° around the [111] axis. This rotation around the body diagonal of the FCC lattice results in the repeating unit as ABCABC instead of ABC as in NaCl. This is schematically illustrated in Fig.4-16 where the two different octahedral sites are designated by "O" and "O_r". The octahedral sites between the first A, B layers is different from that of the second A, B layers. Hence, in calcite, the repeating unit must extend to the third A, B layers as shown in Fig.4-16. The actual crystal structure of calctie is shown in Fig.4-18 where a rhombohedral crystal structure is illutrated.

The other polymorph for $CaCO_3$ is aragonite which has an orthorhombic crystal system. This is shown in Fig.4-19. Regarding the close packing of calcium atoms, the difference of aragonite from calcite is in the stacking sequence of calcium and the arrangement of carbonates. In calcite, calcium has the ABCABC stacking sequnce whereas the sequnce is ABAB in aragonite. In aragonite, an octahedral layer can be considered as the splitting of

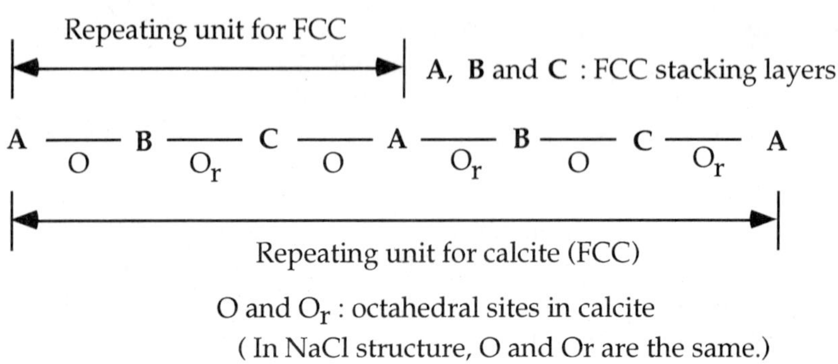

Figure 4-16 Comparison of NaCl structure with calcite structure

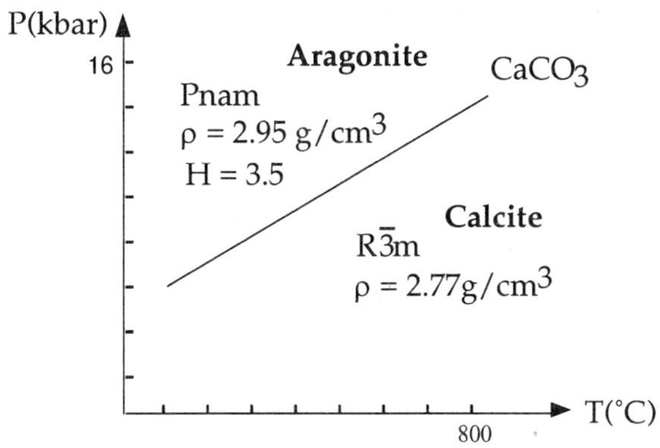

Figure 4-17 Polymorphic formation of $CaCO_3$ with temperature and pressure

the octahedral layer into two sublayers in that the triangular carbonates are different from each other by 180° rotation around the c-axis, which is perpendicular to the close packing layer of calcium. One sublayer of carbonates shown as "O_1" in Fig.4-19 is closer to the upper Ca layer and the other shown as "O_2" in Fig.4-19 is closer to the lower Ca layer. To show this splitting, the crystal structure is elongated along the c-axis, compared to the crystal structure in Fig.4-20.

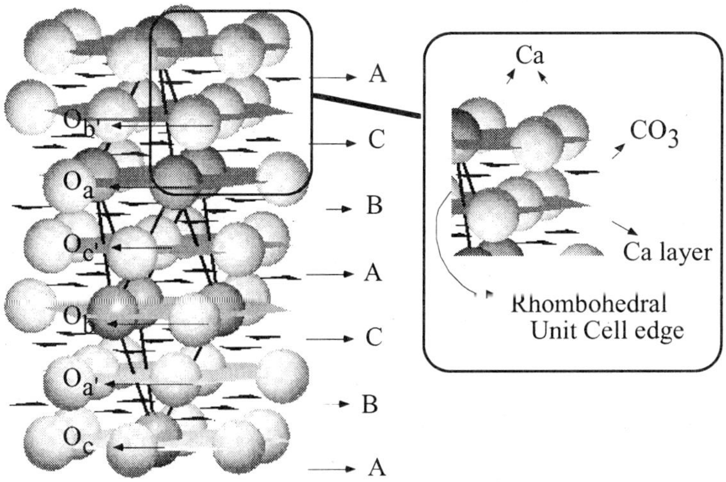

Figure 4-18(a) Rhombohedral crystal structure of calcite

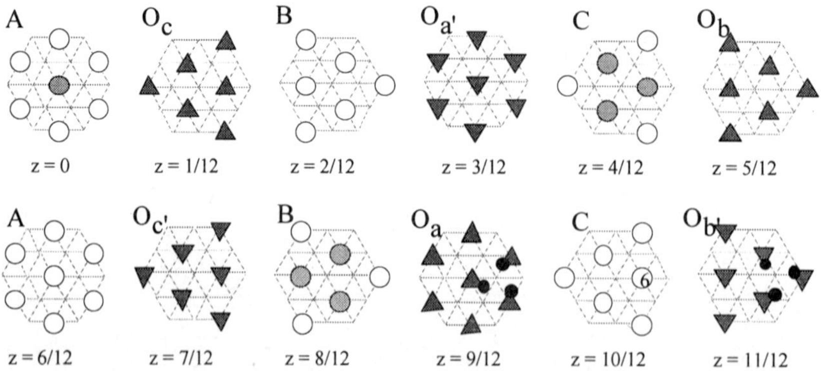

Figure 4-18(b) Rhombohedral crystal structure of calcite with z-notation

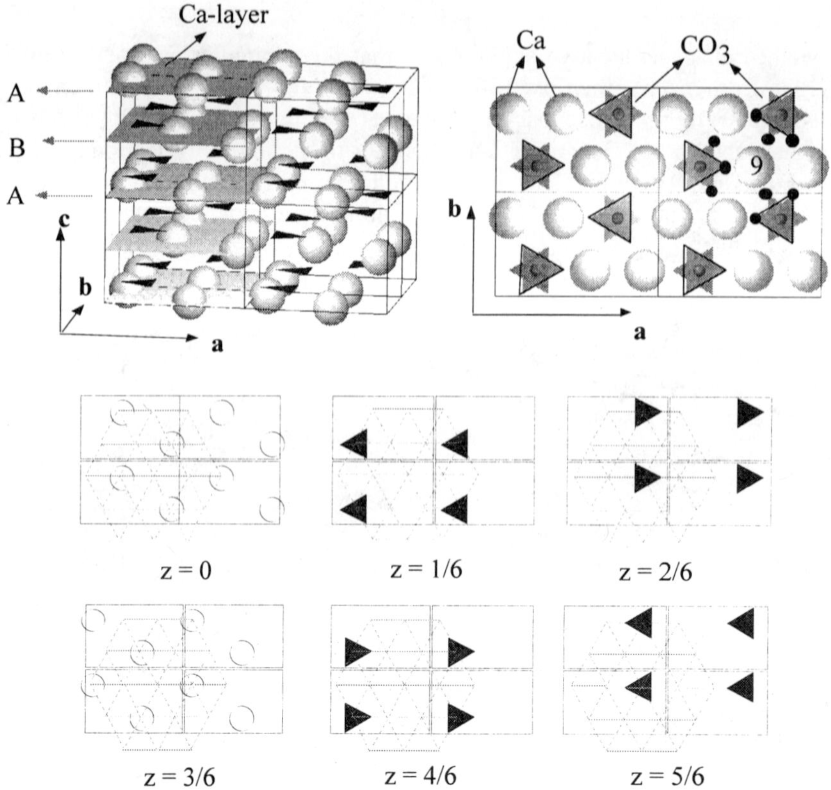

Figure 4-19 Crystal structure of aragonite

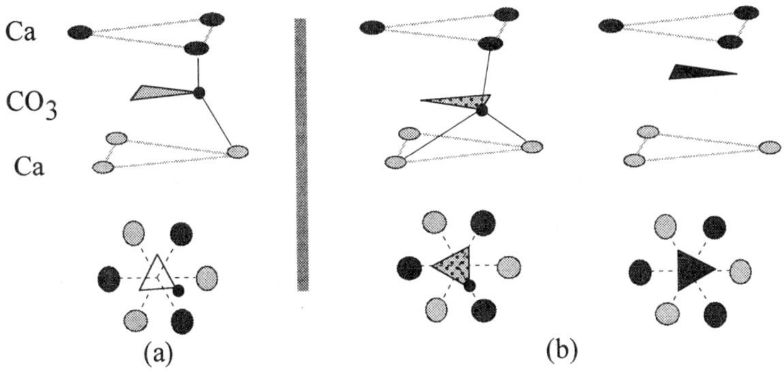

Figure 4-20 Coordination arrangements of carbonates with calcium
(a) in calcite and (b) in aragonite

1-3-1 Other carbonate compounds with calcite and aragonite structures

Carbonates of Mg(0.65Å), Zn(0.74), Ni(0.72) and Fe(0.76) have the calcite structure while carbonates of Sr(1.13) and Ba(1.35) adopt the aragonite structure. The numbers in parantheses represent the radii of +2 ions of the elements. The radius of Ca^{+2} is 0.99Å which is on the boundary between these two group of metal elements. This relationship between the cation radii and the selection of calcite and aragonite structure for the carbonates can be attributed to the same factor that accounts for polymorphism in CsCl with pressure.

As shown in Fig.4-20, oxygen has coordination number 2 in the calcite structure whereas it has coordination number 3 in the aragonite structure if its coordination with the carbon atom is neglected. In order to achieve charge neutrality, the coordination number of the cation in the calcite and aragonite structures must be 6 and 9, respectively since the charge for Ca is +2 and oxygen is –2/3. These numbers are identified on Cation 6 in Fig.4-18 for calcite and Cation 9 in Fig.4-19 for aragonite. The vertices indicated by dots represent the oxygen atoms coordinated to the cations. In light of Pauling's rule, small metal elements take the calcite structure with lower coordination number, 6. The large metal elements adopt the aragonite structure with coordination number 9 for the cations. As for Ca, its radius falls into the category of metal elements which choose the calcite structure. As pressure increases, the relative size of Ca vs oxygen increases since oxygen anions are more susceptible to the compressive effect of the pressure. Therefore, $CaCO_3$ takes the aragonite structure. Dolomite ($Ca_{0.5}Mg_{0.5}CO_3$) and alstonite($Ca_{0.5}Ba_{0.5}CO_3$) are good examples which illustrate the effect of the cation radii. Dolomite has the calcite structure and alstonite the aragonite structure.

Unlike the displacive transformation in NaCl with pressure, the phase transformation for $CaCO_3$ with pressure is reconstructive, since Ca has the

CCP arrangement in calcite and the HCP arrangement in aragonite as illustrated in Fig.s' 4-18 and 4-19. For this reason, aragonite exists in nature along with calcite although NaCl with the CsCl structure does not exist.

1-4. TiO_2

There are three polymorphs of TiO_2. These polymorphs can be described as a framework of octahedra of TiO_6 linked by corner- or edge-sharing. As shown in Chapter 3, rutile is based on close packing of oxygen although its crystal structure is distorted from the ideal close packing. Brookite also has the close packing of oxygen although the distribution pattern of Ti ions among the octahedral sites is different from that of rutile. Also the octahedra of TiO_6 in brookite is a little distorted from the ideal octahedra. Data on the close packed structrues of rutile and brookite are given in Table 4-2 where it can be seen that rutile and brookite have higher densities than anatase, although the average distance of Ti-O in anatase is the shortest among the polymorphs of TiO_2.

Table 4-2 Densities and atomic distances of polymorphs of TiO_2[3]

	Rutile	Anatase	Brookite
$Ti\text{-}O_1$	1.988	1.964	1.87
$Ti\text{-}O_2$	1.988	1.964	2.04
$Ti\text{-}O_3$	1.944	1.937	1.99
$Ti\text{-}O_4$	1.944	1.937	1.94
$Ti\text{-}O_5$	1.944	1.937	1.92
$Ti\text{-}O_6$	1.944	1.937	2.00
Avg. Distance	1.96	1.946	1.959
Density	4.26	3.84	4.17

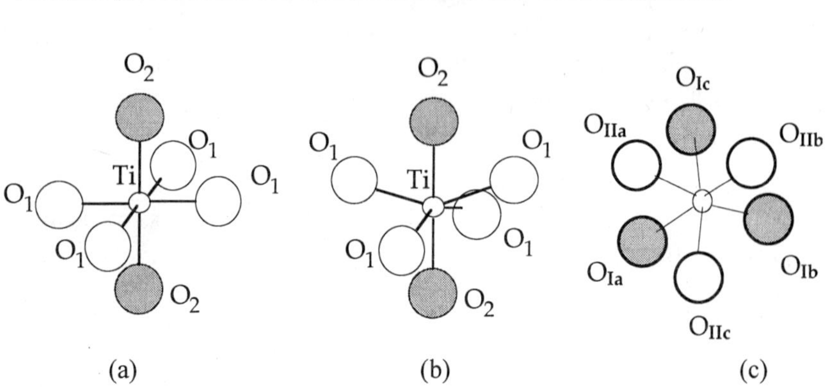

Figure 4-21 TiO_6 octahedron for (a) rutile, (b) anatase and (c) brookite

The lattice structure is detemined by how the octahedra of TiO_6 are linked. The octahedra of TiO_6 are connected by edge sharing in anatase and by a combination of edge and by corner sharing in rutile and brookite. Anatase has the tetragonal structure, while distortions of the crystal structures of rutile and brookite from the ideal close packing result in tetragonal and orthorhombic crystal systems, respectively. The octahedron units for rutile, anatase and brookite are shown in Fig.4-21. Rutile has a regular octahedron that is a little stretched diagonally along the body direction. Hence the distance of Ti-O_1 in Fig.4-21(a) is longer than that of Ti-O_2. Still the angle O_1-Ti-O_2 is 90°. In anatase, this octahedron of rutile is further changed by puckering two sets of Oxygen O_1's along the tetragonal 4 fold rotational symmetry axis. In anatase, therefore, the angle O_1-Ti-O_2 is no longer 90°. In both rutile and anatase, oxygens are symmetrically classified into two groups as O_1 and O_2 in that the line connecting Ti and O_2 is in the digonal direction of the body. In brookite, any line connecting Ti and oxygen does not form a body diagonal since the angle O-Ti-O is not 90° or 180°. Also all Ti-O distances are different from one another as shown in Table 4-2.

1-4-1 Brookite

Like rutile, brookite can be described as closest packing of oxygen anions with Ti^{+4} cations distributed among the octahedral sites. In brookite, oxygen anions take the closest packed structure of –ABACA– sequence while rutile has the HCP structure of oxygen anions. Due to this difference in the packing, the arrangements of Ti in rutile and anatase are different. Octahedral empty sites are arrayed in a zig-zag pattern in each octahedral layer in anatase whereas, in rutile, octahedral sites are arranged in a straight line as shown in Fig.4-22(a). In the notation of the sequence in Fig.4-22(a), capital letters designate the position of oxygen anions and lower case letter the positions of Ti^{+4} ions. In rutile, only c and c' types of octahedral sites are formed. In anatase, there are four types c, c', b and b'. Hence it is expected that the minimum number of TiO_6 octahedra should be 4. Actually the number of octahedra in the unit cell is 8. The rectangle in Fig.4-22(c) implies the two lattice vectors, a_2 and a_3. The unit cell for anatase is formed by displacing the rectangle through 4 octahedral layers. Since the rectangle contains two octahedra, the unit cell consequently contains 8 octahedra. The unit cell dimension of brookite is $a_1 = 9.18$, $a_2 = 5.45$, and $a_3 = 5.15$Å. Conventionally the notation for lattice vectors in the orthorhombic lattice follows the rule that:

$$a > b > c. \text{ (in the orthorhombic lattice)}$$

Hence the lattice vectors, a_1, a_2 and a_3 in Fig.4-22 correspond to a, b and c in the conventional lattice vectors.

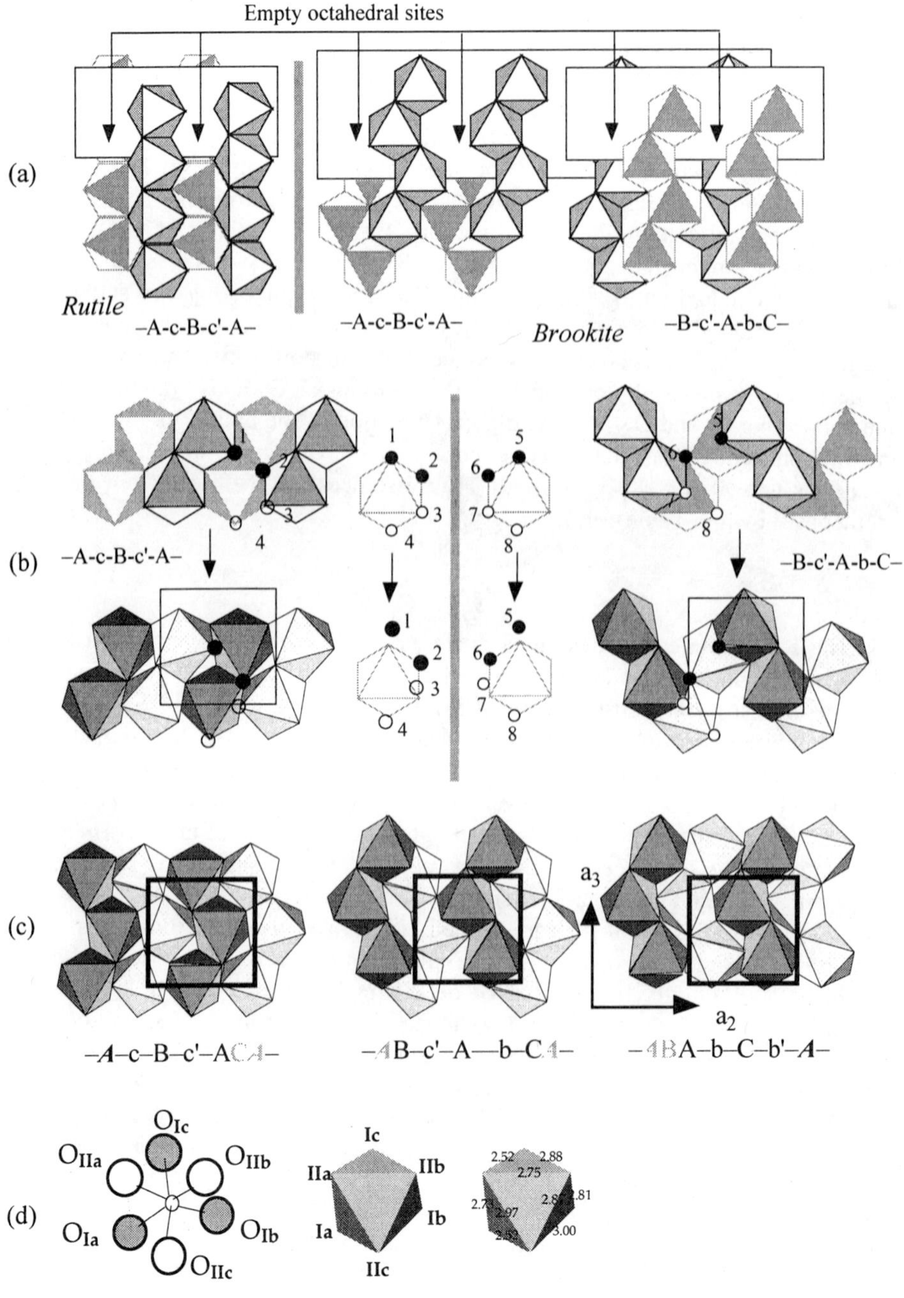

Figure 4-22 Crystal structure of Brookite

In rutile, octaheral chains are rotated along the chain axis to reduce the repulsion between oxygens which make contact but do not share common cations. A similar distortion of the regular octahedra of TiO_6 occurs in anatase in the manner shown in Fig.4-22(b). The distances O_1-O_2, O_3-O_4, O_5-O_6, and O_7-O_8 increase while those of O_2-O_3 and O_6-O_7 decrease. This distortion results in the shape of octahedra in which all the distances of Ti-O are different. Also the distance between adjacent oxygens are all different as shown in Fig.4-22(d).

Since the relative position in oxygen packing is represented by the sequence –ABACA–, the resulting sequence of Ti^{+4} position is –cc'bb'c–. Here c and c' designate the same relative position but octahedra type c' is the mirror image of c across the oxygen layer. In this notation for octahedra types, the same type octahedra are connected to the adjacent ones by edge sharing in a chain mode as shown in Fig.4-22(a). Across the layer, however, octahedra types c and c' and types b and b' are connected to the adjacent octahedra by corner sharing, while types c' and b or types b' and c are connected by corner sharing and edge sharing. This is shown in Fig.4-22(b) and (c). With the packing sequence of –cc'bb'cc'bb'–, each octahedra in brookite has 6 corner sharings and 3 edge sharings. This can be compared to the sharing mode for rutile which has 2 edge sharings and 8 corner sharings. As seen below, anatase, the third polymorphs of TiO_2, has only 4 edge sharings with 4 corner sharings. If the sharing mode is related to the number of neighboring octahedra for Ti^{+4}, anatase has the least number of neighboring octahedra and rutile has the largest number. These configurations of octahedra in polymorphs of TiO_2 agree with the density data in Table 4-2 which show the highest density for rutile and the least for anatase, although anatase has a lower average distance of Ti-O than the other polymorphs.

1-4-2 Anatase

Among the polymorphs of TiO_2, anatase is the most distorted from the regular octahedra. The average distance in Table 4-2 illustrates this aspect of anatase. Although it has the highest distortion, its shape is symmetrical compared to that of brookite, which is shown in Fig.4-23(I). The octahedra are connected by corner sharing so as to form a mesh type octahedral layer as shown in Fig.4-23(II) which is viewed from the top of the octahedral layer. The octahedral layers are stacked by sharing the shorter edges. By so doing, the same position of the octahedra layer repeat every four layers. Hence the unit cell is determined by the three lattice vectors, a_1 and a_2 and c. Here a_1 and a_2 are parallel to the octahedral layer and are perpendicular to each other with the same length. c is perpendicular to the layer and extends four octahedra. The unit cell of anatase is shown in Fig.4-23(III). The unit cell dimension is a_o = 3.76 and c_o = 9.51Å. It is a tetragonal lattice and its space group is $I4_1/amd$.

Fig.4-24 shows the phase diagram for TiO_2. Although TiO_2-II with α-PbO_2 structure, has been obtained only in the laboratory, rutile, brookite, and anatase exist in nature. Both brookite and anatase are low temperature forms of TiO_2 while rutile is the high temperature form. The phase transition from

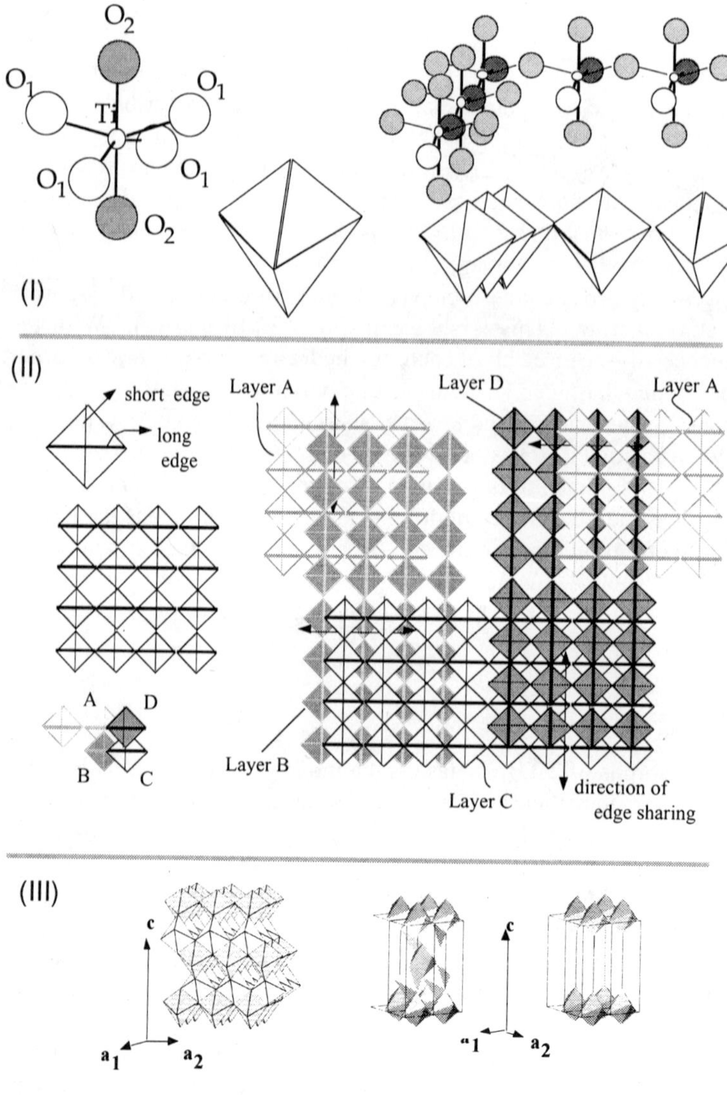

Figure 4-23 Crystal structure of Anatase

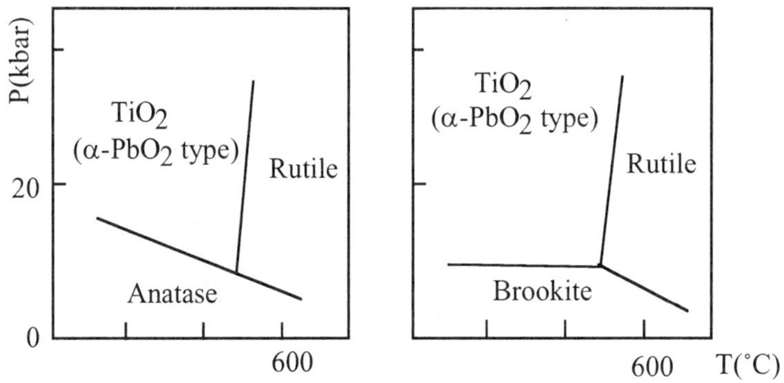

Figure 4-24 Phase diagram of TiO_2

anatase or brookite to rutile occurs at around 500~600°C under pressure and is observed to be irrevisible. Rutile has not been successfully transformed into anatase or brookite in experiments.

Usually the higher temperature form has a more open structure and higher symmetry than the lower temeperature form in polymorphs of crystals. In the case of TiO_2, anatase and brookite have more open structure than rutile. With respect to lattice energy, rutile and anatase have similar lattice energies according to the theoretical calculation by Bollnow. Hence open structure does not mean the greater lattice energy. Also, regarding symmetry, anatase($I4_1/amd$, space group number 141) has even higher symmetry than rutile($P4_2/mnm$, space group number 136). In this respect, it is more appropriate to state that the high temperature form of polymorphs has a more homogeneous environment for each atom than the low temperature form instead of stating that it has the more open structure and higher symmetry. This condition mainly describes the overall crystal strcuture not that of each atom, although generally this homogeneous environment results in a more open structure and higher symmetry. The bond lengths and bond angles between Ti–O's for rutile and anatase illustrate this. In rutile all the bond lengths are similar and the bond angle between two Ti-O bonds is 90° whereas there is a large bond length difference between shorter and longer Ti–O bonds. This partly explains why rutile is the high temperature form of the TiO_2 polymorphs.

2. Displacive and Reconstructive transformation in polymorphs

2-2. $BaTiO_3$

In chapter 3, the crystal structure of $BaTiO_3$ was simply derived on the basis of the close packing of Ba and O. The derived unit cell is cubic as shown in Fig.

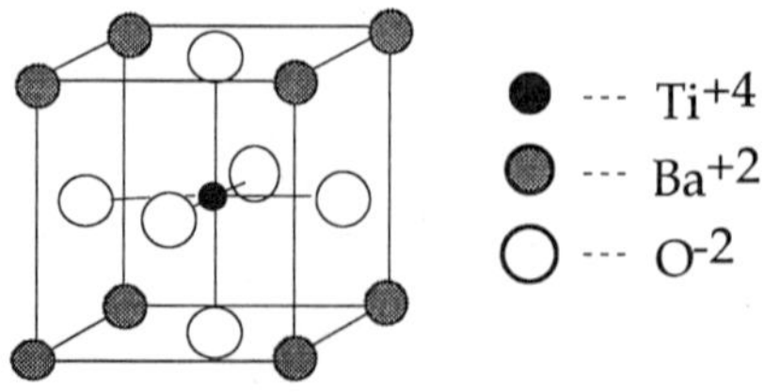

Figure 4-25 B-type unit cell structure of cubic $BaTiO_3$

4-25 where Ti is located at the center of the cube, with oxygens at face-centers and Ba at the cube corners. The length of the cube is 4.00A°. This is a B-type unit cell.

Sometimes $BaTiO_3$ is described with an A-type unit cell in that Ba is located at the center of the cube, oxygens are at the middle of cube edges and Ti's are at cube corners. It is more convenient to describe the cubic structure with the A-type unit cell with respect to the relationship between hexagonal and cubic polymorphs of $BaTiO_3$. The A-type unit cell can be further depicted as TiO_6 octahedra linked by corner sharing with Ba at the center of unit cell. This is shown in Fig.4-26 (b) and (c). Rotating the A-type unit cell reveals how TiO_6 octahedra are formed from the close packing of oxygens. As shown in Fig. 2, Octahedron A-4 consists of A and B layers of oxygens. Octahedra C-2, C-3, and C-4 are formed between B and C layers of oxygens. Octahedra B-1, B-2, and B-3 are made of oxygens of layers C and A. To show that cubic $BaTiO_3$ is based on an FCC structure of oxygens, TiO_6 octahedra are schematically drawn without Ba in Fig.4-26(c). This simplified layer of TiO_6 octahedra is stacked in 3 layers and compared with perpective view of cubic $BaTiO_3$ in that B-type unit cell is shown using the same numbering as in Fig.4-26(b).

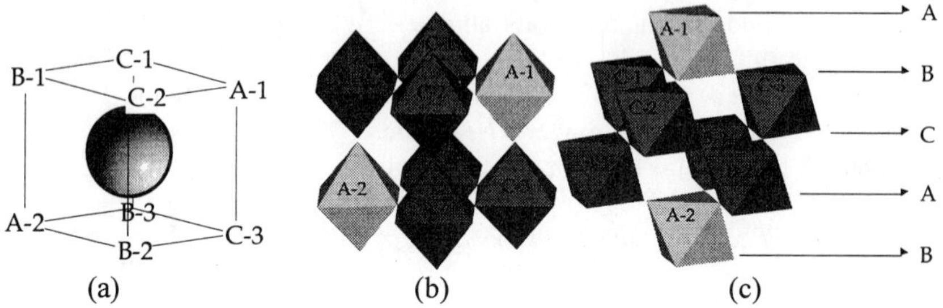

Figure 4-26 A-type unit cell of $BaTiO_3$

One of polymorphs of $BaTiO_3$ is hexagonal structure. Its unit cell has the dimension of a = 5.735A° and c = 14.05A°. Its space group is $C\frac{6_3}{m}mc$. Regarding the stacking of oxygen layers, it has a sequence of ABCACBA. Its perspective view is shown in Fig.4-27. This picture illustrates the resemblance between cubic and hexagonal structures. The three TiO_6 octahedra layers from the top are the same as those of the cubic $BaTiO_3$ as shown in Fig.4-26(a). However, the next three TiO_6 octahedra layers are mirror images of the top 3 layers. This mirror image can be better illustrated in the side and top views shown in Fig.4-28 and Fig.4-29.

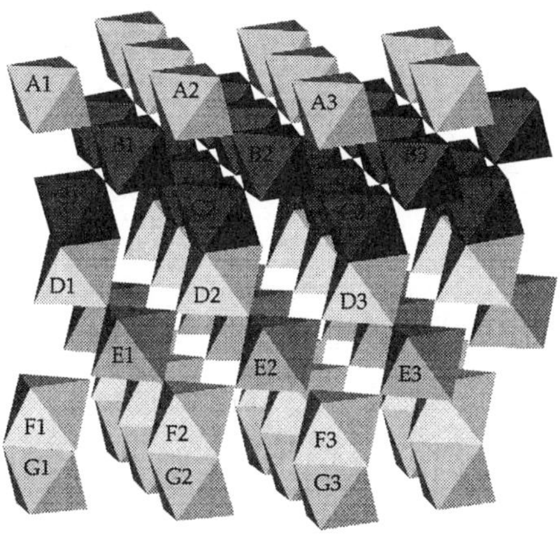

Figure 4-27 Perspective view of hexagonal BaTiO3

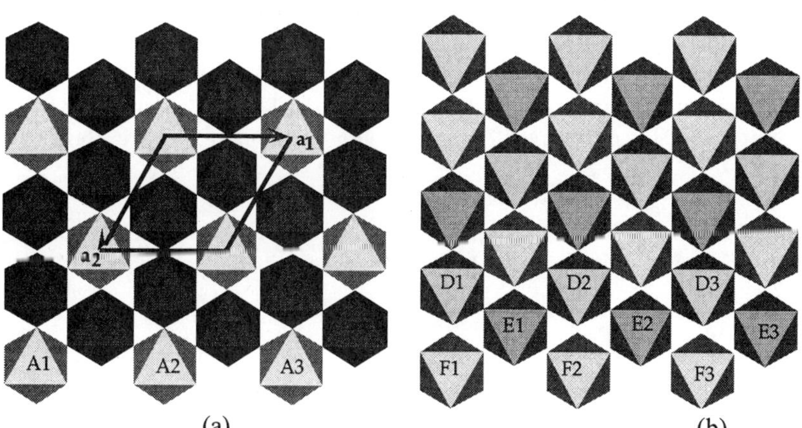

Figure 4-28 View of hexagonal $BaTiO_3$ along c-axis
(a) upper and (b) lower three TiO6 octahedra layers

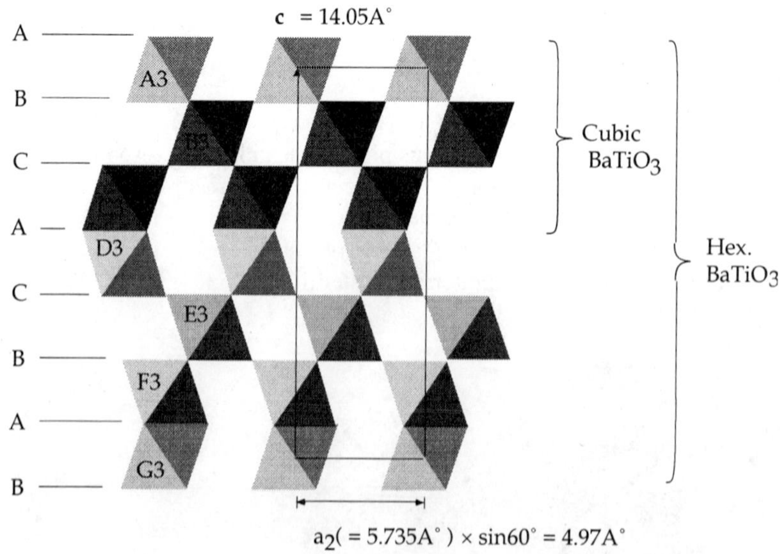

Figure 4-29 View of hexagonal $BaTiO_3$ along a_1 axis

The side view along the a_1 axis shown in Fig.4-29 reveals more distinctively the stacking sequence of oxygen layers in hexagonal $BaTiO_3$ as ABCACBA. Due to this sequence, the oxygen layer A forms a mirror plane. Hence A3, B3, and C3 are mirror images of F3, E3, and D3 through the mirror plane crossing C3 and D3. Through this mirror plane, TiO_6 octahedra are linked by face sharing not by corner sharing. This significant difference between cubic and hexagonal $BaTiO_3$, therefore, is the face sharing between TiO_6 octahedra resulting in a Ti_2O_9 group. This Ti_2O_9 group in hexagonal $BaTiO_3$ can explain its difference in physical properties from cubic $BaTiO_3$. One of distinctive feature of $BaTiO_3$ is phase transformations with temperature. This is shown in Fig.4-30.

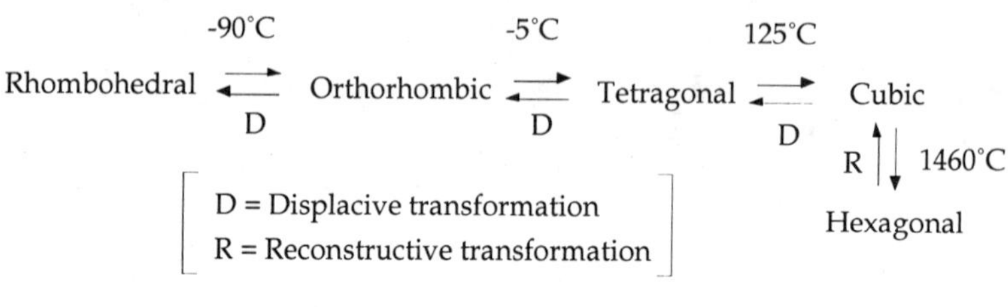

Figure 4-30 Phase transformation of $BaTiO_3$ with temperature

As shown in Fig.4-30, the transitions between polymorphs are displacive except between hexagonal and cubic structures. Hence cubic $BaTiO_3$ is easily transformed to tetragonal $BaTiO_3$ which also transforms to the orthorhombic structure at low temperature around -5°C. If the orthorhombic structure is further cooled to -90°C, it transforms to the rhombohedral structure. Heating of rhombohedral $BaTiO_3$ reverses the transformation, finally resulting in the cubic crystal structure above 125°C. In contrast to cubic $BaTiO_3$, the hexagonal form does not transform to tetragonal or other polymorphs of $BaTiO_3$. Even at room temperature the hexagonal structure is retained once it is formed at high sintering temperature.

Since 1940, there has been tremendous interest in $BaTiO_3$ due to its ferroelectric properties along with high dielectric constant at room temperature, where the most stable structure is the tetragonal crystal structure. Characteristics of tetragoanal $BaTiO_3$ are high dielectric constant and spontaneous polarization even without an external electric field. These are attributed to the off-center movement of Ti within the TiO_6 octahedron. Since, in the cubic structure the activation energy for this off-center movement of Ti is overcome by thermal energy, the averaged equilibrium position of Ti is at the center of the cube. This is the case above 120°C. However, at room temperature, the thermal energy is less than the activation energy and the off-centered position in the tetragonal is retained, resulting in polarization. In Fig.4-31, up and down polarizations are shown, depending on the direction of the

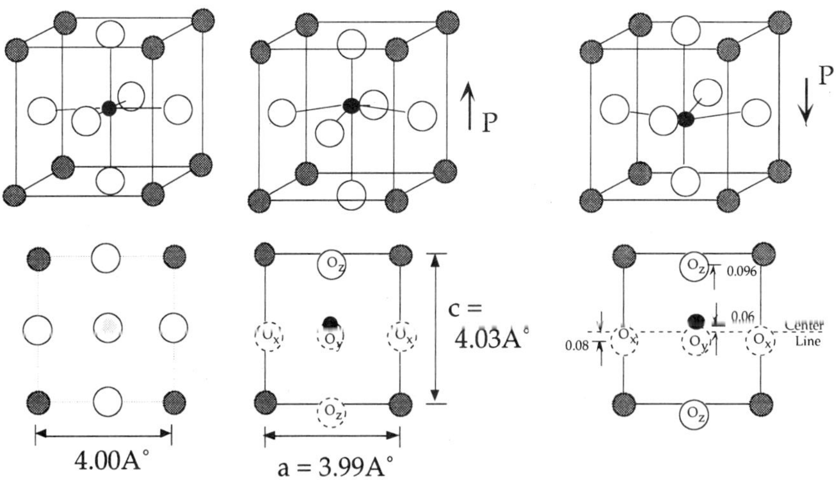

Figure 4-31 Crystal structure change in transition from cubic to tetragonal $BaTiO_3$

shift of Ti. These up and down polarizations can be reversed by applying an external electric field in the opposite direction. The formation of spontaneous polarization under no electric field and its reversibility by an external electric field is the defining characteristic of ferroelectric materials. The name 'ferro' originates from 'ferromagnetic' which indicates the reversibility of magnetic moments under an external magnetic field. Detailed coordinates of elements in tetragonal $BaTiO_3$ are illustrated in Fig.4-31. This figure shows that oxygens above and below Ti in the direction of the c-axis move in opposite directions. Also oxygens in parallel positions with Ti in the ab plane (shown as O_x and O_y) move in the same direction as Oz but for a smaller distance. This concerted movement of elements during the transition from cubic to tetragonal $BaTiO_3$ illustrates why hexagonal $BaTiO_3$ does not show any phase transitions.

Fig.4-32 shows the TiO_9 group in hexagonal $BaTiO_3$(a) and its side view(b) along with the tetragonal structure(c). In contrast to cubic structure, the concerted movements of Ti and oxygens in one direction are restricted by the other TiO_6 octahedron in the TiO_9 group. This alignment of TiO_6 octahedra in cubic $BaTiO_3$ in one direction may be the important aspect for ferroelectric properties. This structural aspect can be compared with those of TiO_2 polymorphs that are not ferroelectric. In TiO_2 polymorphs, TiO_6 octahedra are linked by edge sharing and corner sharing. This mixed sharing mode of TiO_6 octahedra results in a structure in which the concerted movement of cations and anions is difficult for spontaneous polarization to be realized. This structural difference between TiO_2 polymorphs and cubic $BaTiO_3$ is one example of the effect of crystal structure on physical properties.

2-3. ZrO_2

Zirconia is one of the most important ceramic materials for its structural,

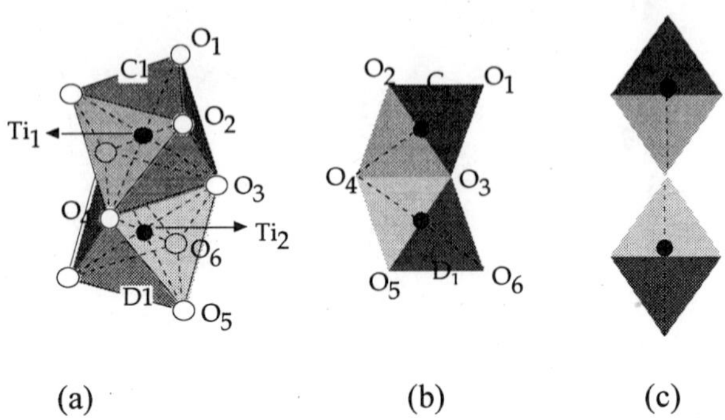

Figure 4-32 Comparison of (a) and (b) TiO_9 groups in hexagonal $BaTiO_3$ with (c) tetragonal $BaTiO_3$ in a concerted mode along the c-axis

$$\underset{\text{Monoclinic}}{ZrO_2} \overset{1150^\circ C}{\rightleftharpoons} \text{Tetragonal} \overset{2100^\circ C}{\rightleftharpoons} \text{Cubic}$$

Figure 4-33 Polymorphism of ZrO_2

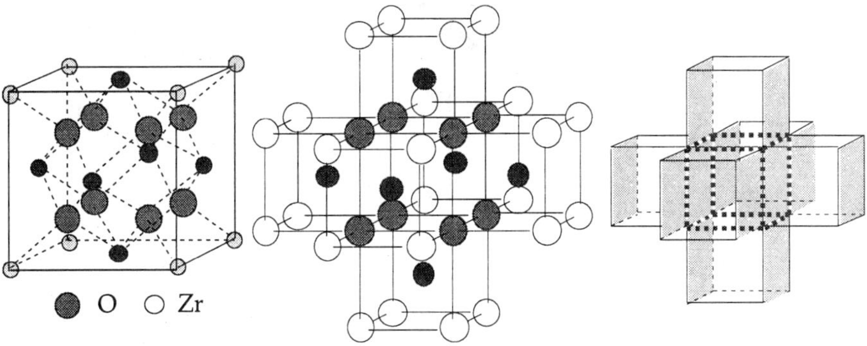

(a) (b) (c)

Figure 4-34 (a) Fluorite type crystal structure of cubic ZrO_2 (b) Edge sharing of ZrO_8 cubes and (c) schematic drawing of edge shared cubes

electrical and refractory applications. Zirconia has three polymorphs which adopt the cubic, tetragonal and monoclinic structures, as illustrated in Fig.4-33. Above 2100°C, its structure is cubic with the fluorite type structure. As shown in Fig.4-34(a), Zr occupies FCC lattice positions and O occupies all of tetrahedral sites. From a different perspective, it can be considered as an arrangement of ZrO_8 cubes by edge sharing. This is shown in Fig.4-34(b) where the dotted oxygens form a cube having no Zr.

2-3-1 Tetragonal ZrO_2

Around 2100°C, cubic zirconia transforms to the tetragonal form by displacive transformation as illustrated in Fig.4-35. During the displacive transformation, oxygens lying on the same line along the c-axis move upward or downward by 0.15 of a unit cell length. This upward and downward movement of oxygens alternates along the a-axis and b-axis. In the same cube of ZrO_8, oxygens facing in the diagonal direction on the ab-plane move together either downward or upward. Hence the coordination of Zr does not change. There is no bond breaking. The space group of cubic zirconia is Fm3m with the four fold rotational axis going through the center of cube face. During the transformation of cubic to the tetragonal phase, this four fold rotational axis disappears while the 4-fold screw axis lying on the cube edge along c-axis remains. The space

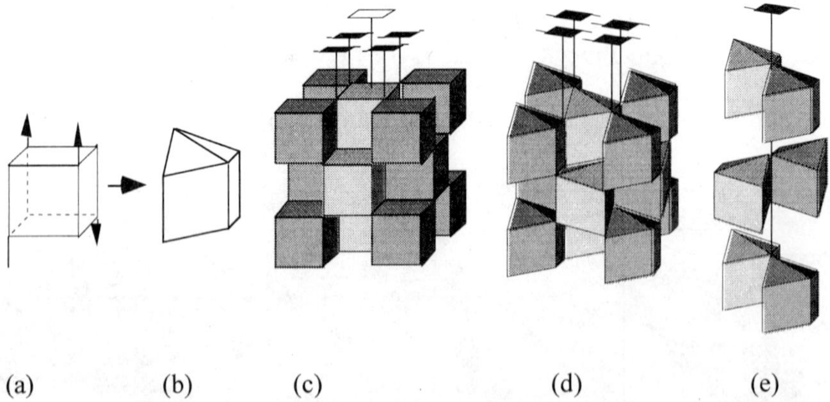

Figure 4-35 Displacive movement of oxygens and symmetry change after the transformation (Only four fold axes are shown.)

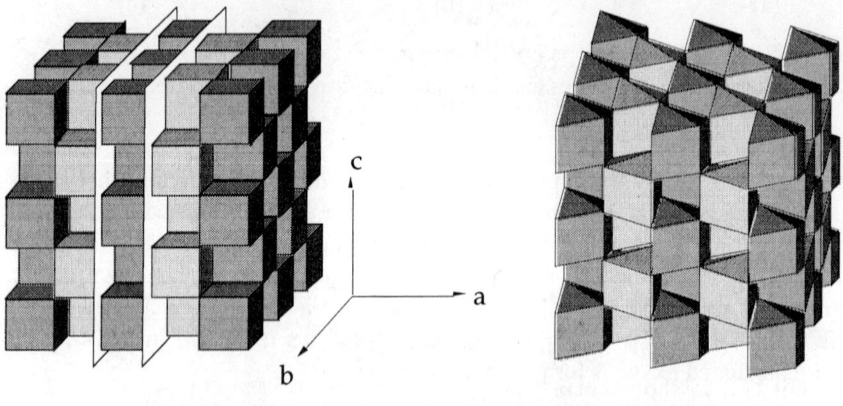

Figure 4-36 Overall crystal structure of cubic and tetragonal ZrO_2

group of tetragonal ZrO_2 is $P4_2/nmc$. Also the unit cell dimension changes after the transformation. The unit cell dimension of the cubic phase is $a_o \approx 5.07Å$ and that of tetragonal ZrO_2 is a = 3.64 and c = 5.27Å. Cubic ZrO_2 has four formula units $(ZrO_2)_4$ and tetragonal two formula unit $(ZrO_2)_2$. The overall structures for cubic and tetragonal ZrO_2 are illustrated in Fig.4-36(a) and (b), respectively.

2-3-2 Monoclinic ZrO_2

(i) Stereographic representation of monoclinic ZrO_2

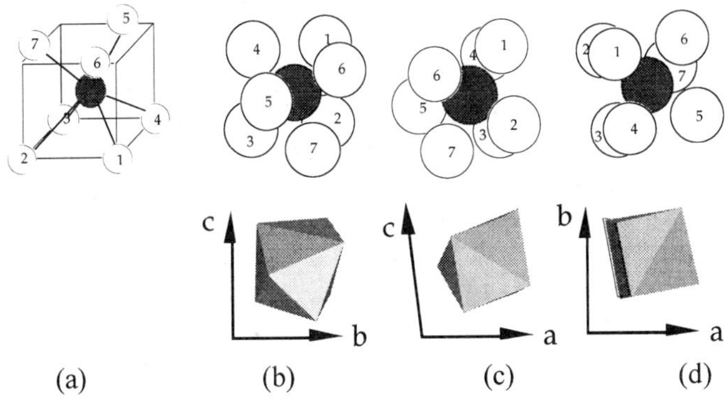

Figure 4-37 ZrO_7 polyhedra in monoclinic ZrO_2
(a) Schematic illustration of ZrO_7 in light of ZrO_8 cube
(b), (c) and (d) View of polyhedra along c* (perpendicular to ab-plane) a, and b-axes, respectively

Monoclinic ZrO_2 is a rare case in that the cation (Zr^{+4}) has coordination number of seven. As a comparison with the ZrO_8 cube, the seven coordinated ZrO_7 in monoclinic ZrO_2 is depicted simply as shown in Fig.4-37(a). In this depiction, four oxygen atoms occupy the four corners of the cube and the other three oxygen atoms are on the other side of the cube. The actual arrangement of oxygens around Zr are further distorted from the schematic description shown in Fig 4-36(a). The actual polyhedra of ZrO_7 are shown in Fig.4-37(b), (c), and (d) with the monoclinic lattice axes shown as a reference.

The overall crystal structure of monoclinic ZrO_2 is shown in Fig.4-37. The repeating unit is the group of four ZrO_7 polyhedra shown in Fig.4-38(c). The unit cell is shown in Fig. 4-41(a) and (b) with solid lines. The cell dimensions are a= 5.169A° , b= 5.232A°, c = 5.341A°, $\alpha = \gamma = 90°$ and β = 99°15'. The space group of monoclinic ZrO_2 is $P2_1/c$. For the connection mode of ZrO_2 polyhedra, it is helpful to characterize the ZrO_7 polyhedra in terms of faces. Let's call the face formed by oxygen atoms 1, 2, 3 and 4 in Fig.4-37(a) a 'rectangular face' while that formed by oxygen atoms 5, 6 and 7 is called a 'triangular face.' Characteristics in the connection mode of the polyhedra ZrO_7 is that each rectangular face is connected to four neighboring rectangular faces along the a and c axes, which is illustrated by polyhedra b4, b5, d4, d5 and d6 in Fig.4-38(f). Likewise, each triangular face is connected to two neigboring two triangular faces along the c-axis as shown by polyhedra a4, a5, c4, c5 and c6 in Fig.4-38(e). With this configuration, a-type and c-type polyhedra construct a polyhedra layer with triangular faces toward the positive side of the a-axis. b-type and d-type polyhedra form another layer with triangular faces toward the negative side of the a-axis. Therefore, two types of polyhedral layers altenate along the a-axis as shown in Fig.4-38(b). Dotted lines B in Fig.4-38(b) indicate a plane which is parallel to the bc plane. Across this plane two types of polyhedral layers are connected by edge sharing of rectangular faces, which can be seen in Fig.4-38(f). Also dotted line A indicate a plane across which two types of polyhedra layer are connected by edge and corner sharing of triangular faces. A more schematic description of face connections across Planes A and B is illustrated in Fig.4-39. The connection mode across Plane B in Fig.4-39(II) is almost the same as that in the tetragonal structure although each polyhedra is slightly rotated along the c-axis. The dotted line in Fig.4-39(I) indicates the original bonding between Ti and O. Hence these bonds are broken during the phase transformation from tetragonal to monoclinic or are reconstructed in the reverse transformation from monoclinic to the tetragonal phase. Unlike other reconstructive phase formation, however, the phase transformation is not slow, and has been characterized as a martensitic phase transformation. Orignially the martensitic transformation was defined in the iron system which shows a phase transition from the face centered cubic structure to the body centered tetragonal. Now the martensitic transformation is not confined to the iron system, but represents a group of phase transformations which have certain features such as:

(i) There is a special crystallographic relationship between parent phase and product phase. The boundary between the parent and product phases is called the habit plane since this plane is always the same.
(ii) Displacement of atoms during the transformation is usually not longer than the atomic distance. Also the displacement of atoms is concerted which means that atoms move together in a similar manner or direction. Hence it is sometimes called a military movement.
(iii) The transformation accompanies a shearing of a discrete volume of the material. This is identified by the fact that a straight line on the parent phase is broken into three line segments in the product phase, even though they are continuous.
(iv) The transformation is very rapid reaching the speed of sound in some cases.

It should be noted here that these features are not sufficient for defining the martensitic phase transformation since some other phase transformations can also satisfy these features. The purpose of this list of special features in the martensitic transformation is to illustrate the relationship of the crystal structures of monoclinic and tetrgonal ZrO_2 to these features.

First, feature (ii) can be related to bond breaking through displacement of atoms in a straight line as shown in Fig.4-39(I). Although bonds are broken in the side of the triangular face of the polyhedral layer during the transformation of the tetragonal to monoclinic phase, the rectangular face of the polyhedra still

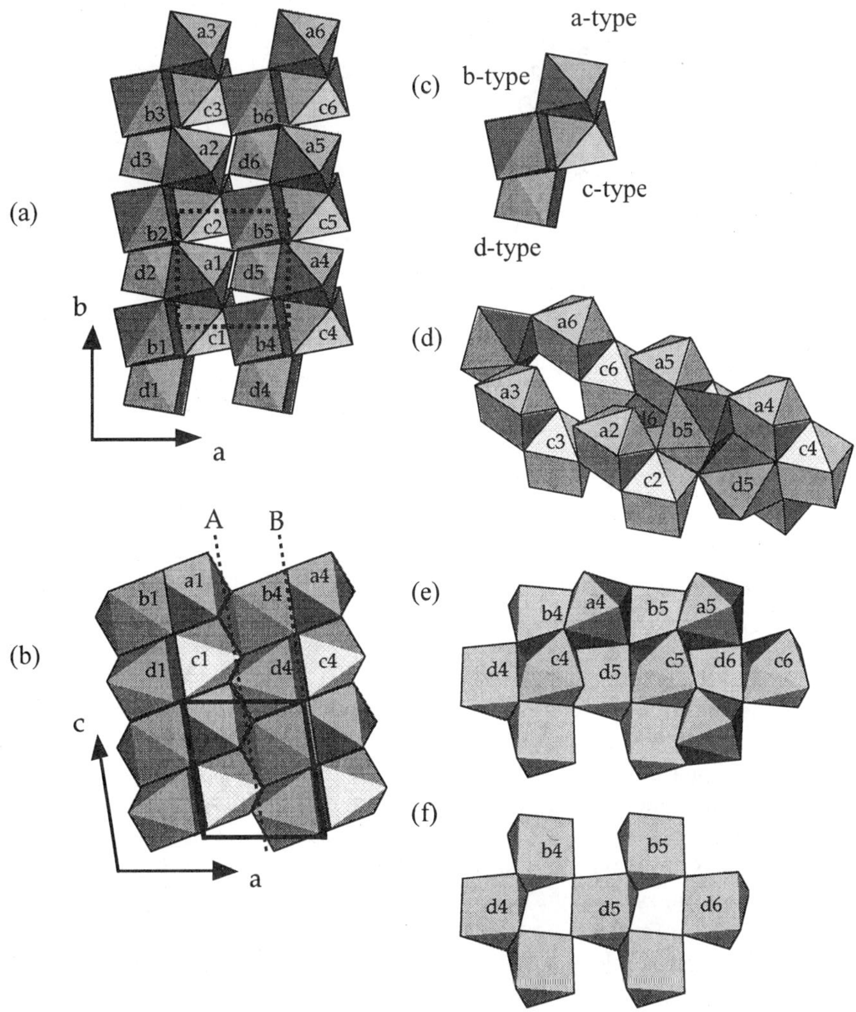

Figure 4-38 (a) View of monoclinic ZrO_2 above the ab plane,
(b) View along b-axis, (c) Basic unit of monoclinic ZrO_2,
(d) Side view 45° above from the ab plane,
(e) and (f) view along a-axis from slightly below the ab-plane

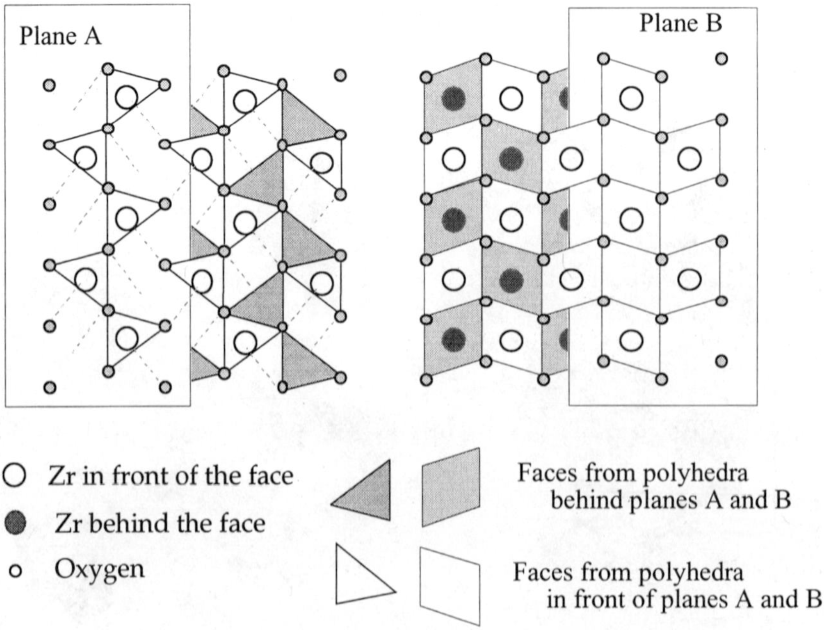

Figure 4-39 Connection mode of triangular and rectangular faces of ZrO_7 polyhedra across (a) Plane A and (b) B in Fig. 4-41(b)

retains the original configuration of tetragonal ZrO_2. With respect to feature(iii), the shift in the polyhedra layer of the monoclinc phase from that of the tetragonal phase, which is illustrated in Fig.4-40, can indicate shearing in a certain portion of the ZrO_2 during the transformation. Also, the match of the two phases can be related to feature (i). In a strict sense, the two phases do not match precisely although they appear to match each other in Fig.4-40. To accomodate this mismatch, the product phase undergoes the formation of twins or slip planes. In martensitic transformations within metallic systems, both twin and slip are possible, but in ceramics such as ZrO_2, slip is not possible. In the monoclinic phase, the prominent twin is that which occurs across the (100) plane. The deformation direction of this twin is [001]. This is shown in Fig.4-41. Another characteristic of this twin is the concerted atomic movement in the opposite directions.

Finally feature(iv) is utilized in the toughening of ceramics by adding particles of tetragonal ZrO_2 to other brittle ceramics. During the phase transformation from tetragonal to monoclinic phase, there is a volume increase. This volume change prevents the crack tip in the brittle ceramics from growing further since pressure generated at the crack tip, forces a change in the tetragonal to the monoclinic phase which has the larger volume by 4%. The transition time is very short. However, this 3 – 4% volume expansion causes cracking during sintering of ZrO_2. Therefore, the monoclinic phase is detrimental to ZrO_2 itself although it is used to increase toughness of other brittle ceramics. To avoid the formation of the monoclinic phase, a high temperature form, cubic ZrO_2, is stabilized by the addition of oxides such as CaO and Y_2O_3. The addition of these oxides hinders the atomic movements required for the transformation from cubic to the tetragonal phase and, hence, prevent the formation of the monoclinic phase.

With respect to the symmetry of the crystal structures, the low temperature form, monoclinic ZrO_2, has a lower symmetry than the high temperature form, tetragonal ZrO_2, although it has the larger volume. This agrees with the general trend between high and low temperature forms of polymorphs. The volume increase and coordination number change from 8 to 7 during the phase transformation is attributed to the covalent bonding character in monoclinic ZrO_2.

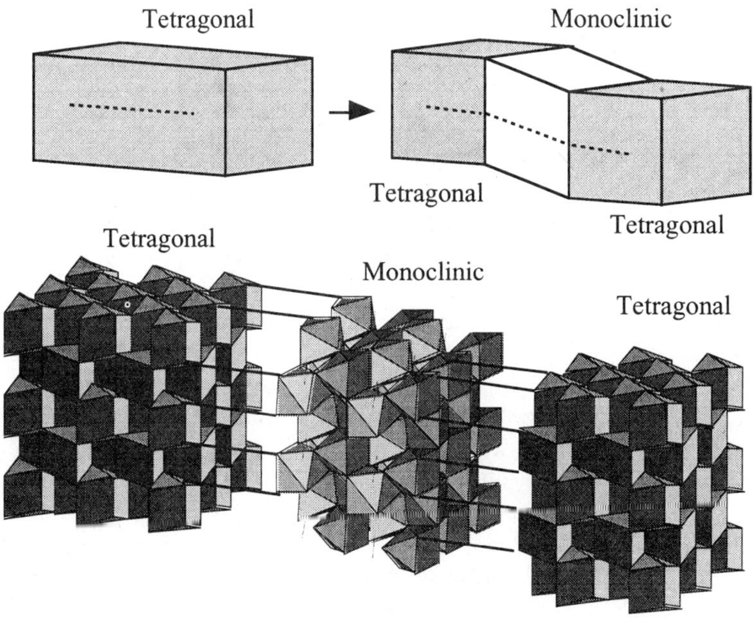

Figure 4-40 Schematic illustration of the boundary between monoclinic and tetragonal phases

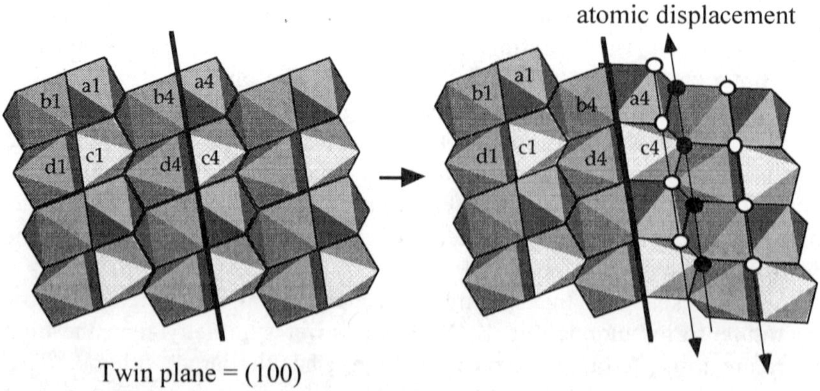

Figure 4-41 Twin formation in monoclinic ZrO_2 with twin plane (100)

(ii) Symmetric representation of space group $P2_1/c$ for monoclinic ZrO_2

The space group for the monoclinic ZrO_2 is $P2_1/c$. Unlike cubic, tetragonal, or orthorhombic lattices, it is difficult to describe crystals with the monoclinic lattice structures. The atomic coordinates of the oxygen and zirconium atoms in the monoclinic ZrO_2 are shown in Fig.4-42(a) with projection onto the ac plane. With these coordinates, the polyhedra and their connections as shown in Fig.4-38 are not easily perceived. In order to show the relationship of atomic positions with the stereographic representation given by these coordinates, a view of the repeating unit shown in Fig.4-38(c) is also shown in Fig.4-42(b). Although the representation of the unit cell with these coordinates does not readily reveal the 3-dimensional perspectives such as polyhedra connection or coordniation, its advanctage over the stereographic representation is in that the symmetry elements can easily identified within the unit cell.

The primary symmetry elements in monoclinic ZrO_2 are the 2 fold screw axis and c-glide plane while an inversion center is the secondary symmetry element. The position of these symmetry elements is shown in Fig.4-43, which is the planar representation of the space group $P2_1/c$. Here, it seems worthwhile to explain the symmetry representation of a space group since monoclinic ZrO_2 has both types of translational symmetry, a screw axis and a glide plane. These two types of translational symmetry elements cannot be describerd in the same way as the point group symmetry elements which are represented by a circular configuration. The symmetry elements of a space group are depicted within the framework of the corresponding lattice type, more exactly within the unit cell of that lattice type. It is the conventional rule to illustrate the corresponding unit cell by its chracteristic faces. For example, the monoclinic unit cell is represented by three faces ab, bc, and ca as shown in Fig.4-43. These faces are called projection faces since symmetry elements are projected onto these faces.

There are three symmetry elements in the space group $P2_1/c$. These elements are represented in different ways. In Fig.4-43(a), the arrow indicates the direction of the glide plane at a height of 1/4 of the lattice vector b above the projection face(ac plane). To distinguish the glide plane from a rotational symmetry element, which also sometimes is represented by an arrow, another line segment is attached at the other end of the arrow. This line segment usually represents one of the two lattice vectors from the projection face. The small circle in Fig.4-43 represents the inversion center. In the symmetric representation of a space group, a dotted line represents a glide plane whose glide direction is perpendicular to the corresponding projection face while a broken line implies that the glide direction is parallel to the projection face. Hence the dotted line in Fig.4-43(b) indicates that the glide direction of the glide plane is perpendicular to the projection face(ab plane), whereas the broken line in Fig.4-43(c) indicates that the direction of the glide is parallel to the projection face(bc plane). As notation for a glide plane, 'c' does not itself represent the plane. Since the direction of glide is along the c-axis, the plane must be either the ac or bc plane. The determination of the plane is dependent on the position of 'c' in the space group notation. In the notation for the space group $P2_1/c$, 'c' is underneath 2_1. This means that the glide plane is perpendicular to the 2-fold screw axis. Since it is conventional to designate the b-axis as the 2-fold rotational axis, the glide plane must be the ac plane. This fact can be verified from the planar symmetry representation in Fig.4-43. The notation $P2_1/c$, therefore, provides sufficient information from which to construct the planar symmetry representation shown in Fig.4-43.

As seen in Fig.4-43(b) and (c), the screw axis is represented by a half arrow if the screw axis is parallel to the projection face. Sometimes there is a numerical fraction attached, which indicates how high it is located above the projection face in terms of fraction of the lattice vector which is perpendicular to that face. If there is no numerical fraction attached to arrow or any other symmetry element, it means that that symmetry is located at the projection face.

Although Fig.4-43(a) and (b) are sufficient to illustrate all the symmetry elements for the space group P21/c, it is conventional to list all three projection faces with symmetry elements as in Fig.4-43. Also, in the insert of Fig.4-43, symmetry elements are illustrated on the 3-dimensional unit cell in order to help the reader to understand the planar representation of a space group. This is done in order to relate the symmetry representation of the space group to that of the unit cell, using the coordinates of monoclinic ZrO_2 shown in Fig.4-42. By way of illustration, let's superimpose the symmetry representation of the

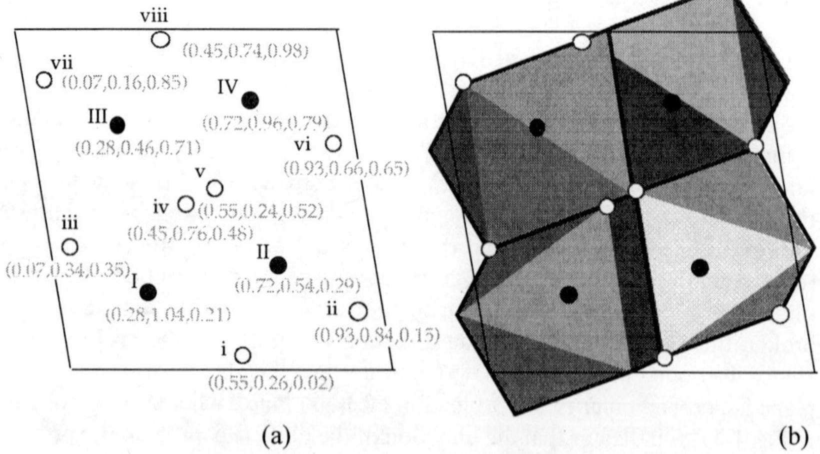

Figure 4-42 (a) Coordinates of atoms in the unit cell of monoclinic ZrO_2 and (b) its equivalent to stereographic represention in Fig.4-38(b)

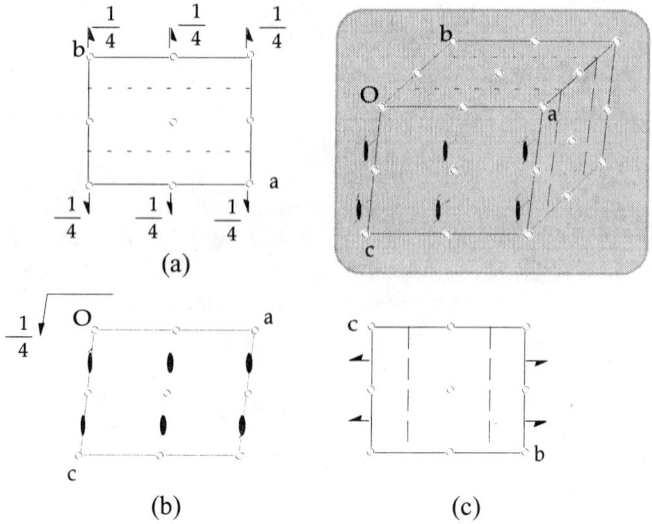

Figure 4-43 Planar illustration of space group $P2_1/c$

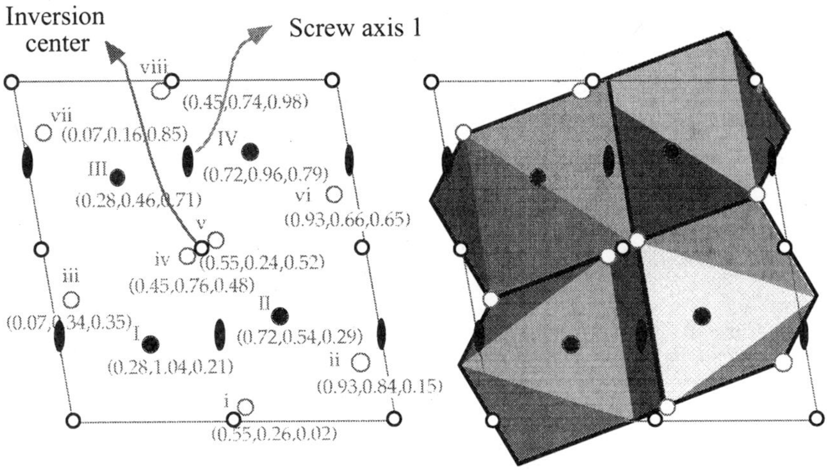

Figure 4-44 Symmetric representation of the unit cell of monoclinic ZrO_2

space group $P2_1/c$ in Fig.4-43(b) on the coordinate representation of the unit cell of monoclinic ZrO_2 in Fig.4-42(a). The superimposed condition is shown in Fig.4-44 which reveals that screw axis 1 moves Zr(I) to Zr(II) and O(i) to O(iv). Also O(iv) moves to O(v) by the inversion center at the center of the unit cell and Zr(I) moves to Zr(IV) by the inversion center. Therefore it is shown that the coordinate representation of the unit cell is a better representation of the symmetric aspects of monoclinic ZrO_2 than the stereographic representation.

SECTION 2 MODIFICATION OF IDEAL CRYSTAL STRUCTURES

An ideal structure, i.e. one with no irregularity in the crystal, rarely exists in nature. Due to the entropy effect, such an ideal structure is not obtainable even at 0°K. Also, irregularities in crystal structures can result from compositional inhomogeneities. These irregularities are interchangeably used with the term defects which can connote harmfulness to ideal physical properties. In fact, however, defects are often intentionally developed for desired physical properties. In this respect, the irregularities in crystal structures can more aptly be expressed as imperfections.

A. Defects in crystal structure

1. Defect Structures

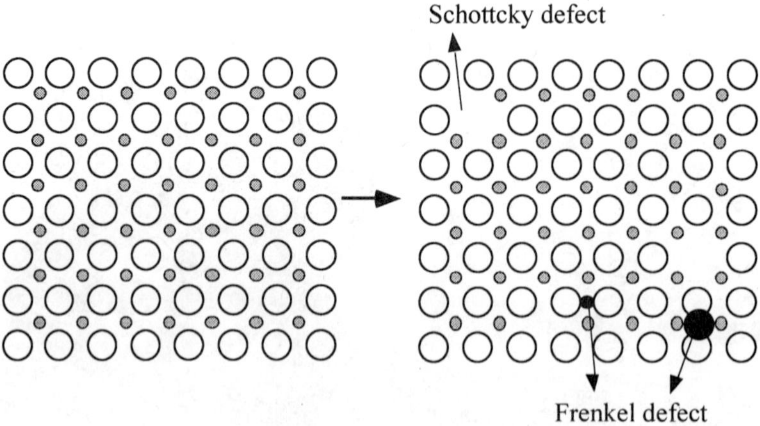

Figure 4-45 Vacancy formation in ionic crystals

The smallest imperfection in a crystal structure is a vacancy or the location of an atom on an interstitial site among the regular lattice positions. These are called point defects since the difference from the ideal structure is in just one atom. In ionic crystals such as substances with the MX formula, point defects can be developed by

i) removing both M and X
ii) moving M or X to interstitial position

The first case is called a Schottcky defect and the second a Frenkel defect. This is illustrated in Fig.4-45. Since anions are larger than cations, it is more difficult to move anions to interstitial positions than to move cations to interstitial positions.

A two dimensional defect in a crystal is a dislocation which forms a line of missing atoms in the crystal. In metals, point defects and dislocations are common. In covalent or ionic crystals, however, it is usually very difficult to match atoms surrounding the dislocation, in contrast to metals. Dislocations are, therefore rare, in covalent and ionic crystals although points defects can be readily formed.

1-1. Defect structure of FeO (Wüstite)

FeO is an example of a compound that can be used to illustrate the effect of point defects on magnetic property. FeO has the NaCl structure and the crystal structure can be depicted as having Fe^{+2} ions distributed among the octahedral sites of close packed oxygens. As explained in chapter 1, Fe^{+2} has a net magnetic dipole moment. Although its atomic magnetic moment is not zero, the overall magnetic moment for the FeO crystal is zero due to the opposite alignment of atomic magnetic moments within the crystal lattice. This is shown in Fig.4-46 where, to illustrate the alignment of magnetic dipole moment, only Fe atoms are shown. Since FeO has the NaCl structure, Fe is located at octahedral sites within the FCC structure.

Fe atoms in the same (111) plane have the same direction of magnetic moments whereas the magnetic moments of atoms in the nearest (111) plane are of opposite direction. This alternating arrangement of magnetic moments along the <111> direction results in zero net magnetic moment for the FeO crystals, a

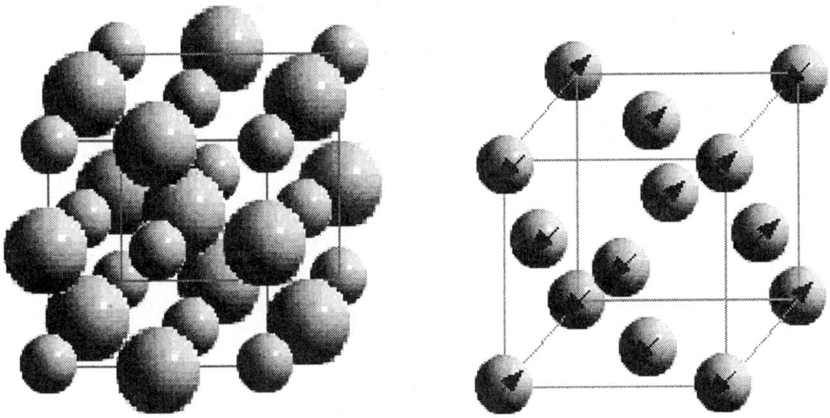

Figure 4-46 FeO structure and arrangement of magnetic moments of Fe atoms

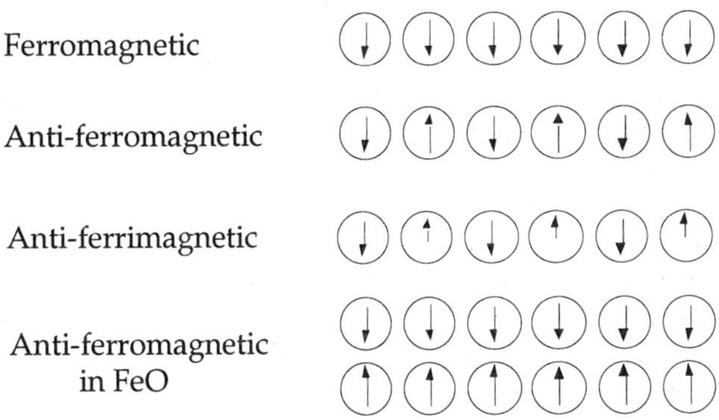

Figure 4-47 Magnetic moments from different atomic arrangement of magnetic dipoles

condition which is called antiferromagnetic. As shown in Fig.4-47, antiferromagnetic behavior is different from paramagnetic since the atomic magnetic dipole is not zero.

While the ideal FeO crystal is antiferromagnetic due to the opposite arrangement of Fe ions, experimental results show that $Fe_{1-x}O$ is also antiferromagnetic. $Fe_{1-x}O$ is an Fe-deficient structure that is formed under reducing conditions.[Roth,1960] As shown in Fig.4-45 for ionic crystals, cation vacancies are mainly formed rather than interstitial anions in the case of $Fe_{1-x}O$. With this defect structure, Fe_{1-x} is a better formula than FeO_{1+y} which could imply the presence of excess interstitial oxygens. Considering the purely geometric aspect of atomic magnetic moments, an Fe^{+2} vacancy would induce a ferrimagnetic moment since it destroys the magnetic moment balance between magnetic dipoles in opposing directions. However, the defects structure developed in iron deficient FeO resembles the structure of Fe_3O_4 (magnetite), in which the Fe ion takes a +3 charge. This increased magnetic moment compensates for the magnetic moment decrease caused by the Fe vacancy. Other dipole-dipole interaction mechanism between nearby Fe atoms also aid in compensating for the magnetic moment decrease from the Fe vacancy.

2. Substitution in crystals

Ions or atoms on regular lattice sites can be substituted for by "similar" ions or atoms of the correct size and charge. The limiting conditions for substitution to occur are :

i) Ion (or atom) sizes must be within ±15% of each other.
ii) Ionic charge differences should not be too great

In this respect Na^+ may substitute for Ca^{+2} but not for Y^{+3}. Likely also Ca^{2+} and Y^{3+} may substitute for Zr^{4+} but not Li^+ or Si^{4+}. Also substitution must be compatible with the existing coordination structure. In ZrO_2, Ca^{2+} substitutes for Zr^{4+} with coordination number 8. In the spinel structure, $MgFe_2O_4$, Fe^{2+} can substitute for Mg^{2+} with coordination number 6. In silicate minerals, Al^{3+} replaces Si^{4+} in tetrahedral sites that have coordination number 4. Si cannot substitute for Al with six coordination, since the radius of Si^{+4} is 0.41Å whereas Al^{+3} has an ionic radius of 0.53Å with coordination 6. With coordination number 4, Al^{+3} has ionic radius of 0.39Å. Hence it can readily locate in the tetrahedral sites of silicates where it does not change radius. However it can also fit into the octahedral sites within oxygen closest packing. In silicates, therefore, $Al^{3+} \rightarrow Si^{4+}$ but not Si^{4+} for Al^{3+}.

$$Al^{3+}\ (CN = 4)\ ;\ r_{Al^{3+}} = 0.39A°.\ \ r_{Al^{3+}}/r_{O^{2-}} = 0.28 \rightarrow CN = 4$$

$$Al^{3+}\ (CN = 6)\ ;\ r_{Al^{3+}} = 0.53A°.\ \ r_{Al^{3+}}/r_{O^{2-}} = 0.38 \rightarrow CN = 6$$

2-1. Structural change with substitution

If the atoms replacing and being replaced are very similar in size and charge, the resulting crystal structure is a solid solution in which the average concentration of replacing atoms in the crystal is the same as the chemical formula. Extrinsic atoms replacing similar atoms in a crystal lattice cause structural change if these are of different in size and charge. If different in charge, the substitution is accompanied by formation of defects, such as vacancies or interstitials, or charge difference must be compensated by other ions or by change in valence.

2-2. Isomorphous substitution

Isomorphs are a groups of materials having crystal structures with the same

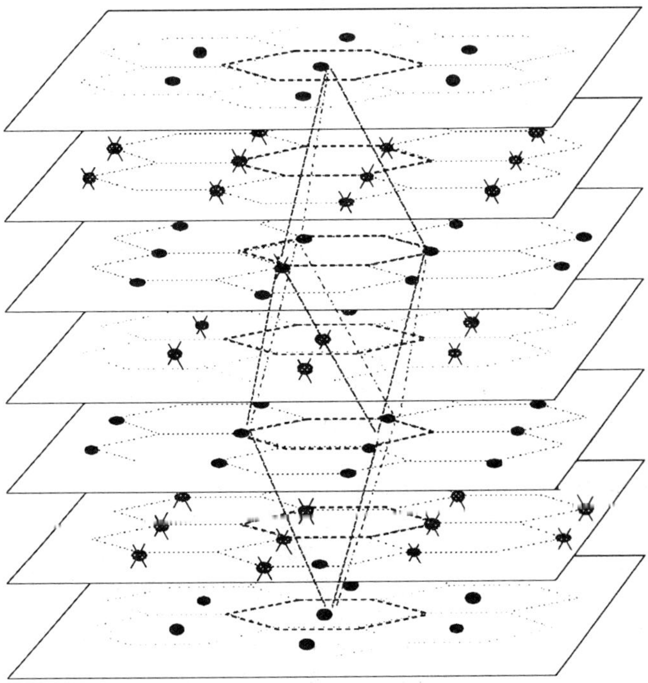

● Ca ✱ Mg

Figure 4-48 Arrangement of Ca and Mg in dolomite(Refer to Fig.4-18 for calcite)

symmetry which can be represented by the same crystallographic parameters with slight differences in magnitude. In this respect, $CaCO_3$ and $MgCO_3$ are only isomorphs. A particular mixture of $CaCO_3$ and $MgCO_3$ leads to dolomite, a mineral in which half of the calcium in calcite are replaced by magnesium. The crystal structure of dolomite is very similar to that of calcite but not isomorphous since a c-glide plane is not possible due to the alternating arrangement of Ca and Mg along c-axis. This is shown is Fig.4-48.

$$CaCO_3 \rightarrow \quad (Ca_{0.5}Mg_{0.5})CO_3 \rightarrow CaMg(CO_3)_2$$

Calcite→ Same structure → Dolomite

2-3. Other structural changes on substitution by extrinsic atoms

If a small portion of ions or atoms are replaced in isomorphous substitution, the cations or anions or atoms will be randomly distributed on the same sublattice forming a solid solution. This occurs without formation of a compound and there is little change in lattice parameter. Hence, there can be relatively limited solid solubility. This type isomorphous substitution can, however, cause significant change in physical properties, and is often used to accomplish such desired changes. The examples given below of $BaTiO_3$, $NiFe_2O_4$ and ZrO_2 illustrate how small substitution can be used to effect desired electric, magnetic, and mechanical property changes.

2-3-1 $BaTiO_3$

As discussed, $BaTiO_3$ has the perovskite structure to which $CaTiO_3$ and $SrTiO_3$ belong. $SrTiO_3$ is cubic while $CaTiO_3$ is orthorhombic in which TiO_6 octahedra are tilted from the ideal cubic structure. The Ti – O distances in $BaTiO_3$, $CaTiO_3$, and $SrTiO_3$ are 2.0, 1.92, and 1.95Å, respectively. The TiO_6 octahedron in $BaTiO_3$ is roomy for the Ti ion movement needed for phase transition while Ti is tightly bound within the TiO_6 octahedra in $CaTiO_3$ and $SrTiO_3$.

The differences between these crystal structures can be inferred from the differences in ionic radii of Ba^{+2}, Ca^{+2} and Sr^{+2} which are 1.35, 0.99, and 1.13Å, respectively. The Sr ion is the right size to fit into the center of the cube formed by eight TiO_6 octahedra while the Ca ion needs a little smaller space which is satisfied by tilting of the TiO_6 octahedra.

With respect to substitution, Sr easily replaces Ba due to similar size ionic radii whereas Ca is much smaller than Ba. Fig.4-49 shows the change in dielectric constant for a solid solution of $(Ba_{1-x}Sr_x)TiO_3$ with Sr. As shown previously, $BaTiO_3$ undergoes specified phase transitions with temperature. The dielectric constant also changes with temperature and has a maximum associated with the tetragonal/cubic transition. In terms of crystal structure, the solid solution phase on the right side of the peak is cubic and is tetragonal on the left side. Fig.4-49 illustrates also that an increase in Sr concentration shifts the phase transition point to lower temperatures. As mentioned previously the cubic phase is favored by $BaTiO_3$. This preference means that $BaTiO_3$ remains a cubic phase even when the transition point shifts to lower temperatures. The substitution of Pb for Ba results in a shift in the phase transition to higher temperature due to the large size of the Pb ion opposite direction. By substituting Sr or Pb for Ba, a desired dielectric constant magnitude and temperature profile can be obtained.

$BaTiO_3 \quad \rightarrow \quad (Ba_{1-x}Sr_x)TiO_3$ — A site substitution

If Ba is replace by La^{+3} which has the same size as Sr^{+2} but with +3 charges, there must be charge compensation. The charge difference is compensated by a change in the oxidation state of Ti from +4 to +3 on the B site. Hence substitution of La for Ba results in B site charge distribution.

$(Ba^{2+}{}_{1-x}La^{3+}{}_x)(Ti^{4+}{}_{1-x}Ti^{3+}{}_x)O_3$ — A and B sites

Similar to A site substitution, Ti ions can be replaced by ions of similar charge and size such as Sn.

$Ba(Ti_{1-x}Sn_x)O_3$ — B site

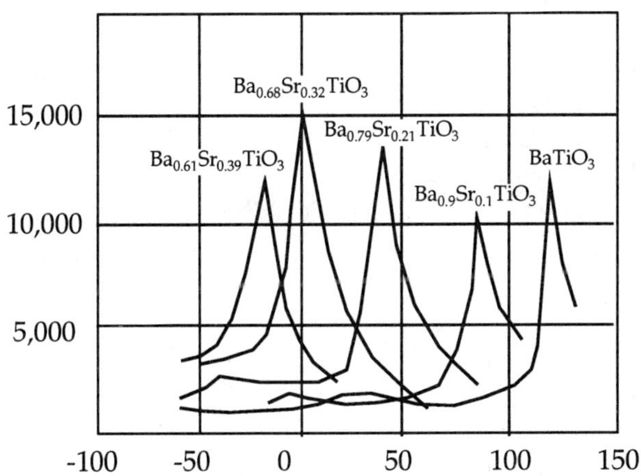

Figure 4-49 Effect of Sr addition on Dielectric constant change of $BaTiO_3$ with temperature

2-3-2 Substitution in magnetite (Fe_3O_4) structure

Magnetite has the inverse spinel structure in that Fe^{+2} occupies the half portion of octahedral sites available and Fe^{+3} occupies the other half. The remaining half of the Fe^{+3} ions occupy tetrahedral sites. The chemical formula for the compound can be represented as Fe^{+3} [Fe^{+2} Fe^{+3}] O_4 where the bracket designates octahedral sites. Since, in the spinel ferrite structure, octahedral and tetrahedral sites are anti-ferromagnetically coupled, the total magnetic moment is expressed as

$$\mu = \mu_{octa} - \mu_{tetra}$$

where μ is the magnetic moment of the metallic atom. The non-paired d-electron structure of Fe^{+2} is 4 and that of Fe^{+3} is 5. The net dipole moment of magnetite is $4\mu_B$(Bohr magnetron) assuming each non-paired d-electron has a magnetic moment of μ_B.

$$\mu_{Fe_3O_4} = \mu_{Fe+3} + \mu_{Fe+2} - \mu_{Fe+3} = 4\ \mu_B$$

Basically, octahedral sites are occupied by metallic ions with +3 charges and tetrahedral sites are occupied by ions with +2 charges. However, Fe^{+2}, Ni^{+2}, Cu^{+2} and Mn^{+2} prefer the octahedral sites, resulting in an inverse spinel structure. Zn^{+2}, however, prefers the tetrahedral site and its ferrite shows a normal spinel structure. The substitution of other metal atoms for Fe atoms in ferrites results in a change in magnetic properties due to the preference of different metal ions for either tetrahedral or octahedral sites. This causes a change in the magnitude of the anti-ferromagnetic interaction between the octahedral and tetrahedral sites. For example, if Fe^{+2} is replace by Mn^{+2}, the resulting structure is 20% inverse spinel structure.

$$Fe^{+3}\ [\ Fe^{+2}\ Fe^{+3}\ ]\ O_4 \rightarrow Mn^{+2}{}_{0.8}\ Fe^{+3}{}_{0.2}\ [\ Mn^{+2}{}_{0.2}\ Fe^{+3}{}_{0.8}\ Fe^{+3}\ ]\ O_4$$

$$\begin{aligned}\mu &= \mu_{octa} - \mu_{tetra} \\ &= 0.2\ \mu_{Mn+2} + 0.8\ \mu_{Fe+2} + \mu_{Fe+3} - 0.8\ \mu_{Mn+2} - 0.2\ \mu_{Fe+3} \\ &= 5.2\ \mu_B\end{aligned}$$

The result of the substitution of Mn^{+3} for Fe^{+3} is to increase the magnetic moment since 20% of Mn^{+2} replaces Fe^{+2} in octahedral sites. This is similar to the effect of substitution of other transition metal atoms for Fe in magnetite.

2-3-3 Substitution in ZrO_2

ZrO_2 is an excellent refractory and mechanically strong oxide material, and it is also chemical inert even at high temperatures. A major problem in the application of ZrO_2 as a engineering materials, however, is the difficulty in processing due to the volume increase during transformation from tetragonal to the monoclinic phase. This is detrimental to the structural integrity of the material. The tetragonal - monoclinic transition is so rapid that it is impossible to prevent the volume change by controlling the cooling rate. Fig.4-50 illustrates the detrimental effect of the volume change on the sintering process especially cooling of the sintered ZrO_2 body.

The sample on the left is a stabilized ZrO_2 with TiO_2 and on the right is pure ZrO_2. By adding TiO_2, ZrO_2 does not show a phase transformation and remains cubic even at room temperature. Here stabilization means that the cubic phase is stable even at room temperature. Common compounds that are added to stabilize the cubic phase of ZrO_2 are CaO and Y_2O_3. The concentration limit for the stabilization of the cubic phase of ZrO_2 are 14 – 20 and 7 - 50%, respectively. The radius of the Ca and Y^{+3} ions are 0.99 and 0.92 while that for Zr^{+4} is 0.80. In terms of charge and size, therefore, Y^{+3} is more similar to Zr but it can be easily replaced by both Ca or even Mg or Ce. This makes for a wide compositional range of possible substitution.

Like other substitution by foreign atoms with different charges, charge deficiency must be compensated if Y or Ca replaces Zr. As illustrated in Fig.4-51, there are two ways to compensate for this charge difference, namely:

i) Y goes to Zr positions in the ZrO_2 crystal lattice and a vacancy of oxygen is formed

$$ZrO_2 \rightarrow (Zr^{4+}{}_{1-x}Y^{3+}{}_{x})O_{2-0.5x} + 0.5xV_O^{-2}$$

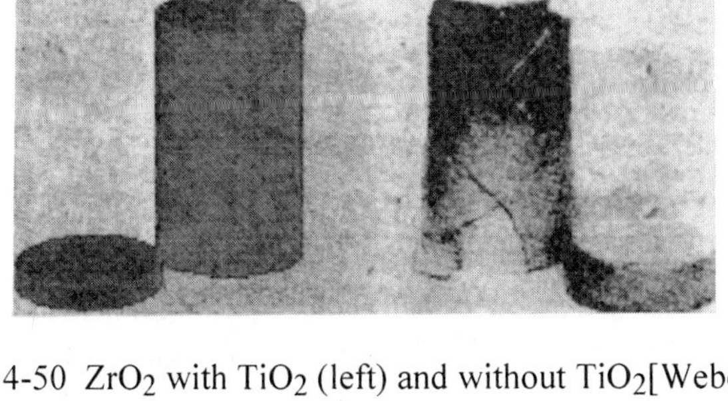

Figure 4-50 ZrO_2 with TiO_2 (left) and without TiO_2[Weber,1956]

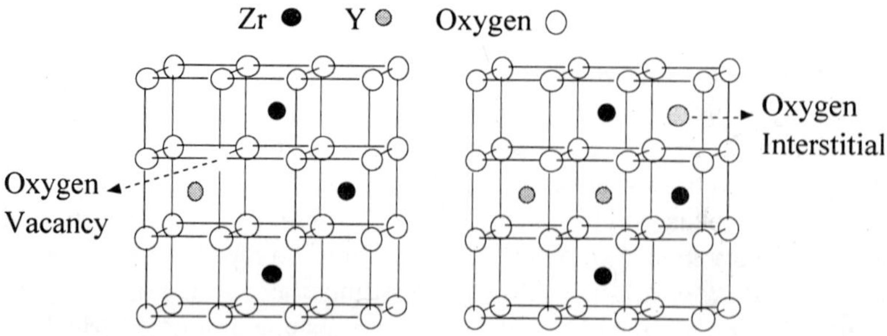

Figure 4-51 Mechanism of defects formation in Y substituted ZrO_2

ii) Y occupies both Zr position and empty interstitial sites at the center of the O_8 cube are occupied by both Y and oxygen

$$ZrO_2 \rightarrow (Zr^{4+}{}_{1-x}Y^{3+}{}_x)O_2 + xY_i + xO_i$$

Experimental data from X-ray diffraction analysis of the unit cell volume as well as pycnometric density data confirms the first mechanism of Y substitution for Zr in ZrO_2.

2-3-4 Substitution in silicates

Similar to $AlPO_4$, substitution in silicates can be summarized primarily as replacement of Si^{+4} by Al^{+3}. In a broad sense, the concept can be expanded to include the structural changes to the SiO_2 unit which take place with addition of other cations. The main building unit in silicates is the SiO_4 tetrahedron. Most substitution in silicates is carried out by replacing Si^{+4} by Al^{+3} thereby changing the configuration of the SiO_4 tetrahedral skeleton. Unlike the solid solution in substituted materials, substitution in silicates has defined rules such that there is a constant ratio between Si and the substituting elements. In orthoclase and albite, one quarter of Si is replaced by Al and the charge deficiency caused by this substitution is compensated by alkali metals such as K and Na. In anorthite, half of the Si is replaced by Al and charge is balanced by introducing Ca. In these materials, the number of charge compensating cations (Ca, Na, and K) is a quarter of the number of Si and Al ions. These materials are in the group of feldspar whose silica framework is based on coesite. In coecite, the number of possible sites for cations such as Ca, K and Na in feldspars is one quarter of the tetrahedral sites for Si and Al.

$$SiO_2(\text{Coesite}) \rightarrow K^+{}_{1/4}(Si^{4+}{}_{3/4}Al^{3+}{}_{1/4})O_2 \rightarrow KAlSi_3O_8(\text{Orthoclase})$$

$$\rightarrow \quad Na^{+}_{1/4}(Si^{4+}_{3/4}Al^{3+}_{1/4})O_2 \rightarrow NaAlSi_3O_8\text{(Albite)}$$
$$\rightarrow \quad Ca^{+}_{1/4}(Si^{4+}_{1/2}Al^{3+}_{1/2})O_2 \rightarrow CaAl_2Si_2O_8\text{(Anorthite)}$$

For nepheline, the ratio of K and Na ions is one half that of Al and Si. This violates the rule for the structure of feldspar which should have the ratio one quarter. Hence, nepheline cannot have the feldspar structure. Rather it is based on the tridymite structure in which half the Si is replace by Al with K and Na occupying interstitial sites between SiO_4 tetrahedra.

$$\underset{\text{(Tridymite)}}{SiO_2} \rightarrow \underset{\text{(Nepheline)}}{K^{+}_{1/8}Na^{+}_{3/8}(Si^{4+}_{1/2}Al^{3+}_{1/2})O_2} \rightarrow KNa_3Al_4Si_4O_8$$

In spodumene, the ratio of charge compensating cations to Si and Al is one third. Therefore it does not adopt the crystal structure of feldspar or nephelene. In spodumene, Si is not replaced by Al. Rather, Al and Li are arranged parallel to the chain of SiO_4 tetrahedra. Hence the first chemical formula shown below is somewhat in error since Al is not located in Si sites. Rather the chemical formula for spodumene must be expressed by the second formula.

$$SiO_2 \rightarrow Li^{+}_{1/3}(Si^{4+}_{2/3}Al^{3+}_{1/3})O_2$$
$$SiO_2 \rightarrow Li^{+}Al^{3+}(Si_2O_6) \rightarrow LiAlSi_2O_6\text{(Spodumene)}$$

The criteria for the structure of spodumene is that the number of Si ion is equal to the number of other cations. The actual crystal structure illustrates why the criteria mentioned above work for the feldspar group, nephelene and spodumene.

2-4. Hybrid substitution

Aluminum phosphate is a good example of hybrid substitution. Half of the Si ions in cristobalite are replaced by P^{+5} and the other half by Al^{+3} ions. Overall charge is balanced by replacing two of Si^{+4} ions by P^{+5} and Al^{+3}.

$$\underset{\text{Cristobalite}}{2(SiO_2)} \rightarrow AlPO_4$$

Since the framework of SiO_2 is not destroyed, $AlPO_4$ shows the same transition as the $\alpha \rightarrow \beta$ cristobalite transformation. Hence the high temperature form of $AlPO_4$ looks like β-cristobalite and the low temperature form like α-cristobalite. The low temperature form of $AlPO_4$ is shown in Fig.4-52. AlO_4

Figure 4-52 View from 4-fold axis (a) b-Cristobalite and (b) low temperature form of $AlPO_4$

tetrahedra are connected to 4 of the PO_4 tetrahedra and Si is substituted alternately by Al and P. The average distance of Si - O in low cristobalite is 1.59A°. The average distances of Al – O and P – O in the low temperature form of $AlPO_4$ are 1.70 and 1.56A°, respectively. Although the silica framework does not change, the tetrahedra undergo changes in volume and shape due to differences in the charge and radius of the Si^{+4}, Al^{+3} and P^{+5} ions.

2-5. Omission substitution in pure substances

This type substitution occurs when an ion or atom is missing from a host lattice for charge compensation purposes. This requires a change in valence state for some other ions in the host lattice . Because of the requirement for charge balance, these structures occur only where variable valence cations are present, such as in FeO. Under oxidative conditions, some Fe^{+2} atoms of FeO are converted to Fe^{+3} resulting in Fe vacancies.

$$FeO \rightarrow Fe^{2+}{}_{1-y}\, Fe^{3+}{}_{y}\, O_{1+0.5y} \rightarrow Fe^{2+}{}_{1-x}\, Fe^{3+}{}_{0.67x}\, O + 0.33x\, V_{Fe^{2+}}$$

Hence, this is a vacancy substitution on lattice atomic position

2-6. Substitutional Order-Disorder transformation

In a complex compound, there typically exists discrete sets of equivalent positions in the crystal lattice that are occupied by the same or similar kinds of atoms. If the sites areoccupied by atoms that have similar size and charge atoms. If the sites areoccupied by atoms that have similar size and charge characteristics, but are different chemically, they can readily switch positions. At low temperature, however, there will be little switching or mixing due to the activation barrier. This activation barrier is overcome at high temperature and switching of dissimilar ions or atoms readily occurs between the equivalent lattice sites. $MgFe_2O_3$ is an example of this order-disorder transformation.

$$\underbrace{Fe^{3+}}_{A}\ \underbrace{(Mg^{2+}Fe^{3+})}_{B}\ O^4 \leftrightarrow \underbrace{Fe^{3+}{}_{1-x}Mg_x}_{A}\ \underbrace{(Mg_{1-x}Fe^{3+}{}_{1+x})}_{B}\ O_4$$

ordered state disordered state

At low temperature Mg occupies the octahedral sites designated as B. At high temperature Mg^{2+} can occupy either of the octahedral or tetrahedral sites designated as A.

SECTION 3 CRYSTAL STRUCTURES OF THE SILICATES

Silicates are the most abundant materials in the earth's crust. Silicate minerals form the basis for most traditional ceramic raw materials, including whitewares, glass, and refractories. The structure and chemistry of silicates are therefore of great importance for these uses. Silicates are also of interest with respect to their crystal structures since they exhibit the most varied structures among materials. This is partially illustrated in the polymorphism of silica (SiO_2). Typical polymorphs of silica(SiO_2) are cristobalite and tridymite. The former has the zincblende structure and the latter has the wurtzite structure. The density of cristobalite is slightly higher than that of tridymite. Silica is the basic structure from which other silicate minerals are developed. Simply stated, the silicate minerals are modifications of silica with substitution of Si by other elements, along with the polymorphism in each silicate mineral.

A. Classification of silicate

In the silicate minerals, the fundamental unit of crystal structure building is the $[SiO_4]^{4-}$ polyhedral as shown in Fig.4-53.

The coordination number of the Si^{4+} ion is 4. In silicates, SiO_4 tetrahedra are linked together by corner sharing not by edge or face sharing. Pauling's third rule can provide the basics for understanding the characteristic corner sharing of SiO_4 tetrahedra in silicate structures. This corner sharing of SiO_4 can be compared to TiO_6 and $CsCl_8$ units which are connected by edge sharing and face sharing, respectively, as illustrated in Fig.4-54. The characteristics of these polyhedral connections are listed in Table 4-3. Silicate

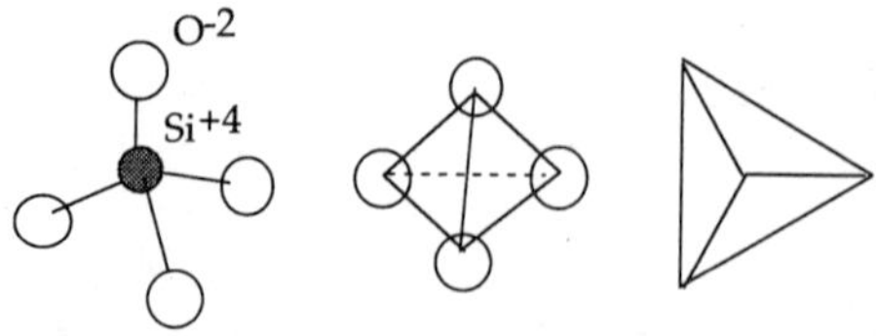

Figure 4-53 Structure of $[SiO_4]^{4-}$ polyhedral and its 2-dimensional representation

structures are, therefore, largely determined by how the tetrahedral units are connected, especially when other cations are mixed with Si in the silicates.

While Pauling's third rule is concerned with one component systems, Pauling's fourth and fifth rules are concerned with the multicomponent system. Hence, the fourth and fifth rules seem more relevant than the third rule to understanding of silicate structures, although the third rule provides the basis for the corner sharing of SiO_4 units.

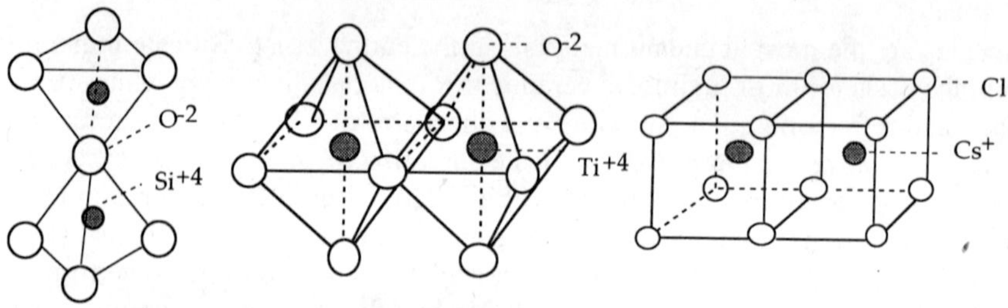

Figure 4-54 The arrangements of polyhedra in (a) SiO2 - corner sharing of tetrahedra $[SiO_4]$, (b) TiO_2 - face sharing of octahedral $[TiO_6]$ and (c) CsCl - face sharing of $[CsCl_8]$

Table 4-3 Characteristics of polyhedra connection in SiO_2, TiO_2, CsCl

Sharing mode	Corner	Edge	Face
Crystal structure	SiO_2	TiO_2*	CsCl
CN of cation	$CN(Si^{4+}) = 4$	$CN(Ti^{4+}) = 6$	$CN(Cs^{+}) = 8$
Ratio of radii of cation and anion	$\frac{r_{Si^{+4}}}{r_{O^{-2}}} = 0.29$	$\frac{r_{Ti^{+4}}}{r_{O^{-2}}} = 0.48$	$\frac{r_{Cs^{+}}}{r_{Cl^{-}}} = 0.89$
Cationic radius	$r_c = 0.41Å$	$r_{Ti}{}^{4+} = 0.68Å$	$r_{Cs}{}^{+} = 1.69$
Bond strength	EBS = 1.0	EBS = 2/3	EBS = 1/8

* Anatase and brookite (Rutile has edge and corner sharing.)

Pauling's fourth rule states that

In a crystal structure containing different cations, those of high valency and small coordination number tend not to share polyhedron elements with each other.

And the fifth rule states that

The number of essentially different kinds of constituents in a crystal tend to be small.

In a strict sense, the fourth and fifth rules are merely extentions of the third rule. Hence, in a crystal with different polyhedral units, those with high charge and low CNs, i.e., poorly shielded polyhedra, tend not to be associated with each other. The number of different polyhedral units in a crystal tends to be small, i.e., similar ions in a structure will be similarly coordinated. By applying Pauling's rules, schematic illustration of silicates structures are possible with other factors such as composition, charge balance, size effect and geometric relationships taken into account.

1. Basic patterns in silicates

Silicates are variations of the silica frame structures caused by the addition of modifier cations. Therefore, they are generally classified according to the way in which the SiO_2 tetrahedra are arranged. Fig.4-55 illustrates these tetrahedral arrangements, resulting in would be island, chain, sheet or 3-dimensional framed structures.

In the island silicate structures, all the tetrahedra are seperated from each other and the oxygen anions of the tetrahedra are connected not to silicon but to other modifier cations. Hence the –4 charge on a SiO_4 tetrahedron in the

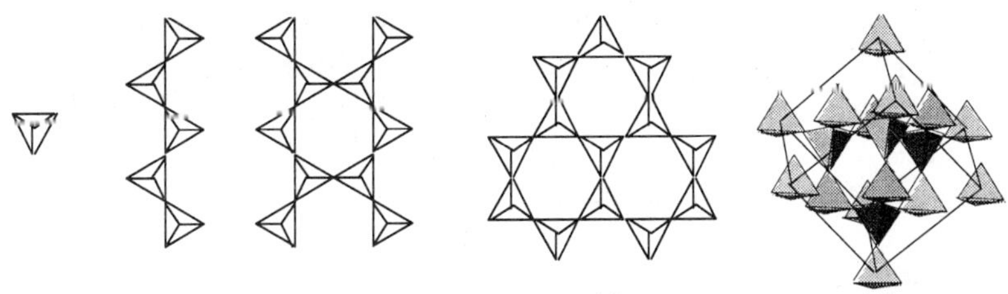

(a) (b) (c) (d)

Figure 4-55 Arrangements of silica tetrahedron in (a) island, (b) single chain, (b) double chain (c) Sheet and (d) frame structures island silicates must be balanced by positive charges from these modifier cations. If the charge on the cation is 'm', the number of modifier cation per Si atom must be 4/m since each Si cation develops –4 charge around the tetrahedron. A general formula for the island silicate with two components is given by $X^{+m}{}_{4/m}\ Si\ O_4$. Typical island silicates are phenacite(Be_2SiO_4) and Zircon($ZrSiO_4$). In the case of phenacite, 'm' in the general formula is two. Hence there must be two Be atoms per one Si atom. As for zircon, 'm' is 4 and the number of Zr atoms per Si atom must be one. Hence the chemical formula, Be_2SiO_4 and $ZrSiO_4$ for phenacite and zircon, respectively, agree with the general formula for the island structures. More generally, the condition of island silicates can be stated as

$$x \times y \geq 4z \text{ in general formula of } R^{+x}{}_y\, Si_zO_n \tag{2}$$

$$\text{or} \quad \frac{\text{Total charges on the other cations}}{\text{the number of Si}} \geq 4 \tag{3}$$

Here x and y represent the charge on the modifier cation and the number of the cation in the formula, respectively. And z represents the number of silicon in the chemical formula. In a simpler form, Eq.(3) becomes

$$\frac{\sum\limits_{R \neq Si} \sigma_R}{N_{Si}} \geq 4 \tag{4}$$

For the single chain silicates, each tetrahedron is connected two adjacent tetrahedra. Due to these connections, each tetrahedron develops –2 charges which must be balanced by positive charges provided from the modifier cations. If the charge on the cation is 'm', the number of modifier cation per Si atom in the single chain silicates must be 2/m since each Si cation develops –2 charge around the tetrahedron. Hence the general formula for two-component single chain silicates is given by $X^{+m}{}_{2/m}\ Si\ O_3$. Typical examples of single chain silicates are enstatite($MgSiO_3$) and Orthoferrosilicate($FeSiO_3$). Therefore the condition of single chain silicate formation can be stated as

$$x \times y = 2z \text{ in the general formula of } R^{+x}{}_y\, Si_zO_n \tag{5}$$

or, the condition expressed for single chain silicates in the form of Eq.(4) is

$$\frac{\sum\limits_{R \neq Si} \sigma_R}{N_{Si}} = 2 \tag{6}$$

Here the summation extends over all cations except Si^{+4} and N_{Si} is the number of Si in the chemical formula. Likewise the correlation between compositions and silicate types can be derived for double chain, sheet and framed silicates.

$$\frac{\sum_{R \neq Si} \sigma_R}{N_{Si}} = 1.5 \qquad \text{for double chain silicates} \qquad (7)$$

$$\frac{\sum_{R \neq Si} \sigma_R}{N_{Si}} = 1.0 \qquad \text{for layered silicates} \qquad (8)$$

$$\frac{\sum_{R \neq Si} \sigma_R}{N_{Si}} < 1.0 \qquad \text{for framed silicates} \qquad (9)$$

More detailed descriptions for the correlation between chemical composition and silicate types will be given as each type silicate is described.

Other factors influencing the actual crystal structures of silicates are charge and CN of the modifier cations, bonding scheme of the O^{2-} ions to the modifier cations, and isomorphous substitution of cations or anions. The conditions for the isomorphous substitution are:

i) Ion (or atom) sizes must be within 15% of each other.
- Na^+ substitute for Ca^{2+}; not for Ba^{2+}

ii) Ionic charge differences should not be too great.
- Ca^{2+} for Zr^{4+} , Y^{3+} for Zr^{4+} , But not Li^+ for Zr^{4+} or Si^{4+}
Only very limited solubility possible.

iii) Substitution must be compatible with existing CN structure.
$Ca^{2+} \rightarrow Zr^{4+}$: CN = 8,
$Fe^{2+} \rightarrow Mg^{2+}$: CN = 6 ,
$Al^{3+} \rightarrow Si^{4+}$: CN = 4

Note that

Al^{3+} (CN = 4); $r_{Al^{3+}} = .39$Å $\quad \frac{r_{Al^{+3}}}{r_{O^{-2}}} = 0.28 \rightarrow CN = 4$

Al^{3+} (CN = 6); $r_{Al^{3+}} = .53$Å $\quad \frac{r_{Al^{+3}}}{r_{O^{-2}}} = 0.38 \rightarrow CN = 6$

Therefore $Al^{3+} \rightarrow Si^{4+}$

but never $Si^{4+} \rightarrow Al^{3+}$(CN=6) in silicate structures.

B. Types of Silicate Minerals

1. Island type silicates

1. Orthosilicates (Island type silicates)

The silicates in this group have isolated group structures that have no bridging oxygens. Therefore, [SiO_4] units are separated from one another and bonded by modifier cations. Olivine(Mg_2SiO_4) minerals are in this group. Mg may be substituted with Fe, Mn, and Ca. Orthosilicate minerals may be conveniently classified according to the charge on the modifier cations that connects the [SiO_4] units and by the complexity of the structure, viz.:

(i) Divalent cations (Be^{2+}, Zn^{2+}, Mg^{2+}, Fe^{2+}, Ca^{2+} . . .)
$R_2^{2+}SiO_4$ – phenacite, olivine, Ca orthosilicate minerals.

(ii) Trivalent cations (Al^{3+} . .)
$R_2^{3+}O^{2-} \cdot SiO_4$ – alumina silicates.

(iii) Quadri-valent cations (Zr^{4+}, Th^{4+} . .)
$R^{4+}SiO_4$ – Zircon minerals.

(iv) Complex:
$R_3^{2+}R_2^{3+}(SiO_4)_3$ – Garnets
$Al_2(F_2)SiO_4$ or $Al_2(OH,F)_2SiO_4$ – Topaz

In the cases of (i) and (iii), the general chemical composition given by $R^{+m}_{4/m}SiO_4$ agrees with actual minerals such as phenacite, olivine, and zircon. However, in the cases of (ii) and (iv), the number of modifier cations more than compensates for the negative charges on the $(SiO_4)^{-4}$ tetrahedra. In the case of (ii), some modifier cations are connected not only to $(SiO_4)^{-4}$ tetrahedra but also to other cation polyhedra, but the island configuration is still retained.

In case(iv), Topaz has OH^- and F^- as anions. Since Si exist in the form of $(SiO_4)^{-4}$, OH^- and F^- are connected to modifier cations. Hence the positive charges neutralized by these anions cannot contribute to compensating the negative charge on $(SiO_4)^{-4}$. Considering this reducing effect of anions on the net positive charge on modifier cations, Eq(2) is further modified as

$$x \times y - m \geq 4z \text{ in the general formula of } R^{+x}_y Si_z(OH)_m O_n \quad (10)$$

or

$$\frac{\sum_{R \neq Si} \sigma_R - N_{OH,F}}{N_{Si}} \geq 4 \qquad (11)$$

Here $N_{OH,F}$ is the number of $(OH)^-$ and F^- in the chemical formula. For Topaz, therefore, the equality Eq.(a1) holds. The same consideration of OH^- and F^- applies to other types of silicates. Eq.(5), (6) and (7) become the following

	Single chain type	Double chain type	Layered type	Framed type
$\frac{\sum_{R \neq Si} \sigma_R - N_{OH,F}}{N_{Si}}$	2	1.5	1	<1

1-1-1 Divalent Group : $R_2^{+2}SiO_4$

(i). Phenacite - $[Be_2SiO_4]_6$

Phenacite is an important Be mineral. The basic unit in phenacite can be considered as a combination of one SiO_4 and two BeO_4 tetrahedra which forms a 3-wing-fan. This is shown in Fig.4-56(I) and (II). Coordination numbers for Be and Si are 4. These three tetrahedra are connected to one oxygen. Hence each oxygen is connected to two Be and one Si. The EBS of O is given as

$$+1(\text{Si–O}) + 0.5\ (\text{Be-O}) + 0.5\ (\text{Be-O}) = 2$$

which is equal to the negative charge on the oxygen anion.

This 3-wing-fan basic unit repeats along one direction resulting in the column shown in Fig.4-56(I) and (II). If each basic unit of 3 tetrahedra is called **a**, **b** and **c**, depending on the relative position of SiO_4 tetrahedra, **a** represents a basic uint in which SiO4 is located in the 3 o'clock direction while **b** and **c** indicate the position of SiO4 tetrahedra in the 7 and 11 o'clock positions, respectively. With this nomenclature, the columns of the basic units can be named in the same way. Hence column **a** means that the basic unit type **a** is on the top.

Since the relative position of SiO_4 tetrahedra in the basic unit rotates by 120° around the column axis along this column, the sequnce of basic units along the column can be –abcabc– or –cbacba–. To distinguish these two types of rotational modes, columns with right handed rotation are called **a**, **b**, or **c** column while columns with left handed rotation are called by **a**', **b**', and **c**' columns. Regarding the connection mode between columnns, columns **a**, **b**, and **c** are connected by another column of basic units designated **a**', **b**' and **c**' in Fig.4-56(IV). These columns **a**', **b**' and **c**' are shifted by half the length of the tetrahedra in **a**, **b**, and **c** along the direction of the column. While the column of basic units forms a 3 fold axis, the combination of 6 columns shown in Fig.4-56(III) makes a 6 fold screw axis.

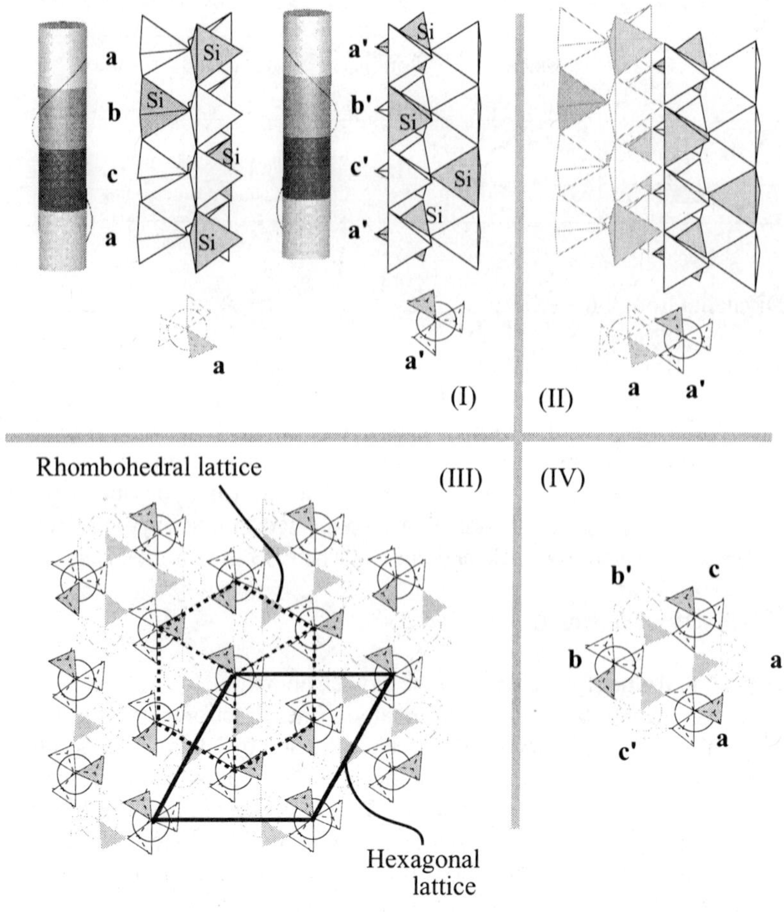

Figure 4-56 (I) Basic units consisting of SiO_4 and BeO_4 in phenacite and
(II) connection configuration between columns
(III) and (IV) views along the columns with schematic notation

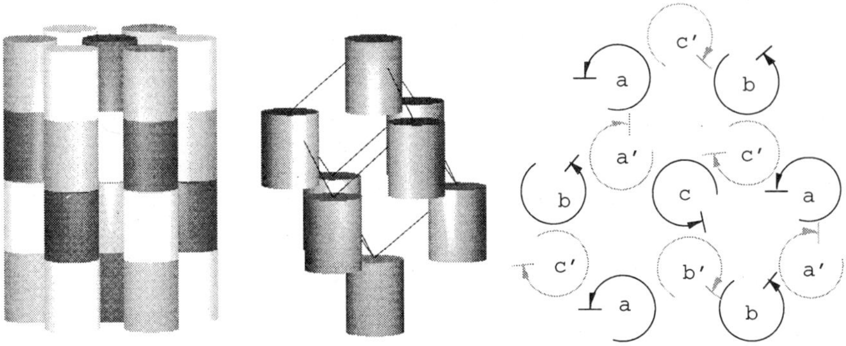

(I) (II) (III)

Figure 4-57 (I) Crystal structure of phenacite with right handed columns only (II) representation of rhombohedra lattice with only SiO_4 tetrahedra and (III) schematic representation by circular arrow notation for the columns (SiO_4 tetrahedra in black and BeO_4 in gray and white color)

The crsytal lattice of phenacite is rhombohedral. The unit cell of the rhombohedral lattice of phenacite is shown in Fig.4-57(I) and (II) with only right handed basic units. Hence only columns are illustrated in Fig.4-57(I) while the left handed columns are omitted for better illustration of the rhombohedral lattice. The whole crystal structure of phenacite can be described by repetition of six basic units along the 3 fold screw axis as shown schematically in Fig.4-57(III). Bold dotted lines in Fig.4-56(III) represent the rhombohedral lattice while the bold lines designate the corresponding hexagonal lattice in the phenacite crystal structure. In this hexagonal lattice notation, the cell dimension of phenacite is a_o = 12.42Å and c_o = 8.24Å. There are 18 formular units (Be_2SiO_4) in the unit cell of this hexagonal lattice since the hexagonal lattice indicated by the bold line includes 4 formula units and the c-axis extends three tetrahedral layers. The space group for phenacite is $R\overline{3}$. This structure occurs for small cations like Be^{+2} with slightly covalent bonding.

ii) Willemite

Willemite with structural formula, $[Zn_2SiO_4]_6$, is similar to phenacite and occurs as flat shaped crystals in Zn-containing enamels and glazes. The ionic radius of Zn^{+2} is 0.74Å(CN=6). Even after the effect of 4 coordination is considered, the radius seems to be large compared to Be^{+2}(0.31Å). This illustrates the covalent

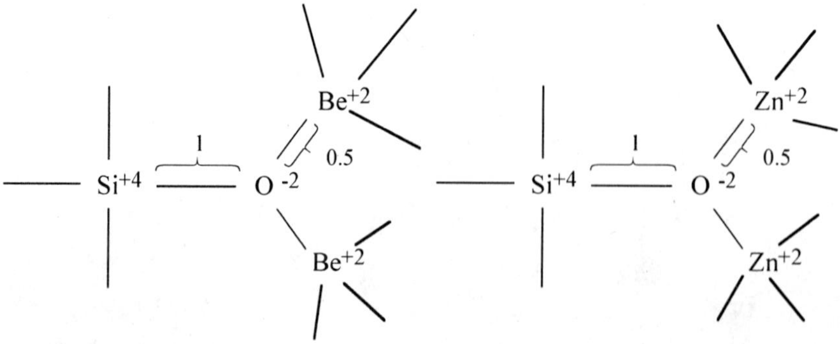

Figure 4-58 Bonding scheme for O^{2-} ions in phenacite and willemite in terms of EBS.

character of Be–O and Zn–O in this structure. Hence, this structure occurs for the more covalent oxides. The cell dimension of willemite is a_o = 13.93Å and c_o = 9.33Å, which is a little larger than that of phenacite, reflecting the larger ionic radii of Zn^{+2}. With respect to EBS for oxygen anions in phenacite and willemite, EBS of Si–O for both phenacite and willemite is 1.0 since they have the same crystal structure. Also, EBS of Zn–O or Be–O is 0.5 as shown in Fig.4-58.

(b) Olivine

Olivine[$(Mg_{1-x}Fe_x)SiO_4$] is the mineral group representing the solid solution of Fayalite($Fe_2^{+2}SiO_4$) and forsterite($Mg_2^{+2}SiO_4$). The chemical formula for this group is given by $R_2^{+2}SiO_4$ which agrees with Eq.(a1) for the island type silicates. Also monticellite($MgCaSiO_4$) is in this group. In the crystal structure of olivine, oxygens take the hcp structure. The crystal structure of olivine(monticellite) is shown in Fig.4-59 where Ca^{2+} and Mg^{2+} ions occupy half of the octahedral sites and Si^{+4} ions occupy one eighth the tetrahedral sites. For illustrative purposes, the regular structure of octahedra are shown, although in the actual crystal structure these octahedra are distorted as in brookite. Ca^{2+} cations take the outside octahedral sites(M2 type) in the octahedra chain while Mg^{2+} cations take the center positions(M1 type). Connection between tetrahedra and octahedra is established through the basal plane of a tetrahedra in the upper polyhedra layer, which sits on the triangle formed by three octahedra as indicated 'B' in Fig.4-59. The other type tetrahedra with inverse shape, points to a common point of three octahedra in the lower layer, indicated by 'A' in Fig.4-59.

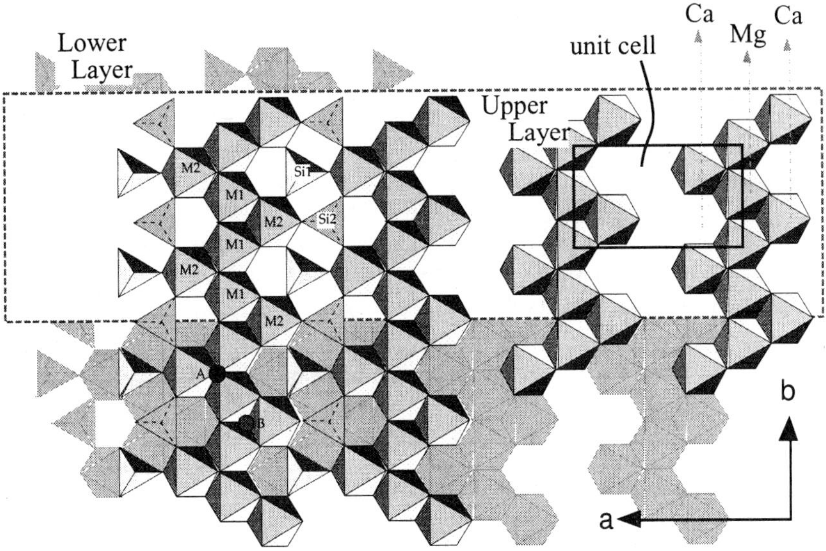

Figure 4-59 Crystal structure of monticellite($MgCaSiO_4$)

Fayalite and olivine are important iron bearing minerals. Forsterite - $[Mg_2SiO_4]_4$ is a refractory mineral and an important source of MgO. The space group for the olivine group is Pnma. Hence it is an orthorhombic lattice. The rectangle in Fig.4-59 contains 4 octahedra which correspond to two formula units. Since the a-axis extends two octahedral layers, there are four formular units in the unit cell. Also, it is easily seen that the c-axis is perpendicular to the oxygen layer while axes b and c are parallel to this layer. The unit cell dimensions for forsterite and fayalite are given as :

a =10.20Å, b = 5.98Å, c = 4.76Å, for forsterite
a =10.48Å, b = 6.11Å, c = 4.82Å, for fayalite

1-1-2 Trivalent Group

(i) Al-Silicates

The chemical formula for this group is expressed as $Al_2O{\bullet}SiO_4$ or Al_2SiO_5. Al^{3+} ions may be in 6, 5 or 4 CN. The three known mineral types are Kyanite, Andalusite, and Sillimanite which all have the same chemical formula- $Al_2O{\bullet}SiO_4$. Among these, kyanite has the highest density and sillimanite the lowest. This can be understood from the phase diagrm in Fig.4-60, which shows that kyanite is formed at high pressure while sillimanite is formed at high temperature. Characteristics of these polymorphs are listed in Table 4-4. Compared to olivine, aluminum silicates have cations which are more than sufficient to balance the negative charges on the SiO_4 tetrahedra. Hence, not all O^{2-} ions are bonded to Si^{4+} ions and AlO_6 octahedra are formed in a chain structure with SiO_4 tetrahedra in between the chains of AlO_6 octahedra.

Alumina silicates are important refractory minerals. A large volume change accompanies the transition of kyanite to mullite which on heat treatment may cause cracking. Therefore, kyanite must be precalcined to convert it to one of the other forms prior to its use as a refractory. Mullite has the chemical formula of $3Al_2O_3{\bullet}2SiO_2$ and density of $\rho = 3.03$ g/cm^3. Its structure is close to that of sillimanite, but has not been accurately determined due to variability with composition and pressure. It is a high temperature product of all three minerals, and typically does not exist in nature.

Table 4-4 Comparison of three alumina silicates in orthosilicates group

Kyanite	Andalusite	Sillimanite
Al^3 : CN = 6	Al^{3+} : CN = 5, 6	Al^{3+} : CN = 4, 6
Trinclinic	Orthorhombic	Orthorhombic
$P\bar{1}$; Z = 4	Pnnm; Z = 9	Pbnm; Z = 4
ccp O^{2-} ions	Not close packed	Not close packed
$\rho = 3.63$ g/cm^3	$\rho = 3.15$ g/cm^3	$\rho = 3.08$ g/cm^3

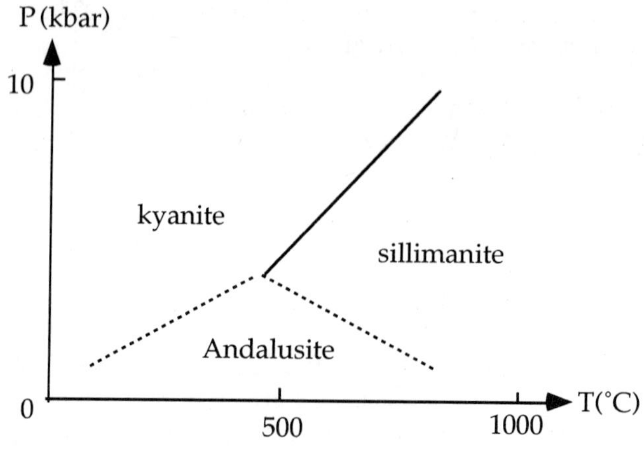

Figure 4-60 Phase diagram for alumina silicates in orthosilicate group

(ii) Kyanite

In kyanite, oxygen anions are in a CCP structure with Al^{+3} and Si^{+4} ions occupying the octahedral and tetrahedral sites, respectively, as are shown in Fig.4-61(I) and (II). In type A, one fifth of the octahedral sites are occupied by Al^{+3} and a quarter of the tetrahedral sites are occupied by Si^{+4}. Si^{+4}s do not share any oxygens. Hence kyanite is a island type silicate. In type B, three

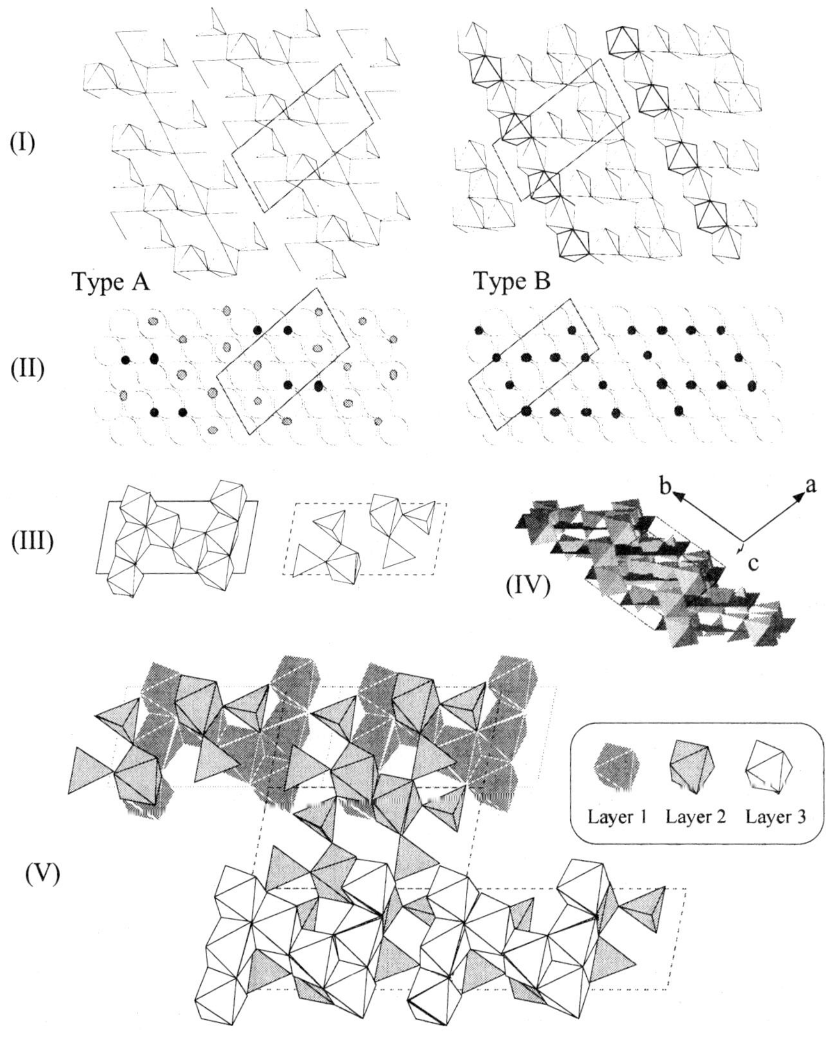

Figure 4-61 Crystal structure of kyanite

fifths of the octahedral sites are occupied by Al^{+3}, while Si^{+4} ions do not occupy any tetrahedral sites in type B. The crystal structure is a connection of polyhedral layers of type A and B in the sequence of –ABABAB–. The unit cell of kyanite is shown in Fig.4-61(IV) and the connections between polyhedral layers are illustrated in Fig.4-61(V) with 3 layers. The parellelogram in Fig.4-61(I), (II) and (III) represents the repreating unit in each layer. The short edge of this parallelogram corresponds to the c-axis of the crystal lattice of kyanite. The lattice vector a and b are the repeating directions of the basic planar unit represented by the parallelogram. Although oxygens have the CCP structure, the lattice generated is triclinic due to the inclined repeating directions from the basal plane of the oxygen packing. The inversion center can be easily identified in each of the basic planar units in Fig.4-61(III) which is located at the center of the plane. Inversion centers exists at lattice points, at the middle of the unit cell edge, at the face-centers and at the body-centers of the unit cell. The space group for kyanite crystal is $P\bar{1}$ and its cell dimensions are given as:

$$a = 7.12Å,\ b = 7.84Å\ \text{ and } c = 5.57Å$$
$$\alpha = 89.98^\circ\ \ \beta = 101.12^\circ\ \text{ and } \gamma = 106.01^\circ$$

There are three different types of octahedra, one in type A and two in type B. Since their enviroments are different, they are not symmetrically equivalent. With respect to the shape of the octahedra in kyanite, it is distorted from regular as in brookite and olivine. Regarding the arragement of tetrahedra, the configuration in type A along the c-axis, which is the short edge of the parallelogram, is similar to that of olivine. However, due to the smaller ratio of Si/O in kyanite, less than that of olivine(0.25), it is more crowded in the type A layer. Also no SiO_4 tetrahedra exists in the type B layer, resulting in the triclinic structure.

(iii) Silimanite

In kyanite, all the Al cations occupy octahedral sites. In constrast, half of Al take tetrahedral sites in silimanite. In a simple way, the crystal structure of sillimanite can be depicted as in Fig.4-62. Octahedral chains run along the edge and center of the unit cell and tetrahedral chains run in the same direction and connect to the octahedral chains. In the tetrahedral chains, SiO_4 and AlO_4 alternate. In order to show the connection mode between octahedral and tetrahedral chains, some tetra and octahedra are omitted in Fig.4-63(a). A cross section of the octahedra and tetrahedra chains is shown in Fig.4-63(b). The space group of sillimanite is Pbnm. Its lattice type is orthorhombic. The unit cell dimension is given as

$$a = 7.49Å\ \ b = 7.67Å\ \text{ and }\ c = 5.77Å$$

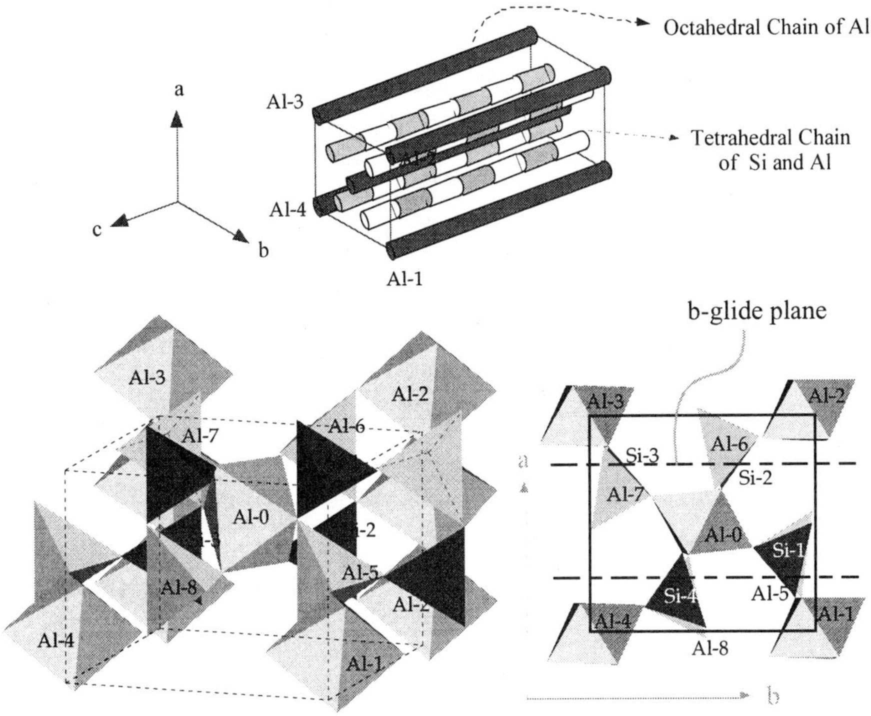

Figure 4-62 Crystal structure of Sillimanite

(iv) Andalusite

In andalusite, Al^{+3} takes coordination numbers of 6 and 5. With coordination number 5, Al^{+3} forms a bipyramid of AlO_5. Due to the 5 coordination of Al^{+3}, tetrahedra no longer form a chain like in sillimanite. The configuration of octahedral chains is the same as in sillimanite, which can be seen in the cross sectional view along the c-axis, but the directions of the axes a and b are switched. The change in the magnitude of axes a and b can be understood by comparing the distances between O2–O5 and O2–O3. Octahedra 2 and 5 in sillimanite are connected by bipyramids B2 and B5 along their pyramidal axes. With respect to this bipyramid of AlO_5, these octahedral chains are connected to one another by SiO_4 tetrahedra or AlO_5 bipyramid.

The space group for andalusite is Pnnm, and its lattice type is, therefore, orthorhombic. Its unit cell dimensions are :

a = 7.79Å b = 7.89Å and c = 5.56Å

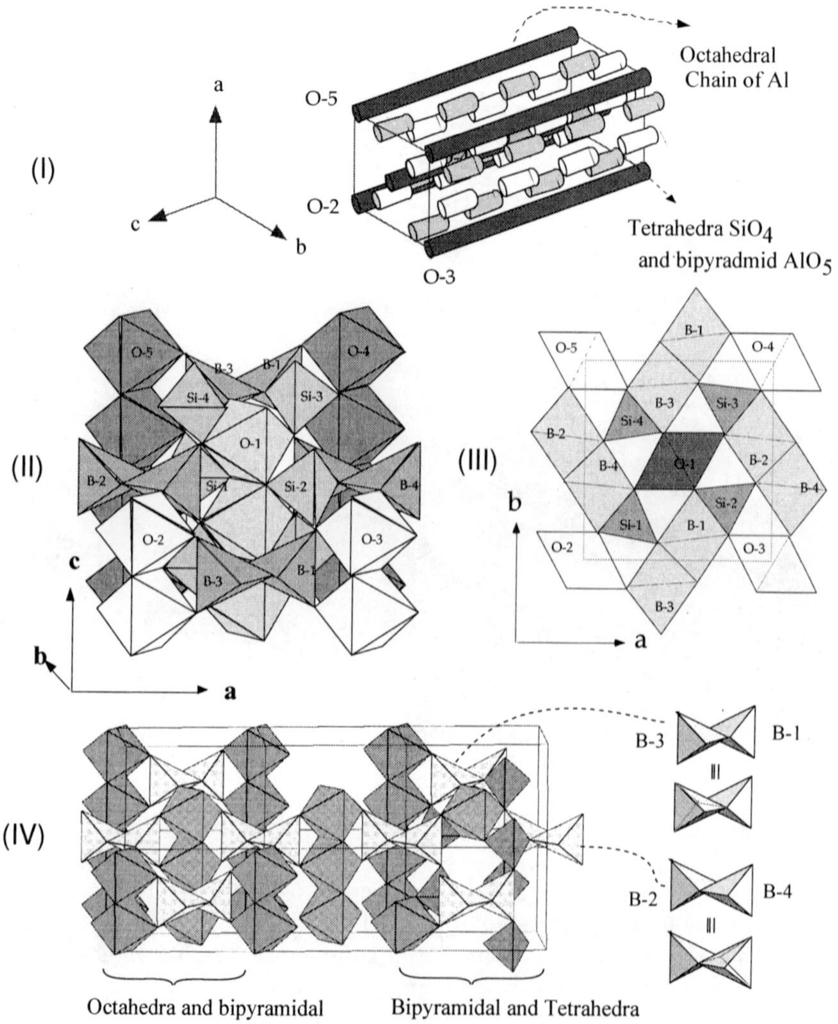

Figure 4-63 Crystal structure of Andalusite

1-1-3 Quadrivalent Group

Zircon($ZrSiO_4$)

The Zr ion in Zircon[$ZrSiO_4$] has eight coordination with O^{-2} ions. The ZrO_8^{-4} polyhedra unit is a distorted dodecahedron as shown in Fig.4-64. Oxygens ions are grouped into two types, O_S and O_L. They are not symmetrically equivalent. The edge formed by two O_L type oxygens is used for edge sharing with SiO_4. Each ZrO_8^{-4} dodecahedron is connected to four adjacent neigbouring ZrO_8^{-4} dodecahedron and two SiO_4^{-4} tetrahedra by edge sharing. SiO_4^{-4} tetrahedra are connected to two by edge sharing. Hence, each oxygen anion is connected to 2 Zr^{+4} and 1 Si^{+4}. In zircon, O^{2-} ions take a fairly close packed structure.

The space group of zircon is $I4_1/amd$. Hence it has the body centered tetragonal lattice. The unit cell dimension is a = 6.616Å and c = 6.015Å with four formular units in the unit cell. The four fold rotational symmetry can be illustrated by using columnar representation for the chain of alternating ZrO_8 and SiO_4 as shown in Fig.4-64. There are two types of columns, type A and B. They are 90° rotated image of each other around the c-axis. These columns alternate along the a and b axes. Hence the length of the lattice vector $\mathbf{a_o}$ is equal to twice the distance between adjacent columns.

Zircon is used in ceramic pigments, as a gemstone and is an important source of ZrO_2. Similar minerals are thorvite, $[ThSiO_4]_4$. Among phosphates, YPO_4, and $CePO_4$ have this structure.

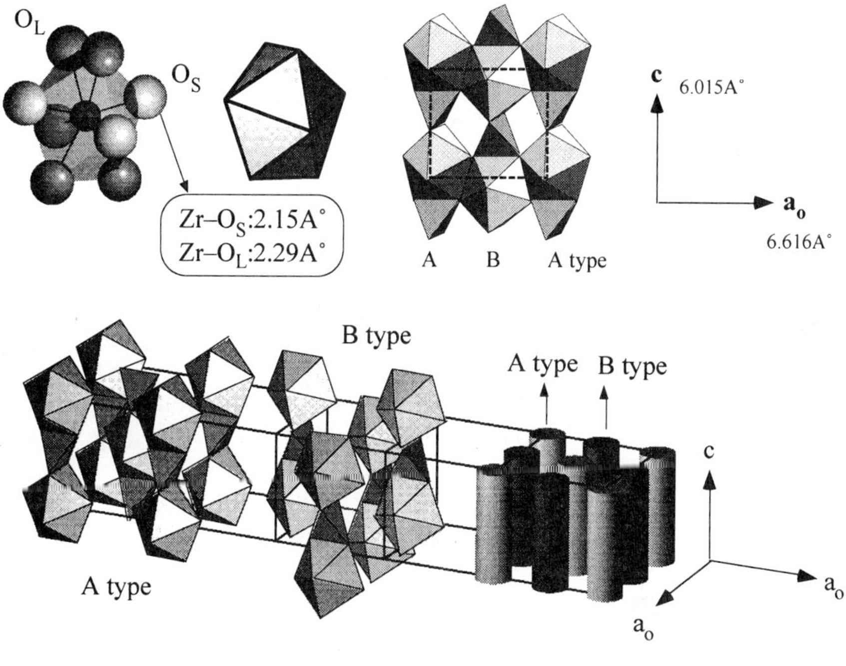

Figure 4-64 Crystal structure of zircon

1-1-4 Complex Group

Garnet Minerals :

The garnet minerals, [Ca^{2+}_3 Al^{3+}_2 $(SiO_4)_3$], crystallize in the cubic or isometric system and belong to the space group Ia3d with 8 formula units in the unit cell. The structural formula can be represented as A^{2+}_3 B^{3+}_2 $(SiO_4)_3$, where A and B refer respectively to 8 and 6 coordinated cation sites. This chemical formula satisfies Eq.(a1) for the island type silicates. The A sites are typically occupied by large divalent cations such as Ca^{2+}, while the B sites house smaller trivalent cations such as Al^{3+}. Occupancy of the sites A, B and Si can be by one of elements in the following :

	A_3^{2+}	B_2^{3+}	$(SiO_4)_3$
CN	8	6	4
Possible Elements	Mg^{2+}, Ca^{2+}, Fe^{2+}, Mn^{2+}	Cr^{3+}, Al^{3+}, Fe^{3+}, Y^{3+}	Al^{3+}, Fe^{3+}, Ti^{3+}

Since there are 8 formula units in the unit cell, the total number of atoms in the unit cell is 160. Illustration of the unit cell of the garnet structure with atoms is so difficult that it is better described by polyhedra. The A sites are surrounded by 8 oxygens in irregular coordination polyhedra, with oxygens roughly in a BCC array. For example, in grassularite, CaO_8 forms irregular cubes in that two opposite planes are twisted by about 17 - 25° while SiO_4 and AlO_6 are regular tetrahedron and octahedron, respectively. Fig.4-65 illustrate the three perspectives of the twisted octahedra in garnet.

The CaO_8 cube is linked to two SiO_4 tetrahedra and 4 AlO_6 octahedra by edge sharing which is very similar to the connection mode of ZrO_8 in zircon. This connection mode of the cube with adjacent neigboring tetrahedra and octahedra is shown in Fig.4-66(I). Each AlO_6 octahedra is surrounded by 6 CaO_8 cubes with edge sharing, and by two SiO_4 tetrahedron with corner sharing. Fig.4-66(II) illustrates this connection mode for AlO_6. The overall crystal

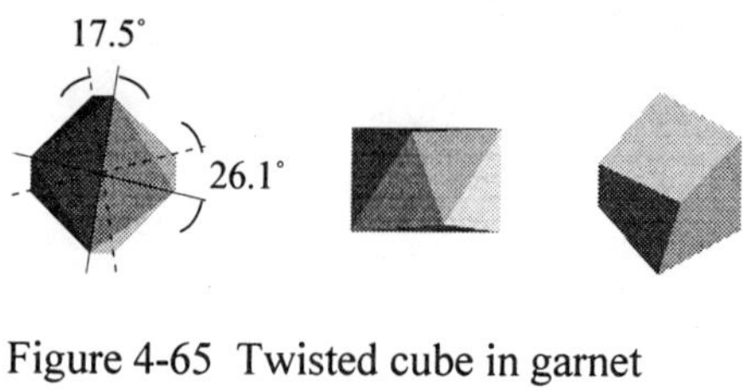

Figure 4-65 Twisted cube in garnet

structure of garnet is shown in Fig.4-67 with projection along the x axis. The structure of garnet consists of alternating SiO_4 tetrahedra and BO_6 octahedra that share corners to form a continuous 3-dimensional network. Alternatively it can be considered as linear chains of alternating SiO_4 tetrahedra and CaO_8 cubes linked by AlO_6 octahedra. Each SiO_4 tetrahedra is connected to two adjacent CaO_8 cubes by edge sharing and to four AlO_6 octahedra by corner sharing. In zircon, the ratio of Zr/Si is 1 which is equal to the ratio of Ca/Si in garnet. The same ratio of Zr/Si and Ca/Si can be attributed to the alternating sequence of Ca and Si in the chain of octahedra and tetrahedra with AlO_6 octahedra between the chains. The x-projection at x = 0 in Fig.4-67 shows the similarity of garnet structure with that of zircon. Hence SiO_4 is separated from other SiO_4 by CaO_8 cubes in the vertical direction and by AlO_6 octahedra in the horizontal direction or vice versa in the garnet structure.

The distance of all Si-O in SiO_4 tetrahedron is the same(1.651Å) and that of Al–O is 1.927A° whereas eight Ca-O bonds degenerate into four of length 2.326Å and four of 2.483Å. The structural arrangement is such that the atomic population of {100} and {111} planes is depleted.

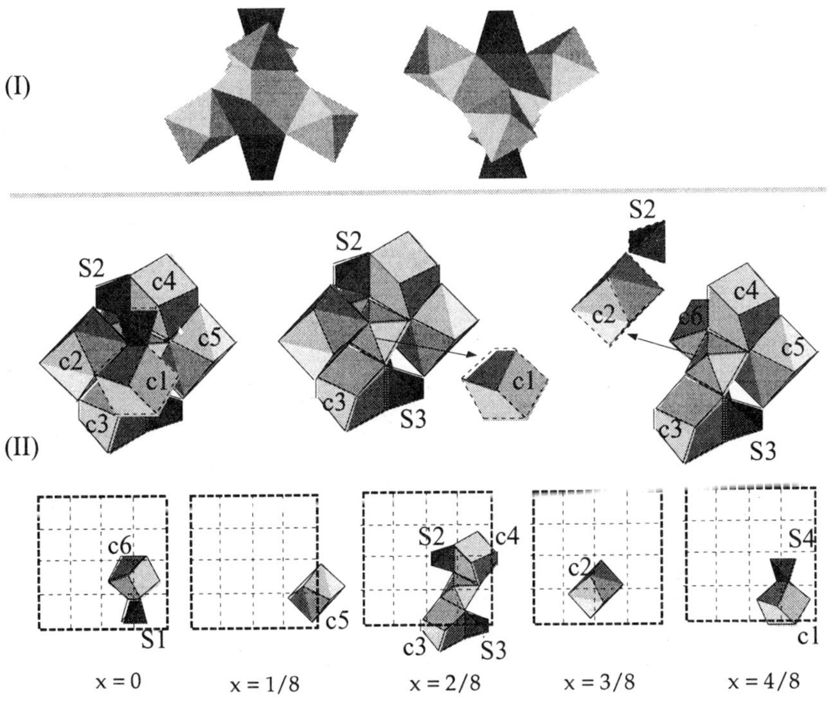

Figure 4-66 Connection modes of (I) octaheda(AlO_6) and (II) cube(CaO_8) with neigboring polyhedra in garnet

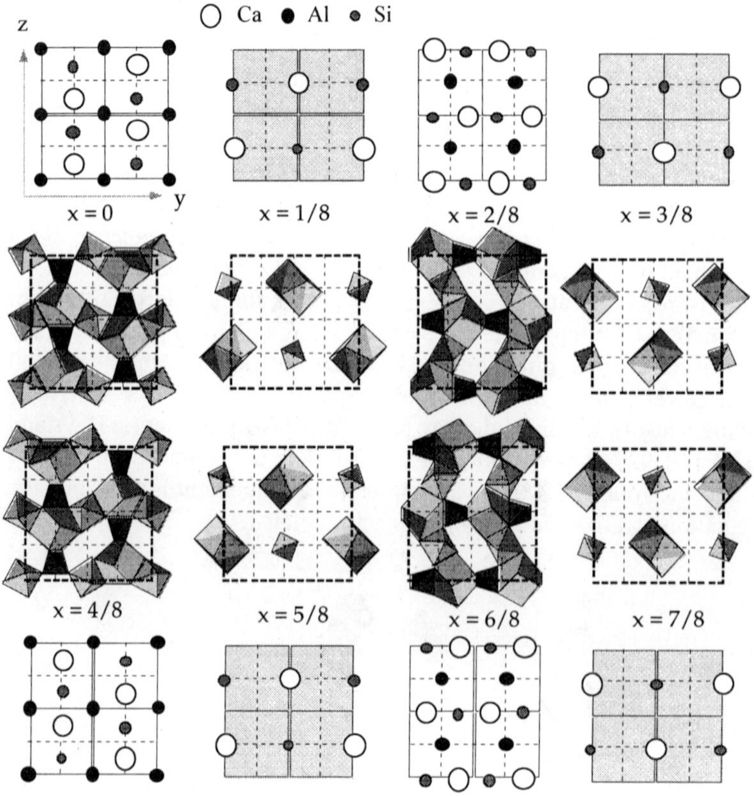

Figure 4-67 x-projection profile of garnet crystal structure

The garnet group includes a wide variety of naturally occurring minerals that are all silicates and have characteristic $(SiO_4)_3$ groups, although partial substitution of $(OH)_4{}^{4-}$ groups occurs in such hydrous garnets as grossularite $[Ca_3Al_2(SiO_4)_{3-m}](OH)_{4m}$ where m = 0~3. It is however, possible to synthesize crystals with the garnet structure that are totally free of silicon in the tetrahedral sites. These crystals form the basis of a very large family of synthetic garnets, with very interesting optical, magnetic and electrical properties. Garnet materials are especially good candidates for magnetic applications such as microwave devices owing to the antiparallel interaction between A and B sites. Here, the garnet structure can be compared to spinel structure in respect to coexistence of and antiparallel interaction between tetrahedra and octahedra with magnetic elements such as Fe, and Mn. The magnetic properties of garnets can be controlled by allocating different cations on A, B and C sites.

i) Natural Mineral

Grossular	$[Ca_3Al_2(SiO_4)_3]_8$	Cubic, Ia3d; Z = 8
Pyrope	$[Mg_3Al_2(SiO_4)_3]_8$	
Almadine	$[Fe_3Al_2(SiO_4)_3]_8$	
Andradite	$[Ca_3Fe_2(SiO_4)_3]_8$	

ii) Synthetic Garnet Minerals

$Ca_3Al_2(OH)_{12}$ – Synthetic hydro-garnet
$3Y_2O_3{\bullet}5Fe_2O_3$-YIG – Microwave magnetic materials
$3Y_2O_3{\bullet}5Al_2O_3$-YAG – Laser material

It is evident for the synthetic hydro-garnet that H_2O substitutes for SiO_4 as $(OH)_4^{4-}$ groups, with $4H^+$ replacing Si^{4+}. For the YIG[Yttrium Iron Garnet] and YAG[Yttrium Aluminum Garnet] systems the reactions become

$$4[3Y_2O_3{\bullet}5Fe_2O_3] \rightarrow [Y_3^{3+}Fe_2^{3+}(Fe^{3+}O_4)_3^{5-}]_8 \text{ and}$$
$$4[3Y_2O_3{\bullet}5Al_2O_3] \rightarrow [Y_3^{3+}Al_2^{3+}(Al^{3+}O_4)_3^{5-}]_8$$

where Fe^{3+} and Al^{3+} respectively replace Si^{4+} in the tetrahedral sites. Note that all cations are in their highest valence states. This eliminates electron exchange between cations and leads to high insulation resistance and low dielectric losses. YIG materials are, therefore, suitable and much used for microwave applications.

1-2. Pyrosilicates

The pyrosilicates are a group of silicates that have only one bridging oxygen between two SiO_4 tetrahedra. This can be depicted as an island type silicate with SiO_7 as the island. Three oxygen anions of the SiO_4 tetrahedra or six oxygen anions of SiO_7 unit must be connected to other cations. To balance the negative charges on the oxygen anions, there must be +3 charges per one Si. Therefore the chemical formula for this group of silicates is given as $R_2^{+3}Si_2O_7$ for trivalent cations and $R_3^{+2}Si_2O_7$ for divalent cations, or

$$\frac{\sum_{R\neq Si}\sigma_R - N_{OH,F}}{N_{Si}} = 3 \qquad (10)$$

Thorvetite ($Sc_2Si_2O_7{\cdot}H_2O$) belongs to this group. Also silicate slag(Ca_2MgSiO_5) is in this group. Both thorvetite and silicate slag satisfy the charge balance condition for the one bridge between two SiO_4 tetrahedra. However, these type silicates are rare.

1-3. Ring Silicates

Ring silicates are formed by two bridging oxygens in each tetrahedron forming 3-, 4- and 6-membered rings as shown in Fig.4-68. Five-membered ring silicates are unknown. These rings are not connected to other SiO_4 tetrahedra. Hence this group of ring silicates is another island type with the rings as the islands.

Since there are two unbridged oxgyen anions per each SiO_4 tetrahedra, the negative charges on these ions must be balanced by connections to other cations. Hence the condition for 3-membered ring silicates is given as:

$$\frac{\text{Total charges on the other cations}}{\text{the number of Si}} = 2 \qquad (11)$$

Benitoite, $BaTiSi_3O_9$, is an example of a three-membered ring silicate. The total positive charges from Ba^{+2} and Ti^{+4} is equal to the negative charges, -6, on the three-membered ring of $[Si_3O_9]^{6-}$. This satisfies Eq.(11).

Only a few four-membered ring silicates are known to exist. One is axinite, $Ca_3Al_2BSi_4O_{15}(OH)$. In relation to Eq.(11), axinite illustrates a unique feature of B^{+3}. B^{+3} forms triangular BO_3^{-3} units which are usually separated from SiO_4^{-4} tetrahedra.

This separation causes the BO_3^{-3} unit to function as an OH^- ion, thereby reducing the effective positive charges on other cations which contribute to neutralizing the negative charges on SiO_4^{-4}. Due to this reducing effect of BO_3^{-3} on the positive charges on other cations, Eq.(11) becomes :

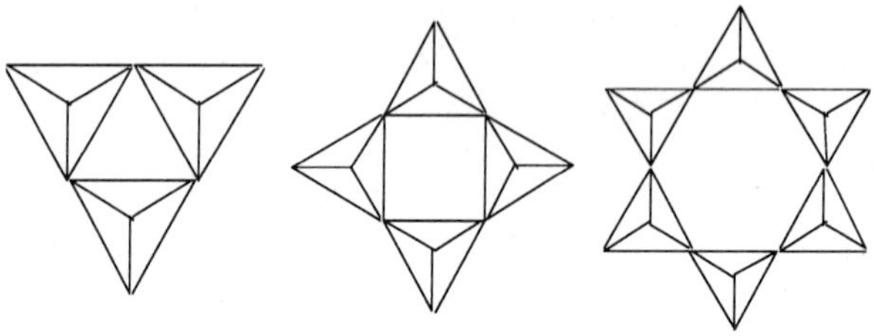

(a) (b) (c)

Figure 4-68 (a) 3, (b) 4 and (c) 6 membered rings of the $[Si_3O_9]^{6-}$, $[Si_4O_{12}]^{8-}$, $[Si_6O_{18}]^{12-}$ ring silicates

$$\frac{\sum\limits_{R \neq Si,B} \sigma_R - 3N_B - N_{OH,F}}{N_{Si}} = 2 \qquad (12)$$

Here N_B is the number of boron cations in the chemical formula. Typical six-membered ring silicates are beryl($Be_3Al_2Si_6O_{18}$), cordierite($Al_3Mg_2(Si_5Al)O_{18}$) and tourmaline($NaMg_3Al_6(BO_3)_3[Si_6O_{18}](OH)_4$). The chemical formula for cordierite implies that one tetrahedra in the six-membered ring is replaced by AlO_4^{+5}. This means that only three Al^{+3} cations can contribute to neutralizing the negative charges on the Si tetrahedra. The replacement of Si by Al also increases the negative charge on the tetrahedra. By considering this change Eq.(12) for ring silicates becomes :

$$\frac{\sum\limits_{R \neq Si,Al(t),B} \sigma_R - 3N_B - N_{OH,F} - N_{Al(t)}}{N_{Si,Al}} = 2 \qquad (13)$$

Here Al(t) indicates Al^{+3} cation in the tetrahedra. $N_{Al(t)}$ is the number of Si^{+4} replaced by Al^{+3}. The left side of Eq.(13) represents the criteria correlating chemical formula not only for ring silicates but also for other type silicates.

4. Single Chain Silicates (Pyroxenes)

The pyroxenes consist of silica chains connected by modifying cations between the silica chains. The typical formula is $AB(Si_2O_6)$ or $BSiO_3$. Cation in A sites have 8 coordination but may not be occupied. B sites are typically occupied by Fe^{2+}, Ni^{2+}, Li^{+}, Al^{3+} cations. In this structure, Si may have minor substitution by Al^{3+} or Fe^{3+}. This is typified by pyroxenes with the structure depicted in Fig.4-69. In single chain silicates, each tetrahedra has a net -2 charge which must be balanced by other cations. Hence the chemical formula for this structure must satisfy the following:

$$\frac{\text{Total charges on the other cations}}{\text{the number of Si}} = 2 \qquad (14)$$

$$\text{or} \quad \frac{\sum\limits_{R \neq Si,Al(t),B} \sigma_R - 3N_B - N_{OH,F} - N_{Al(t)}}{N_{Si,Al}} = 2 \qquad (15)$$

This equation is the same as Eq.(11) which represents the ring silicates. The difference between the ring and chain silicates is in the fact that the number of Si in the chemical formula for single chain silicates is 1 or 2 while for ring silicates in the chemical formula for single chain silicates is 1 or 2 while for ring silicates there are more than 3 Si ions. Hence ring silicates and single chain silicates can be distinguished by their chemical formula although they both satisfy the same condition expressed in Eq.(11).

In pyroxenes, the direction of the silica(SiO_4) chain is along the c-axis of the unit cell. These chains extend parallel to the basal plane of the tetrahedra,

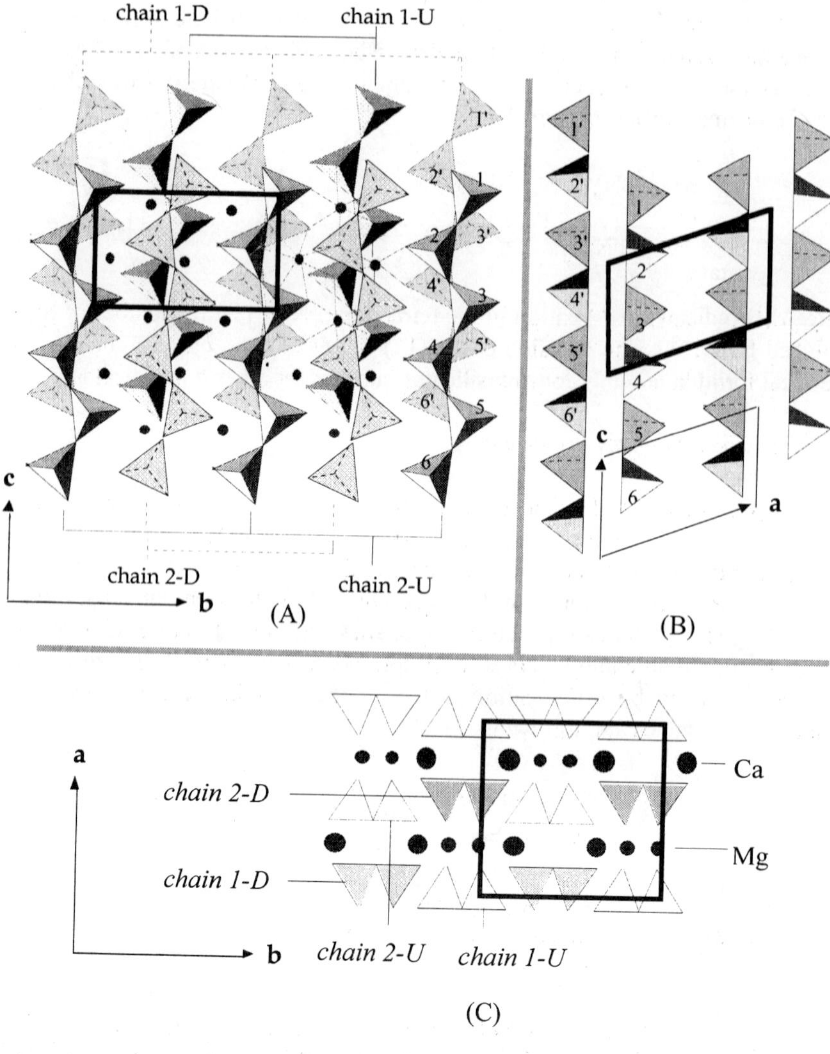

Figure 4-69 Perspective of single chain silicate(diopside, $CaMgSiO_6$),

(A) along the a-axis (B) the b-axis and (C) the c-axis

resulting in a layer of tetrahedra chains. The b-axis in pyroxene is in this direction of repetition of the chains in parallel with the basal plane. Chain layers also stack in the direction perpendicular to the basal plane and this stacking direction is the a-axis of the pyroxene lattice. The configuration of chains is related to the cations which connect the silica chains. Also, depending on the configuration of these silica chains, the structures are orthorhombic or monoclinic. In monoclinic pyroxene, adjacent chain layers are shifted from each other along the c-axis. In orthorhombic pyroxene, there is a shift of adjacent chain layers along the a-axis. This shift of chain layers, however, results in no net movement from the lattice point in every four shifts as shown in Fig.4-70. Therefore, the unit cell of orthorhombic pyroxene is about twice the unit cell of the monoclinic pyroxene. The number of formula units in the unit cell is 4 for monoclinic pyroxene while that for orthorhombic is 16. The reason is that the basic unit in the monoclinic pyroxene is usaully Si_2O_6 while it is SiO_3 in orthorhombic pyroxene. However, in the cases of hypersthene ($FeMgSi_2O_6$) and clinoenstatite ($MgSiO_3$), the basic units are reversed. The formula unit per unit cell, therefore, is not a good measure to compare monoclinic and orthorhombic pyroxenes. The characteristic of these two types of pyroxene are summarized in Table 4-5.

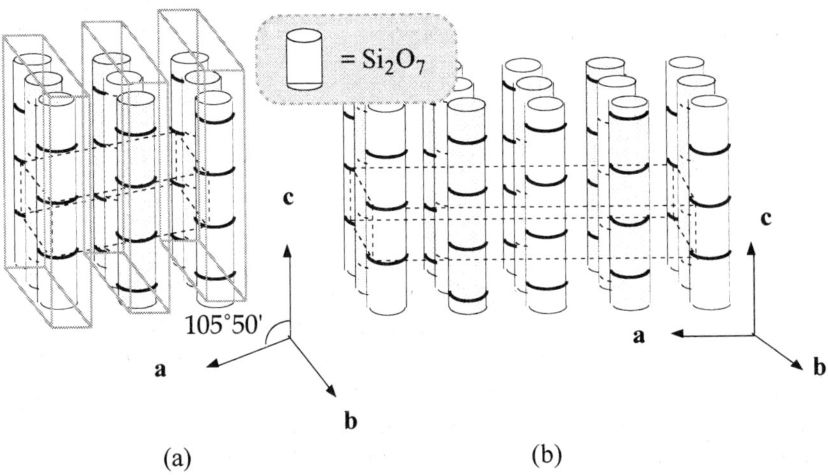

Figure 4-70 Schematic illustration of (a) monoclinic structure for diopside and (b) orthorhombic structure of enstatite

(Small cylinder is equivalent to two tetrahedra. Shaded cylinders are downward tetrahedra and white cylinders are upward tetrahedra.)

Table 4-5 Two types of single chain silicates

	Diopside	Enstatite
Lattice type	Monoclinic	Orthorhombic
Space group	C2/a	Pbca
Chemical formula	$(CaMg)Si_2O_6$	$MgSiO_3$
Formula units in the unit cell	Z = 4	Z = 16
Unit cell dimension	a =9.75, b=8.93, c=5.25Å β = 105°50'	a =18.2, b=8.8, c=5.2Å β = 105°50'
Other Minerals	Jaedite: $(NaAl)Si_2O_6$ Spodumene: $(LiAl)Si_2O_6$ Clinoenstatite:$MgSiO_3$	Orthoferrosilicate: $FeSiO_3$ Hypersthene:$(MgFe)(SiO_3)_2$

5. Double Chain Silicates (Amphiboles)

Structures of amphiboles can be schematically drawn as in Fig.4-71. The negative charges on tetrahedra in double chain silicates are 1.5 per Si. Hence, with respect to Eq.(11), a similar equation can be expressed for double chain silicates as :

$$\frac{\sum_{R \neq Si, Al(t), B} \sigma_R - 3N_B - N_{OH,F} - N_{Al(t)}}{N_{Si,Al}} = 1.5 \qquad (16)$$

A typical double chain silcate mineral is tremolite, $Ca_2Mg_5(Si_8O_{22})(OH)_2$. The crystal structure of tremolite is shown in Fig.4-71 and the unit cell is shown in bold lines. The space group of tremolite is C2/m. Its lattice type is monoclinic and the cell dimensions are a = 9.85Å, b = 18.1Å, c = 5.3Å and β = 104°40'. Like pyroxene, the c-axis runs along the chain direction. The b-axis lies parallel with the basal plane of the tetrahedra. Double chains stack along the direction of the a-axis. There are two formula units in the unit cell. In order to illustrate this fact, the chemical formula may be expressed as $[R_2^{3+}R_5^{2+}(Si_8O_{22})(OH)_2]_2$. In relation to Eq.(16), the crystal structure illustrates that $(OH)^-$ is not connected to Si and reduces the effective positive charges on other cations which compensate the negative charges on the Si tetrahedra. Anthophyllite, $(Mg,Fe)_7(Si_8O_{22})(OH)_2$, and hornblende,

$Ca(Mg,Fe)_4Al(Si_7Al)O_{22}(OH)_2$, belong to this group. These all satisfy Eq.(16). Also they are among the most important asbestos minerals. Pyroxene and amphibole minerals are of fibrous habit and may be distinguished by their cleavage direction. The coordination numbers for R^{3+}(or R'^{2+}), R^{2+} and Si are 8, 6 and 4, respectively.

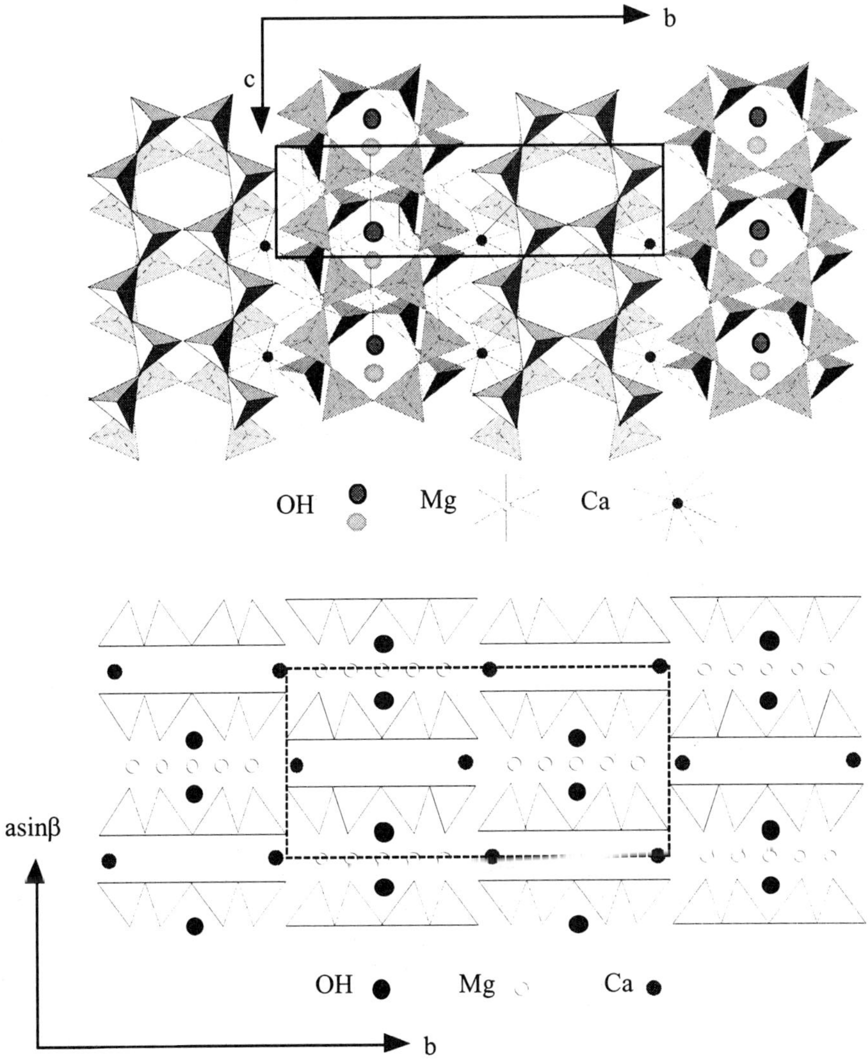

Figure 4-71 Crystal structure of tremolite

6. Sheet or Layer Structures

Here it is useful to distinguish between the terms 'sheet' and 'layer.' Although both terms imply a planar extension of some repeating unit, the sheet is homogeneous, while a layer can be a single sheet or multiple sheets of different type. Another difference between sheet and layer is in their connections to adjacent sheets or layers. For example, if there are two sheet types, A and B and their sequence is –ABAB–, the bonding modes between A–**B** and **B**–A would be different. If that is the case, sheet **B** has different bonding schemes with two adjacent A sheets. However, if –[AB]– is grouped as a layer, the bonding scheme is always the same for any two adjacent layers. Layer silicates are so called because they consist of tetrahedral sheets of $(Si_2O_5)^{2-}$, shown in Fig.4-72, interlayered with other types of sheets such that there is always a difference between intralayer and interlayer bonding modes.

In the silicate sheets, there is one non bridging O^{2-} per [SiO4] unit, which forms the apex of pyramid of the tetrahedra sheet. The other bridging oxygens form the hexagonal basal plane with OH^- at the center of each hexagonal ring of oxygens. This tetrahedra sheet is bonded usually to a modified brucite ($Mg(OH)_2$) or gibbsite ($Al(OH)_3$) sheet. Brucite and Gibbsite have the same closest packing of OH^- with Mg and Al located at the octahedral sites. Due to the electric charge difference between Al^{+3} and Mg^{+2} ions, two thirds of the octahedral sites are filled in gibbsite while all of the octahedral sites are filled in brucite. Fig.4-74 shows how brucite and gibbsite sheets are stacked over the tetrahedral sheet. Layer silicates with the gibbsite sheet are called trioctahedral

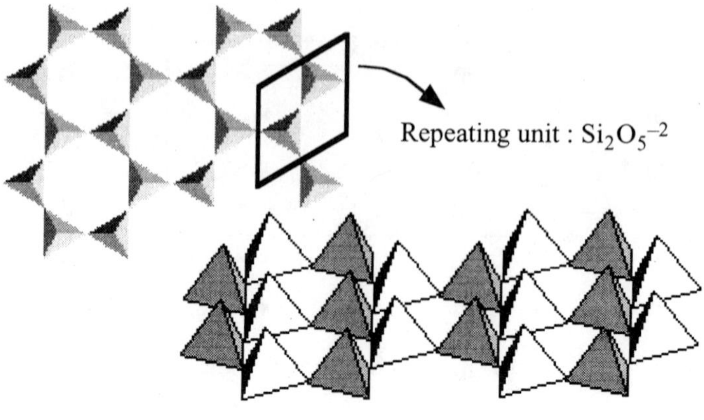

Figure 4-72 Perspective view of Silica tetrahedra sheet

silicates since all the three octahedral sites per basic unit of Si_2O_5 tetrahedra sheet are filled. Likewise, layer silicates with the gibbsite sheet are called dioctahedral silicates since two thirds of the octahedral sites per basic unit of Si_2O_5 tetrahedra sheet are filled. Silicates with the brucite structure, are called trioctahedral silicates. The structures of brucite and gibbsite are shown and compared in Fig.4-73.

Another classification of layered silicates depends on whether the $(Si_2O_5)^{2-}$ layers are stacked on both sides of the brucite or gibbsite octahedral sheets or just on one side. If one tetrahedra sheet is stacked on the octahedra sheet, the silicate is called a single layer silicate. If both sides of the octahedra sheet are stacked by tetrahedra sheets, the silicate is called a double layer silicate. Fig.4-74 illustrates four possible silicate types regarding single, double, trioctahedra, and dioctahedra silicates.

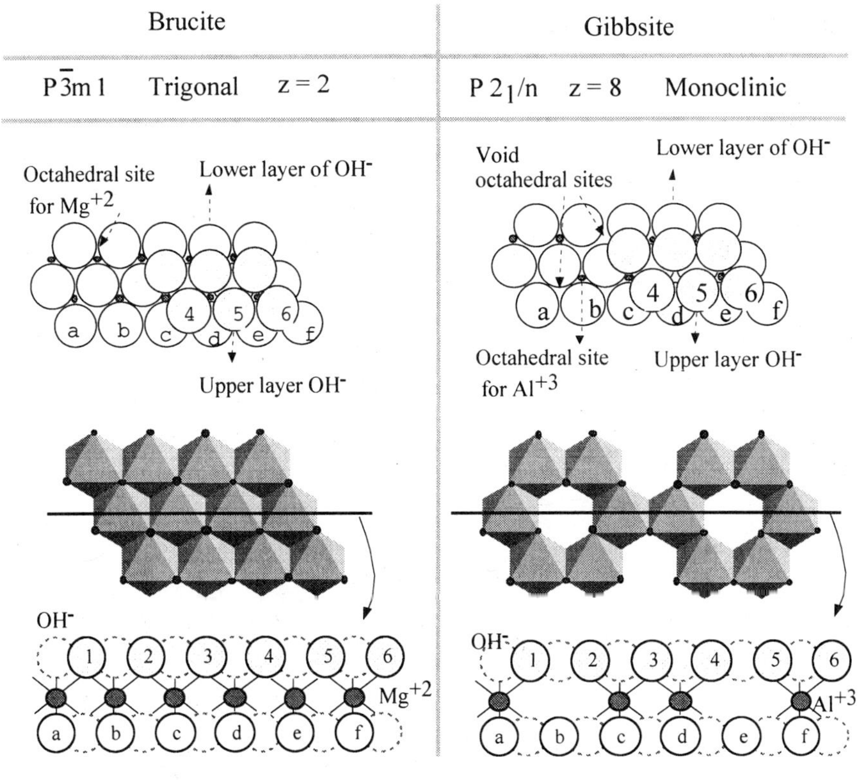

Figure 4-73 Illustration of structures of brucite and gibbsite

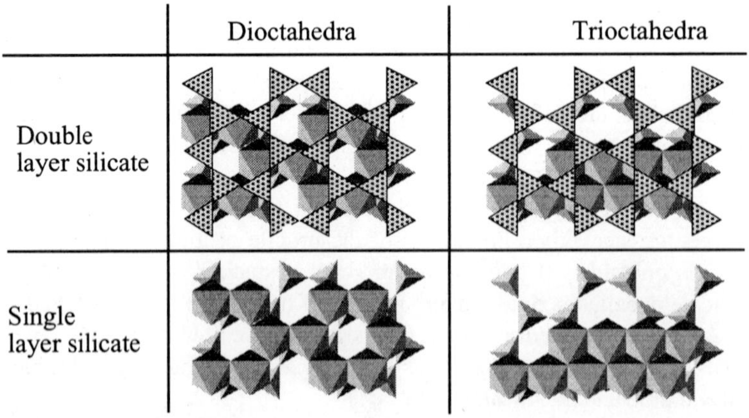

Figure 4-74 Four types of layered silicates according to the modes of combining tetrahedra sheet and octahedra sheet

6-1. Single Layer Silicates

Single layer silictes are made up of layers which consist of a single (Si_2O_5) sheet with [SiO_4] units in hexagonal array and a single sheet of brucite or gibbsite as shown in Fig.4-75. Apical oxygens of tetrahedra sheet point in the same directions and bond to the gibbsite or brucite sheet so as to form a layer. However, octahedra sheet in the lower layer and tetrahedra sheet in the upper layer are bonded by weak van der Waals forces. Therefore the layer structure can be cleaved at this interlayer. Also stresses do exist between the octahedral and tetrahedral sheets within the layer because of different anion spacing and mismatch of the two type sheets. This stress usually cause single or double layer silicates to exist only as fine crystals.

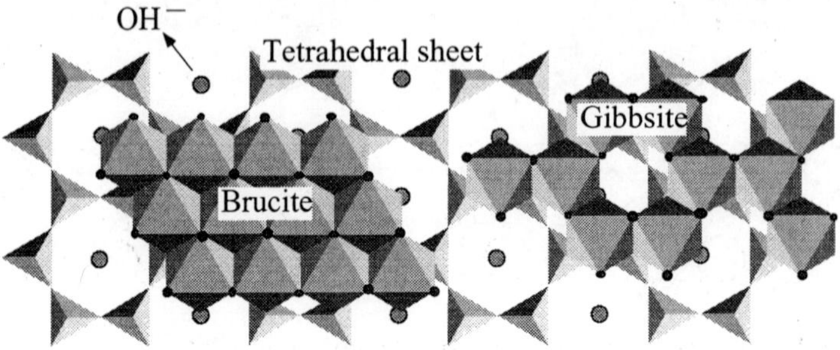

Figure 4-75 Stacking of brucite(or gibbsite) sheet over silicate sheet

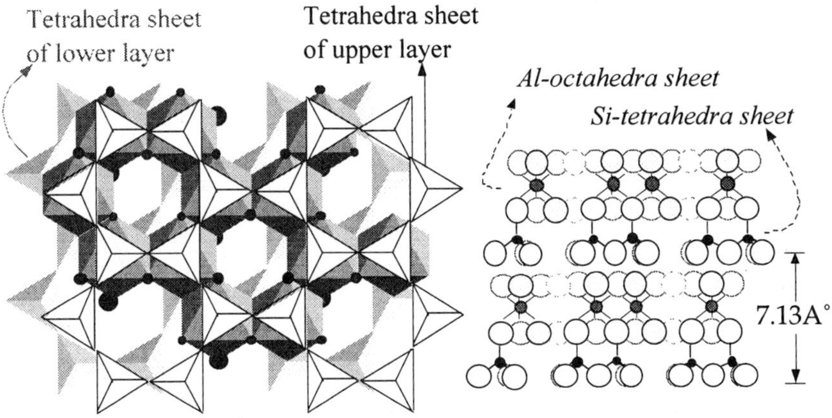

Figure 4-76 Crystal structure of kaolin

6-1-1 Kaolinite $[(OH)_4Al_2Si_2O_5]_2$

Kaolinite is a dioctahedral single layer silicate with gibbsite as the octahedral sheet. The crystal structure of kaolinte is shown in Fig.4-76. Kaoline is the main clay mineral and typically very pure. Little substitution of Al for Si takes place and it exists as very fine particles mainly due to interlayer stresses. The O-OH arrangement in the basal plane results in different polytypes. Polymorphs such as dickite, nacrite, halloysite represent different stacking arrangements and repeat distances. Stacking arrangements will be explained in detail in discussing polytypism of layer silicates. Kaolinite is used as a main constituent of ceramic whitewares, in paper and in medicinal formulations.

6-1-2 Chrysotile $[(OH)_4Mg_3Si_2O_5]_2$

Chrysotile is a main asbestos mineral with a fibrous morphology. This fibrous morphology among layer silicates, which typically have plate morphologies, can be explained by its crystal strcuture. Chrysotile is a trioctahedral single layer silicate, analogous to kaolin in which the gibbsite layer is replaced by a brucite layer. This replacement is the main cause of the morphology difference between kaolin and chrysotile. In kaolin, the misfit between tetrahedral and octahedral sheets is small. In chrysotile, Al^{+3} (0.50A°) in octahedra sheets is replaced by Mg^{+2}(0.65A°) with a larger ionic diameter resulting in a larger misfit between the octahedral and tetrahedral sheets. The resulting stresses are accomodated by a larger curvature and the formation of fibers (~ 100Å diameter). This morphological difference is illustrated schematically in Fig.4-77.

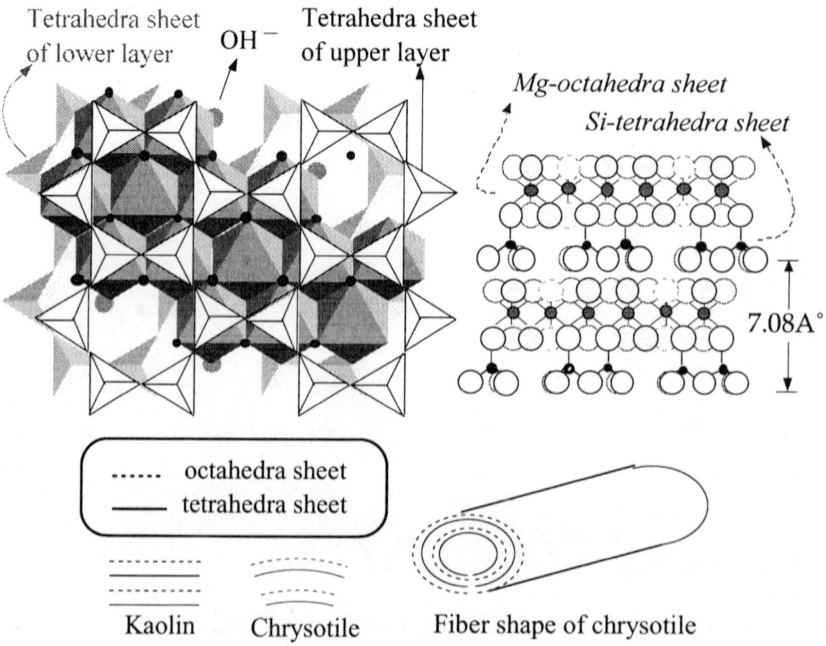

Figure 4-77 Crystal structure of chrysotile and its morphological difference from kaolin

6-2. Double Layer silicates

Double layer silicates have one more silica sheet on the other side of the brucite sheet which results in a sandwich type structure. Hence double means two silica sheets on both sides of a brucite (or gibbsite) sheet. Si is sometimes partially replaced by +3 cations such as Al^{+3} or Fe^{+3} in the tetrahedral sheet. The electric charge unbalance caused by this substitution of Al^{+3} or Fe^{+3} for Si^{+4} is compensated by cations such as Na^+, K^+ or Ca^{+2} which locate between the two layers. Hence double layers which involve the substitution of Si by lower charge cations are typically bonded by electrostatic forces through interlayer cations. However, this ionic bonding is still weak although it is stronger than van der Waals bonding.

6-2-1 Double layer silicates without interlayer cations

These are "sandwich" type structures in which the brucite or gibbsite octahedral layer is bonded symmetrically to two $(Si_2O_5)n$ tetrahedral layers.

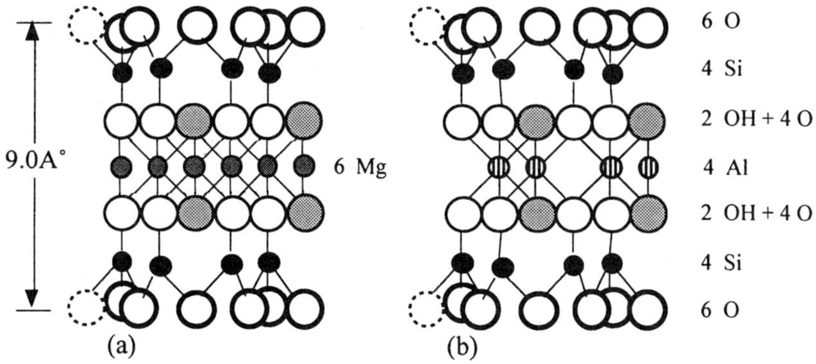

Figure 4-78 Unit cell structure of double layer silicates
(a) Talc(trihedral) and (b) Pyrophilite(dihedral)

Atomic arrangements in talc and phrophillite are shown in Fig.4-78. In this figure, OH^- hydroxyl groups are shown by gray circles. It is revealed in Fig.4-78 that the hydroxyl groups(OH^-) are connected to Al^{+3} but not to Si^{+4}. Hence it does not contribute to compensating the negative charge of Si^{+4}.

Talc and pyrophillite are the typical double layer structures with brucite and gibbsite as the octahedral layers, respectively. Talc is used in electrical porcelains, floor and wall tiles, and as cosmetic and pharmaceutical ingredients.

6-2-2 Double layer silicates with interlayer cations

(i) Montmorillonite Clays

The structure is based on that of talc or pyrophillite but the structures are not electrically neutral. This is due to excessive substitution in the layers. This has partial fraction of cations per unit silica Si_4O_{10} unit.

(ii) MIcas

The micas are derived from pyrophillite or talc by substitution of 1/4 Si^{4+} ions by Al^{3+} ions in tetrahedral sites (layers). K^+ or Na^+ ions are stuffed between layers to balance the charge. This has one cation per unit silica Si_4O_{10}. Mica minerals exist as thin sheets that are readily sheared along the plane. The substitution scheme is illustrated in Table 4-6

Table 4-6 Substituted double layer from talc and pyrophillite

Partial substitution of one Si^{+4} per formular unit$(Si_2O_5)_2$		
Talc derivative - $(OH)_2Mg_3(Si_2O_5)_2$	Hectorite $(OH)_2(Mg_{2.67}Li_{0.33})(Si_2O_5)_2$ ↓ $Na^+_{0.33}$ – Surface adsorbed Na ions	
Pyrophillite derivative - $(OH)_2Al_2(Si_2O_5)_2$	Montmorillonite $(OH)_2(Al_{1.67}Mg_{0.33})(Si_2O_5)_2$ ↓ $Na^+_{0.33}$	
	Beidellite $(OH)_2Al_{2+x}(Si_{2-x/2}Al_{x/2}O_5)_2$ ↓ $Na^+_{x/2}$	
Full substitution of one Si^{+4} per formular unit$(Si_2O_5)_2$		
Talc derivative - $(OH)_2Mg_3(Si_2O_5)_2$	Pyrophillite derivative - $(OH)_2Al_2(Si_2O_5)_2$	
Phlogopite	Muscovite	Biotite
$(OH)_2Mg_3(Si_3Al)O_{10}$	$(OH)_2Al_2(Si_3Al)O_{10}$	$(OH)_2(MgFe)_3(Si_3Al)O_{10}$
↓ K^+	↓ K^+	↓ K^+
Surface adsorbed K^+ ions		

* Note that Adsorbed ions give high base exchange capacity to the clays.
Montmorillonite minerals add green strength to ceramic bodies.

6-2-3 Chemical formula for layer silicates

In layer silicates, SiO_4^{+4} tetrahedra are connected to two adjacent tetrahedra leaving one oxygen anion of the tetrahedra unbridged. Therefore there is one negative electric charge per the Si ion. Like Eq.(16), the criteria for layer silicates can be expressed with respect to its chemical formula as :

$$\frac{\sum_{R\neq Si,Al(t),B} \sigma_R - 3N_B - N_{OH,F} - N_{Al(t)}}{N_{Si,Al(t)}} = 1 \qquad (17)$$

Let's apply Eq.(17) to muscovite which has the chemical formula given as $KAl_2(AlSi_3)O_{10}(OH)_2$. There is no boron. With $N_{Si,Al(t)} = 4$, $N_{OH} = 2$, $N_{Al(t)} = 1$, the total cationic charge except for Si, Al(t) and B is +7. By inserting the values of these terms into the left side of Eq.(17), the result is 1. Hence Eq.(17) illustrates from the chemical formula that muscovite must be a sheet silicate. All the chemical formula in Table 4-4 satisfy Eq.(17). Even for vermiculite, with chemical formula, $(Mg,Fe^{3+}Al)_3(Al_2Si_2)O_{10}(OH)_2{\bullet}3.5H_2O$, Eq.(17) is effective. However, if the chemical formula is given without showing any site selection for cations such as in $H_2MgFe^{3+}Al_3Si_2O_{12}$, excluding water molecules, the first step for applying Eq.(17) is to convert hydrogen into hydroxyl groups(OH), and calculate according to Eq.(17) assuming only Si takes tetrahedral sites, viz:

$$\frac{\sum_{R\neq Si,Al(t),B} \sigma_R - 3N_B - N_{OH,F} - N_{Al(t)}}{N_{Si,Al(t)}}$$

$$= \frac{14 - 0 - 2 - 0}{2} = 6$$

If one Al in the chemical formula takes a tetrahedral site, then Eq.(17) becomes

$$= \frac{11 - 0 - 2 - 1}{3} = \frac{8}{3}$$

If two Al in the chemical formula takes tetrahedral sites, then Eq.(17) becomes

$$= \frac{8 - 0 - 2 - 2}{4} = 1$$

Among these values, 8/3 is not significant since it does not satisfy any of silicate types. Hence the possible structure type for the chemical formula given is an island type or sheet type silicate. If it is a sheet type silicate, half the tetrahedral sites must be substituted by Al for Si, which agrees with the actual crystal structure for vermiculite. Hence, the objective of Eq.(17) is not only to identify the structure type of silicates but also to show the strong relationship between chemical composition and crystal structure. Eq.(17) reveals, therefore, that silicate minerals crystallize in a systematic way.

6-3. Polytypism

Structures with polytypism occur when the polymorphs differ only in the stacking of the two-dimensional sheets. The best example is polymorphs with

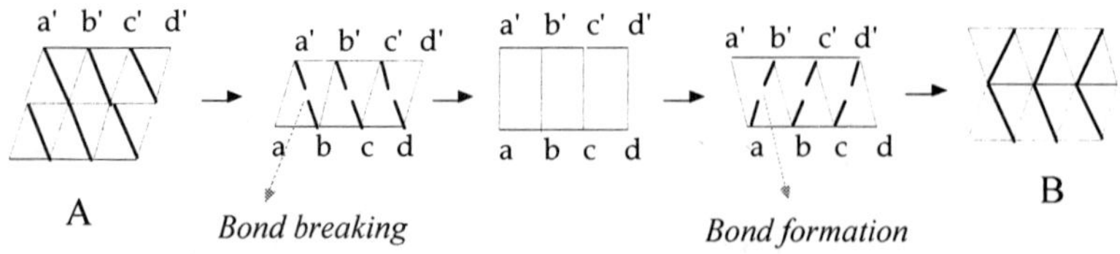

Figure 4-79. Transformation of crystal structures in polytypism

FCC and HCP structures like SiO_2 and ZnS. In this case, the transformation can be depicted as the change of sequence from ABCABC to ABABAB.

ZnS : Zincblende (Cubic : FCC) ↔ Wurtzite (Hexagonal: HCP)
SiO_2 : Tridymite (Cubic : FCC) ↔ Cristobalite (Hexagonal : HCP)

As shown in Fig.4-79, conversion of structure A to structure B, requires an alternating layer in A with its mirror image. Hence many physical properties such as density are almost the same. However, these polytypes are not easily converted to each other since this conversion requires breaking of bonds. In this respect, polytypism is a special case of reconstructive transformations. Hence this conversion is slow, it being a one-dimensional polymorphism.

Prominent examples of polytypism are layered silicates such as kaolinite and mica. In these layered silicates, each layer consists of two or three sheets of silica and other oxides. Although the atomic arrangements in each sheet are the same throughout the layers, the relative positions of these sheets to each other result in the formation of several polytypes. Silicate sheets can stack in many different ways since the bonding forces between layers are not primary chemical bonds but van der Waals or hydrogen bonds. Hence there are 12 polytypes in kaolin due to this reason.

6-3-1 Polytypism in single layer silicates

There are two main factors that change the relative positions of neighboring layers resulting in two polytypes formed in single layer silicates. The first factor is how the octahedral sheet sits on the tetrahedral sheet in the same layer. There are two sitting modes for octahedral sheet over the tetrahedral sheet. This is illustrated in Fig.4-80. The difference between mode I and mode II is in the shape of the octahedra sheet. The octaheda in mode II is the 180°-rotated image of that in mode I.

The other factor is the interlayer bonding mode which is characterized by how the tetrahedral sheet of one layer is positioned relative to the octahedral sheet of the next layer. This interlayer bonding mode is determined

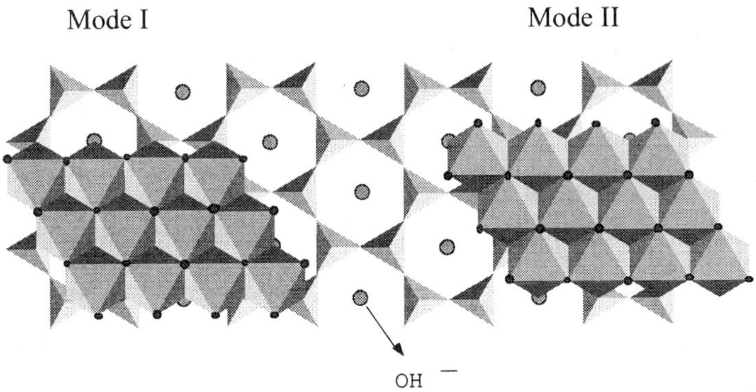

Figure 4-80 Two modes for intralayer bonding between tetrahedra sheet and octahedral sheets

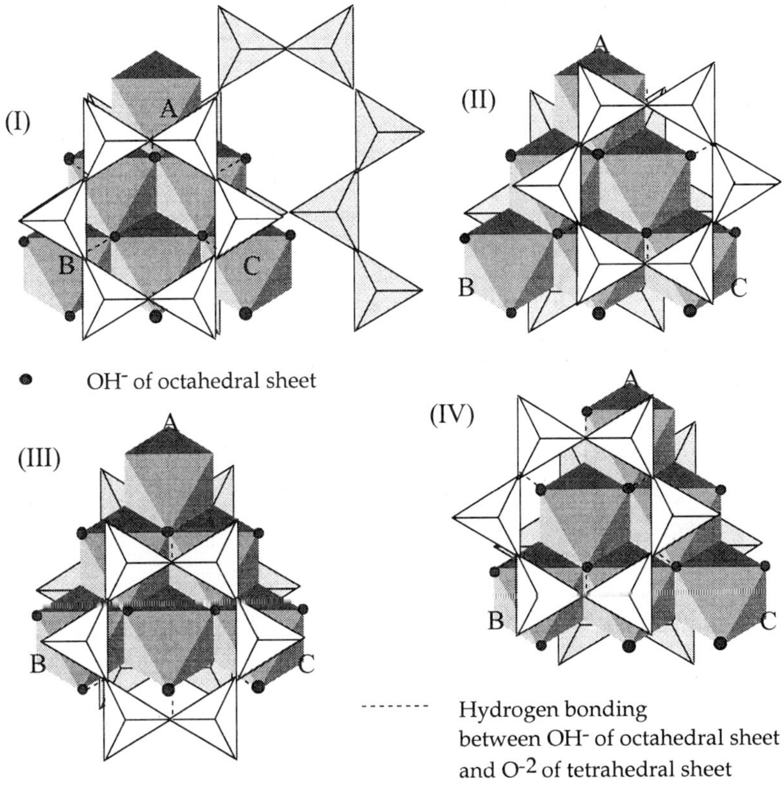

Figure 4-81 Configurations of interlayer shift due to hydrogen bonding (I) no shift and (II), (III) and (IV) shifts with a 1/3 distance of each edge of the hexagonal ring

by the hydrogen bonding between OH- groups in the octahedral sheet of the lower layer and oxygen ions in the basal plane of the tetrahedra sheet of the upper layer. The interlayer bonding configuration is shown in Fig.4-81(I) in which tetrahedra sheet of the first layer overlap with the tetrahedra sheet of the next layer. With this hydrogen bonding structure, there are three more possible modes of stacking tetrahedra sheet over the next lower layer. This is shown in Fig.4-81(II), (III) and (IV). By combining the shift within the layer and the interlayer shift, many polytypes of single layer silicates are possible.

6-3-2 Polytypism in 2:1 double layer silicates

In the double layer silicates like mica, two tetrahedra sheets of the adjacent two layers, which face each other, are superimposed exactly by matching the hexagonal rings of each tetrahedral sheet. Unlike single layer silicates, therefore, shifting of one layer relative to the next layer is determined by the intralayer stagger which is the shift of the upper tetrahedral sheet relative to the lower sheet, within the same layer. There are 3 possible intralayer shifts that are shown in Fig.4-82 (b), (c) and (d). If the octahedral sheet changes its configuration from mode I to mode II, illustrated in Fig.4-80, the intrashift in

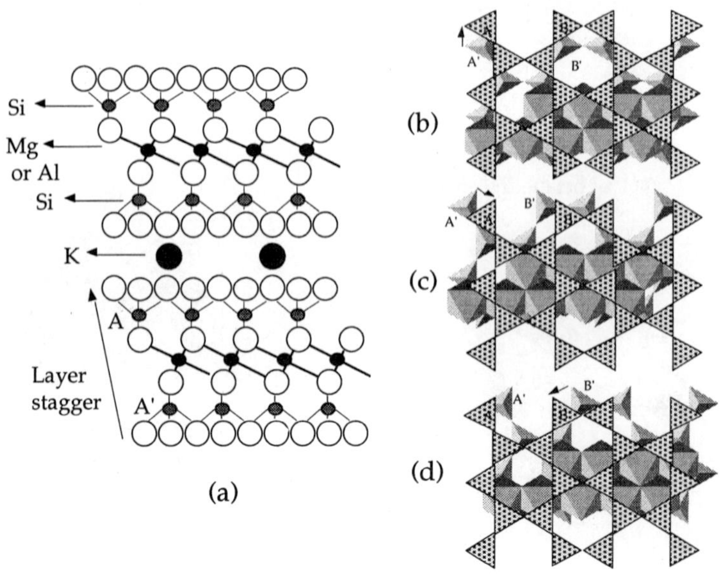

Figure 4-82 (a) View of mica from the right side of the layer (b), (c) and (d) giving three possible configurations between tetrahedra sheets in the same layer of mica

double layer silicates reverses. Hence 6 intrashift modes are possible in double layer silicates like micas.

6-3-3 Nomenclature of polytypes in layered silicates

Many polytypes in layer silicates can exist since unlimited combinations of intralayer or interlayer shifting is possible along the perpendicular direction to

Notation	Shift Mode	Notation	Shift Mode
3T $P3_1$ (or $P3_2$) $P3_112$ (or $P3_212$)	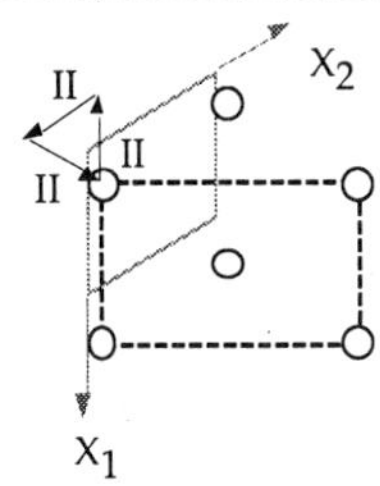	2Or $Ccm2_1$ $Ccm2_1$	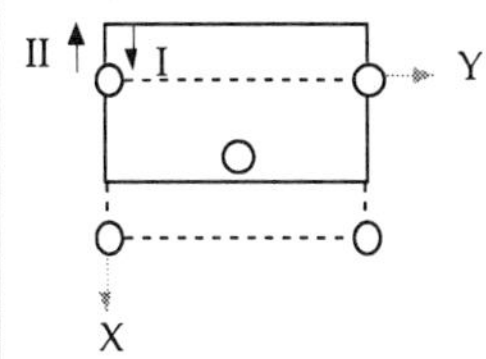
6H $P6_1$ (or $P6_5$) $P6_122$ (or $P6_522$)	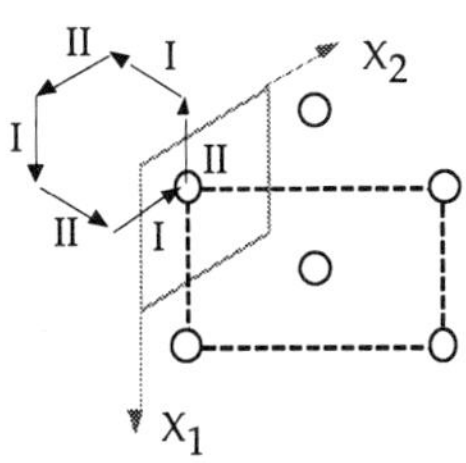	1M Cm C2/m	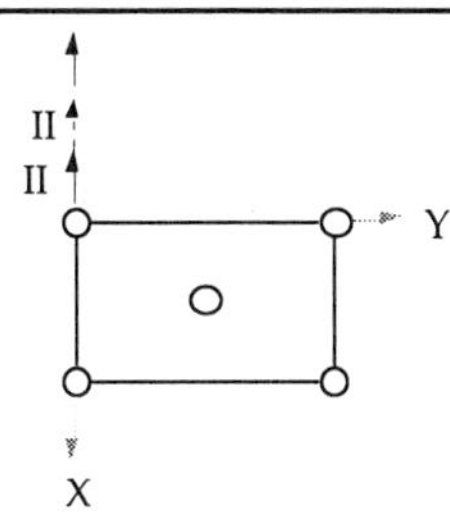
$2M_1$ Cc Cc	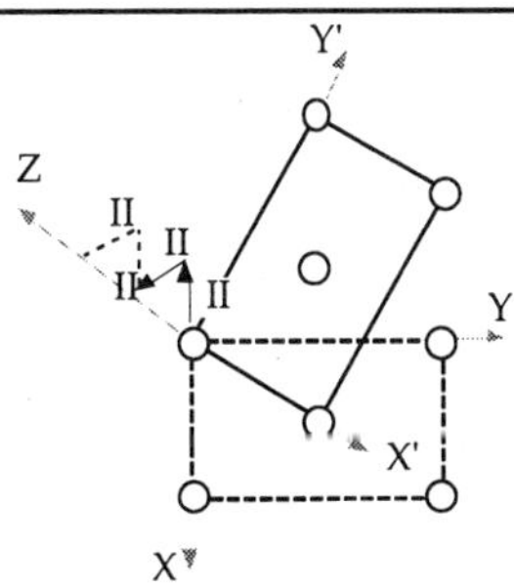	$2M_2$ Cc Cc	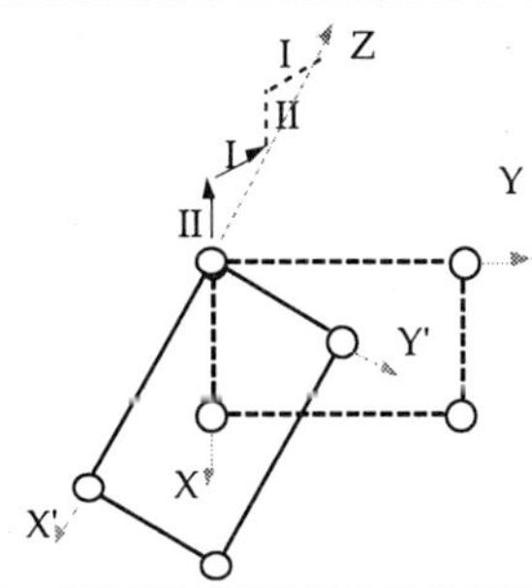

Notes for notation : 3T - Ramsdell's notation
$P3_1$ - Space group (sinlge layer silicate)
$P3_112$ - Space group (double layer silicate)
Circles represent the center of hexagonal ring.

Figure 4-83 Polytypes of layered silicates and their notations with space groups the layer. However, if the pattern of interlayer or intralayer shifting is repeated in a regular mode, the possible polytypes are reduced. Some typical examples are illustrated in Fig.4-83. Three notations for each polytype are labeled on the left side of the pattern. The first is Randell's notation. The others are space group notations which describe the symmetry and unit cell configuration of the single and double layer silicates, respectively. In Randell's notation, alphabetic letters designate the crystal lattice. M, Or, T, and H indicate monoclinic, orthorhombic, trigonal and hexagonal. The number 'n' before these letters indicates that the layer repeats in every 'n' layers. The number after the letters differentiate repeating modes with the same symmetry and repeating layers. For example, $2M_1$ and $2M_2$ in Fig.4-83 have the same monoclinic crystal lattice and layer shift which repeats in every 2 layers. However they have different repeating modes as shown in Fig.4-83.

7. Framework Silicates

This is a group of silicate minerals in which all 4 corners of the $[SiO_4]^{4-}$ tetrahedral unit are shared. Because of the high charge on the Si^{4+} ion, these tetrahedra do not share edges or faces but usually only corners. This results in a relatively open structure. This aspect of pure silica is explained in Chap.3 with respect to anion-anion repulsion on the basis of closest packing of oxygen anions.

7-1. The silica minerals

The silica minerals exhibit different orientation and stacking of $[SiO_4]^{4-}$ units, resulting in four different structures which are quartz, tridymite, cristobalite, and coesite. Among these polymorphs, quartz has already been described. Hence, only the crystobalite, tridymite and coecite, structures will be described. Opal ($SiO_2 \cdot nH_2O$) is a silica mineral which includes H_2O within its structure. In addition to its use as a gemstone, opan is a main component of diatomaceous earth.

7-1-1 Cristobalite

Cristobalite exist in both α and β forms in which the $[SiO_4]^{4-}$ - units are linked in 4 member rings. β-cristobalite has the Fd3m structure based on ZnS (cubic). Silicon locates in the carbon positions in the diamond-like structure and oxygen resides between silicons.

Among the polymorphs of silica, cristobalite best illustrates the close packing of oxygens. In close packed oxygens, Si must occupy one quarter of the tetrahedral sites to balance the charges. In Fig.4-85(a) is shown the most

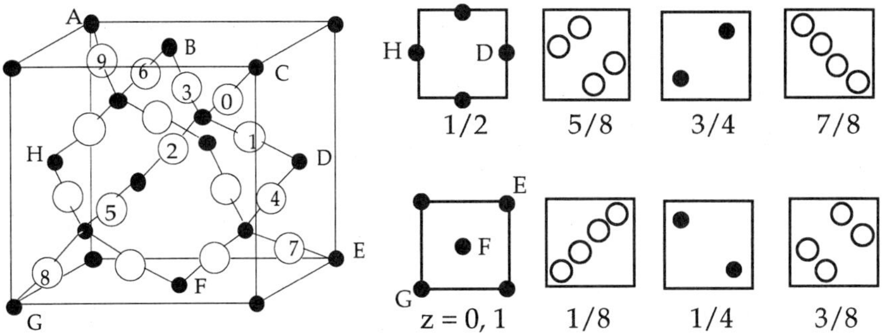

Figure 4-84 Crystal structure of β-cristobalite where silicon is shown as black circles and oxygen as open circles.

probable positions for Si within the oxygen layer assuming the lattice energy is lowest when any Si has the same distance to the nearest Si. The configuration of silica tetrahedra resulting from this occupation of Si among tetrahedral sites within oxygen layers is that shown in Fig.4-85(b). In this configuration, oxygens are degenerated into two groups, A and B. Oxygens in group A form the basal plane of the silica tetrahedra with apices pointing upward. Oxygens in group B form silica tetrahedra with apices downward.

As shown in Fig.4-85, two layers of oxygen form one layer of silica tetrahedra. The crystal structure of SiO_2 which has the close packing of oxygen can be described by connecting the silica tetrahedra layer shown in Fig.4-85(b).

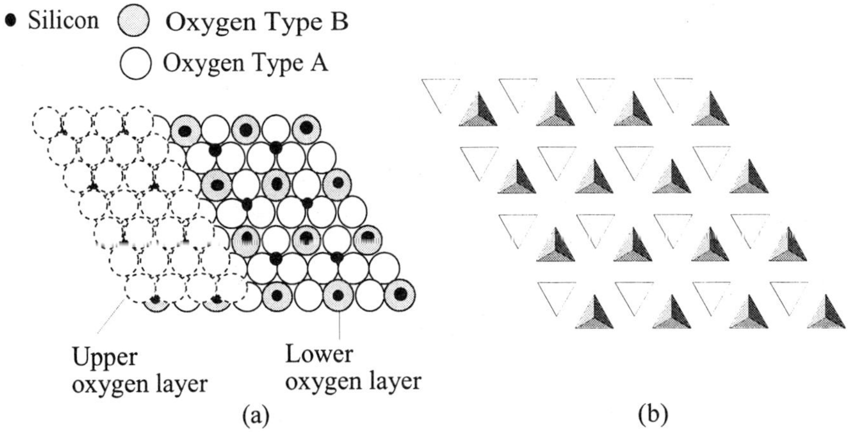

Figure 4-85 (a) Occupation of one quarter of tetrahedral sites by Si^{+4} and (b) Its polyhedra illustration

Since there are no connections between the two types of silica tetrahedra, upward and downward, within the layer, the anionic repulsion between these two tetrahedra sheets makes the structure unstable. In order to avoid this repulsion, actual crystal structures of cristobalite and tridymite adopt closest packed oxygen anions with half of the tetrahedra sheet removed, resulting in an alternating sequence of downward and upward tetrahedral sheets as the basic unit, as shown in Fig.4-86(I) There are two possible connection modes for the basic unit of two tetrahedral sheet. The first as shown in Fig.4-86(II) is generated by overlapping one basic unit on top of the other basic unit. This is the crystal structure of cristobalite. The second, shown in Fig.4-86(III) is generated by overlapping two basic units, which are mirror images of each other across the basal plane of the tetrahedra sheet. This is the crystal structure of tridymite. Regarding the oxygen layer sequence, the first has the sequence of –ABCABC–. Hence cristobalite takes the cubic lattice whereas tridymite has the hexagonal lattice. To illustrate the equivalence of stacking mode in Fig.4-86 for cristobalite, the unit cell with tetrahedra is shown in Fig. 4-87(a) and its rotated image in Fig. 4-87(b). With repespect to the atomic arrangement in the critobalite crystal, the stacking mode in Fig.4-86(III) is described in detail in Fig.4-88 using the same numbering as in Fig.4-84.

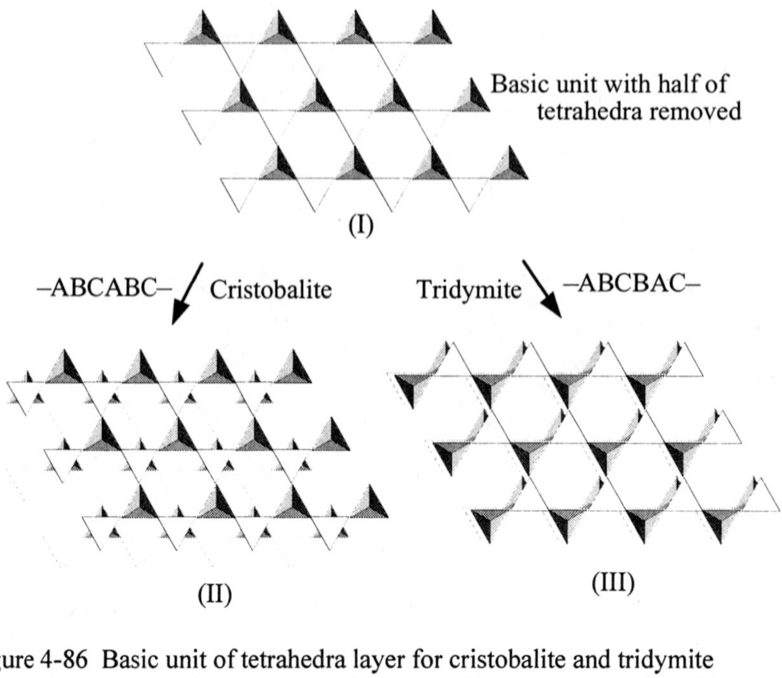

Figure 4-86 Basic unit of tetrahedra layer for cristobalite and tridymite
(Upper layer is shown by solid line and lower layer by dotted line.)

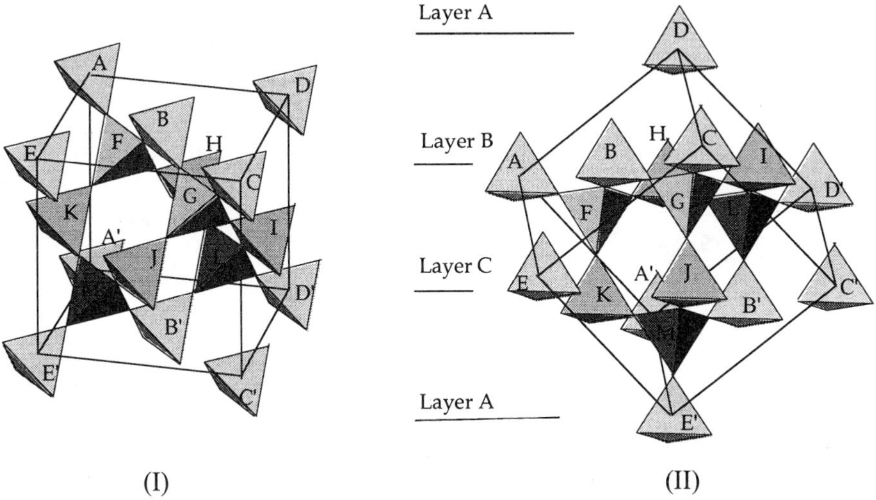

Figure 4-87 Perspective of unit cell of cristobalite (I) showing the cubic lattice and (II) its image rotated by 45° counterclockwise on the page to illustrate the stacking of the basic unit as in Fig.4-86.

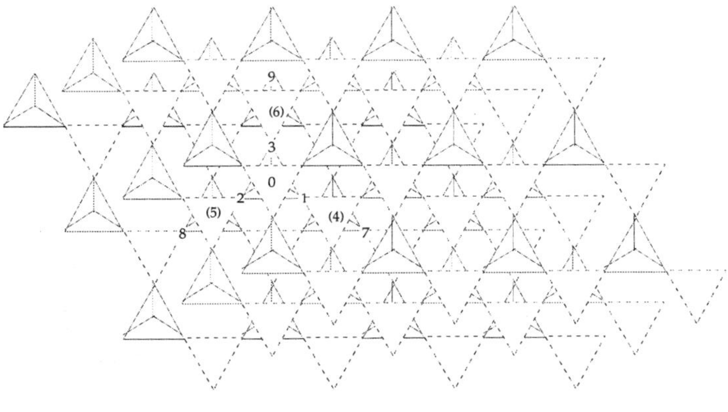

Figure 4-88 Stacking of silica tetrahedra layer with the same numbering as in Fig.4-84 for cristobalite

7-1-2 Tridymite

Tridymite has the regular array of [SiO_4] units arranged with alternate apices up and down. Its space group is P6/m mc. The crystal structure is illustrated in Fig.4-89. Tetrahedra of same number with and without apostrophe('), are mirror images to each other, like tetrahedra 2 and 2'. The hexagonal unit cell

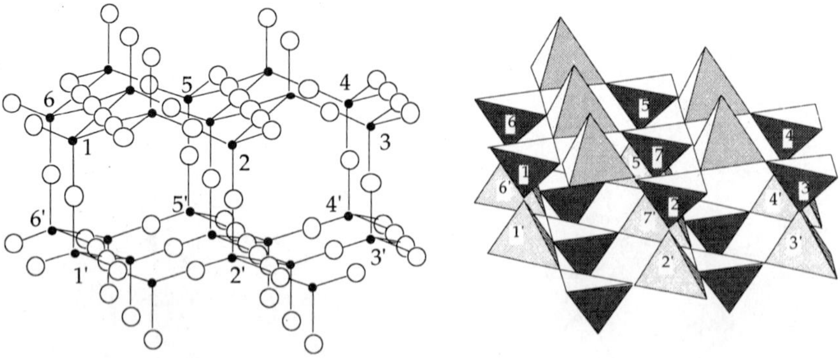

Figure 4-89 Crystal structure of tridymite

extends four tetrahedra along the vertical direction to the tetrahedra layer. The unit cell dimensions for tridymite are a = 5.03Å and c = 8.22Å.

7-1-3 Polymorphic transformation in cristobalite and tridymite

The crystal structures shown in Fig. 4-87 and 4-93 represent the high temperature forms of cristobalite and tridymite, respectively, which are usually called β-forms. The polymorphic transformation from β to α form or from α to β form is a displacive transforamtion. As in quartz, the displacive transformations in cristobalite and tridymite occur by rotation of tetrahedra. Fig.4-90 illustrates this rotational transformation from β to α form of cristobalite with the same numbering as in Fig. 4-87. The lines in Fig.4-90(a) indicate the rotational axes. This displacive movement elongates the cubic lattice horizontally and shortens it in the vertical direction. Due to the rotations around these axes, cristobalite loses its 3 fold rotational symmetry axis which runs through the body diagonal of the cubic lattice. The lattice type of α–cristobalite is the tetragonal lattice and no longer a face centered lattice. Hence the space group changes from F3dm to $P4_12_1$ and the unit cell becomes smaller as indicated by the dotted line in Fig.4-90(b). The unit cell dimensions are a = 4.97Å and c = 6.92Å. The lattice vector 'a' in α-cristobalite is equal to the half of the face diagonal of the cubic lattice for β-cristobalite.

Tridymite also shows displacive transformation which changes β-tridymite from a hexagonal lattice to α-tridymite with the orthorhombic lattice or vice versa. These displacive transformations between high temperature and low temperature forms are so rapid that heat quenching does not prevent them.

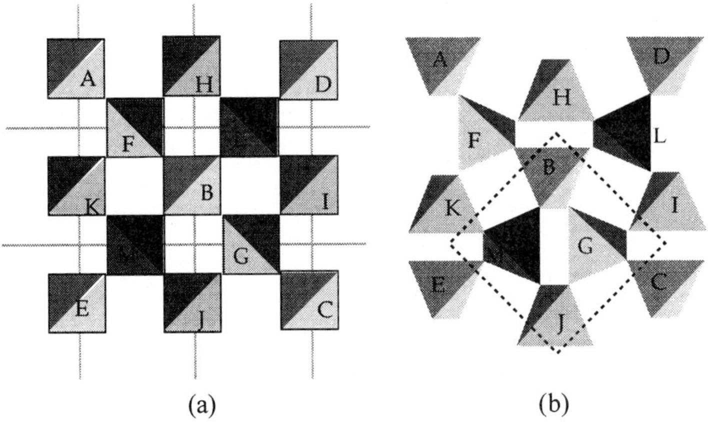

Figure 4-90 Displacive transformation of (a) β-cristobalite to (b) α-cristobalite by rotation around the axes shown by lines

However, reconstructive transformations like β-quartz → β-tridymite or β-quartz → β-cristobalite need external energy input or catalyst assist to aid the transformation. The transformation of β-quartz → β-tridymite is aided by Li, Na, or K carbonate additions. β-quartz → β-cristobalite is aided by Ca, Mg, or Ba carbonate. The carbonates provide a liquid phase at the transformation temperature, which speeds up the diffusion and nucleation processes. Conversely, ions such as Al^{3+} and Ti^{3+} inhibit the quartz conversion.

7-1-4 Coecite

A four membered ring of silica tetrahedra is the basic unit for coecite. This is shown in Fig.4-91. Another basic unit is formed by inverse symmetry of the basic unit. In Fig.4-91, tetrahedron 1' is the inverse image of tetrahedron 1. Each basic unit forms one layer of tetrahedra as shown in Fig.4-92. The ab-plane of the coecite crystal coincides with the plane of this layer and the c-axis is perpendicular to this plane also. The crystal class of coecite is monoclinic. The layers alternate in the sequence a-b-a-b with respect to the basic unit. Since the second layer a-b shifts by half the length of the a-axis relative to the next a-b layer, the unit cell of coecite consists of four layers, a-b-a-b.

Trydimite and cristobalite are high temperature forms of silica. Channels run through all the silica minerals with quartz having small channels into which only Li^+ ions fit. Tridymite and Cristobalite have larger channels

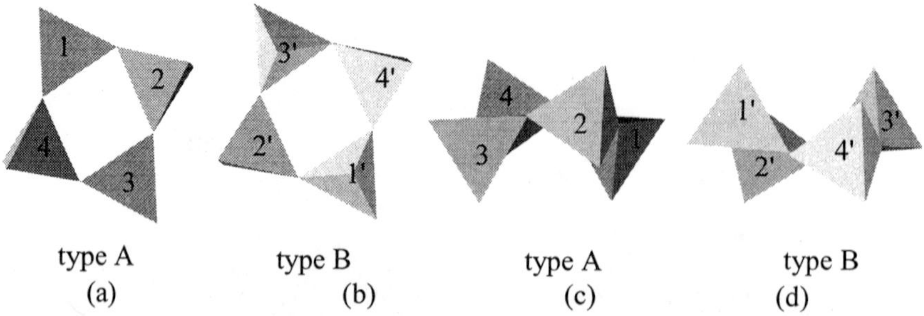

Figure 4-91 (a) and (b) Two types of basic units for coecite and (c) and (d) their side views

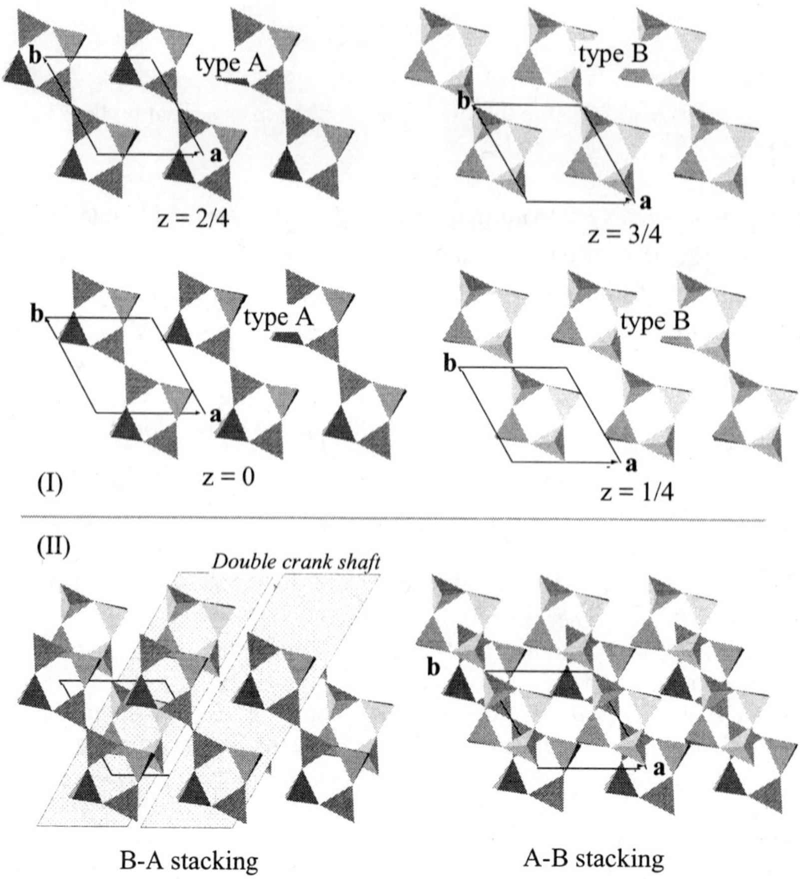

Figure 4-92 (I) Layers of silica tetrahedra along the c-axis and (II) Stacking of two layer types of tetrahedra, A–B and B–A

into which Na and K ions can fit. Coesite has very large channels, which forms the base for the feldspar and zeolite minerals. However, the volumetric density of the channels in coecite is smaller than in tridymite and cristobalite. This can be compared to the density of those silica polymorphs as given in Table 4-1.

7-2. The feldspar minerals

Feldspars are framework silicate minerals (based on the open structure of coesite) in which 1/4 to 1/2 the $[SiO_4]^{4-}$ units are replaced by $[AlO_4]^{5-}$ units. The charge deficiency is made up by alkali (Li^+, Na^+, K^+) or alkaline earth cations (Ca^{2+}, Ba^{2+}) which fit into the channels in the framework structure of coecite (SiO_2). Although the portion of the replacement of Si by Al varies from 1/4 to 1/2, the ratio of total amount of cations (M^+, or M^{2+}) to the sum of Si and Al is always 1/4. This can be illustrated by the fact that one vacant site in coecite for charge compensating cations such as Na^+ and Ca^{++} is developed from four of silica (and alumina) tetrahedra. The following illustrates how silica can become a feldspar mineral with respect to chemical formula change.

SiO_2 (Coecite) $\rightarrow Ca_{1/4}\begin{bmatrix} Si_{1/2} \\ Al_{1/2} \end{bmatrix} O_2 \rightarrow CaAl_2Si_2O_8$(Anorthite)

$\rightarrow Na_{1/4}\begin{bmatrix} Si_{3/4} \\ Al_{1/4} \end{bmatrix} O_2 \rightarrow KAlSi_3O_8$ (Albite)

$\rightarrow K_{1/4}\begin{bmatrix} Si_{3/4} \\ Al_{1/4} \end{bmatrix} O_2 \rightarrow KAlSi_3O_8$ (Orthoclase)

Plagioclase $Ab_x\ An_{1-x}$ (Ab - Albite, An - Anorthite)

The basic unit of feldspar is the linear chain of four silica tetrahedra as shown in Fig.4-93. One or two of Si is replaced by Al. In anorthite, two of the Si are replaced by Al and in albite and orthoclase one Si is replaced by Al.

7-2-1 Ca-feldspar ($CaAl_2Si_2O_8$: Anorthite)

The crystal structure of anorthite is shown in Fig.4-93. The basic unit of anothite can be considered as a linear chain of four alternating SiO_4 and AlO_4 tetrahedra. The direction of the length of this chain is along the b-axis.

These basic chain units are interconnected in such a way that SiO_4 connects to AlO_4 and not SiO_4 to SiO_4 or AlO_4 to AlO_4 . One type of basic unit is connected to the other type resulting in the repeating unit shown in Fig.4-93(c). The rectangle in Fig.4-93(c) shows the unit cell in the ab-plane. The length of the a-axis is about twice that of the basic unit while the length of the b-axis is 1.25 times the length of the basic unit. By arrangement of the

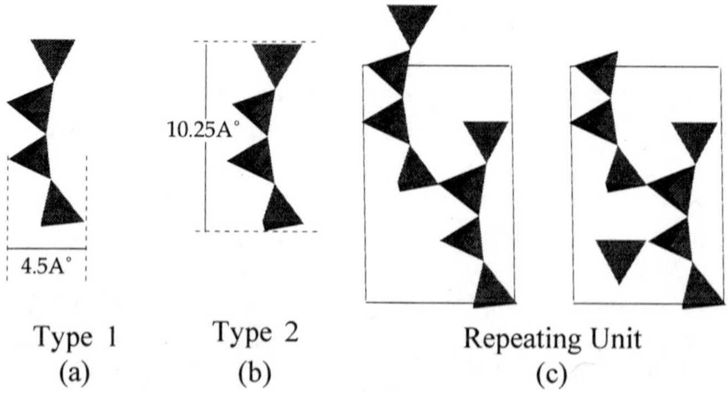

Figure 4-93 Two types of basic unit in anorthite and repeating unit (Odd number - SiO_4, Even number - AlO_4)

repeating units along the a and b-axes, a two dimensional lattice is formed as shown in Fig.4-94. Anorthite is formed by the stacking of four layers of this two dimensional lattice along the c-axis. In each layer, however, the basic units have different configurations. If the layers are named as A, A', B and B', A and A' have the same sequence of Si-Al-Si-Al in the basic unit. In B and B', the sequence is reversed to give Al-Si-Al-Si. In A, the basic unit shows a concave curve toward the a-axis while in A', the curve is convex towards the a-axis. This change in curvature is the same between B and B'. The configuration of the basic unit repeats in every four layers as A-A'-B-B'. Therefore the four layers of two-dimensional lattice shown in Fig.4-94 form the unit cell of anorthite. Since each repeating unit in the two dimensional lattice of anorthite has eight tetrahedra, the unit cell of anorthite has 32 tetrahedra. These layers in anorthite are shown in Fig.4-94. Also, the image of anorthite projected along the a-axis is shown in Fig.4-95(b). With respect to symmetry, layer A is the inversion image of layer A'. Also Layer B and B' are inversion images of one another. Hence inversion centers exist between A and A' and B and B' not between A and B' and A' and B. These inversion centers are shown in Fig.4-95(b) which is seen along the a-axis. The inversion centers located in between B and B' are shown in Fig.4-95(a) which is a view seen perpendicular to the ab-plane. The inversion centers are located at the origin and halfway from the origin in each direction along the a, b and c-axis and combinations of these directions. The crystal structure data for anorthite are a= 8.18A°, b= 12.88A°, c=14.169A° and α= 93.17, β = 115.85, γ = 91.22°. Its space group is Triclinic $P\bar{1}$.

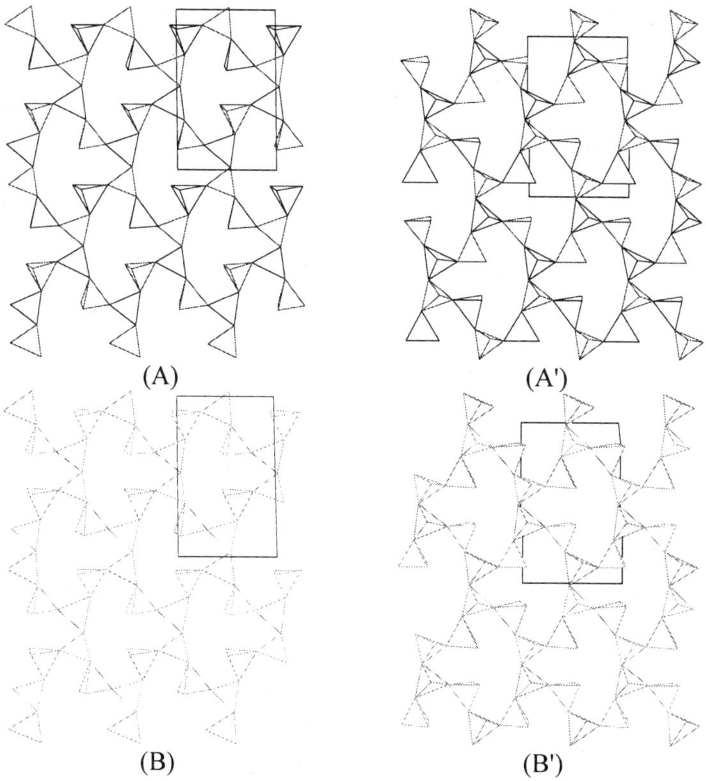

Figure 4-94 Four types of tetrahedra sheet in anorthite

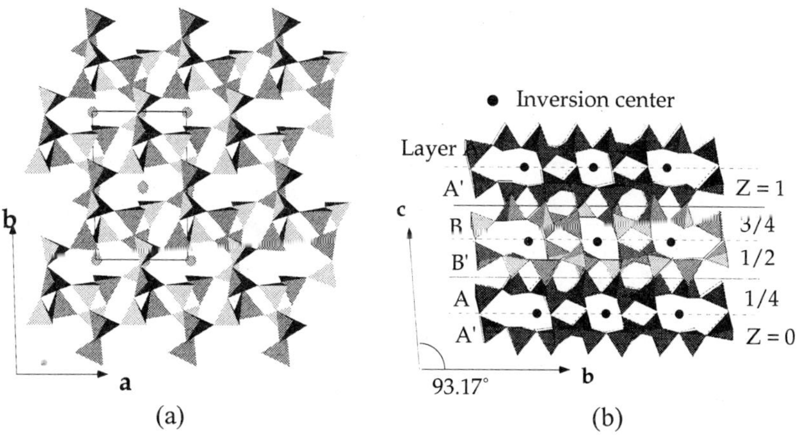

Figure 4-95 Crystal structure of anorthite (a) viewed along the c-axis and (b) viewed along the a-axis

(i) Description of feldspar by Double crankshaft

It has become conventional to describe feldspar and coecite structures by a double crankshaft. With respect to the crystal structure shown in Fig.4-94, a double crank shaft can be described as formed between layer A and B' and layer B and A'. This is shown in Fig.4-96. A double crankshaft runs along the a-axis. In other words, feldspar consists of two double crankshafts with distinctive configurations formed by layers A and B' and B and A'. The double crankshaft

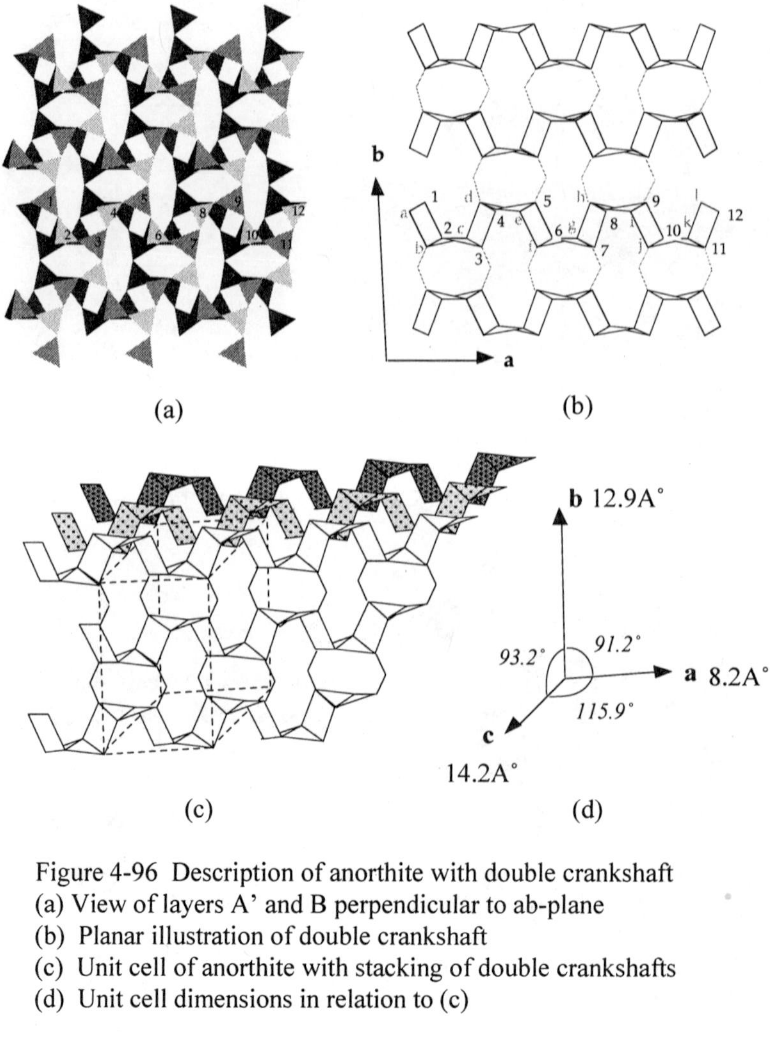

Figure 4-96 Description of anorthite with double crankshaft
(a) View of layers A' and B perpendicular to ab-plane
(b) Planar illustration of double crankshaft
(c) Unit cell of anorthite with stacking of double crankshafts
(d) Unit cell dimensions in relation to (c)

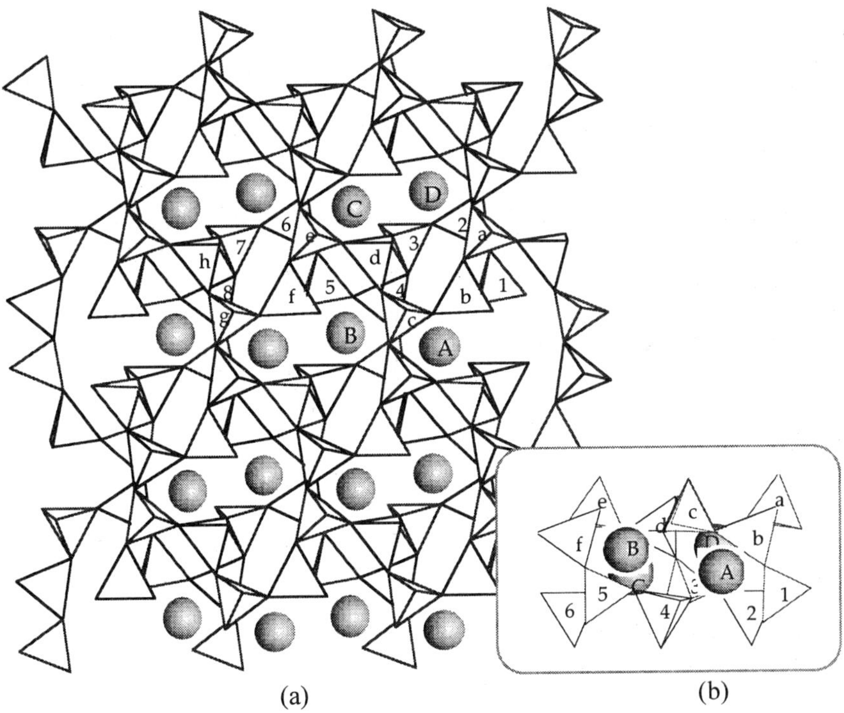

Figure 4-97 Locations of Ca^{+2} in anorthite
(a) View of layers A-A' perpendicular to ab-plane and (b) along b-axis

is configured in Fig.4-96(a) and (b), and atomic positions and bonding between them are illustrated in Fig.4-96. The advantage of the double crankshaft description of feldspar is its simplicity. In Fig.4-96(c) and (d), the double crankshaft shape of feldspar is drawn in a schematic way. This figure shows the positions of Si ions, the plane connecting the four Si's and the resulting unit cell of the triclinic lattice.

(ii) Locations of charge compensating cations (Ca^{+2} in anorthite)

In anorthite, Ca^{+2} ions are located between Layers A and A' and between B and B'. In other words, Ca^{+2}'s reside between two adjacent double crankshafts of the silica chain not within the double crankshaft. Four Ca's are held by sixteen tetrahedra (1-8 and a-h) as shown in Fig.4-97(b). Only one empty space for charge compensating cations is provided by the four tetrahedra. This ratio of one cation M^{+} (or M^{++}) per four tetrahedra is a structural characteristic of feldspar. If two of the four tetrahedra are AlO_4 like in anorthite, then M^{++} must be placed into the vacant space as shown in Fig.4-97(b) to compensate for the negative charges developed with the replacement of Si^{+4} by Al^{+3}. If one of the four tetrahedra is AlO_4, then M^{+} instead of M^{++} must be placed into this vacant space. If M is sodium, the feldspar formed is albite and if M is potassium it is orthoclase. If the portion of Si replaced by Al is not exactly half or quater but in between 0.25 and 0.5, the vacant spaces are filled with a mixture of M^{+} and M^{++}. This is the case for the plagioclase and alkali feldspars. In plagioclase, Ca^{++} and Na^{+} are distributed throughout the vacant spaces of the feldspar framework and in alkali feldspar, Na^{+} and K^{+} are so distributed. Since Ca and Na have very similar cation sizes, they easily form discrete crystallites within the feldspar framework as in plagioclase. K, however, does not form discrete crystallites with Ca due to size difference although it forms a crystallite with Na. The phase diagram for the feldspars shown in Fig.4-98 illustrate these points.

7-2-2 Albite

Albite has the same framework of tetrahedra and distribution of charge compensating cations within the framework as anorthite. However, the difference is in the configuration of the basic unit. In albite, Al occupies the end

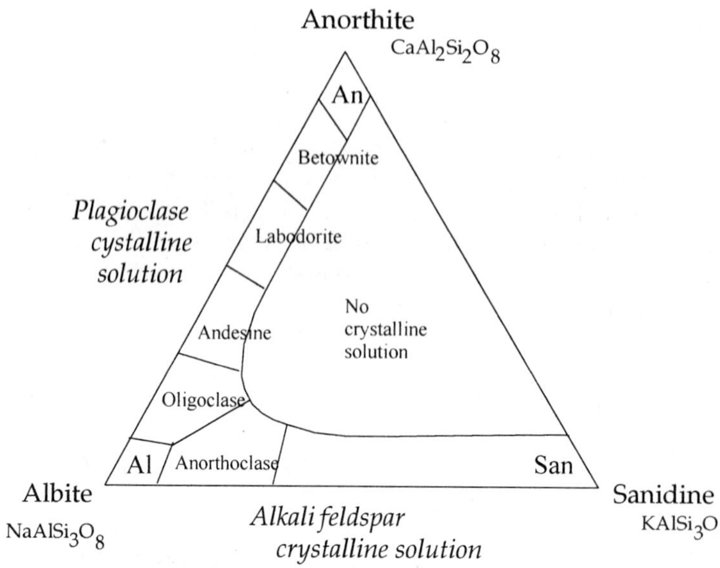

Figure 4-98 Phase diagram of feldspar

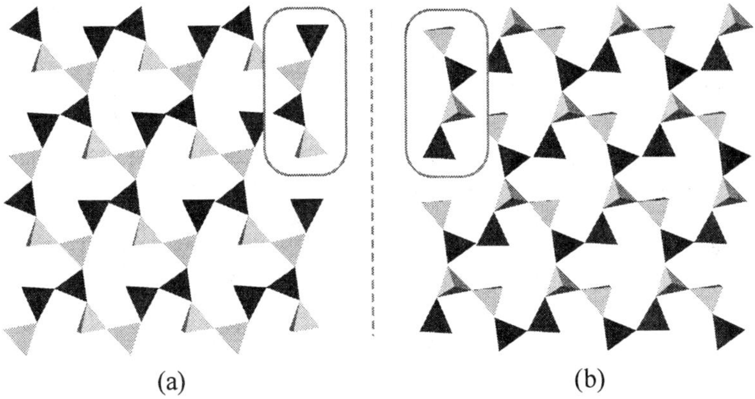

Figure 4-99 (a) Layer A and (b) Layer A' with basic units in the inserts

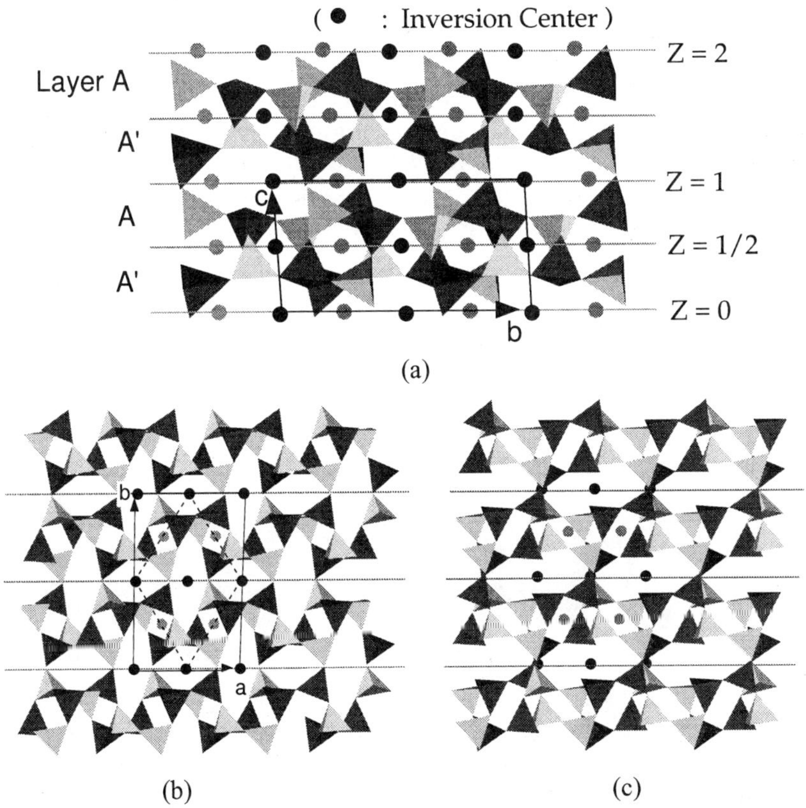

Figure 4-100 View of low albite along (a) the a-axis (b) along the c-axis with A(top)-A'(bottom) and (c) A'-A perpendicular to ab-plane

tetrahedra in the basic unit as shown in Fig.4-99. Due to this configuration, the stacking sequence of layers along the c-axis is A-A'-A-A' instead of the A-A'-B-B' of anorthitie. Hence the cell dimension along the c-axis is shortened by half compared to anorthite. The double crankshaft configuration is formed by A-A' in which Layer A is over Layer A'. Cations are located in between A'-A. In order to make the unit cell structure similar to that of anorthite, the unit cell of albite is not primary but rather C-face centered. This is shown in Fig.4-100. Therefore the unit cell of albite has two formula units, $NaAlSi_3O_8$. The cell dimensions of low albite are a=8.138, b=12.789, c=7.156, $\alpha = 94°20'$, $\beta = 116°34'$ and $\gamma = 87°39'$. The space group is Triclinic $P\bar{1}$.

7-2-3 K-feldspars ($KAlSi_3O_8$) – Monoclinic and Triclinic

These feldspars are formed when the charge deficiency in the framework silicates is balanced by K. Sanidine is the high temperature form with random distribution of Al and Si. This random distribution of Al and Si results in no difference among tetrahedra. A common form of K feldspar is microcline. Microcline has the ordered (AlO_4) tetrahedra which are crystallographically distinguished from silica tetrahedra. It has the same structure as that of albite and is a low temperature form whose space group is triclinic $P\bar{1}$. Since K^+ is

Microcline } Triclinic $P\bar{1}$

Orthoclase
Sanidine } Monoclinic $C\frac{2}{m}$

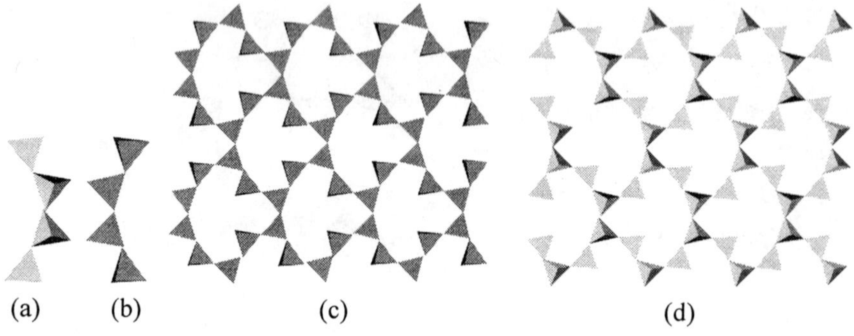

(a) (b) (c) (d)

Figure 4-101 (a) and (b) basic units of microline (c) layer A and (d) layer A'

(a)

(b)

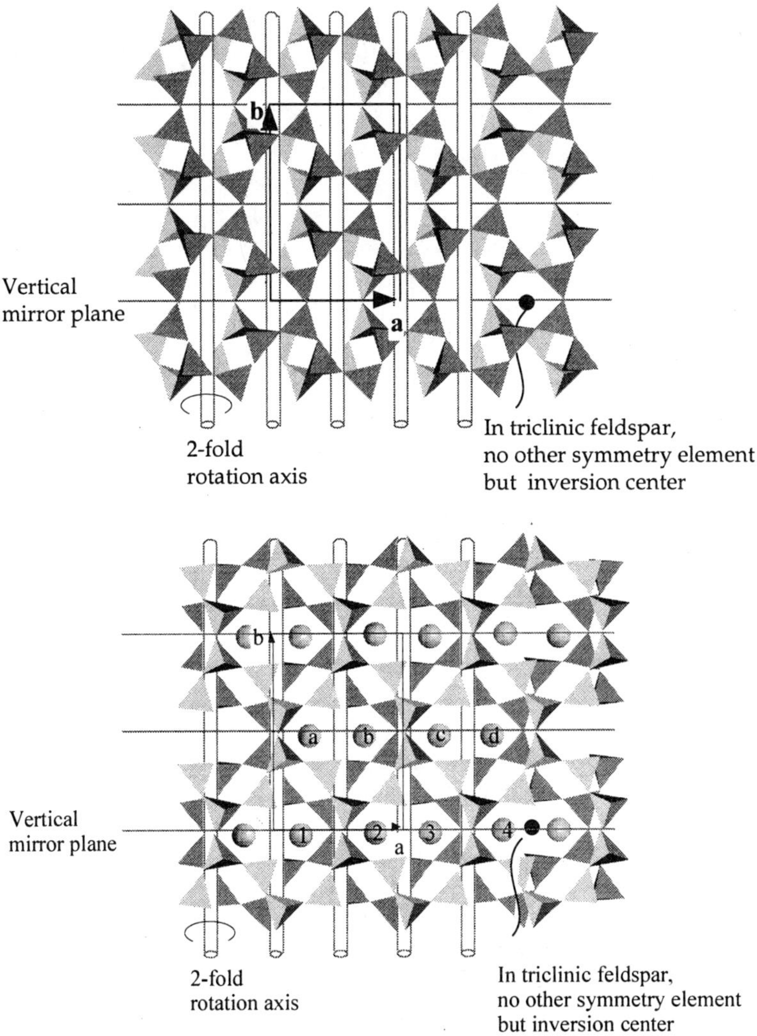

Figure 4-102 Illustration of symmetry elements in (a) A-A' and (b) A'-A (A-A' forms double crankshaft configuration.)

larger than Na^+, the unit cell dimension of microline is a little larger than that of albite. Its cell parameters are a=8.56, b=12.964, c=7.215, α = 90°39', β = 115°50' and γ = 87°42'.

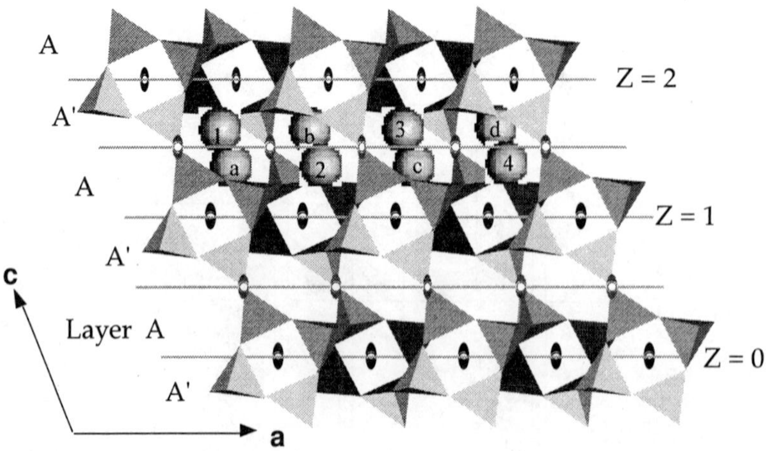

Figure 4-103 Illustration of K^+ positions and symmetry elements viewed along b-axis

Orthoclase is intermediate in randomness between sanidine and microcline. Its space group is monoclinic $C\frac{2}{m}$. Its monoclinic crystal structure is illustrated in Fig.4-102 and Fig.4-103. The basic unit of orthoclase is the same as that of anorthite or albite. However, the difference is in its symmetry. The basic unit has a vertical mirror plane in the middle along the length. Due to this symmetry, the two dimensional lattice has a rectangular unit cell as shown in Fig.4-101(c) and (d). Also the stacking sequence along the c-axis of these two dimensional lattices are A-A'-A-A' in which A and A' are inverse images. Hence orthoclase has not only an inversion center but also a 2-fold rotation axis along the b-axis and a mirror plane perpendicular to the b-axis. This is shown in Fig.4-102. The location of K^+ ion is shown in Fig.4-103. Similar to anorthite and albite, K^+ resides between the double crankshaft of the tetrahedral chain. K-feldspar is an essential ingredient in ceramic porcelain bodies.

7-2-4 Plagioclase Feldspars ($Na_{1-x}Ca_x(Si_{3-x}Al_{1+x})O_8$) – Triclinic

Isomorphous substitution is common between Na and Ca Feldspars resulting in several intermediate minerals being formed. Na-feldspar is used similar to K-feldspar as a main ceramic raw material and as flux component in whiteware bodies and glasses.

7-2-5 Feldspathoid Minerals

Anhydrous framework silicates, similar to feldspars. These minerals are based on other silica polymorphs than coecite on which feldspar is based. As mentioned in the introduction to silica polymorphs, the main difference between feldspathoid from feldspar is in the alkali (or alkali earth) content. Consequently, the chief chemical difference lies in the SiO_2 content. Feldspathoids such as Eucryptite, Nepheline and Kalsilite contain only about 2/3 as much SiO_2 as alkali feldspars, hence they are better fluxes. Li minerals are based on the quartz structure. Na and K minerals are used as fluxes for glass manufacture and ceramic processing. The Li minerals are also source of Li. However, those like Leucite and Spodumene are not derived from silica polymorphs, but have a different crystal structure. Leucite belongs to the zeolite group and spodumene is derived from diopside in which Ca and Mg are replaced by Al and Li, respectively. In spodumene, the coordination of Al is six rather than four. Hence spodumene is in the group of pyroxene which is not a framework structure of silica. Fig.4-104(b) and (c) illustrate the derivative

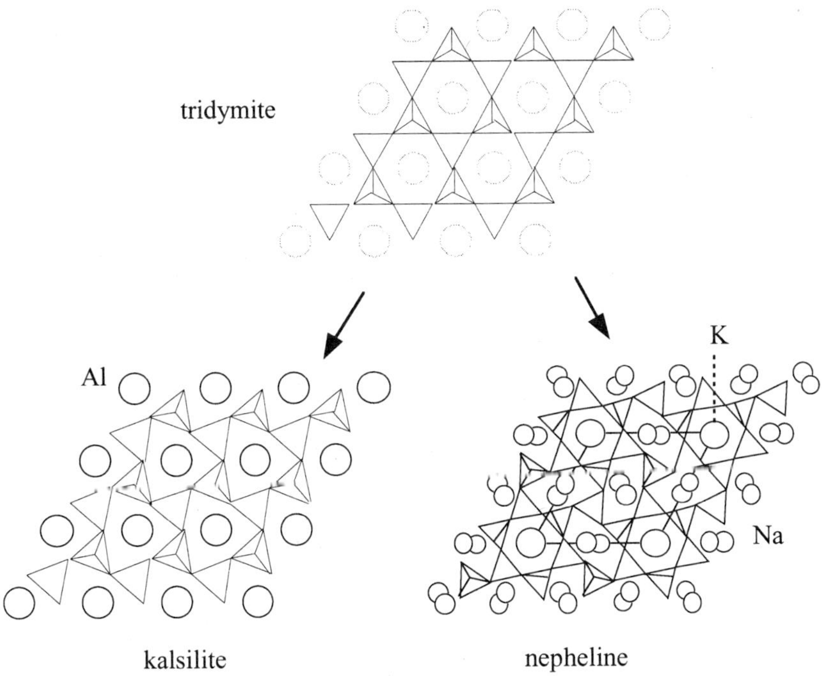

Figure 4-104 Derivative structure of kalsilite and nepheline from tridymite

structures of tridymite for kalsilite and nepheline which has the distorted hexagonal ring of tridymite as in Fig.4-104(a).

SiO_2 (Quartz)	→	$Li_{1/2}$	$\begin{bmatrix} Si_{1/2} \\ Al_{1/2} \end{bmatrix} O_2$	→	$LiAlSiO_4$ (Eucryptite)
SiO_2 (Tridymite)	→	$K_{1/8}\,Na_{3/8}$	$\begin{bmatrix} Si_{1/2} \\ Al_{1/2} \end{bmatrix} O_2$	→	$KNa_3Al_4Si_4O_{16}$ (Nepheline)
SiO_2 (Tridymite)	→	$K_{1/2}$	$\begin{bmatrix} Si_{1/2} \\ Al_{1/2} \end{bmatrix} O_2$	→	$KAlSiO_4$ (Kalsilite)
SiO_2	→	$K_{1/3}$	$\begin{bmatrix} Si_{2/3} \\ Al_{1/3} \end{bmatrix} O_2$	→	$LiAlSi_2O_6$ (Leucite)
$CaMgSi_2O_6$ (Diopside)	→	LiAl	Si_2O_6	→	$LiAlSi_2O_4$ (Spodumene)

7-3. Channeled frame silicates - Zeolites

Zeolites form a large group of hydrous silicates and are similar to feldspars. But the zeolites contain large channels which may absorb cations or trap water. Zeolites are characterized by the extent of Al^{3+} substitution for Si^{4+}. Listed below are the typical zeolite minerals. These show the variation in water content in the structure and the difference in the extent of substitution of Si with Al.

Analcite	$NaAlSi_3O_8 \cdot H_2O$
Natrolite	$Na_2Al_2Si_3O_{10} \cdot 2H_2O$
Heulandite	$CaAl_2Si_7O_{18} \cdot 6H_2O$

Due to the large channels, zeolites are used as molecular sieves and catalysts. As molecular sieves, the water in the channels can be removed by heating, and the channels may be filled by other chemical species. The absorption process of chemical species is selective, depending on the size of the channels and of the molecules. Hence, the channels function as "sieves" for different size molecules. Also zeolites can be used as ion exchangers. Ca^{2+} ions in water are readily replaced in the channels by Na^+ ions (and vice versa). The reaction is reversible as shown in Eq.(18). This cation exchange is the basis for water softening.

$$2Na\left(AlSi_2O_6\right) \cdot H_2O + Ca^{2+} \Leftrightarrow Ca^{2+}\left(AlSi_2O_6\right)_2 \cdot 2H_2O + Na^+ \qquad (18)$$

Regeneration takes place by reverse treatment with (hot) brine (NaCl) solution.

7-3-1 Crsystal structures of zeolites

Another useful property of the zeolites in addition to channel sizes is their rigidity during dehydration, so that they do not shrink or change shape. However, not all zeolites have this property. Usually the useful zeolites are synthetic since channel size can be made appropriate for selectivity of the chemical species in catalysis. The channel sizes can be controlled by controlling the chemical composition of zeolites as well as the reaction conditions. With respect to crystal structure, channel size is varied by the number of ring tetrahedra in the structure – 4, 5, 6, 8, 12 ring tetrahedra.

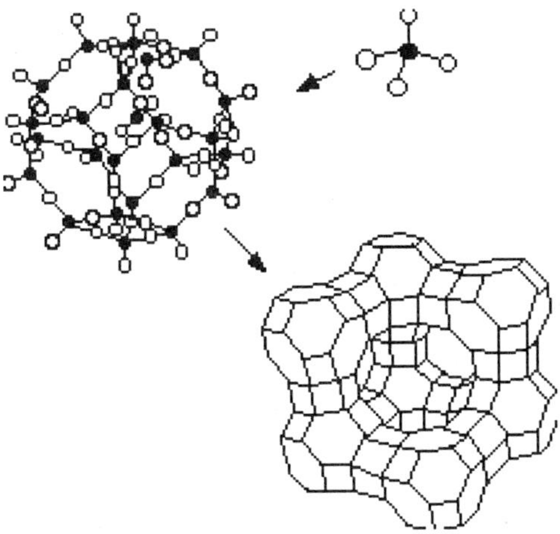

Figure 4-105 Schematic drawing of framework of zeolites from silica tetrahedron

<u>Crystal Structure States</u>

A. <u>Crystalline State</u> - - - exists when there is a periodic arrangement of the atoms or ions in a solid such that a unit of repetition (the unit cell) can be defined which by transition forms the crystal, e.g., NaCl, MgO and diamond.

B. <u>Amorphous</u> (Non-Crystalline State) - - - exists where there is a lack of long range order or periodicity in the solid, may be ordering over only small regions, e.g., glasses, polymer (most of it) and liquids.

C. <u>Polymorphics</u> - - - represents changes in crystal structure brought about by external parameters such as pressure, temperature, e.g., quartz, opal → fused silica glass.

Crystal Structure Analysis

1. Crystal Morphology
   Measurement of faces, angles, zone relationship on actual crystals.
2. Chemical Composition
   Density Determination / X-ray Fluorescence / Wet Chemical Analysis
3. X-ray Diffraction Analysis
   Unit Cell Dimension / Atomic Plane Orientation and Spacing / Atom-atom Distances (Ionic Radii) / Phase Analysis
4. Electron Diffraction Analysis
   Same as 3.
   Microsamples less convenient
5. Spectroscopic Analysis
   NMR (Nuclear Magnetic Resonance) / IR (Infrared) / UV (Ultraviolet)

Appendix 1

Special functions in quantum mechanics

I. Legendre Equation

$$(1-x^2)y''-2xy'+n(n+1)y=0 \tag{1}$$

The solution of this equation is called Legendre Polynomials

$$P_n = \frac{1}{2^n n!}\left(\frac{d}{dx}\right)^n \sum_{t=0}^{T} \frac{(-1)^t n!}{t!(n-t)!} x^{2n-2t}$$

$$= \frac{1}{2^n n!} \sum_{t=0}^{T} \frac{(-1)^t n!}{t!(n-t)!} \frac{(2n-2t)!}{(n-2t)!} x^{n-2t} \tag{2}$$

Here T is the largest number not greater than n/2

The proof that Eq.(2) is the solution for Legendre Equation expressed as Eq.(1) is done for the case of n as even number. Since the proof in case of odd 'n' is similar, this will be omitted.

The first and second derivative of Pn are as follow:

$$P_n' = \frac{1}{2^n n!} \sum_{t=0}^{T-1} \frac{(-1)^t n!}{t!(n-t)!} \frac{(2n-2t)!}{(n-2t-1)!} x^{n-2t-1} \tag{3}$$

$$P_n'' = \frac{1}{2^n n!} \sum_{t=0}^{T-1} \frac{(-1)^t n!}{t!(n-t)!} \frac{(2n-2t)!}{(n-2t-2)!} x^{n-2t-2} \tag{4}$$

Here, notice that the summation for both derivatives extends to t = T-1 since, when t = T, derivative of x^{n-2t} in the summation is equal to zero. If n =

2T + 1, the summation for the first derivative of Pn extend to t = T while the summation for the second derivative extends to t = T – 1. This is the difference between odd number and even number cases for n. The rest of process of the proof is the same.

$$P_n' = \frac{1}{2^n n!} \sum_{t=0}^{T} \frac{(-1)^t n!}{t!(n-t)!} \frac{(2n-2t)!}{(n-2t-1)!} x^{n-2t-1} \qquad \text{if } n = 2T+1 \tag{3'}$$

$$P_n'' = \frac{1}{2^n n!} \sum_{t=0}^{T-1} \frac{(-1)^t n!}{t!(n-t)!} \frac{(2n-2t)!}{(n-2t-2)!} x^{n-2t-2} \tag{4'}$$

Inserting Eq.(2), (3) and (4), Eq.(1) becomes

$$= (1-x^2) \frac{1}{2^n n!} \sum_{t=0}^{T-1} \frac{(-1)^t n!}{t!(n-t)!} \frac{(2n-2t)!}{(n-2t-2)!} x^{n-2t-2} \tag{1a}$$

$$-2x \frac{1}{2^n n!} \sum_{t=0}^{T-1} \frac{(-1)^t n!}{t!(n-t)!} \frac{(2n-2t)!}{(n-2t-1)!} x^{n-2t-1} \tag{1b}$$

$$n(n+1) \frac{1}{2^n n!} \sum_{t=0}^{T} \frac{(-1)^t n!}{t!(n-t)!} \frac{(2n-2t)!}{(n-2t)!} x^{n-2t} \tag{1c}$$

If (1a) is expanded with respect to the order of x in the summation

$$(1-x^2) \frac{1}{2^n n!} \sum_{t=0}^{T-1} \frac{(-1)^t n!}{t!(n-t)!} \frac{(2n-2t)!}{(n-2t-2)!} x^{n-2t-2}$$

$$= \frac{1}{2^n n!} \sum_{t=0}^{T-1} \frac{(-1)^t n!}{t!(n-t)!} \frac{(2n-2t)!}{(n-2t-2)!} x^{n-2t-2} \tag{1aa}$$

$$- \frac{1}{2^n n!} \sum_{t=0}^{T-1} \frac{(-1)^t n!}{t!(n-t)!} \frac{(2n-2t)!}{(n-2t-2)!} x^{n-2t} \tag{1ab}$$

To make the order of x in the summation of Eq.(1aa), t is replaced by t'-1

$$\frac{1}{2^n n!} \sum_{t'=1}^{T} \frac{(-1)^{t'-1} n!}{(t'-1)!(n-t'+1)!} \frac{(2n-2t'+2)!}{(n-2t')!} x^{n-2t'} \tag{1aa'}$$

Note that the summation is from t'=1 to T. Since the summation does not change with the change of t' to t

$$\frac{1}{2^n n!}\sum_{t=1}^{T}\frac{(-1)^{t-1}n!}{(t-1)!(n-t+1)!}\frac{(2n-2t+2)!}{(n-2t)!}x^{n-2t} \qquad (1aa'')$$

To change the range of summation of Eq.(1aa"), it becomes

$$\frac{1}{2^n n!}\sum_{t=1}^{T-1}\frac{(-1)^{t-1}n!}{(t-1)!(n-t+1)!}\frac{(2n-2t+2)!}{(n-2t)!}x^{n-2t} \qquad (1aa\text{-}1)$$

$$+\frac{1}{2^n n!}\frac{(-1)^{T-1}n!}{(T-1)!(n-T+1)!}\frac{(2n-n+2)!}{(n-n)!}x^{0} \qquad (1aa\text{-}2)$$

Likewise Eq(1ab) becomes

$$-\frac{1}{2^n n!}\sum_{t=0}^{T-1}\frac{(-1)^{t}n!}{t!(n-t)!}\frac{(2n-2t)!}{(n-2t-2)!}x^{n-2t}$$

$$=-\frac{1}{2^n n!}\sum_{t=1}^{T-1}\frac{(-1)^{t}n!}{t!(n-t)!}\frac{(2n-2t)!}{(n-2t-2)!}x^{n-2t} \qquad (1ab\text{-}1)$$

$$-\frac{1}{2^n n!}\frac{(-1)^{0}n!}{0!n!}\frac{2n!}{(n-2)!}x^{n} \qquad (1ab\text{-}2)$$

Making the range of the summation from t=1 through T-1, Eq.(1b) and (1c) become

$$-2x\frac{1}{2^n n!}\sum_{t=0}^{T-1}\frac{(-1)^{t}n!}{t!(n-t)!}\frac{(2n-2t)!}{(n-2t-1)!}x^{n-2t-1} \qquad (1b)$$

$$=-2\frac{1}{2^n n!}\sum_{t=1}^{T-1}\frac{(-1)^{t}n!}{t!(n-t)!}\frac{(2n-2t)!}{(n-2t-1)!}x^{n\ 2t} \qquad (5b\text{-}1)$$

$$-2\frac{1}{2^n n!}\frac{(-1)^{0}n!}{0!n!}\frac{2n!}{(n-1)!}x^{n} \qquad (5b\text{-}2)$$

$$n(n+1)\frac{1}{2^n n!}\sum_{t=0}^{T}\frac{(-1)^t n!}{t!(n-t)!}\frac{(2n-2t)!}{(n-2t)!}\;x^{n-2t} \qquad (1c)$$

$$= \; n(n+1)\frac{1}{2^n n!}\;\frac{(-1)^T n!}{T!(n-T)!}\frac{(2n-2T)!}{(n-2T)!}\;x^{n-2T} \qquad (5c\text{-}1)$$

$$+\; n(n+1)\frac{1}{2^n n!}\sum_{t=1}^{T-1}\frac{(-1)^t n!}{t!(n-t)!}\frac{(2n-2t)!}{(n-2t)!}\;x^{n-2t} \qquad (5c\text{-}2)$$

$$+\; n(n+1)\frac{1}{2^n n!}\;\frac{(-1)^0 n!}{0!n!}\frac{2n!}{n!}\;x^n \qquad (5c\text{-}3)$$

Combining Eq.(1aa), (1ab), (5b), and (5c)

$$(1-x^2)y'' - 2xy' + n(n+1)y$$

$$= \frac{1}{2^n n!}\sum_{t=1}^{T-1}\frac{(-1)^{t-1} n!}{(t-1)!(n-t+1)!}\;\frac{(2n-2t+2)!}{(n-2t)!}\,x^{n-2t} \qquad (6aa\text{-}1)$$

$$+\;\frac{1}{2^n n!}\frac{(-1)^{T-1} n!}{(T-1)!(n-T+1)!}\;\frac{(2n-n+2)!}{(n-n)!}\,x^0 \qquad (6aa\text{-}2)$$

$$-\frac{1}{2^n n!}\sum_{t=1}^{T-1}\frac{(-1)^t n!}{t!(n-t)!}\frac{(2n-2t)!}{(n-2t-2)!}\,x^{n-2t} \qquad (6ab\text{-}1)$$

$$-\frac{1}{2^n n!}\;\frac{(-1)^0 n!}{0!n!}\;\frac{2n!}{(n-2)!}\;x^n \qquad (6ab\text{-}2)$$

$$-2\frac{1}{2^n n!}\sum_{t=1}^{T-1}\frac{(-1)^t n!}{t!(n-t)!}\frac{(2n-2t)!}{(n-2t-1)!}\,x^{n-2t} \qquad (6b\text{-}1)$$

$$-2\frac{1}{2^n n!}\;\frac{(-1)^0 n!}{0!n!}\;\frac{2n!}{(n-1)!}\;x^n \qquad (6b\text{-}2)$$

$$+\; n(n+1)\frac{1}{2^n n!}\;\frac{(-1)^T n!}{T!(n-T)!}\frac{(2n-2T)!}{(n-2T)!}\;x^{n-2T} \qquad (6c\text{-}1)$$

$$+\ n(n+1)\frac{1}{2^n n!}\sum_{t=1}^{T-1}\frac{(-1)^t n!}{t!(n-t)!}\frac{(2n-2t)!}{(n-2t)!}\,x^{n-2t} \qquad (6c\text{-}2)$$

$$+\ n(n+1)\frac{1}{2^n n!}\frac{(-1)^0 n!}{0!n!}\frac{2n!}{n!}\,x^n \qquad (6c\text{-}3)$$

By arranging the terms in the same order for x and the summations together

(6aa-2) + (6c-1)

$$=\frac{1}{2^n n!}\frac{(-1)^{T-1}n!}{(T-1)!(T+1)!}\frac{n!(n+1)(n+2)}{0!}x^0-n(n+1)\frac{1}{2^n n!}\frac{(-1)^{T-1}n!}{T!T!}\frac{n!}{0!}\,x^0$$

$$=\frac{1}{2^n n!}\frac{(-1)^{T-1}n!n!}{T!T!}\left\{\frac{T(2T+1)(2T+2)}{(T+1)}-2T(2T+1)\right\}=0$$

(6ab-1) + (6b-2) + (6c-3)

$$=-\frac{1}{2^n n!}\frac{(-1)^0 n!}{0!n!}\frac{2n!}{(n-2)!}\,x^n$$

$$-2\frac{1}{2^n n!}\frac{(-1)^0 n!}{0!n!}\frac{2n!}{(n-1)!}\,x^n+n(n+1)\frac{1}{2^n n!}\frac{(-1)^0 n!}{0!n!}\frac{2n!}{n!}\,x^n$$

$$=\frac{1}{2^n n!}\frac{(-1)^0 2n!}{0!n!}x^n\{-n(n-1)-2n+n(n+1)\}=0$$

(6aa-1) + (6ab-1) + (6b-1) + (6c-2)

$$=\frac{1}{2^n n!}\sum_{t=1}^{T-1}\frac{(-1)^{t-1}n!}{t!(n-t)!}\frac{(2n-2t)!}{(n-2t)!}\frac{(2n-2t+2)(2n-2t+1)t}{(n-t+1)}\,x^{n-2t}$$

$$-\frac{1}{2^n n!}\sum_{t=1}^{T-1}\frac{(-1)^t n!}{t!(n-t)!}\frac{(2n-2t)!}{(n-2t)!}\frac{(n-2t)(n-2t-1)}{1}\,x^{n-2t}$$

$$-2\frac{1}{2^n n!}\sum_{t=1}^{T-1}\frac{(-1)^t n!}{t!(n-t)!}\frac{(2n-2t)!}{(n-2t)!}\frac{(n-2t)}{1}\,x^{n-2t}$$

$$+ n(n+1)\frac{1}{2^n n!}\sum_{t=1}^{T-1}\frac{(-1)^t n!}{t!(n-t)!}\frac{(2n-2t)!}{(n-2t)!}x^{n-2t}$$

$$= \frac{1}{2^n n!}\sum_{t=1}^{T-1}\frac{(-1)^t n!}{t!(n-t)!}\frac{(2n-2t)!}{(n-2t)!}x^{n-2t}\{-2(2n-2t+1)t$$

$$-(n-2t)(n-2t-1)-2(n-2t)+n(n+1)\} \tag{7}$$

Since $-2(2n-2t+1)t-(n-2t)(n-2t-1)-2(n-2t)+n(n+1)=0$,

Eq.(7) is equal to zero. Hence Eq.(1) is equal to zero when Pn and its derivatives are inserted. Therefore, it is proved that the polynomial expressed by Eq.(2) is the solution for Legendre equation in Eq.(1).

II. Associated Legendre Equation

In quatum mechanics, what is encounted is the associated Legendre equation which is the same except for a extra term, $-\frac{m^2}{1-x^2}$ y, in the equation. Eq.(8) is called the associated Legendre Equation. Its solution is based on the Legendre polynomial expressed in Eq.(2). The solution for the associated Legendre equation is expressed as Eq.(9). There appears two scripts in the representation for the associated Legendre polynomial which are n and m.

$$(1-x^2)y'' - 2xy' + \left(n(n+1)-\frac{m^2}{\left(1-x^2\right)}\right)y = 0 \tag{8}$$

$$P_n^m = (1-x^2)^{m/2}\frac{d^m}{dx^m}P_n \tag{9}$$

The proof that Eq.(9) is the solution for the associated Legendre equation starts from differentiate m times the Legendre Equation. To make the proof is easier, Leibnitz formula expressed as Eq.(10) is applied.

$$\frac{d^m}{dx^m}A(x)B(x) = \sum_{t=0}^{m}\left[\frac{m!}{(m-t)!t!}\frac{d^{m-t}}{dx^{m-t}}A(x)\frac{d^t}{dx^t}B(x)\right] \tag{10}$$

$$(1-x^2)P_n'' - 2xP_n' + n(n+1)\ P_n = 0 \tag{11}$$

Differentiate Eq.(8) m times

$$\frac{d^m}{dx^m}\left\{(1-x^2)P_n'' - 2xP_n' + n(n+1)\ P_n\right\} = 0 \tag{12}$$

According to Leibnitz formula, Eq.(12) becomes

$$0 = \sum_{t=0}^{m}\left[\frac{m!}{(m-t)!t!}\frac{d^{m-t}}{dx^{m-t}}P_n''\frac{d^t}{dx^t}(1-x^2)\right] \tag{13a}$$

$$+\sum_{t=0}^{m}\left[\frac{m!}{(m-t)!t!}\frac{d^{m-t}}{dx^{m-t}}P_n'\frac{d^t}{dx^t}2x\right] \tag{13b}$$

$$+\sum_{t=0}^{m}\left[\frac{m!}{(m-t)!t!}\frac{d^{m-t}}{dx^{m-t}}P_n\frac{d^t}{dx^t}n(n+1)\right] \tag{13c}$$

Since the terms in the summation survives when $t \le 2$ in (13a), $t \le 1$ in (13b), $t \le 0$ in (13c), Eq.(12) becomes

$$0 = \sum_{t=0}^{2}\left[\frac{m!}{(m-t)!t!}\frac{d^{m-t}}{dx^{m-t}}P_n''\frac{d^t}{dx^t}(1-x^2)\right]$$

$$+\sum_{t=0}^{1}\left[\frac{m!}{(m-t)!t!}\frac{d^{m-t}}{dx^{m-t}}P_n'\frac{d^t}{dx^t}(-2x)\right]$$

$$+\sum_{t=0}^{0}\left[\frac{m!}{(m-t)!t!}\frac{d^{m-t}}{dx^{m-t}}P_n\frac{d^t}{dx^t}n(n+1)\right] \tag{14}$$

$$= (1-x^2)\frac{d^m}{dx^m}P_n'' \quad 2mx\frac{d^{m-1}}{dx^{m-1}}P_n'' - m(m-1)\frac{d^{m-2}}{dx^{m-2}}P_n''$$

$$- 2x\frac{d^m}{dx^m}P_n' - 2m\frac{d^{m-1}}{dx^{m-1}}P_n'$$

$$+ n(n+1)\frac{d^m}{dx^m}P_n \tag{15}$$

Replacing $\frac{d^m}{dx^m}P_n$ by 'u', Eq.(15) becomes

$$0 = (1-x^2)u'' - 2mxu' - m(m-1)u - 2xu' - 2mu + n(n+1)u \tag{16}$$

$$0 = (1-x^2)u'' - 2(m+1)xu' + (n-m+1)(n+m)u \tag{17}$$

If a new function is defined as

$$v = \left(1-x^2\right)^{m/2} u = \left(1-x^2\right)^{m/2} \frac{d^m}{dx^m}P_n \quad \text{and} \quad u = \frac{v}{\left(1-x^2\right)^{m/2}} \tag{18}$$

The first derivative of v(x) with respect to x is

$$v' = -mx\left(1-x^2\right)^{m-2/2} u + \left(1-x^2\right)^{m/2} u' \tag{19}$$

Inserting Eq.(19) for u and rearranging with respect to u'

$$u' = \frac{1}{\left(1-x^2\right)^{m/2}}\left\{v' + mx\frac{v}{\left(1-x^2\right)}\right\} \tag{20}$$

Differentiate v twice with respect to x

$$v'' = -m\left(1-x^2\right)^{m-2/2} u$$

$$+ mx^2(m-2)\left(1-x^2\right)^{m-4/2} u - 2mx\left(1-x^2\right)^{m-2/2} u' + \left(1-x^2\right)^{m/2} u'' \tag{21}$$

Inserting Eq.(18) and (20) for u and u' in Eq.(21)

$$v'' = -m\left(1-x^2\right)^{m-2/2}\frac{v}{\left(1-x^2\right)^{m/2}} + m(m-2)x^2\left(1-x^2\right)^{m-4/2}\frac{v}{\left(1-x^2\right)^{m/2}}$$

$$- 2mx\left(1-x^2\right)^{m-2/2}\frac{1}{\left(1-x^2\right)^{m/2}}\left\{v' + mx\frac{v}{\left(1-x^2\right)}\right\} + \left(1-x^2\right)^{m/2} u'' \tag{22}$$

$$v'' = -m\frac{v}{\left(1-x^2\right)} + m(m-2)x^2\frac{v}{\left(1-x^2\right)^2}$$

$$- 2mx\frac{1}{\left(1-x^2\right)}\left\{v' + mx\frac{v}{\left(1-x^2\right)}\right\} + \left(1-x^2\right)^{m/2}u'' \tag{23}$$

Rearranging Eq.(23) with respect to u"

$$\left(1-x^2\right)^{m/2}u'' = v'' + m\frac{v}{\left(1-x^2\right)} - m(m-2)x^2\frac{v}{\left(1-x^2\right)^2}$$

$$+ 2mx\frac{1}{\left(1-x^2\right)}\left\{v' + mx\frac{v}{\left(1-x^2\right)}\right\}$$

$$u'' = \frac{1}{\left(1-x^2\right)^{m/2}}\left[v'' + m\frac{v}{\left(1-x^2\right)} - m(m-2)x^2\frac{v}{\left(1-x^2\right)^2}\right.$$

$$\left. + 2mx\frac{1}{\left(1-x^2\right)}\left\{v' + mx\frac{v}{\left(1-x^2\right)}\right\}\right]$$

$$u'' = \frac{1}{\left(1-x^2\right)^{m/2}}\left[v'' + m\frac{v}{\left(1-x^2\right)} + \frac{2mxv'}{\left(1-x^2\right)}v' + \frac{m(m+2)x^2}{\left(1-x^2\right)^2}v\right] \tag{24}$$

Inserting Eq(18), (20) and (24) for u, u' and u" in Eq.(17)

$$(1-x^2)u'' - 2(m+1)xu' + (n-m+1)(n+m)u$$

$$= (1-x^2)\frac{1}{\left(1-x^2\right)^{m/2}}\left[v'' + \frac{m}{\left(1-x^2\right)}v + \frac{2mx}{\left(1-x^2\right)}v' + \frac{m(m+2)x^2}{\left(1-x^2\right)^2}v\right]$$

$$- 2(m+1)x\frac{1}{\left(1-x^2\right)^{m/2}}\left\{v' + mx\frac{v}{\left(1-x^2\right)}\right\}$$

$$+ (n-m+1)(n+m)\frac{v}{\left(1-x^2\right)^{m/2}} \tag{25}$$

$$= \frac{1}{\left(1-x^2\right)^{m/2}}\left[(1-x^2)v'' + mv + 2mxv' + \frac{m(m+2)x^2}{\left(1-x^2\right)}v \right.$$

$$\left. - 2(m+1)x\left\{v' + mx\frac{v}{\left(1-x^2\right)}\right\} + (n-m+1)(n+m)v\right] \quad (26)$$

$$= \frac{1}{\left(1-x^2\right)^{m/2}}\left[(1-x^2)v'' + 2mxv' + \frac{m(m+2)x^2}{\left(1-x^2\right)}v \right.$$

$$\left. - 2(m+1)x\left\{v' + mx\frac{v}{\left(1-x^2\right)}\right\} + n(n+1)v - m^2v\right] \quad (27)$$

Since

$$2mxv' + \frac{m(m+2)x^2}{\left(1-x^2\right)}v - 2(m+1)\,x\left\{v' + mx\frac{v}{\left(1-x^2\right)}\right\} = -\,2xv' - \frac{m^2x^2}{\left(1-x^2\right)}v$$

$$\text{Eq.(27)} = \frac{1}{\left(1-x^2\right)^{m/2}}\left[(1-x^2)v'' - 2xv' - \frac{m^2x^2}{\left(1-x^2\right)}v + n(n+1)v - m^2v\right]$$

$$= \frac{1}{\left(1-x^2\right)^{m/2}}\left[(1-x^2)v'' - 2xv' - \frac{m^2}{\left(1-x^2\right)}v + n(n+1)v\right] \quad (28)$$

The equation in the bracket is the associated Legendre Equation. Therefore, it is shown that v which is equal to the function in Eq.(8), is the solution for the associated Legendre Equation.

III Laguerre Equation

$$xy''(x) + (1-x)\,y'(x) + ny(x) = 0$$

The solution of this equation is called Laguerre polynomial and is expressed as

$$y(x) = L_n(x) = L_n = \sum_{t=0}^{n}(-1)^t \frac{n!}{(n-t)!t!t!}x^t$$

IV Associated Laguerre Equation

Like Legendre equation, the associated Laguerre eqaution is the same as Laguerre equation except for a extra term ky'(x) and has the solution which is based on the Laguerre equation.

$$xy''(x) + (k+1-x)\,y'(x) + ny(x) = 0 \tag{29}$$

$$L_n^k = (-1)^k \frac{d^k}{dx^k} L_{n+k} = \sum_{t=0}^{n}(-1)^t \frac{(n+k)!}{(n-t)!(k+t)!t!}x^t \tag{30}$$

The proof that the polynomial in Eq.(29) satisfies the associated Laguerre equation expressed in Eq.(30) is carried out by recurrsion relationships between L_n^k and its derivatives. Therefore, the proof begins with proving the recurrsion relationships. The recurrsion relationships used to prove that the Eq.(30) satisfies the associated Laguerre equation in Eq.(29) are

$$(n+1)L_{n+1}^k = (2n+k+1-x)L_n^k - (n+k)L_{n-1}^k \tag{31}$$

$$xL_n^{k\,\prime} = nL_n^k - (n+k)L_{n-1}^k \tag{32}$$

Recurrsion relationship 1

The first recurssion relationship is to be proved by showing that Eq.(d) is equal to zero when polynomials in Eq.(a), (b) and (c) are put to Eq.(d).

$$L_{n+1}^k = \sum_{t=0}^{n+1}(-1)^t \frac{(n+1+k)!}{(n+1-t)!(k+t)!t!}x^t \tag{a}$$

$$L_n^k = \sum_{t=0}^{n}(-1)^t \frac{(n+k)!}{(n-t)!(k+t)!t!}x^t \tag{b}$$

$$L_{n-1}^k = \sum_{t=0}^{n-1}(-1)^t \frac{(n-1+k)!}{(n-1-t)!(k+t)!t!}x^t \tag{c}$$

$$(n+1)L_{n+1}^{k} - (2n+k+1-x)L_{n}^{k} + (n+k)L_{n-1}^{k} \tag{d}$$

Putting Eq.(a), (b) and (c) for L_{n+1}^{k}, L_{n}^{k}, and L_{n-1}^{k} in Eq.(d)

$$(n+1)\sum_{t=0}^{n+1}(-1)^t \frac{(n+1+k)!}{(n+1-t)!(k+t)!t!}x^t \tag{e-1}$$

$$-(2n+k+1-x)\sum_{t=0}^{n}(-1)^t \frac{(n+k)!}{(n-t)!(k+t)!t!}x^t \tag{e-2}$$

$$+(n+k)\sum_{t=0}^{n-1}(-1)^t \frac{(n-1+k)!}{(n-1-t)!(k+t)!t!}x^t \tag{e-3}$$

Rearranging Eq.(e-1) to make the summation extend to t = n

$$(n+1)\sum_{t=0}^{n+1}(-1)^t \frac{(n+1+k)!}{(n+1-t)!(k+t)!t!}x^t$$

$$= \sum_{t=0}^{n+1}(-1)^t \frac{(n+1+k)!(n+1)}{(n+1-t)!(k+t)!t!}x^t$$

$$= \sum_{t=0}^{n}(-1)^t \frac{(n+1+k)!(n+1)}{(n+1-t)!(k+t)!t!}x^t$$

$$+ \sum_{t=0}^{n}(-1)^{n+1} \frac{(n+1+k)!(n+1)}{(n+1-n-1)!(n+1+k)!(n+1)!}x^{n+1}$$

$$= \sum_{t=0}^{n-1}(-1)^t \frac{(n+1+k)!(n+1)}{(n+1-t)!(k+t)!t!}x^t$$

$$+ (-1)^n \frac{(n+1+k)!(n+1)}{(n+1-n)!(k+n)!n!}x^n$$

$$+ (-1)^{n+1} \frac{(n+1+k)!(n+1)}{(n+1-n-1)!(n+1+k)!(n+1)!}x^{n+1}$$

$$= \sum_{t=0}^{n-1}(-1)^t \frac{(n+1+k)!(n+1)}{(n+1-t)!(k+t)!t!}x^t$$

$$+ (-1)^n \frac{(n+1+k)(n+1)}{n!}x^n + (-1)^{n+1}\frac{1}{n!}x^{n+1} \tag{e-1'}$$

Rearranging Eq.(e-2) to make the order of x in the summation be 't'

$$-(2n+k+1-x)\sum_{t=0}^{n}(-1)^t\frac{(n+k)!}{(n-t)!(k+t)!t!}x^t$$

$$= -\sum_{t=0}^{n}(-1)^t\frac{(n+k)!(2n+k+1-x)}{(n-t)!(k+t)!t!}x^t$$

$$= -\sum_{t=0}^{n}(-1)^t\frac{(n+k)!(2n+k+1)}{(n-t)!(k+t)!t!}x^t + \sum_{t=0}^{n}(-1)^t\frac{(n+k)!}{(n-t)!(k+t)!t!}x^{t+1}$$

$$= -\sum_{t=0}^{n-1}(-1)^t\frac{(n+k)!(2n+k+1)}{(n-t)!(k+t)!t!}x^t$$
$$-(-1)^t\frac{(n+k)!(2n+k+1)}{(n-n)!(k+n)!n!}x^n + \sum_{t=0}^{n}(-1)^t\frac{(n+k)!}{(n-t)!(k+t)!t!}x^{t+1} \quad \text{(e-2')}$$

Rearranging Eq(e-3)

$$(n+k)\sum_{t=0}^{n-1}(-1)^t\frac{(n-1+k)!}{(n-1-t)!(k+t)!t!}x^t$$

$$= \sum_{t=0}^{n-1}(-1)^t\frac{(n-1+k)!(n+k)}{(n-1-t)!(k+t)!t!}x^t \quad \text{(e-3')}$$

By combining Eq.'s (e-1'), (e-2') and (e-3'), Eq.(d) becomes

$$\sum_{t=0}^{n-1}(-1)^t\frac{(n+1+k)!(n+1)}{(n+1-t)!(k+t)!t!}x^t + (-1)^n\frac{(n+1+k)(n+1)}{n!}x^n + (-1)^{n+1}\frac{1}{n!}x^{n+1}$$

$$-\sum_{t=0}^{n-1}(-1)^t\frac{(n+k)!(2n+k+1)}{(n-t)!(k+t)!t!}x^t$$

$$-(-1)^t\frac{(n+k)!(2n+k+1)}{(n-n)!(k+n)!n!}x^n + \sum_{t=0}^{n}(-1)^t\frac{(n+k)!}{(n-t)!(k+t)!t!}x^{t+1}$$

$$+ \sum_{t=0}^{n-1}(-1)^t\frac{(n-1+k)!(n+k)}{(n-1-t)!(k+t)!t!}x^t$$

$$= \sum_{t=0}^{n-1}(-1)^t\frac{(n+1+k)!(n+1)}{(n+1-t)!(k+t)!t!}x^t \quad \text{(d'-1)}$$

$$-\sum_{t=0}^{n-1}(-1)^t\frac{(n+k)!(2n+k+1)}{(n-t)!(k+t)!t!}x^t+\sum_{t=0}^{n-1}(-1)^t\frac{(n-1+k)!(n+k)}{(n-1-t)!(k+t)!t!}x^t \quad (d'\text{-}2)$$

$$+\sum_{t=0}^{n}(-1)^t\frac{(n+k)!}{(n-t)!(k+t)!t!}x^{t+1} \quad (d'\text{-}3)$$

$$+(-1)^n\frac{(n+1+k)(n+1)}{n!}x^n+(-1)^{n+1}\frac{1}{n!}x^{n+1}$$

$$-(-1)^t\frac{(n+k)!(2n+k+1)}{(n-n)!(k+n)!n!}x^n \quad (d'\text{-}4)$$

Combining (d'-1) and (d'-2) in order to put together terms having x^t in the summation,

$$\sum_{t=0}^{n-1}(-1)^t\left\{\left(\frac{(n+1+k)!(n+1)}{(n+1-t)!(k+t)!t!}\right)-\left(\frac{(n+k)!(2n+k+1)}{(n-t)!(k+t)!t!}\right)-\left(\frac{(n-1+k)!(n+k)}{(n-1-t)!(k+t)!t!}\right)\right\}x^t$$

$$=\sum_{t=0}^{n-1}(-1)^t\left\{\frac{(n+k)!}{(n-t)!(k+t)!t!}\left[\frac{(n+1)(n+1+k)}{(n+1-t)}-\frac{(2n+k+1)}{1}+\frac{(n-t)}{1}\right]\right\}x^t$$

$$=\sum_{t=0}^{n-1}(-1)^t\left\{\frac{(n+k)!}{(n-t)!(k+t)!t!}\left[\frac{t(k+t)}{(n+1-t)}\right]\right\}x^t$$

$$=\sum_{t=0}^{n}(-1)^t\left\{\frac{(n+k)!}{(n-t)!(k+t)!t!}\left[\frac{t(k+t)}{(n+1-t)}\right]\right\}x^t$$

$$-(-1)^n\left\{\frac{(n+k)!}{(n-n)!(k+n)!n!}\left[\frac{n(k+n)}{(n+1-n)}\right]\right\}x^n \quad (d'\text{-}1,2)$$

To make Eq.(d) simpler, insert (d'-1,2) to Eq.(d')

$$=\sum_{t=0}^{n}(-1)^t\left\{\frac{(n+k)!}{(n-t)!(k+t)!t!}\left[\frac{t(k+t)}{(n+1-t)}\right]\right\}x^t$$

$$-(-1)^n\left\{\frac{(n+k)!}{(n-n)!(k+n)!n!}\left[\frac{n(k+n)}{(n+1-n)}\right]\right\}x^n \quad (d'\text{-}1,2)$$

$$+\ \sum_{t=0}^{n}(-1)^{t}\frac{(n+k)!}{(n-t)!(k+t)!t!}x^{t+1} \tag{d'-3}$$

$$+\ (-1)^{n}\frac{(n+1+k)(n+1)}{n!}x^{n}$$
$$+\ (-1)^{n+1}\frac{1}{n!}x^{n+1}\ -(-1)^{t}\frac{(n+k)!(2n+k+1)}{(n-n)!(k+n)!n!}x^{n} \tag{d'-4}$$

By adding the last term of (d'-1,2) with the first term of Eq.(d'-4)

$$(-1)^{n}\frac{(n+1+k)(n+1)}{n!}x^{n}\ -(-1)^{n}\left\{\frac{(n+k)!}{(n-n)!(k+n)!n!}\left[\frac{n(k+n)}{(n+1-n)}\right]\right\}x^{n}$$

$$=(-1)^{n}\frac{2n+1+k}{n!}x^{n} \tag{e}$$

Eq.(e) cancels out with the last term of Eq.(d'-4). Hence, Eq.(d') becomes in a compact form

$$\sum_{t=0}^{n}(-1)^{t}\frac{(n+k)!}{(n-t)!(k+t)!t!}x^{t+1}$$
$$+\ \sum_{t=0}^{n}(-1)^{t}\left\{\frac{(n+k)!}{(n-t)!(k+t)!t!}\left[\frac{t(k+t)}{(n+1-t)}\right]\right\}x^{t}\ +(-1)^{n+1}\frac{1}{n!}x^{n+1} \tag{d''}$$

(The second integration is equal to zero when $t = 0$)

$$=\ \sum_{t=0}^{n}(-1)^{t}\frac{(n+k)!}{(n-t)!(k+t)!t!}x^{t+1}$$
$$+\ \sum_{t=1}^{n}(-1)^{t}\left\{\frac{(n+k)!}{(n-t)!(k+t)!t!}\left[\frac{t(k+t)}{(n+1-t)}\right]\right\}x^{t}\ +(-1)^{n+1}\frac{1}{n!}x^{n+1} \tag{d'''}$$

By replacing t by t'+1 in the second summation, the second summation plus the last term in Eq.(d''') becomes

$$\sum_{t'=0}^{n}(-1)^{t'+1}\left\{\frac{(n+k)!}{(n-t'-1)!(k+t'+1)!(t'+1)!}\left[\frac{(t'+1)(k+t'+1)}{(n-t')}\right]\right\}x^{t'+1}+(-1)^{n+1}\frac{1}{n!}x^{n+1}$$

$$= \sum_{t'=0}^{n}(-1)^{t'+1}\left\{\frac{(n+k)!}{(n-t'-1)!(k+t'+1)!(t'+1)!}\left[\frac{(t'+1)(k+t'+1)}{(n-t')}\right]\right\}x^{t'+1}$$

$$= \sum_{t'=0}^{n}(-1)^{t'+1}\left\{\frac{(n+k)!}{(n-t')!(k+t')!t'!}\right\}x^{t'+1}$$

$$= -\sum_{t'=0}^{n}(-1)^{t'}\left\{\frac{(n+k)!}{(n-t')!(k+t')!t'!}\right\}x^{t'+1} \qquad \text{(f)}$$

Since the result of summation does not change with summation variable, Eq.(f) is equal to the first summation in Eq.(d''') with negative sign and Eq.(d''') is equal to zero. Hence the recurrsion relation expressed in Eq.(d) is eqaul to zero. This verifies that the the associated Laguerre polynomial, L_n^k, satisfy the first recurrsion relationship in Eq.(31).

Recurrsion relationship 2

Putting Eq.(b) and (c) and derivative of Eq.(b) for L_n^k, L_{n-1}^k and $L_n^{k}{}'$ in Eq.(4')

$$xL_n^{k}{}' - nL_n^k + (n+k)L_{n-1}^k \qquad \text{(32')}$$

$$= \sum_{t=0}^{n}(-1)^t\frac{(n+k)!}{(n-t)!(k+t)!(t-1)!}x^t$$

$$-n\sum_{t=0}^{n}(-1)^t\frac{(n+k)!}{(n-t)!(k+t)!t!}x^t + (n+k)\sum_{t=0}^{n-1}(-1)^t\frac{(n-1+k)!}{(n-1-t)!(k+t)!t!}x^t$$

$$= \sum_{t=0}^{n}(-1)^t\frac{(n+k)!}{(n-t)!(k+t)!(t-1)!}x^t$$

$$-\sum_{t=0}^{n}(-1)^t\frac{(n+k)!n}{(n-t)!(k+t)!t!}x^t + \sum_{t=0}^{n-1}(-1)^t\frac{(n+k)!}{(n-1-t)!(k+t)!t!}x^t \qquad \text{(32'')}$$

Since, when t = n, the last term is zero,

$$\sum_{t=0}^{n-1}(-1)^t\frac{(n+k)!(n-t)}{(n-t)!(k+t)!t!}x^t = \sum_{t=0}^{n}(-1)^t\frac{(n+k)!(n-t)}{(n-t)!(k+t)!t!}x^t$$

Eq.(32") becomes

$$\sum_{t=0}^{n}(-1)^t \frac{(n+k)!t}{(n-t)!(k+t)!t!}x^t - \sum_{t=0}^{n}(-1)^t \frac{(n+k)!n}{(n-t)!(k+t)!t!}x^t + \sum_{t=0}^{n}(-1)^t \frac{(n+k)!(n-t)}{(n-t)!(k+t)!t!}x^t$$

$$\sum_{t=0}^{n}(-1)^t \frac{(n+k)!t}{(n-t)!(k+t)!t!}x^t - \sum_{t=0}^{n}(-1)^t \frac{(n+k)!n}{(n-t)!(k+t)!t!}x^t$$

$$+\sum_{t=0}^{n}(-1)^t \frac{(n+k)!(n-t)}{(n-t)!(k+t)!t!}x^t$$

$$= -\sum_{t=0}^{n}(-1)^t \frac{(n+k)!(n-t)}{(n-t)!(k+t)!t!}x^t + \sum_{t=0}^{n}(-1)^t \frac{(n+k)!(n-t)}{(n-t)!(k+t)!t!}x^t = 0$$

Hence the second recurssion relationship is proved.

Associated Laguerre Equation

$$(n+1)L_{n+1}^{k} = (2n+k+1-x)L_n^k - (n+k)L_{n-1}^k \qquad (31)$$

$$xL_n^{k\prime} = nL_n^k - (n+k)L_{n-1}^k \qquad (32)$$

Differentiate Eq.(32)

$$xL_n^{k\prime\prime} + L_n^{k\prime} = nL_n^{k\prime} - (n+k)L_{n-1}^{k}{}' \qquad (33)$$

From Eq.(32)

$$L_{n-1}^k = \frac{xL_n^{k\prime} - nL_n^k}{(n+k)} \qquad (34)$$

From Eq.(33)

$$L_{n-1}^{k}{}' = \frac{xL_n^{k\prime\prime} + L_n^{k\prime} - nL_n^{k\prime}}{(n+k)} \qquad (35)$$

Inserting Eq.(35) into Eq.(31)

$$(n+1)L_{n+1}^{k} = (2n+k+1-x)L_n^k - (n+k)\frac{xL_n^{k}{}' - nL_n^k}{(n+k)}$$

$$(n+1)L_{n+1}^{k} = (n+k+1-x)L_n^k + xL_n^{k}{}' \tag{36}$$

Replacing n+1 by n in Eq.(36)

$$nL_n^k = (n+k-x)L_{n-1}^k + xL_{n-1}^{k}{}' \tag{37}$$

Inserting Eq.(35) and Eq.(36) into Eq.(37)

$$nL_n^k = (n+k-x)\frac{xL_n^{k}{}' - nL_n^k}{(n+k)} + x\frac{xL_n^{k}{}'' + L_n^{k}{}' - nL_n^{k}{}'}{(n+k)}$$

$$n(n+k)L_n^k = (n+k-x)\left(xL_n^{k}{}' - nL_n^k\right) + x\left(xL_n^{k}{}'' + L_n^{k}{}' - nL_n^{k}{}'\right)$$

$$n(n+k)L_n^k = (n+k-x)\left(xL_n^{k}{}' - nL_n^k\right) + x\left(xL_n^{k}{}'' + L_n^{k}{}' - nL_n^{k}{}'\right)$$

$$0 = n(n+k)L_n^k + (n+k)xL_n^{k}{}' - x^2L_n^{k}{}' + nxL_n^k + x\left(xL_n^{k}{}'' + L_n^{k}{}' - nL_n^{k}{}'\right)$$

$$0 = (n+k)xL_n^{k}{}' - x^2L_n^{k}{}' + nxL_n^k + x\left(xL_n^{k}{}'' + L_n^{k}{}' - nL_n^{k}{}'\right)$$

$$0 = (n+k)L_n^{k}{}' - xL_n^{k}{}' + nL_n^k + \left(xL_n^{k}{}'' + L_n^{k}{}' - nL_n^{k}{}'\right)$$

$$0 = kL_n^{k}{}' - xL_n^{k}{}' + nL_n^k + xL_n^{k}{}'' + L_n^{k}{}'$$

$$0 = xL_n^{k}{}'' + (k-x+1)L_n^{k}{}' + nL_n^k \tag{38}$$

Eq.(38) is the same as the associated Laguerre Equation. Therefore it is proved that the equation expressed in Eq.(30) satisfies the associated Laguerre Equation and it is the solution for that.

Appendix 2

Quantum mechanical background for the simple expression of the lattice energy of ionic crystals

Regarding wave functions for electrons, ionic crystal is the simplest model case among crystals since electrons can be described as in the atomic states. As shown in Fig.A-1, the wave function for Na and F in atomic state is described as $\Psi_{NaF} = \Psi_{Na}\Psi_{F}$ since there is no interaction between Na – F. In ionic state, the valence electron of Na is transferred to F resulting in the wave function of Na – F as $\Psi_{NaF} = \Psi_{Na^+}\Psi_{F^-}$ since the valence electron transferred to F is confined to F. Even in crystalline state, the wave function for Na and F can be approximated as $\Psi_{NaF} = \Psi_{Na^+}\Psi_{F^-}$.

The SWE shown in Eq.(31) in Chapter 1 can be applied to the ionic crystal to relate it to the macroscopic data such as lattice energy which will be explained in the following section. In another expression, the SWE is expressed as $H\Psi = E\Psi$ in which H is an operator generating a energy value from the wave function and equal to $\frac{h^2}{8\pi^2 m}\left(\frac{\partial^2}{\partial^2 x}+\frac{\partial^2}{\partial^2 y}+\frac{\partial^2}{\partial^2 z}\right) - V$. Since the there is no interaction between electrons of Na and F in the atomic state, the term, V, in the operator H can be expressed as

$$H \equiv \sum_{i\neq j}\left(\frac{h^2}{8\pi^2 m}\left(\frac{\partial^2}{\partial^2 x_i}+\frac{\partial^2}{\partial^2 y_i}+\frac{\partial^2}{\partial^2 z_i}\right) - \frac{e^2}{r_{Na,i}} + \frac{e^2}{r_{i,j}}\right) + \sum_{i\neq j}\left(\frac{h^2}{8\pi^2 m}\left(\frac{\partial^2}{\partial^2 X_i}+\frac{\partial^2}{\partial^2 Y_i}+\frac{\partial^2}{\partial^2 Z_I}\right) - \frac{e^2}{R_{F,I}} + \frac{e^2}{R_{I,J}}\right) \quad \text{(A-1)}$$

Here (x,y,z) and (X,Y,Z) indicate the coordinates relative to Na and F, respectively. $\sum_{i\neq j}$ indicates a summation over electrons and inequality sign implies that $r_{i,j}$ is a distance between electrons 'i' and 'j' which are different. On the assumption that there is no interaction between Na and F, (x,y,z) and (X,Y,Z) are independent from each other. Hence the wave function for Na–F

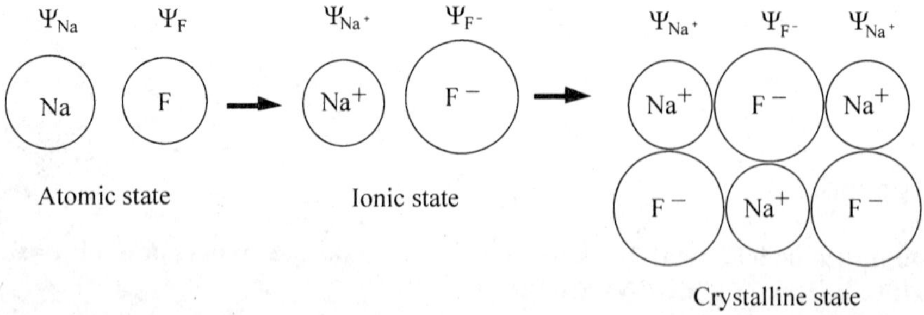

Figure A-1 Change of Na and F from atomic state to crystalline state with electron transfer

can be expressed as $\Psi_{Na,F} = \Psi_{Na}(r_1, r_2, ..., r_9)\Psi_F(R_1, R_2, ..., R_{11})$. By applying this wave function in the SWE with the operator in Eq.(A-1),

$$H\Psi_{Na,F} =$$

$$\sum_{i\neq j}\left(\frac{h^2}{8\pi^2 m}\left(\frac{\partial^2}{\partial^2 x_i}+\frac{\partial^2}{\partial^2 y_i}+\frac{\partial^2}{\partial^2 z_i}\right)-\frac{e^2}{r_{Na,i}}+\frac{e^2}{r_{i,j}}\right)\Psi_{Na}(r_1,..)\Psi_F(R_1,...)$$
$$+\sum_{i\neq j}\left(\frac{h^2}{8\pi^2 m}\left(\frac{\partial^2}{\partial^2 X_i}+\frac{\partial^2}{\partial^2 Y_i}+\frac{\partial^2}{\partial^2 Z_i}\right)-\frac{e^2}{R_{F,i}}+\frac{e^2}{R_{i,j}}\right)\Psi_{Na}(r_1,..)\Psi_F(R_1,..) \quad \text{(A-2)}$$

This is abbreviated into

$$H_{Na}\Psi_{Na}(r_1,r_2,...,r_9)\Psi_F(R_1,R_2,...,R_{11}) + H_F\Psi_{Na}(r_1,r_2,...,r_9)\Psi_F(R_1,R_2,...,R_{11})$$

$$= (E_{Na}\Psi_{Na})\Psi_F + (E_F\Psi_F)\Psi_{Na} = (E_{Na}+E_F)\Psi_{Na,F} \qquad \text{(A-3)}$$

This is called a production rule which means that the wave function for multiple electrons having no interactions between them can be expressed as a product of individual wave function. In the same token, the wave functions for ionic state of NaF can be expressed as

$$\Psi_{NaF} = \Psi_{Na^+}\Psi_{F^-} \qquad \text{(A-4)}$$

if Na^+ and F^- are separated and have no interactions. Regarding the SWE for the ionic states, the energy change in Na and F can be expressed as

$$H\Psi_{Na^+} = \left(E_{Na} - IE_{Na}\right)\Psi_{Na^+} \text{ and } H\Psi_{F^-} = \left(E_F - EA_F\right)\Psi_{F^-} \qquad \text{(A-5)}$$

In a crystalline state shown in Fig.A-1, ions make contact with each other. If the wave functions of NaF crystal assumes the form of Eq.(4), the operator for the SWE takes the form of Eq.(A-1). However, the operator H in Eq.(A-1) must include the interaction between electrons of different atoms and the interaction between nuclei since ions are not any longer separated so that the interaction between ions cannot be ignored. These interactions give rise to the potential terms included as the summation terms in Eq.(A-6)- b, c, and d. Eq.(A-6)-b is the attractive electrostatic interaction between electrons on F^- with Na^+ and electrons on Na^+ with F^-. Eq.(A-6)-c indicates the repulsive interactions between nuclei of Na^+ and F^-. Eq.(A-6)-d represents the interaction between electrons on different nuclei.

$$H^{o}_{NaF,crystal} =$$

$$\sum_{I}\left(H_{Na^+_I} + H_{F^-_I}\right) \qquad \text{(A-6)-a}$$

$$+ \sum_{I\neq J}\left(\frac{e^2}{r_{J-j,Na^+_I}}\right) + \sum_{I\neq J}\left(\frac{e^2}{r_{I-i,F^-_J}}\right) \qquad \text{(A-6)-b}$$

$$- \left[\sum_{I,J}\left(\frac{e^2}{r_{Na^+_I,F^-_J}}\right) + \sum_{I\neq J}\left(\frac{e^2}{r_{Na^+_I,Na^+_J}}\right) + \sum_{I\neq J}\left(\frac{e^2}{r_{F^-_I,F^-_J}}\right)\right] \qquad \text{(A-6)-c}$$

$$- \sum_{I\neq J}\left(\frac{e^2}{r_{I-i,J-j}}\right) \qquad \text{(A-6)-d}$$

Here I or J can represent coordinates of nuclei while i and j connote the electrons. Hence I-i and J-j imply the i-th electron in the I-th nucleus and j-th electron J-th nucleus. $H_{Na_I^+}$ and $H_{F^-_J}$ are operators working on I-th Na^+ and J-th F^- ions which are expressed as

$$H_{Na^+_I} \equiv \frac{h^2}{8\pi^2 m}\sum_{i\neq i'}\left(\nabla_i^2 - \frac{e^2}{r_{Na^+_I,i}} + \frac{e^2}{r_{i,i'}}\right)$$

$$H_{F_I} \equiv \frac{h^2}{8\pi^2 m}\sum_{i\neq i'}\left(\nabla_i^2 - \frac{e^2}{r_{F_I,i}} + \frac{e^2}{r_{i,i'}}\right)$$

If the operator in Eq.(A-6) is applied to the wave function of NaF crystal expressed as Eq.(A-4), the SWE is

$$H^{o}_{NaF,crystal}\,\Psi_{Na^+}\Psi_{F^-} = E^{o}\,\Psi_{Na^+}\Psi_{F^-}$$

$$= \sum_{I,J}\left(E_{Na_I} + E_{F_J} + EA_{F,I} - IE_{Na,J}\right)\Psi_{Na^+}\Psi_{F^-} + E'\,\Psi_{Na^+}\Psi_{F^-} \qquad \text{(A-7)}$$

The result with the operator of Eq.(A-6)-a on $\Psi_{Na^+}\Psi_{F^-}$ is a simple arithmetic summation of atomic energy of Na (E_{Na}) and F(E_F) and corresponding ionization energy of Na (IE_{Na}) and electron affinity of F(EA_F). These values are known. For the operators in Eq.(A-6)-b, c, and d, however, there is no simple solution. To obtain the energy term resulting from these operators on $\Psi_{Na^+}\Psi_{F^-}$, it is assumed that ions are regarded as point charges. Then the energy term, U, is approximately set as

$$U \approx \sum_{I,J}\left(-\frac{e^2}{r_{Na^+_I,Na^+_{I'}}} - \frac{e^2}{r_{F^-_J,F^-_{J'}}} + \frac{e^2}{r_{Na^+_I,F^-_J}}\right) \qquad \text{(A-8)}$$

$$E^{o} = \sum_{I,J}\left(E_{Na_I} + E_{F_J} + EA_{F,I} - IE_{Na,J}\right) + U \qquad \text{(A-9)}$$

Here the only coordinates required are those of nucleus which are measured from experimental data such as X-ray diffraction. Therefore the energy of ionic crystal is easily calculated according to Eq.(A-9). Those energy values in the summation of Eq.(A-9) are pertaining to atomic states. Hence only the energy term concerned in Eq.(A-9) is U which represents characteristic of the ionic crystal. This energy depends on the crystal structure and represents it. Since the binding energy is the energy gained from making a bond, those atomic energy terms in Eq.(A-9) must be subtracted. Hence the bonding energy (E_b) is expressed as

$$E_b = \sum_{I,J}\left(E_{F_J} - IE_{Na,J}\right) + U \qquad \text{(A-10)}$$

In other types of crystal, mathematical simple expressions for the binding energy like Eq.(A-10) is not possible like in ionic crystals. The first reason is that the valence electrons involved in a bonding cannot be localized as in purely ionic crystal. Hence the wave function in a crystal cannot be explicitly expressed as in Eq.(A-4). Also the potential terms in the SWE cannot be formulated in a simple way as in Eq.(A-7) and (A-8). The total energy term of the SWE for a non-ionic crystal cannot be formulated with those parameters which are well characterized theoretically like ionization energy and electron affinity.

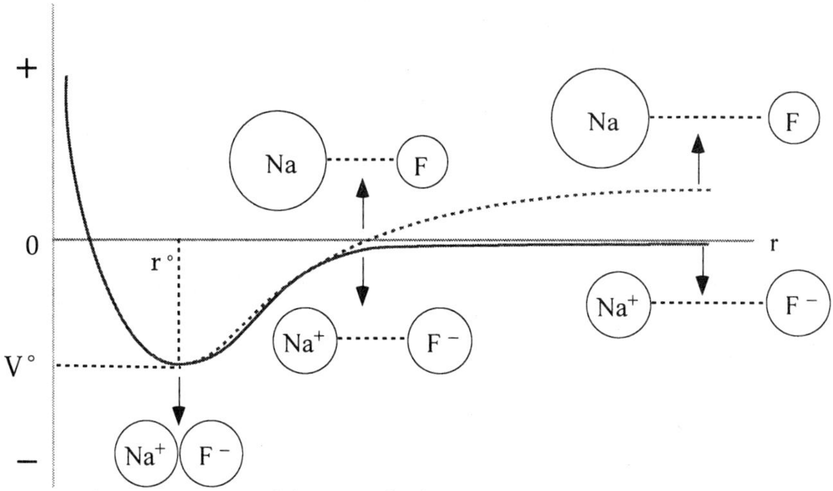

Figure A-2 Potential curves for ionic state and atomic state with bond distance

BIBLIOGRAPHY

General References:

Adams D. M., *Inorganic Solids,* John Wiley and Sons, New York (1974).

Bloss F. D., *Crystallography and Crystal Chemistry*, Holt Rheinhart and Co., New York (1971).

Chalmers B., Holland J., Jackson K. and Brody R., *Modules of Crystallography,* EMMSE, Penn. State University, State Park, PA (1968).

Hauth W. E., Jr., "Crystal Chemistry in Ceramics," *Am. Ceram. Soc. Bull.*, Jan. - June, 1951.

DeKock R. L. and H. B. Gray, *Chemical Structure and Bonding*, The Benjamin Cummins Publishing Co., Menlo Park, CA (1980).

Evans R. C., *An Introduction to Crystal Chemistry, Second Edition*, Cambridge University Press, London (1964).

Guinier A., *The Structure of Matter*, Edward Arnold, Co., Baltimore, MD (1984).

Hench L. L. and West J. K. *Principles of Electronic Ceramics*. John Wiley & Sons, Inc. New York (1990).

James E. Huheey, *Inorganic Chemistry, Second Edition*, Harper and Row, New York (1978).

Kelly A. G. and G. W. Groves, *Crystallography and Crystal Defects*, Addison-Wesley Co., Ltd., London (1970).

Nye J.F. 1957. *Physical Properties of Crystals*. Oxford University Press.

Pauling L. 1940. *The Nature of the Chemical Bond.* Cornell University Press, Ithaca.

Ralls K. M., Courtney T. H. and Wulff J., *Introduction to Materials Science and Engineering*, Chaps. 2, 3 and 4 (Atomic Structure, Chemical Bonding, Molecular Orbitals) John Wiley and Sons, Ltd., New York (1976).

Solid State Chemistry, Eds. A. K. Chatham and P. Day, Clarendon Press, Oxford (1992).

Windle A., *A First Course in Crystallography*, G. Bell and Sons, Ltd., London (1971).

Yanagida H., Koumoto K., and Miyayama M. *The Chemistry of Ceramics*. John Wiley & Sons, Inc., West Sussex, England (1996).

References for Chapter 1:

Azaroff L. V. and Brophy J. J. *Electronic Processes in Materials*, McGraw-Hill Inc., New York (1963)

Burdett J.K. *Molecular Shapes. Theoretical Models of Inorganic Stereochemistry.* Wiley, New York (1980).

Cochran W. *The Dynamics of Atoms in Crystals, The Structures and Properties of Solids 3* (Ed B. R. Coles), Edward Arnold (Publishers) Limited, London (1973).

Coles B. R. and Caplin A. D. *The Electronic Structures of Solids*. Edward Arnold (Publishers) Ltd., New York (1976).

Dunitz. J. *X-ray Analysis and the Structure of Organic Molecules* Cornell University Press, Ithaca. Authoritative review (1980).

Galasso F. S. *Structure and Properties of Inorganic Solids: International Series of Monographs in Solid State Physics, Vol. 7*, Pergamon Press, Inc., Oxford (1970).

Gay Peter. *The Crystalline State*, Hafner Publishing Co., Inc., New York (1972).

Jaffe H. W., *Introduction to Crystal Chemistry*, Cambridge University Press, Cambridge (1988).

Kitaigorodsky A. I. *Molecular Crystals and Molecules.* Academic Press, New York. (1973).

Krebs H. *Fundamentals of Inorganic Crystal Chemistry.* McGraw Hill, London (1968).

Interrante L. V., Caspar L. A., Elllis A. B. *Materials Chemistry: An Emerging Discipline*, American Chemical Society, Washington, D.C. (1995).

Rose R. M., Shepard L. A. and Wulff J. *The Structure and Properties of Materials, Vol. IV, Electronic Properties*, John Wiley & Sons, Inc., New York (1966).

Shannon, R.D. and Prewitt C.T. "Effective ionic radii in oxides and fluorides" *Acta Cryst.* B25 925-945 (1969.)

Shannon R. D. "Revised effective ionic radii and systematic studies of interatomic distances in halides and chalcogenides" *Acta Cryst.* A32, Zs 1-767 (1976).

Wade K. "Structural and bonding patterns in cluster chemistry." *Adv. Inorg. Chem. and Radiochem.* 18 1-66 (1976).

Wells A.F. *The Third Dimension in Chemistry.* Clarendon Press, Oxford (1956).

Wells A.F. *Models in Structural Inorganic Chemistry.* Oxford University Press, New York (1970).

Wells A.F. *Structural Inorganic Chemistry. 4th ed.* Clarendon Press, Oxford (1975).

References for Chapter 2:

Boisen N.B., Jr. and Gibbs G.V. "A derivation of the 32 crystallographic point groups using elementary group theory," *Amer. Mineral*, 61:145-165 (1976).

Brown P. J. and Forsyth J. B., *The Crystal Structure of Solids: The Structures and properties of Solids 2* (ed. B. R. Coles), Edward Arnold (Publishers) Limited, London (1973).

Buerger M.J. *Elementary Crystallography.* Wiley, New York (1956).

Burns G. and Glazer A.M. *Space Groups for Solid State Scientists.* Academic Press, New York (1978).

Cotton F.A. *Chemical Applications of Group Theory.* Wiley Interscience, New York (1971).

DentGlasser L.S. *Crystallography and Its Applications.* VanNostrand Reinhold, New York (1977).

Fischer W., Burzlaff H., Hellner E., and Donnay J.D.H. *Space Groups and Lattice Complexes.* National Bureau of Standards Monograph 134, Washington D.C. (1973).

Frank F.C. and Kasper J.S. "Complex alloy structures regarded as sphere packings. I. Definitions and basic principles." *Acta Cryst.* 11 184- 190 (1958).

Holser W.T. "Point groups and plane groups in a two-sided plane and their subgroups." *Z. Krist.* 110 266-281 (1958b).

Lockwood E.H. and Macmillan R.H. *Geometric Symmetry.* Cambridge University Press, London (1978).

O'Keefe M. and Hyde B.G. "Plane nets in crystal chemistry" *Phil. Trans.*, Roy. Soc. London 295 553-622 (1980).

Phillips F.C. *An Introduction to Crystallography. 4th ed.* Oliver and Boyd, Edinburgh (1971).

Schneer C.J. (Ed.). *Crystal Form and Structure.* Dowden, Hutchinson & Ross, Stroudsburg (1977).

Wachtman, J. B. *Characterization of Materials*, Butterworth-Heinemann, Stoneham, MA (1993).

Wooster W.A. *Tensors and Group Theory for the Physical Properties of Crystals.* Clarendon Press, London (1973).

Wyckoff R.W.G. *Crystal Structures.* 2nd ed., vols. 1-6. Interscience, New York (1963-1971).

References for Chapter 3:

Bhagavantan S. *Crystal Symmetry and Physical Properties.* Academic Press, New York (1996).

Bradley C.J. and Cracknell A.P. *The Mathematical Theory of Symmetry in Solids. Representation Theory for Point Groups and Space Groups.* Clarendon Press, Oxford. Comprehensive treatise (1972).

Buerger M.J. *Crystal Structure Analysis.* Wiley, New York (1960).

Bunget I. and Popescu M. *Physics of Solid Dielectrics*, Elsevier Science Publishing Company, Inc. New York (1984).

Kelly A. and Groves G.W. *Crystallography and Crystal Defects.* Addison-Wesley, Reading, Mass (1970).

Ladd M.F.C. and Palmer R. A. *Structure Determination by X-Ray Crystallography*, Plenum Press, New York (1977).

Lipson H. and Cochran W. *The Determination of Crystal Structures.* 3rd ed. Bell, London (1966).

Loeb A.L. *Space Structures.* Addison-Wesley, Reading, Mass. (1976)

McKie D. and McKie C. *Crystalline Solids.* Nelson, London (1974).

Megaw H.D. *Ferroelectricity in Crystals.* Methuen, London (1957).

Megaw H.D. *Crystal Structures. A Working Approach.* Saunders, Philadelphia (1973).

Pearson W. B. *The Crystal Chemistry and Physics of Metals and Alloys.* Wiley-Interscience, New York (1972).

Smith J V. *Geometrical and Structural Crystallography* John Wiley & Sons, Inc., New York (1982).

Wells A. F. *Three-Dimensional Nets and Polyhedra.* Wiley, New York (1977).

Weyl H. *Symmetry.* Princeton University Press, Princeton (1952).

Wenninger M. *Polyhedron Models.* Cambridge University Press, London (1971).

Wang R., Bradley W.F., and Steinfink H. "The crystal structure of alunite." *Acta Cryst.* 18 249-252 (1965).

Winter J.K. and Ghose S. "Thermal expansion and high-temperature crystal chemistry of the Al2SiOs polymorphs" *Amer. Mineral.* 64 573-586 (1979).

Yagi T., Mao H.-K, and Bell P.M. "Structure and crystal chemistry of perovskite-type $MgSiO_3$." *Phys. Chem. Minerals* 3 97-110 (1978).

References for Chapter 4:

Birle J. D., Gibbs G. V., Moore P. B. and Smith J. V. "Crystal structures of natural olivines." *Amer. Mineral.* 53 807-824 (1968).

Bragg Sir L., Claringbull G. F., and Taylor W. H. *Crystal Structures of Minerals.* Cornell University Press, Ithaca (1965).

Brindley G.W. and Brown G. (Eds.) *Crystal Structures of Clay Minerals and Their X-Ray Identification.* Mineralogical Society, Monograph no. 5, London (1980).

Burnham C.W. "Refinement of the crystal structure of sillimanite." *Z. Krist.* 118 337-360 (1963).

Britton D. and Dunitz J.D. "A complete catalogue of polyhedra with eight or fewer vertices." *Acta Cryst.* A29 362-371 (1973).

Chabot B., Cenzual K., and Parthe E. "Nested polyhedra units: A geometrical concept for describing complicated cubic structures." *Acta Cryst.* A37 6-11 (1981).

Fischer W. and Koch E. "Limiting forms and comprehensive complexes for crystallographic point groups, rod groups and layer groups." *Z. Krist.* 147 255-273 (1978).

Fischer W. and Koch E. "Geometrical packing analysis of molecular compounds." *Z. Krist.* 150 245-260 (1978).

Frye K. *Modern Mineralogy*, Prentice-Hall, Inc., New Jersey (1974).

Hazen R.M. and Finger L.W. "Crystal structure and compressibility of zircon at high pressure." *Amer. Mineral.* 64 196-201 (1979).

Hellner E. "The frameworks (Bauverbande) of the cubic structure types." *Structure and Bonding,* 37 61-140. (Ed. J.D.Dunitz et al., Springer, Berlin.) (1979).

King R.B. "Chemical applications of topology and group theory. I. Coordination polyhedra. II. Metal complexes of planar unsaturated carbon systems." *J. Amer. Chem. Soc.* 91 7211-16, 7217-23 (1969).

King R. B. "Chemical applications of topology and group theory. V. Polyhedral clusters and boron hydrides." *J. Am. Chem. Soc.* 94 95-103 (1972).

Cundy A.M. and Rollett A.P. *Mathematical Models. 2nd ed.* Clarendon, Oxford (1961).

Deer W.A., Howie R.A., and Zussman J. *Single-chain Silicates, Vol. 2A, 2nd ed. of Rock-forming Minerals*. Wiley, New York (1978).

Geller S. "Crystal chemistry of the garnets." *Z. Krist.* 125 1-47 (1967).

Gibbs G.V. "The polymorphism of cordierite-I: The crystal structure of low cordierite." *Amer. Mineral.* 51 1068-1087 (1966).

Hurlbut C. S., Jr. and Klein C. *Manual of Mineralogy. 19th ed. Wiley*, New York (1977).

Koch E. and Fischer W. "Types of sphere packings for crystallographic point groups, rod groups and layer groups." *Z. Krist.* 147 21-38 (1978a).

Koch E. and Fischer W. "Types of sphere packings for crystallographic point groups, rod groups and layer groups." *Z. Krist.* 148 107-152 (1978b).

Lima-de-Faria J. "Rules governing the layer organization of inorganic crystal structures." *Z. Krist.* 148 1 -5 (1978b).

Meier W.M. and Olson D.H. *Atlas of Zeolite Structure Types.* Polycrystal Book Service, Pittsburgh (1978).

Moore P. B. "Structural hierarchies among minerals containing octahedrally coordinating oxygen. I. Stereoisomerism among cornersharing octahedral and tetrahedral chains." *Neues Jahrb. Mineral. Mh.* 163-73 (1970a).

Novak G.A. and Gibbs G.V. "The crystal chemistry of the silicate garnets." *Amer. Mineral.* 56 791-825 (1971).

Nyman H. and Andersson S. "The elongated rhombic dodecahedron in alloy structures." *Acta Cryst.* A35 305-308 (1979).

Ohashi Y. and Finger L.W. "The role of octahedral cations in pyroxenoid crystal chemistry. I. Bustamite, wollastonite, and the pectoliteschizolite-serandite series." *Amer. Mineral.* 63 274-288 (1978).

Pannhorst W. "Structural relationships between pyroxenes." *Neues Jahrb. Miner. Abh.* 135 1-17 (1979).

Papike J.J. and Cameron M. "Crystal chemistry of silicate minerals of geophysical interest." *J. Geophys. Res.* 14 37-80 (1976).

Perrotta A.J. and Smith J.V. "The crystal structure of kalsilite, $KAISiO_4$." *Mineral. Mag.* 35 588-595 (1965).

Povarennykh A. S. *Crystal Chemical Classification of Minerals.* Plenum, New York (1972).

Ribbe P.H., Gibbs G.V., and Jones N.W. "Cation and anion substitutions in the humite minerals. "*Mineral. Mag.* 37 966-975 (1968).

Robinson K., Gibbs G.V., and Ribbe P.H. "The crystal structures of the humite minerals, IV. Clino- and titanoclinohumite." *Amer. Mineral.* 58 43-49 (1973).

Smith J. V. *Feldbsar Minerals.* SDrinzer. Heidelberg (1974).

Smith J.V. "Enumeration of 4-connected 3-dimensional nets and classification of framework silicates. I. Perpendicular linkage from simple hexagonal net." *Amer. Mineral.* 62 703-709 (1977).

Smith J.V. and Bennett J.M. "Enumeration of 4-connected 3dimensional nets and classification of framework silicates. IV. The infinite sets of ABC-6 nets; the Archimedean and a-related nets." *Amer. Mineral.* 66 777-788 (1981).

Smith J.V. and Yoder H.S. "Experimental and theoretical studies of the mica polymorphs. "*Mineral. Mag.* 31 209-234 (1956).

Takeuchi Y. and Donnay G. "The crystal structure of hexagonal $CaA1_2$ Si208." *Acta Cryst.* 12 465-70 (1959).

Verma A. R. and Krishna P. *Polymorphism and Polytypism in Crystals.* Wiley, New York (1966).

Zoltai T. "*Classification of silicates and other minerals with tetrahedral structures.*" *Amer. Mineral.* 45 960-73 (1960).

INDEX

Materials Crystal Chemistry